Electron Impact Ionization

Edited by
T. D. Märk and G. H. Dunn

Springer-Verlag Wien GmbH

Prof. Dr. Tilmann D. Märk
Institut für Experimentalphysik, Universität Innsbruck, Austria

Prof. Dr. Gordon H. Dunn
Joint Institute for Laboratory Astrophysics,
National Bureau of Standards and University of Colorado, Boulder, Colorado, U.S.A.

With 156 Figures

Library of Congress Cataloging in Publication Data. Main entry under title: Electron impact ionization. Includes index. 1. Electron impact ionization. I. Märk, T. D. (Tilmann D.), 1944– . II. Dunn, G. H. (Gordon H.), 1932– . QC702.7.E38E44. 1985. 537.5'32. 84-23595.

ISBN 978-3-7091-4030-7   ISBN 978-3-7091-4028-4 (eBook)
DOI 10.1007/978-3-7091-4028-4

Preface

It is perhaps surprising that a process which was one of the first to be studied on an atomic scale, and a process which first received attention over seven decades ago, continues to be the object of diverse and intense research efforts. Such is the case with the (seemingly) conceptually simple and familiar mechanism of electron-impact ionization of atoms, molecules, and ions. Not only has the multi-body nature of the collision given ground to theoretical effort only grudgingly, but also the variety and subtlety of processes contributing to ionization have helped insure that progress has come only with commensurate work: no pain − no gain.

Modern experimental methods have made it possible to effectively measure and explore threshold laws, differential cross sections, partial cross sections, inner-shell ionization, and the ionization of unstable species such as radicals and ions. In most instances the availability of experimental data has provided impetus and guidance for further theoretical progress.

If it is surprising that a field so old is still a contemporary, vital research area, it is then all the more astonishing that no book has provided a comprehensive review on the varied aspects of the kinetics and dynamics of electron impact ionization. There is a vast literature of many hundreds (perhaps thousands) of original research papers. Books by Field and Franklin, by Reed, and by Peterkop, as well as shorter treatments in chapters of books by Massey and Burhop, by McDaniel, by Hasted and by others, are limited in scope or perspective and are not comprehensive in their discussion of the electron-impact ionization process. Indeed, since electron-impact ionization is so important in the study, analysis, and modeling of physical/chemical systems ranging from bodies in intergalactic space to analytical mass spectrometers in the laboratory, many limited treatments of the process have been found in books on these "user" fields.

For some time we have felt that continued orderly progress in understanding this important collision mechanism would be enhanced if a comprehensive treatment of the process were available in a single reference book. Similarly, those in the "user" fields would also benefit by having a resource that was not limited in scope and perspective. With these goals in mind, this work was undertaken. We have sought and obtained participating authors who are experts in the field and who are distinguished by their original contributions to their areas.

Quantum mechanical, semiclassical, and semiempirical methods for calculating ionization cross sections are considered in Chapters 1−2. Chapter 3 deals with experimental and theoretical aspects of the threshold behavior of ionization cross sections. Summaries of today's knowledge about differential, partial, innershell, and total ionization cross sections make up Chapters 4−7. Chapter 8 deals with the active subject of electron-ion ionization; and the final chapter, Chapter 9, deals with applications of quantitative knowledge of the electron ionization process in various fields: mass spectrometry, plasma diagnostics, astrophysics, fusion research, aeronomy, gaseous electronics, and radiation physics. It is our hope and expectation that researchers and graduate students in these "user" areas as well as those working to further understand the intricacies of this fascinating process will find this book an important source of information.

We are grateful to the contributing authors both for the quality of their manuscripts and for their patience: some in waiting for the process to come to completion and some in bearing the harassment of the editors to complete their parts. We also commend the people at Springer Verlag Wien-New York for the quality and speed with which they put together the finished product, and for their forbearance.

January 1985

T. D. Märk
G. H. Dunn

Contents

1

Quantum Theoretical Methods for Calculating Ionization Cross Sections

S. M. Younger

Lawrence Livermore National Laboratory, Livermore, Cal., U.S.A.

1.1 Introduction

The electron impact ionization of atoms and ions is a topic of both fundamental and practical importance in modern atomic physics. From a practical standpoint, estimates of the rates of ionization of highly charged ions in high temperature plasmas are important components of models of the structure and dynamics of laboratory and astrophysical plasmas. Theoretical modeling of the complex ionic transport mechanisms observed in fusion plasma machines demands that rates for the relevant atomic processes be known. Since the ionization rate along with the associated recombination rates govern the charge state structure of plasmas, they are also important for the prediction of the spectral distribution of radiation from hot plasmas.

From a fundamental viewpoint the electron ionization event is one of the more complex of electron-atom interactions. In addition to the usual complexities of electron-atom scattering such as complex coupling effects, resonances, and target polarization, in electron ionization there is a double-continuum electron final state. The problem of two free electrons in the field of an ionized atom is a particularly difficult example of the quantum many body problem and has not yet been formulated in a manner conducive to practical calculation. Although impressive progress has been made over the past two decades on many-electron correlation in bound states of atoms and ions, few of these techniques can be carried over to the multiple continuum problem. Most of the progress in the theoretical calculation of electron ionization cross sections of atoms and ions has been associated with the development of approximation methods which are both numerically manageable and physically realistic.

The need for estimates of the electron ionization cross sections of atoms and ions has led to the use of a wide variety of semi-empirical and semi-classical approximations to the cross section. A detailed discussion of such methods is given in Chapter 2 of th s volume. In the present discussion we will concentrate on what might be termed an "intermediate" level approximation to electron ionization which, while certainly short of a rigorous theory of the process, attempts to incorporate as much physical insight into the theory as possible. Since the direct knock-out electron impact ionization cross section decreases with increasing nuclear charge, Z, roughly as Z^{-4}, the experimental determination of cross sections for highly charged ions is very difficult (Crandall 1981 a). Thus, the burden of providing the bulk of realistic estimates of such data falls to theory at least for the forseeable future. Given the need for information concerning the electron ionization of a wide variety of atoms and ions, we will concentrate on theoretical methods which have been or can readily be applied to the widest spectrum of atomic and ionic targets. For a discussion of more advanced theoretical developments applied mainly to neutral atoms we refer the reader to several review articles (McCarthy and Stelbovics 1980, Weingold and McCarthy 1978, Fano and Inokuti 1976, Rudge 1968, Massey and Burhop 1969, Younger 1982).

1.2 Theory of Electron Ionization of Atoms and Ions

1.2.1 General Theory

It may be said that there currently exists no rigorous theory of the electron impact ionization of atoms and ions. Indeed, developments since the early work of Thomson (1912) have concentrated on increasingly sophisticated approximations to an as yet undefined formal theory. The reason for such a state of affairs, unusual to atomic physics, is the inherently three-body nature of the electron ionization process. In the final state of the electron + ion system one has three free particles interacting via a long range potential, i.e., the Coulomb interaction

$$V_{12}(\vec{r}_1, \vec{r}_2) = \frac{Z_1 Z_2}{|\vec{r}_1 - \vec{r}_2|} \tag{1-1}$$

where Z_i is the charge on the i-th particle with position vector $\vec{r}_i$. Atomic units $(e = m = \hbar = 1)$ are used throughout this chapter unless otherwise indicated.

Although very powerful computational techniques have been developed to deal with atoms containing many *bound* electrons, or even many bound electrons plus a single free electron, the extension of such methods to the double continuum electron case has not yet been satisfactorily accomplished. Fadeev (1961) has recast the three body problem in terms of an infinity of two body problems, but from a practical standpoint this has only transformed a *formally* difficult problem into an *operationally* difficult problem in that Fadeev's equations have not yet been brought into a useful format.

Considerable effort has been devoted to theories based on optical potentials or perturbative approaches of one form or another (McCarthy and Stelbovics 1980). Steady progress has been reported in this direction by several groups, but most of the work has so far been concentrated on the electron ionization of neutral hydrogen and helium. Recently some purely numerical studies have been performed using a finite element technique which provide a visual display of the progress of the ionization event given a well defined set of initial conditions (Bottcher 1982).

Although very important progress has been made in the development of a comprehensive theory of electron ionization during the past two decades, most of the systematic calculations performed on many-electron atoms and ions have been made in some variant of the partial wave approximation. Since it is the intent of this work to discuss the theoretical description of electron ionization for a very broad class of targets, indeed, the entire periodic table insofar as it is possible to make reasonable statements, it is on this method that we shall concentrate most of our attention.

1.2.2 Partial Wave Theories

In the partial wave approximation each free electron with momentum $\vec{K}$ is represented by an expansion wavefunction, $\phi(\vec{K}, r)$, of the form

$$\varphi(\vec{K}, \vec{r}) = \sum_{lm} a_{lm} Y_{lm}(\theta, \phi) F_l(K, r) \tag{1-2}$$

where l and m are the orbital and azimuthal quantum numbers corresponding to the spherical harmonic $Y_{lm}(\theta, \phi)$ which is a solution to the angular part of the Schrödinger equation. The radial function, $F_l(K, r)$, is a solution to the associated radial Schrödinger equation

$$\left[-\frac{1}{2} \frac{d^2}{dr^2} + \frac{l(l+1)}{2r^2} + V(r) \right] F(r) = \frac{K^2}{2} F(r) \tag{1-3}$$

where $V(r)$ is the potential in which the free electron is assumed to move. Scattering approximations based on partial wave expansions are often identified by the potential, $V(r)$, used to describe the free electron. There are three common choices for $V(r)$:

Plane Wave Born Approximation $V(r) = 0$

The effect of the target on the free electron is ignored in the plane wave Born approximation. The radial functions $F_l(K, r)$ are simply spherical Bessel functions. The plane wave Born approximation has been extensively applied to atoms and ions by Peach (1966a, 1966b, 1968, 1970, 1971) and McGuire (1971, 1977a, 1977b, 1979).

Coulomb Approximation

$V(r) = \dfrac{-Z}{r}$, where Z is a constant representing an effective nuclear charge experienced by the free electron. The $F_l(K, r)$ are regular Coulomb functions, described in standard mathematical handbooks. Moores (1972, 1978) and Moores and Nussbaumer (1970) have applied the Coulomb Born approximation to a number of ionic targets.

An interesting variant of the Coulomb-Born approximation is the so-called "infinite Z" or "scaled hydrogenic" approximation which makes use of the fact that a change in the radial scale of the many-electron Hamiltonian yields a zeroth order Hamiltonian identical to the hydrogenic case plus a perturbation term equal to the interelectronic interaction times $1/Z$. For large Z the perturbation will be small relative to the zeroth order term which is Z-independent. The $Z = \infty$ method ignores the many-electron perturbation entirely and yields estimates of the electron ionization cross section for the nl orbital of any atom or ion based on values computed using only $Z = 1$ Coulomb functions in the matrix element. This method has been applied to ionization of the orbitals $1s - 4f$ and to many excitation processes as well by Sampson, Golden, and Moores (Golden and Sampson 1977, 1980, Sampson and Golden 1979, 1981, Golden et al. 1978, Moores et al. 1980). For atoms and ions where the target potential has significant influence on the partial waves such a simple method understandably fails. A rule of thumb for the application of the scaled hydrogenic approximation is that the ratio Z/N must be larger than two or three for accurate results to be obtained, where Z is the bare nuclear charge and N is the number of electrons in the initial target.

Distorted Wave Approximation

$$V(r) = \frac{-Z}{r} + V_{DW}(r) \tag{1-4}$$

where $V_{DW}(r)$ is some approximation for the potential due to the atomic target electrons. In the close coupling and R-matrix approximations the atom is allowed to "relax" in the presence of the continuum electron yielding a type of "self-consistent-field" solution for the target + electron system. Even in the "frozen core" approximation where the target wavefunction is fixed the potential $V_{DW}(r)$ may be non-local, and energy dependent. The distorted wave approximation has been applied to a number of atoms and ions by Blaha and Davis (1980), Jakubowicz and Moores (1981), Moores (1980), Trefftz (1962), Stingl (1972), and Younger (1980 a, 1980 b, 1981 a, 1981 b, 1981 c, 1982 a, 1982 b, 1982 c, 1982 d, 1982 e, 1982 f, 1982 g).

Peak and Golden (1980) have investigated the use of an "average approximation" for the transition potential by the scattering electron. This method, which is in the spirit of a WKB approximation developed by Riley (1973), reduces the considerable computational costs associated with the distorted wave approximation. Preliminary results of Peek (1980) indicate that such methods may be an accurate and efficient substitute for full partial wave expansion techniques.

The electron ionization event proceeds as follows: The incident electron approaches the target with energy T and interacts with the target electrons via the Coulomb interaction $V_{12}(r_1, r_2)$. An inelastic transition occurs which scatters the free electron into the state T_f and ejects one of the bound target electrons into state T_e. In first order perturbation theory the matrix element describing this interaction is

$$M_d = \left\langle \phi_b \phi_i \left| \frac{1}{r_{12}} \right| \phi_e \phi_f \right\rangle \tag{1-5}$$

where ϕ_b, ϕ_i, ϕ_e, and ϕ_f are the orbital functions describing the bound, incident, ejected, and final scattered electrons, respectively.

The probability $P(bi \rightarrow ef)$ for ionization of the target is proportional to the square of the interaction matrix element

$$P(bi \rightarrow ef) \sim |M_d|^2. \tag{1-6}$$

Inserting the appropriate kinematical factors one arrives at an expression for the electron ionization cross section

$$\sigma(T, T_e, T_f) = \frac{16\, a_0^2}{T} \sum_L (2L+1) |M_d|^2 \tag{1-7}$$

where L is the total angular momentum of the electron + target system. In order to get the total integrated ionization cross section for single ionization[1] σ, it is necessary to integrate over the possible distribution of energy among the final state electrons.

Then

$$\sigma_1(T) = \int_0^{T_{\max}} dT_e\, \sigma(T, T_e, T_f). \tag{1-8}$$

The upper integration limit is given by

$$T_{\max} = \frac{1}{2}(T - I), \tag{1-9}$$

the so called "half-range Born" approximation, where I is the ionization energy.

Since electrons are identical particles it is necessary to consider exchange processes. One can distinguish between two types of exchange in ionization: potential exchange and scattering exchange. Potential exchange denotes exchange between a free electron and a target electron. It is analogous to exchange in a Hartree-Fock bound state calculation and is easily taken account of in a frozen core approximation by including the appropriate non-local potential terms in V_{DW}. The second form of exchange, scattering exchange, describes the possible interchange of the two final state continuum electrons. Considerable attention has been devoted to the problem of scattering exchange, especially since it is closely involved with the three-body nature of the final state. Problems arise even at the most elementary level of the consideration of scattering exchange, however, as will now be shown.

The exchange contribution to the scattering matrix element is

$$M_{ex} = \left\langle \phi_b \phi_i \left| \frac{1}{r_{12}} \right| \phi_f \phi_e \right\rangle, \tag{1-10}$$

so that the total probability for ionization is given by

$$P(bi \rightarrow ef) \sim |M_d + M_{ex}|^2 = |M_d|^2 + |M_e|^2 - \alpha |M_d|\, |M_{ex}|. \tag{1-11}$$

---

[1] The total ionization cross section here is called total integrated ionization cross section to avoid confusion with the total cross section introduced in Eq. 2-2 and used throughout the book.

$|M_d|^2$ is the direct ionization term, $|M_{ex}|^2$ the exchange ionization term, and $\alpha|M_d||M_{ex}|$ the interference term. The phase factor α is related to the relative phases of the direct and exchange matrix elements and depends on the interaction between the two final state continuum electrons. Rudge and Seaton (1964) have examined various choices for α, and the interested reader may examine their paper for a detailed exposition. For our purpose, we simply identify some of the more common choices for α which, to be honest, are chosen as much or more for computational tractability than for physical necessity.

Born-Oppenheimer Approximation: $\alpha = 0$. The exchange matrix element is computed by simply switching the two final state orbitals in the scattering matrix element, i.e., the same orbitals are used in both the direct and the exchange matrix elements. By omitting the interference term altogether one usually obtains total integrated ionization cross sections which are much too large, since the exchange cross section is often comparable to the direct cross section.

Born-Exchange Approximation [Rudge and Schwartz (1966) formulation]: Here α is chosen so as to cancel the imaginary part of the scattering matrix element. The result is that the cross section including exchange is comparable to the direct cross section.

Born-Exchange Approximation [Peterkop (1977) or "maximum interference" formulation]: $\alpha = 1$. In the maximum interference exchange approximation the interference term is maximized and the total exchange contribution minimized.

These two formulations of the Born exchange approximation usually yield comparable cross sections, with the Peterkop theory yielding slightly smaller values than the Rudge and Schwartz formulation. Since most theoretical predictions of direct knockout electron ionization cross sections are large compared to experimental data, it might seem practical to use the Peterkop formulation of exchange, which minimizes the total integrated ionization cross section. On the other hand the Rudge and Schwartz formulation has the correct form for electron ionization from systems with very large effective nuclear charges. One usually finds the differences between the two approximations to be smaller than the descrepancies between theory and experiment. Since either choice is essentially arbitrary, and both are within the confines of the Born exchange approximation, which is itself only a first order perturbation theory treatment, detailed elaboration on the choice for α is not particularly illuminating.

One point which must be made regarding the Born-exchange approximation is the significance of the choice of the potentials in which the final state partial waves are calculated. In the Born-Oppenheimer approximation the same final state orbitals are used to compute the direct and exchange matrix elements. In the Born-exchange approximation, however, it is essential that overlapping orbitals in a matrix element be computed in the *same* potential so that orthogonality is maintained between orbitals on opposite sides of the interaction matrix element (Jakubowicz and Moores 1980). The slow ejected wave in the exchange matrix element is computed in the potential of the initial target, an N-electron system, and the fast final state partial waves are computed in the potential of the residual ion, an $(N-1)$-electron system. Although somewhat nonphysical, this choice does yield total integrated electron ionization cross sections in good agreement with experiment for light atoms and ions.

From a strict perturbation theory point of view one might be tempted to compute *all* of the orbitals in the scattering matrix element in the *same* potential, namely that of the residual ion. Then the entire ionization event would be between eigenstates of one Hamiltonian. Further consideration, however, will reveal the existence of large first order corrections to such orbitals which effectively correct for the nonphysical occupation numbers of the zeroth order system. The use of different potentials for the target and scattering orbitals is an attempt to include such first order corrections in the zeroth order Hamiltonian basis set, albeit in a nonrigorous fashion.

1.2.3 Resonances

Resonances are common in electron-atom scattering, arising from a correspondence of the energy of an N-electron bound state plus a free electron with the energy of an $N + 1$ bound electron system. For example, in a lithium-like ion a resonance will occur when the energy of a $1 s^2 \varepsilon l$ state matches that of the $1 s 2 s^2\,{}^2S$ state. The latter is not really an "energy level" in the traditional sense, since its energy lies above the ionization limit for the lithium-like system. When $E(1 s^2 \varepsilon l) \sim E(1 s 2 s^2)$, there will be strong mixing between the two configurations with a corresponding change in the scattering cross section.

Resonance effects in electron ionization may be grouped into two classes: resonances in the incident electron channel and resonances in the ejected channel. There are resonances in the final scattered channel as well, but in most theories this channel is treated so poorly (as it is in the Born-exchange approximation where ϕ_f is computed in the potential of the initial target) that resonance effects are a relatively minor worry. Resonances in the incident channel will appear as sharp spikes in the total integrated ionization cross section at specific incident electron energies. Resonances in the ejected channel will appear as sharp spikes in the energy differential cross section $\sigma(T, T_e, T_f)$ and as abrupt step increases in the total integrated ionization cross section, as is illustrated in Fig. 1-1 for sodium-like iron. In forming the total integrated cross section for electron ionization, one integrates over all possible ejected energies, thus the same ejected resonance will contribute to the total integrated cross section at all incident energies above its excitation threshold. Resonances in the ejected continuum are often called "excitation-autoionization" resonances since they correspond to excitation to an autoionizing configuration.

Resonances in the ejected continuum result in dramatic enhancements of the total integrated electron ionization cross section of certain atoms and ions, sometimes by more than an order of magnitude over the direct (non-resonance) cross section. Since most ejected channel resonances involve production of a core-excitation, these effects are most important for alkali-like and alkaline-earth-like ions, where large inner shells closely underlie a sparsely occupied valence shell (Bely 1968, Cowan and Mann 1979, Kim and Cheng 1978, Crandall 1981, Falk *et al.* 1982, Hahn 1977, Hansen 1975, Henry 1979, Griffin *et al.* 1981 a, 1981 b). As the number of valence electrons increases, the relative contribution of the ejected resonances to the total integrated ionization cross section dereases with respect to that for ionization from the valence shell. Thus, for sodium-like ions one might find a factor of two enhancement in the average total integrated ionization cross section due to ejected channel resonances, whereas in argon-like such resonances might change the cross section by only a few percent.

Another important determinant of the importance of ejected channel resonances is the particular atomic structure of the target. In potassium-like ions, for example, an important ejected resonance for low ionization stages is $KL\,3\,s^2\,3\,p^6\,3\,d - KL\,3\,s^2\,3\,p^5\,3\,d^2$. This $\Delta n = 0$, $\Delta l = +1$ excitation is very strong and consumes most of the total oscillator strength of 6 allowed for transitions from the $3\,p^6$ subshell. For Ti^{3+} this resonance (actually a set of resonances due to the alternate coupling possibilities) causes more than an order of magnatude jump in the total integrated ionization cross section at $T \sim 40\,eV$. For V^{4+}, the next ion in the potassium isoelectronic sequence, the $KL\,3\,s^2\,3\,p^5\,3\,d^2$ level is bound. Since it still consumes most of the available oscillator strength, however, resonance effects in V^{4+} will be far less dramatic than in neighboring Ti^{3+}.

The usual prescription for including ejected channel resonances in ionization cross section calculations is to first calculate a non-resonant ionization cross section and add to it the cross sections for excitation of the autoionizing resonances. This assumes that all of the atoms excited to the resonance "state" will autoionize. For highly charged ions there is also the possibility of "radiative stabilization" of the autoionizing resonance, where it radiatively decays to a true bound state of the N-electron system. The autoionization rate is roughly constant as Z increases along an isoelectronic sequence but the radiative rate increases as Z^4, thus at large Z only a fraction of the ejected resonance excitations will actually lead to an ionization event. In such cases it is necessary to multiply the "excitation" cross section by an effective branching ratio for autoionization vs. radiative stabilization before adding it to the cross section for direct (non-resonant) ionization. Due to the complexity of the autoionizing states, the accurate calculation of the excitation cross sections and the associated branching ratios can be very difficult task, as is illustrated by the work of Cowan and Mann (1979) for sodium-like iron.

Butler and Moores (1981) have recently examined the effects of incident and ejected channel resonances on total integrated and energy differential electron ionization cross sections using a quantum defect approach. They treat the resonance in a scattering formalism including the possibility of emission of a photon rather than an electron. Their theory allows for interference effects between the ejection of an electron and emission of a photon which cannot be handled in a simple cross section addition approximation. Although early applications of the theory to lithium-like ions yield total integrated cross sections which are in good agreement with the addition technique, in more complex ions where direct ionization and resonance ionization cross sections are comparable one might expect the two approaches to yield substantially different results.

LaGattuta and Hahn (1981) have recently proposed another indirect mechanism for the electron ionization of atoms and ions whereby dielectronic recombination of the target ion and incident electron is followed by double autoionization, resulting in single ionization by a single electron collision. A substantial increase in the total integrated electron ionization cross section of Fe^{15+} due to this mechanism has been predicted. (See Fig. 1-1.) The relative importance of this contribution to electron ionization is determined by the existence of doubly-excited autoionizing resonances at appropriate energies. A systematic examination of the dielectronic recombination double autoionization mechanism with respect to target atomic structure has not yet been performed, but preliminary estimates indicate that it is somewhat less common than the excitation single autoionization indirect ionization process.

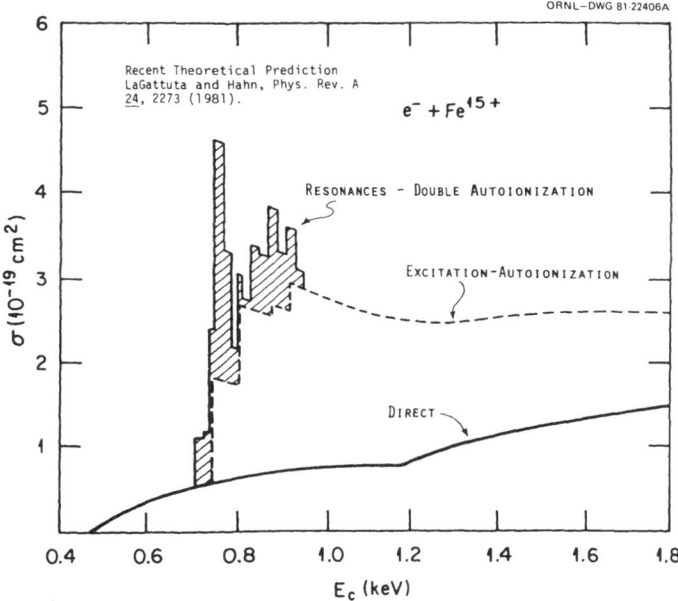

Fig. 1-1. Electron impact ionization cross section of Fe^{15+} as a function of electron energy T showing the relative contribution of direct ionization (dotted line), direct ionization + excitation-autoionization (dashed line) and direct ionization + excitation-autoionization + resonant excitation double Auger ionization (LaGattuta and Hahn 1981)

1.2.4 Target Electron Correlation

One might define electron correlation as the difference between the atomic structure approximation you are using and the non-relativistic truth. The Hartree-Fock approximation represents the best possible single particle description of an atom, since it is variationally optimized, so it is convenient to use it as the reference point for a discussion of electron correlation in the target and the effect of such correlation on the electron ionization cross section.

Except in certain cases of very complex atomic structure, the most important corrections to the Hartree-Fock approximation usually involve pair excitations. For example, in argonlike ions with the ground state $KL\,3\,s^2\,3\,p^6\,^1S$ an important correlation correction involves excitations of the type $3\,p^4\,3\,d^2\,^1S$, of which there are three owing to the different possible couplings between the $3p^4$ and $3d^2$ subshells. The total wave-function, including correlation is then

$$\phi^{\text{Corr}} = C_0\,\phi_0\,(3\,p^6) + \sum_{i=1}^{3} C_i\,\phi_i\,(3\,p^4\,3\,d^2) \qquad (1\text{-}12)$$

where C_i are mixing coefficients. The mixing coefficients as well as the $3\,d$ orbital are optimized such that ϕ^{Corr} minimizes the total energy of the system. In this case, and in most other cases of pair correlation, the resulting $3\,d$ orbital closely overlaps the $3\,p$ subshell. Physically this means that the correlation does not significantly alter the radial charge density of the atom, only its angular distribution.

Pair correlation effects are very important in calculations of optical transition probabilities, excitation cross sections and related atomic quantities. That they are much *less* important for electron ionization can be seen by the following simple argument. The electron ionization event is propagated by the $1/r_{12}$ interaction, a two-body operator. This means that the *total* wavefunctions on each side of the interaction matrix element can differ by *at most* two electrons. One of these electrons must be the scattering electron, which undergoes a transition from the incident to the final scattered state. The other must be the bound orbital, which goes to the ejected wave. If the final state of an argonlike target is assumed to be $3\,p^5$, then including the correlation terms $3\,p^4\,3\,d^2$ in the initial target wavefunction will make the left and right sides of the interaction matrix element differ by too many electrons, i.e.,

$$M = \left\langle 3\,p^4\,3\,d^2\,\phi_i \left| \frac{1}{r_{12}} \right| 3\,p^5\,\phi_e\,\phi_f \right\rangle. \qquad (1\text{-}13)$$

Orthogonality of the initial and final state orbitals causes this matrix element to vanish. One could allow the final target ion to be in the excited state $3\,p^4\,3\,d$, so that the interaction matrix element would be

$$M_i \left\langle 3\,p^4\,3\,d^2\,\phi_i \left| \frac{1}{r_{12}} \right| 3\,p^4\,3\,d\,\phi_e\,\phi_f \right\rangle. \qquad (1\text{-}14)$$

Now the left and right sides of the matrix element differ by only two electrons and the matrix element does not vanish. Note, however, that this matrix element does not add coherently to the zeroth order (non-correlated) matrix element, since in M_i an even symmetry d-orbital is being ionized; whereas in M_0 an odd-symmetry p orbital is being ejected. When the partial cross sections are added, one obtains

$$\sigma_1 = |C_0|^2\,\sigma^0 + \sum_{i=1}^{3} |C_i|^2\,\sigma^i. \qquad (1\text{-}15)$$

Since C_i is approximately 0.2 for argonlike ions while $C_0 \approx 0.95$ the correlation terms will result in only a small change of the total integrated ionization cross section from the non-correlated case. Also, since the $3\,d$ orbital closely overlaps the $3\,p$ orbital the $3\,d$ ionization cross section will be very close to that of the $3\,p$ cross section further reducing the effect of correlation on the total integrated electron ionization cross section.

The above discussion of electron correlation effects is necessarily qualitative, since only a few calculations of electron ionization cross sections including such corrections have so far been made (Younger 1981 c, 1982 e). In cases where the atomic structure is very complex or where violations of the above assumptions occur one might expect a much more complex situation to prevail, but for many systems, particularly highly charged ions, it does not appear as if target electron correlation effects are as important in electron ionization as in other atomic processes. An important exception to this pattern is electron ionization from closed subshells with large orbital angular momenta, which will be discussed in detail below.

1.3 Results of Partial Wave Calculations of Electron Ionization Cross Sections

Having presented a brief outline of partial wave methods of calculating electron impact ionization cross sections, we shall now proceed to examine how well these methods succeed in predicting the total integrated ionization cross sections of real atoms and ions. Since a persistent worry in the simple theories is the extent to which the omitted final state continuum-continuum electron correlation affects the final result, we start with the simplest of test cases, where uncertainty over the accuracy of target wavefunctions, resonance effects, etc. is smallest. Since the same *qualitative* continuum-continuum electron correlation occurs in all cases of electron ionization of atoms and ions, by choosing a target with simple atomic structure we can assign most of the discrepancy between theory and observation to the scattering approximation so that we can get at least a general idea of the adequacy of the partial wave approximation for describing electron ionization.

1.3.1 One Electron Atoms

As a first choice one might select neutral atomic hydrogen for such a test, since its target wavefunction is known analytically and since it has a vanishing asymptotic distorted wave potential in the initial state and no resonance structure in the ejected channel. Neutral hydrogen is among the most difficult of atoms to deal with theoretically, however, since the ionized state consists of three particles (two electrons and a proton) of equal charge magnitude interacting via the long range Coulomb potential. Even for heavier target atoms, where the atomic structure is not as simple, the scattering picture might be cleaner, since the partial waves are more strongly influenced by the higher nuclear charge than by the complex electron-electron interaction. Heroic attempts have been made to accurately compute the electron ionization cross section of hydrogen with results that are sometimes in worse agreement with experiment than is the case for much more complex targets. Even relatively sophisticated calculations such as the close coupling results of Burke and Taylor (1965), the pseudostate calculation of Calloway and Ona (1979), and the calculations of McCarthy (1980) fail to yield accuracy much better than $10-30\%$ for incident electron energies below 100 eV. A word of caution is in order, however, in that the experimental results are uncertain to $10-20\%$ (Kieffer and Dunn 1966; see also Chapters 5 and 7), so that a definitive statement of the discrepancy between theoretical and experimental cross sections is not possible at this time. It is interesting to note that the plane wave Born and the distorted wave Born approximations yield very similar results. This is not surprising since the only significant departure of the distorted wave orbitals from plane waves occur for very low l continuum orbitals which penetrate the $1s$ charge distribution. Since low-l partial waves tend to contribute only a minor part of the total integrated ionization cross section (except at very low incident electron energies) most of the distorted wave partial cross sections will be identical to the plane wave results.

As a second example of electron ionization from a one-electron target consider He^+. In this case all of the partial waves (incident, final, ejected) see a non-vanishing asymptotic charge. Since the helium nucleus has *two* units of charge, all of the partial wave orbitals involved will be strongly influenced by the electron-nucleus interaction. One might then expect simple partial wave methods based on the Coulomb-Born or distorted wave Born approximations to work much better for He^+ than for neutral hydrogen. Almost all of the theoretical approximations applied to He^+ yield total integrated electron ionization cross sections in good agreement with experiment. The plane wave Born results are larger than either the Coulomb-Born or the distorted wave data. This is expected since the attractive potentials in the latter approximations effectively "focus" the partial waves toward the nucleus, increasing their interaction with the target and hence increasing the total integrated cross section. By the same reasoning one might expect the distorted wave cross sections to be larger than the Coulomb-Born, since the former have a stronger effective potential near the nucleus than the latter, which assumes a fixed effective charge equal to the asymptotic charge of the distorted wave potential. (Exceptions may occur for more complex targets where the overlap between partial waves and bound orbitals with complex nodal structures can make the cross section very sensitive to small changes in the distorted waves.) Note that the effect of exchange in the final state is to reduce the total cross section by approximately $0-25\%$. Such a figure is typical of the effect of exchange noted in many other calculations, although in some cases one finds the cross section including exchange to be *larger* than the no-exchange result in the near threshold incident electron energy range (Younger 1980a, 1981b).

1.3.2 Two Electron Atoms

Neutral helium is an interesting test for theories of electron ionization since it has a simple atomic structure which is fairly accurately described by a Hartree-Fock level of approximation and since a large amount of experimental data is available for comparison purposes. A number of theoretical calculations have been made for the electron ionization cross section of neutral helium in variants of the plane wave Born (Bell and Kingston 1969, 1975, Economides and McDowell 1969, Robb *et al.* 1975, Sloan 1965) and distorted wave approximations (Brandson *et al.* 1978, 1979, Madison *et al.* 1977) with generally similar results. For very low incident electron energies ($u < 1.25$, with $u = T/I_n$ and I_n the binding energy of electrons in n-th subshell) the theoretical estimates generally underestimate the experimental results, whereas at higher energies they exceed the measured cross sections by 25% or more.

Several very detailed calculations have been performed on the angular and energy distributions of ejected and scattered electrons following electron ionization of neutral helium (Bell and Kingston 1975, Bransden *et al*. 1978, 1979, Madison *et al*. 1977, Tweed 1980). These calculations generally agree well with available experimental data for incident electron energies above 150 eV but become increasingly inaccurate as the energy is reduced. Since 150 eV is more than six times the ionization energy of neutral helium this is an indication that quantitative agreement of the total integrated ionization cross section with measured values is obtained at significantly lower energies than for differential ionization cross sections. Put another way, a triple integration (two angles, one energy) forgives many sins.

1.3.3 The Neon Isoelectronic Sequence

So far we have shown that although partial wave methods are only qualitatively successful at predicting the electron ionization cross sections of light neutral atoms, they are surprisingly accurate for ions. One might expect that such behavior would extend to moderately heavier systems with simple atomic structures such as neonlike ions but unfortunately this is not the case. The failure of neonlike ions to follow the classical scaling law (which predicts that the quantity $u I^2 \sigma$ is constant along an isoelectronic sequence) was recognized some time ago when measurements of the electron ionization cross section of $Na^+ + e \rightarrow Na^{2+}$ were compared to results for neutral neon (Peart and Dolder, 1968). Further studies of Mg^{2+} confirmed the failure of classical scaling for ten-electron ions (Peart *et al*. 1969).

Theoretical studies of the Ne-sequence have not yet explained the breakdown of classical scaling. Structurally, the neonlike system has one of the simplest of atomic configurations with a completely closed shell ground state. In this case (and also in the case of heliumlike ions) there are no excited states having the same principal quantum number as a ground configuration orbital. Thus target electron correlation effects in the Ne-like ground state should be much weaker than in other noble-gas systems where the lowest nd excited orbital mixes strongly with the ground configuration.

From an *ab initio* standpoint then, a neonlike ion should be an excellent candidate for a successful application of distorted wave methods since it represents a many electron system with a simple atomic structure which should not be significantly affected by ejected channel resonances. Fig. 1-2 illustrates the results of several calculations of the electron ionization cross sections of Ne-like ions. For neutral neon the distorted wave method yields σ_1 in excellent agreement with measurements in the near-threshold incident electron energy range, and is only about 20% too large at $u = 5$. Simple plane wave Born calculations are also in good agreement with experiment, especially at higher energies. For Na^+, however, a qualitatively different situation is found. Here all of the theoretical predictions are in poor agreement with experimental results, especially in the near threshold energy range. Furthermore, there is a significant difference between the results of the plane wave Born, the Coulomb-Born, and variants of the distorted wave approximation. It has been found that the theoretical σ_1 is very sensitive to the details or the potential in which the low-energy *ejected* partial waves are computed (Younger 1981 a).

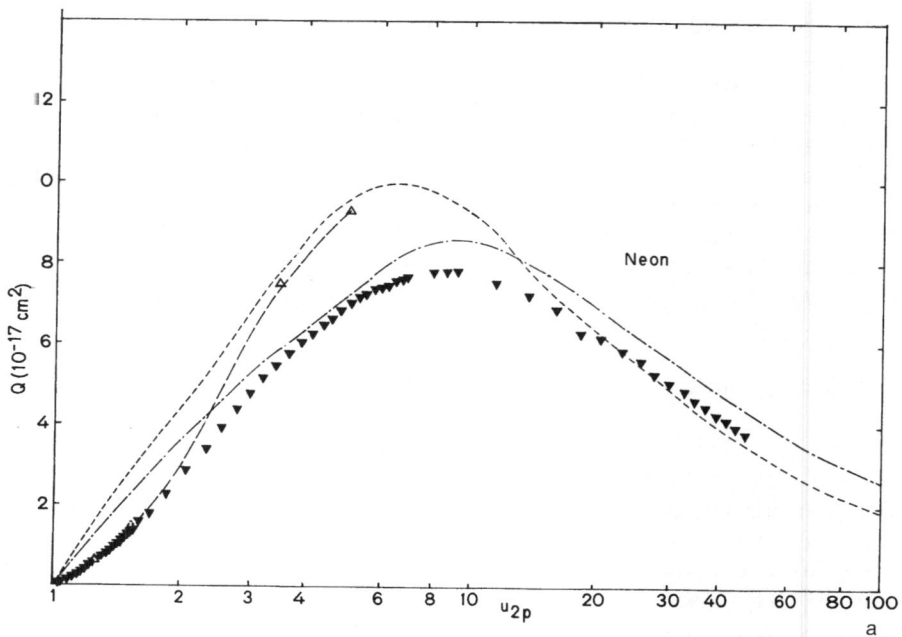

a

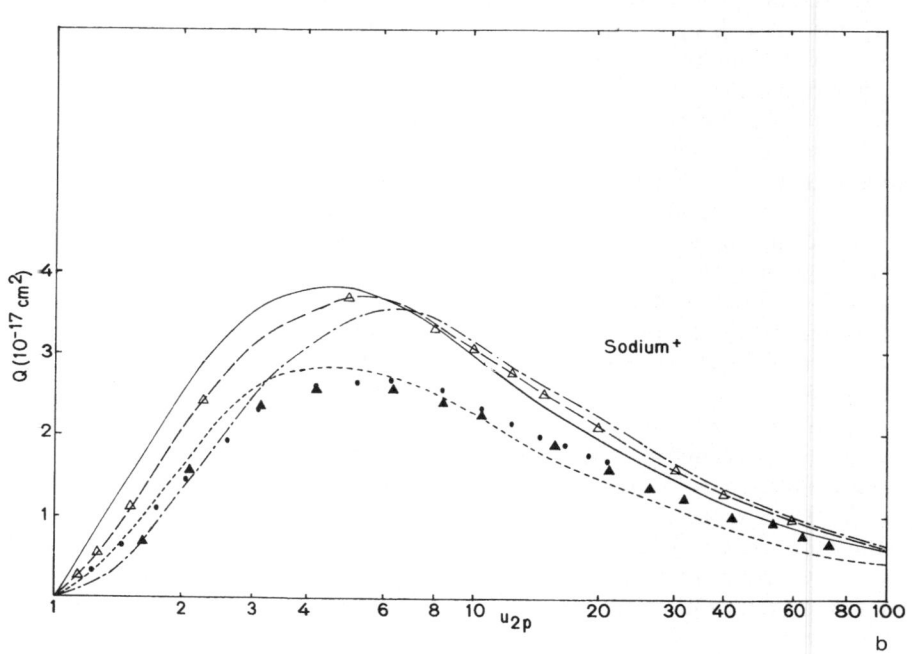

b

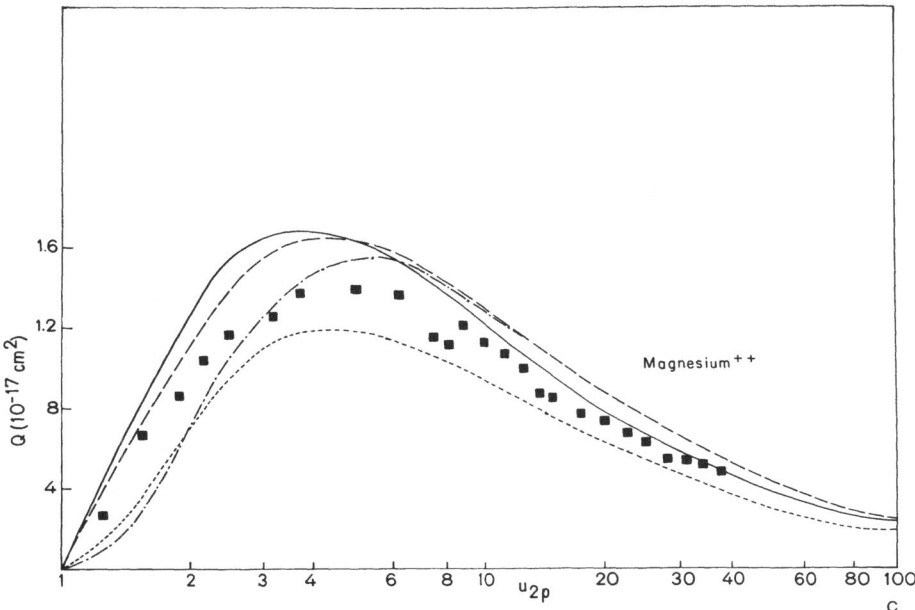

Fig. 1-2. Comparison of the results of various theoretical methods for computing electron impact ionization cross sections of neonlike ions for which classical scaling of the cross section breaks down. ––––– plane wave Born, distorted ejected waves (McGuire 1971, 1977); –·–·– Plane wave Born, Coulomb ejected waves (Peach 1971); ——— Coulomb Born, distorted ejected waves (Moores 1972); — — distorted wave (Younger 1981); Experiments: ▽ Rapp and Englander-Golden 1965; ○ Hooper *et al.* 1966; △ Peart and Dolder 1978; □ Peart *et al.* 1969

Relatively minor modifications of the ejected wave potential can result in factor of two changes in the total integrated cross section. In particular, the inclusion of potential exchange terms in the ejected channel potential was found to be important. For Mg^{2+} the disagreement between theory and experiment is not as pronounced as it is in Na^+, and one might hope for a convergence of predicted and measured results at still higher charge states. Qualitatively, one might argue that for neutral neon the distorted wave potential is not sufficiently strong to significantly modify those partial waves most important in determining an accurate total integrated electron ionization cross section. For Na^+ the potential is strong enough to cause a significant interaction between the partial waves and the ion. For higher charge states this complex interelectronic interaction will be dominated by the effective nuclear charge, resulting in more accurate total integrated cross sections. Such a tentative explanation should be viewed suspiciously, however, since it glosses over several very complex interelectronic interactions such as angular coupling, and since it is based only on comparisons of total integrated cross sections. The possibility of fortuitous agreement of theory and experiment for neutral neon due to cancellation of several important interaction mechanisms cannot be excluded. Still, one does expect the partial wave approximation to become more accurate for higher ionic charges where the scattering potentials approach the pure Coulomb potential limit.

1.3.4 The Argon Isoelectronic Sequence

Argonlike ions differ from heliumlike and neonlike systems in that the $KL\,3\,s^2\,3\,p^6$ ^1S configuration is only a closed *subshell* configuration and not a closed *shell* configuration. Calculations of photon excitation and ionization cross sections have shown that the in-shell pair correlation excitation $3\,p^4\,3\,d^2\,{}^1S$ has a significant influence on transition matrix elements. Furthermore, the dipole excitation channel $3\,p^5\,\varepsilon\,d$ is strongly influenced by a term-dependent exchange interaction between the $3\,p^5$ subshell and the $\varepsilon\,d$ ejected wave. In particular, a large positive coefficient for the $(3\,p, \varepsilon\,d)$ dipole potential exchange term in the wave equation effectively repels the ejected $\varepsilon\,d$ wave from the vicinity of the $3\,p$ subshell. Since the scattering matrix element is dependent on the overlap between the $3\,p$ and $\varepsilon\,d$ orbitals it follows that this exchange interaction will have a strong influence on the cross section. Furthermore, the strong perturbation of the $3\,p^5\,\varepsilon\,d\,{}^1P$ channel will result in non-zero overlap integrals between the $\varepsilon\,d$ wave and the $3\,d$ correlation orbital, which being varationally optimized to minimize the ground state energy, closely overlaps the $3\,p$ ground state orbital. This non-zero overlap of the $3\,d$ and $\varepsilon\,d$ orbitals allows ground state pair correlation to affect the $3\,p \rightarrow \varepsilon\,d$ partial cross section by the introduction of correlation terms within the scattering matrix element itself. This situation contrasts with the usual scenario described in Section 1.2, where target pair correlation was found to have a relatively minor influence on the total electron ionization cross section.

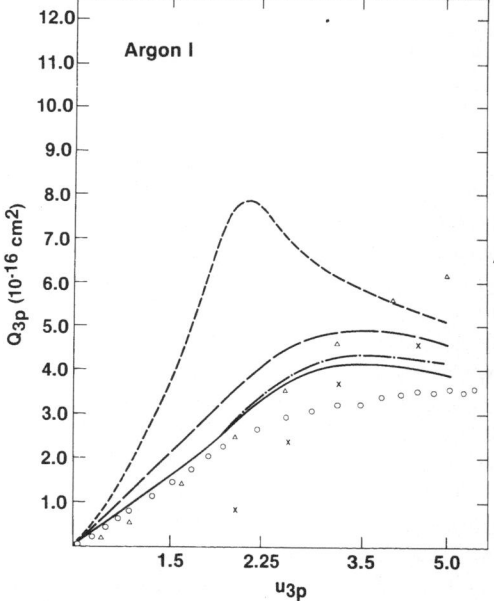

Fig. 1-3. Theoretical approximations for the electron impact ionization cross section of neutral argon. – – – central field distorted waves; — — — distorted wave with term-dependent Hartree-Fock ejected waves; ——— distorted wave with term-dependent Hartree-Fock ejected waves and ground state correlation; —·— solid curve plus contribution due to $3\,s$ subshell (Younger 1982); △ plane wave Born with central field distorted ejected waves (McGuire 1971); × plane wave Born with Coulomb ejected waves (Peach 1971); ○ experiment (Asundi and Kurepa 1963)

Fig. 1-3 illustrates the results of several partial wave calculations of the total electron ionization cross sections of neutral argon (Younger 1982 e). As expected from previous work on the photoionization of neutral argon the use of term-dependent ejected d-waves in electron ionization calculations substantially changes the neutral argon cross section from results obtained with a "configuration averaged" potential. Further improvement of the theory by allowing for target electron correlation yields a total integrated cross section in reasonably good agreement with experiment considering the complexities of the calculation. Note in particular that the distorted wave results correctly predict the location of the cross section maximum with a respectable estimate of its amplitude as well. Based on results of photoionization cross section studies of argonlike ions the influence of the term-dependent potential and target electron correlation on the total integrated electron ionization cross section are expected to diminish rapidly with increasing nuclear charge beyond K^+.

1.3.5 Indirect Mechanisms for Electron Ionization

Recently several dramatic demonstrations of the importance of indirect mechanisms for the electron ionization of ions have been reported (Falk $et\,al.$ 1981, Crandall 1981, 1982). As was discussed in Section 1.2, indirect mechanisms are most important for ions with one or only a few valence electrons outside a many-electron closed shell. Griffin and Bottcher (1981) have studied the location and relative importance of ejected wave resonances in a wide variety of ions and have been able to at least qualitatively explain much of the structure observed in several recent measurements.

Of particular interest from a theoretical standpoint in the case where the matrix elements for direct and indirect ionization are comparable in magnitude. In this situation one might expect coherent interference of the total integrated ionization scattering amplitude to occur with a very noticeable effect on the total integrated ionization cross section. Such a case has been analyzed by Pindzola $et\,al.$ (1982) in calculations of the electron ionization of Ga^+. They found that although the $total$ integrated distorted wave cross sections for electron ionization of Ga^+ with and without resonances are similar, the cross sections differential in the ejected electron energy were quite different, the resonance ionization cross section containing an impressive Feshbach resonance at $T_e \cong 2\,\mathrm{eV}$.

1.3.6 Ionization from Inner Shells

In most of the preceeding discussion we have concentrated on electron ionization of the outermost subshell of the target. In many-electron atoms and ions, however, one can also eject an inner shell electron. Consider, for example, the sodium-like ions Mg^+ and Ar^{7+} (Younger 1981 b). With a ground configuration of $K\,2s^2\,2p^6\,3s$ sodium-like ions have a single valence electron outside of an 8-electron inner L-shell. For Mg^+ the effect of inner shell ionization is to produce a "shoulder" on the total ionization curve for $u_{3s} > 5$. Except for very high incident electron energies ejection of the $3s$ electron is the dominant single ionization mechanism for Mg^+. In Ar^{7+} the situation is quite different. In this ion the stronger nuclear charge results in a contraction of the $2p$ and $3s$ subshells so that for $u_{3s} > 3$ the total integrated electron ionization cross section is dominated by the $2p$-subshell contribution.

Inner shell ionization will be most important in atoms and ions with a sparsely occupied valence shell and a many-electron inner shell with a comparable ionization energy (Hahn 1977, Hansen 1975, Cowan and Mann 1979, Bely 1968, Henry 1979). Since the ionization cross section of a subshell is roughly linear in the number of subshell electrons large inner shell contributions will be found for closed subshells with high orbital angular momentum. From a geometrical point of view, even though an inner shell orbital has a smaller mean radius than a valence orbital the larger occupation number of the inner shell can more than make up for the smaller cross section per inner shell electron.

There are two qualifications which limit the contribution of inner shell ionization to single ionization cross sections. First, as the number of valence electrons increases the threshold energy for ejection of an inner shell electron increases relative to the valence shell ionization energy. Since the ionization cross section goes roughly as I^{-2} more deeply buried subshells will yield smaller relative contributions to the total electron ionization cross section. Second, when a deeply buried electron is ejected from an atom the resulting ion may be left in an autoionizing state, which can emit one or more electrons in a cascade of Auger transitions (Abouaf 1970). We will not discuss multiple ionization here, except to say that such processes are expected to be very important for heavy many electron ions which have many subshells available for multiple ionization cascades.

An interesting side effect of inner shell ionization is its role in the formation of populations of ions in metastable excited states of the final ion (Kupriyanov and Latpov 1965). For example, in ionization of the $2s$ subshell of boronlike ions one can produce substantial numbers of berylliumlike ions in the $K2s2p\ ^3P$ metastable state. For low-Z berylliumlike ions where electron excitation of the triplet states from the berylliumlike singlet ground state is very weak innershell ionization can be an important mechanism for the population of the metastable manifold.

1.3.7 Ionization from Excited States

For the purpose of discussion, one might classify excited states into three groups according to the degree of excitation: First, there are the metastable excited terms within the ground configuration, where they exist. In carbonlike ions for example, one has a $1s^2 2s^2 2p^2\ ^3P$ ground state and excited states ^1D and ^1S. Second, there are metastable terms associated with excited configurations, such as the $1s^2 2s2p$ ^3P term in berylliumlike ions. Third, there are Rydberg excitations such as the $1s^2\ nl$ states in lithium-like ions. Very little information, either theoretical or experimental, is available on the electron ionization of atoms and ions in excited states (Ton-That et al. 1977, Kupriyanov and Latypov 1965, Rosner et al. 1976, Shearer-Inumi and Botter 1974, Vriens et al. 1968). Experimentaly only metastable states usually have lifetimes long enough to allow scattering cross sections to be measured.

Electron ionization from excited terms within the ground configuration produces non-zero cross sections at incident electron energies below the threshold for ionization of the ground state. Such behavior has been observed, for example, in measurements of the electron ionization cross sections of triply ionized rare gases, which have the ground configuration terms ^4S, ^2D, and ^2P in order of increasing energy. Griffin *et al.* have shown that the population of excited terms within the ground configuration can dramatically influence the resonance structure in the ejected channel of Fe^{4+} since several important resonances are associated only with the excited terms (Crandall 1982). Since the population of the excited terms is equivalent to a non-zero temperature of the target, such effects can be viewed as an illustration of the dependence of the electron ionization cross section on the temperature of the target.

Only a few measurements of excited configuration metastable state cross sections have been reported and these are mostly for noble gas neutral atoms and for the $2s$ state of hydrogenic ions (Defrance *et al.* 1981). Theoretical estimates of such cross sections typically yield the same qualitative level of accuracy as is found for the corresponding ground state cases.

Preliminary calculations of electron ionization cross sections from highly excited Rydberg states in Be^+ and O^{5+} using a distorted wave method are in good agreement with the results of scaled hydrogenic theory. At present there are no experimental data available on the electron ionization of highly excited Rydberg orbitals.

1.3.8 Electron Ionization of Heavy Atoms and Ions

Given the modest success of partial wave theories in predicting the single electron ionization cross sections of light atoms and ions it is of interest to see how well such methods perform for more complex targets. Many-electron atoms and ions offer a particularly strong test of distorted wave methods due to the large difference between the bare nuclear charge and the effective asymptotic charge. Unfortunately, only a small number of heavy atoms and ions have been studied experimentally so that comparison data is limited. Belic' *et al.* (1982) describe crossed-beam measurements as well as distorted wave theoretical results for Hg^+.

McGuire (1971, 1977a, 1977b, 1979, 1982) has performed a large number of calculations of electron ionization cross sections of heavy atoms and ions using a plane wave Born approximation with Hartree-Fock-Slater target wavefunctions. Except at low incident electron energies McGuire is generally able to predict the single ionization cross section to within about a factor of two, and sometimes much better.

Calculations for heavy atoms and ions are complicated by the extremely complex atomic structure of the target. In the lanthanides and actinides, for example, many hundreds of energy levels associated with several different configurations occur very near the ground state.

1.4 Discussion

In the preceeding sections we have attempted to give an overview of current theoretical methods used to compute electron impact ionization cross sections for atoms and ions. We have concentrated on methods which have actually been applied to the calculations of cross sections for several different targets, although there has been a great deal of additional work done on the application of advanced theoretical methods to simple atoms such as hydrogen and helium.

The present state of electron ionization theory allows us to make several observations on the kinds of effects encountered in the electron ionization of atoms and ions:

1. For the calculation of total integrated electron ionization cross sections of light neutral atoms partial wave methods are usually accurate to better than 50%.

2. For light ions the Coulomb-Born and distorted wave methods accurately reproduce measured cross sections over a very extended energy range. For light ions the Coulomb-Born and distorted-wave cross sections are very similar. There are currently no experimental measurements of the angular or energy distributions of final state electrons associated with the electron ionization of ions.

3. For moderately complex ions such as argonlike ions, the ejected partial waves may be strongly perturbed by term-dependent potential effects. Target electron correlation can also affect the cross section, particularly in combination with such partial wave potential effects.

4. Inner shell cross sections are sensitive to correlation between the ejected electron and diffuse valence orbitals.

5. Resonances in the ejected electron continuum can produce abrupt increases in the total integrated ionization cross section by an order of magnitude or more. These resonances are most important in alkali-like ions or other ions where a many electron inner subshell occurs in conjunction with a sparcely occupied valence shell. Complex resonant structures can severely perturb the ejected electron energy distribution in complex ions.

6. Heavy atoms and ions suffer from all of the above complexities and can also have large cross sections for multiple ionization by a single electron impact. Relativistic effects may be important even for valence shell ionization, although this point has not yet been systematically explored (Ndefru and Malik 1981).

7. For few-electron isoelectronic sequences the classically scaled electron ionization cross section $u I^2 \sigma$ has been found to vary smoothly with the nuclear charge. For targets with more complex atomic structures, there can be large deviations from classical scaling for the first few ions of the sequence due to target electron correlation and distorted wave potential effects. A guide to the adequacy of simple partial wave methods for electron ionization may be found in the extensive literature on photoionization, where similar problems in dealing with the ejected electron are treated in a simpler computational framework.

8. For sufficiently highly ionized ions the *classically scaled cross section* $u I^2 \sigma$ for direct ionization of a given subshell has been found to be roughly linear in the

number of bound electrons in that subshell, although the cross section, σ, itself may not scale in such a manner.

The past several years have brought major advances in our understanding of the electron ionization of atoms and ions. It is important to recall, however, that most of the theoretical results discussed here were computed in relatively simple partial wave approximations which only crudely account for final state continuum-continuum interaction, target polarization, etc. Partial wave methods can be improved only to a point-further tinkering with the details of the calculation could amount to little more than a complex method of fitting experimental data. Theorists have been able to explain and predict much, but not all, of the observed features of electron ionization. Improvement of these predictions from a qualitative to a quantitative level of accuracy requires a continued attack on the difficult problem of the many body quantum scattering problem.

References

Abouaf, R. (1970): J. Physique *31*, 277 − 283.
Asundi, R. K., Kurepa, M. V. (1965): J. Electron Control *15*, 41 − 50.
Belic', D. S., Falk, R. A., Dunn, G. H.: To be published.
Bell, K. L., Kingston, A. E. (1975): J. Phys. *B 8*, 2666 − 2678.
Bely, O. (1968): J. Phys. B *2*, 23 − 28.
Blaha, M., Davis, J. (1980): Electron Ionization Cross sections in the Distorted Wave Approximation. NRL Memorandum Report 4245, Naval Research Laboratory, Washington, D. C.
Bottcher, C. (1981): J. Phys. *B 14*, L 349 − L 355.
Bransden, B. H., Smith, J. J., Winters, K. H. (1978): J. Phys. *B 11*, 3095 − 3114.
Bransden, B. H., Smith, J. J., Winters, K. H. (1979): J. Phys. *B 12*, 1267 − 1278.
Burke, P. G., Taylor, A. J. (1965): Proc. Roy. Soc. *A 287*, 105 − 122.
Butler, K., Moores, D. L. (1981): XII Intl. Conf. on Phys. Electronic and Atomic Collisions (Datz, S., ed.).
Callaway, J., Oza, D. H. (1979): Phys. Lett. *72 A*, 207 − 208.
Cowan, R. D., Mann, J. B. (1979): Astrophys. J. *232*, 940 − 947.
Crandall, D. H. (1981): Physica Scripta, *23*, 153 − 162.
Crandall, D. H. (1982): To be published.
Crandall, D. H., Phaneuf, R. A., Falk, R. A., Belic, D. S., Dunn, G. H. (1982): Phys. Rev. *A 25*, 143 − 153.
Defrance, P., Claeys, W., Cornet, A., Poulaert, G. (1981): J. Phys. *B 14*, 111 − 117.
Economides, D. G., McDowell, M. R. C. (1969): J. Phys. *B 2*, 1323 − 1331.
Fadeev, L. D. (1961): Sov. Phys. JETP *12*, 1014 − 1019.
Falk, R. A., Dunn, G. H., Griffin, D. C., Bottcher, C., Gregory, D. C., Crandall, D. G., Pindzola, M. S. (1982): To be published.
Fano, U., Inokuti, M. (1976): On the Theory of Ionization by Electron Collisions. Argonne National Laboratory Report ANL-76-80 Argonne, Illinois.
Golden, L. B., Sampson, D. H. (1977): J. Phys. *B 10*, 2229 − 2237.
Golden, L. B., Sampson, D. H. (1980): J. Phys. *B 13*, 2645 − 2652.
Griffin, D. C., Bottcher, C., Pindzola, M. S. (1982): Phys. Rev. *A 25*, 154 − 160.
Griffin, D. C., Bottcher, C., Pindzola, M. S. (1981 b): To be published.
Hahn, Y. (1977): Phys. Rev. Lett. *39*, 82 − 84.
Hansen, J. E. (1975): J. Phys. *B 8*, 2759 − 2770.
Henry, R. J. W. (1979): J. Phys. *B 12*, L 309 − L 313.
Hooper, J. W., Lineberger, W. C., Bacon, F. M. (1966): Phys. Rev. *141*, 165 − 173.
Jakubowicz, H., Moores, D. L. (1980): Comments At. Mol. Phys. *9*, 55.
Jakubowicz, H., Moores, D. L. (1981): To be published.

Kieffer, L. J., Dunn, G. H. (1966): Rev. Mod. Phys. *38*, 1 – 35.
Kim, Y. K., Cheng, K.-T. (1978): Phys. Rev. *A 18*, 36 – 47.
Kupriyanov, S. E., Latypov, Z. Z. (1965): Sov. Phys. JETP *20*, 36 – 41.
LaGattuta, K. J., Hahn, Y. (1981): Phys. Rev. *A 24*, 2273 – 2276.
Madison, D. H., Calhoun, R. V., Shelton, W. N. (1977): Phys. Rev. *A 16*, 552 – 562.
Massey, H. S. W., Burhop, E. H. S. (1969): Electronic and Ionic Impact Phenomena, Vol. 1. Oxford: Oxford Univ. Press.
McCarthy, I. E., Stelbovics, A. T. (1980): In: Theoretical Methods for Ionization in Atomic and Molecular Processes in Controlled Thermonuclear Fusion (McDowell, M. R. C, Ferendeci, A. M., eds.). New York: Plenum Press.
McGuire, E. J. (1971): Phys. Rev. *A 3*, 267 – 279.
McGuire, E. J. (1977 a): Phys. Rev. *A 16*, 62 – 72.
McGuire, E. J. (1977 b): Phys. Rev. *A 16*, 72 – 79.
McGuire, E. J. (1979): Phys. Rev. *A 20*, 445 – 456.
Moores, D. L. (1972): J. Phys. *B 5*, 286 – 298.
Moores, D. L. (1978): J. Phys. *B 11*, L 403 – L 405.
Moores, D. L., Golden, L. B., Sampson, D. H. (1980): J. Phys. *B 13*, 385 – 395.
Moores, D. L., Nussbaummer, H. (1970): J. Phys. *B 3*, 161 – 172.
Ndefru, J. T., Malik, F. J.: To be published.
Peach, G. (1966 a): Proc. Phys. Soc. *87*, 375 – 380.
Peach, G. (1966 b): Proc. Phys. Soc. *87*, 381 – 391.
Peach, G. (1968): J. Phys. *B 2*, 1088 – 1108.
Peach, G. (1970): J. Phys. *B 3*, 328 – 349.
Peach, G. (1971): J. Phys. *B 4*, 1670 – 1677.
Peart, B., Dolder, K. T. (1968): J. Phys. *B 1*, 240 – 244.
Peart, B., Martin, S. O., Dolder, K. T. (1969): J. Phys. *B 2*, 1176 – 1179.
Peek, J. (1980): Private Communication.
Peterkop, R. K. (1977): Theory of Ionization of Atoms by Electron Impact. Boulder, Colorado: Colorado Associated Univ. Press.
Pindzola, M. S., Griffin, D. C., Bottcher, C. (1982): Phys. Rev. *A 25*, 211 – 218.
Rapp, D., Englander-Golden, P. (1965): J. Chem. Phys. *43*, 1464 – 1479.
Riley, M. E. (1973): Phys. Rev. *A 8*, 742 – 753.
Rudge, M. R. H. (1968): Revs. Mod. Phys. *40*, 564 – 590.
Roob, W. D., Roundtree, S. P., Burnett, T. (1975): Phys. Rev. *A 11*, 1193 – 1199.
Rosner, S. D., Gaily, T. D., Holt, R. A. (1976): J. Phys. *B 9*, L 489 – 491.
Rudge, M. R. H., Schwartz, S. B. (1966): Proc. Phys. Soc. *88*, 563 – 578.
Rudge, M. R. H., Seaton, M. J. (1965): Proc. Roy. Soc. London *A 283*, 262 – 290.
Sampson, D. H., Golden, L. B. (1979): J. Phys. *B 12*, L 785 – L 791.
Sampson, D. H., Golden, L. B. (1981): J. Phys. *B 14*, 903 – 913.
Shearer-Izumi, W., Botter, R. (1974): J. Phys. *B 7*, L 125 – L 128.
Sloan, I. H. (1965): Proc. Phys. Soc. *85*, 435 – 442.
Stingl, E. (1972): J. Phys. *B 5*, 1160 – 1174.
Thomson, Sir J. J. (1912): Phil. Mag. *23*, 449 – 457.
Ton-That, D., Manson, S. T., Flannery, M. R. (1977): J. Phys. *B 10*, 621 – 653.
Trefftz, E. (1962): Proc. Roy. Soc. London *A 271*, 379 – 386.
Tweed, R. J. (1980): J. Phys. *B 13*, 4467 – 4479.
Weingold, E., McCarthy, I. E. (1978): (*e, 2 e*) Collisions in Advances in Atomic and Molecular Physics, Vol. 14 (Bates, Sir D. R., Benderson, B., eds.). New York: Academic Press.
Vriens, L. Bonsen, T. F. M., Smit, J. A. (1968): Physica *40*, 229 – 252.
Younger, S. M. (1980 a): Phys. Rev. *A 22*, 111 – 117.
Younger, S. M. (1980 b): Phys. Rev. *A 22*, 1425 – 1428.
Younger, S. M. (1981 a): Phys. Rev. *A 23*, 1138 – 1146.
Younger, S. M. (1981 b): Phys. Rev. *A 24*, 1272 – 1277.
Younger, S. M. (1981 c): Phys. Rev. *A 24*, 1278 – 1285.
Younger, S. M. (1981 d): J. Quant. Spectrosc. Radiat. Transfer *26*, 329 – 337.
Younger, S. M. (1982 a): Comm. At. Mol. Phys. *11*, 193 – 209.
Younger, S. M. (1982 b): J. Res. Nat. Bur. Stds. *87*, 49 – 51.

Younger, S. M. (1982c): J. Quant. Spectrosc. Radiat. Transfer *27*, 541 — 544.
Younger, S. M. (1982d): Phys, Rev. *A 25*, 3396 — 3398.
Younger, S. M. (1982e): Phys. Rev. *A 26*, 3177 — 3186.
Younger, S. M. (1982f): J. Quant. Spectrosc. Radiat. Transfer *29*, 61 — 66.
Younger, S. M. (1982g): Unpublished.

2

Semi-Empirical and Semi-Classical Approximations for Electron Ionization

S. M. Younger and T. D. Märk

Lawrence Livermore National Laboratory, Livermore, Cal., U.S.A.
Institut für Experimentalphysik, Leopold-Franzens-Universität,
Innsbruck, Austria

2.1 Introduction

Until quite recently very little information was available concerning electron impact ionization. Other chapters in this book amply demonstrate the formidable experimental complexities associated with the accurate measurement of electron impact ionization cross sections as well as the still unresolved theoretical picture. In order to fill the void between data required and data available, a large number of attempts have been made to extend our current quantitative knowledge of the electron impact ionization process by means of semi-empirical and semi-classical prescriptions. The general goal of all these methods is to use one or another basic approximation to generate ionization cross section data for a broad class of atoms, molecules and ions in some selected energy region. Each method is designed to fit a specific need, with some methods more generally applicable than others which might be more tightly controlled by empirical evidence.

Ionizing collisions of electrons with atoms and molecules can be classified by comparing the impact electron's velocity with the mean orbital velocity of the target electrons in the (sub-) shell under study, i.e., *fast* collisions and *slow* collisions. For fast collisions, the influence of the incoming electron upon a target can be treated as a sudden and external perturbation. Conversely, for slow collisions, the combined system of incoming electron and target system has to be considered. Thus for cross section data at *high* electron energies one might employ a high energy approximation to the Born approximation (e.g., Bethe 1930, 1932, Inokuti 1971 and Inokuti *et al.* 1978) while for very *low* energies a classical scattering theory might be invoked (e.g., binary encounter approximation (Thomson 1912)).

Some definitions are in order:

The counting cross section for electron ionization is the simple sum of the partial cross sections

$$\sigma_c = \Sigma \ \sigma_z \tag{2-1}$$

where σ_z is the partial ionization cross section, i.e., the cross section for production of a z-times ionized residual ion i by a single collision. The counting cross section, however, is only (directly) accessible by experiment if the number of ionization events is measured (McClure 1953; Rieke and Prepejchal 1969, 1972). In most experiments, however, partial cross sections σ_z are measured or the gross (or total) ionization cross section σ_t is determined by measuring the total ion current produced by all different ionization mechanisms,

$$\sigma_t = \Sigma \ z \cdot \sigma_z. \tag{2-2}$$

Note: It should be mentioned in this conjunction, that some theoreticians use the term total ionization cross section for single partial ionization cross section, because the term "total" is used to indicate that integration over all final state electron momenta has been carried out.

We have selected a group of representative and commonly used semi-empirical and semi-classical methods and illustrate their performance on a set of test cases. In this manner we hope to demonstrate some of the inherent strengths and weakness of the different approaches and perhaps even suggest ways in which they might be improved. For more details see also reviews of Seaton (1962 a), Kurepa (1963), Stafford (1966), Burgess and Percival (1968), Rudge (1968), Vriens (1969), Inokuti (1971), Vainshtein *et al.* (1973), Peterkop (1977), Kato (1977), Inokuti *et al.* (1978), Vriens and Smeets (1980), Itikawa and Kato (1981), Sobelman *et al.* (1981), Casnati *et al.* (1982), Märk (1982 a, b, 1984). Although some of these approximate theories give prescriptions for differential ionization cross sections (e.g., Vriens 1964 b, 1966 a, Flannery 1971), most are concerned with the calculation of either single or double (e.g., Gryzinski 1965 a, b, McFarland 1967, Roy *et al.* 1972, Roy and Rai 1973, Chatterjee *et al.* 1982) partial ionization cross sections or (less commonly) with total (or counting) ionization cross sections. The present chapter is confined to the discussion of partial and total (counting) ionization cross sections for electron impact ionization of atoms, molecules and ions.

2.2 Empirical and Semi-Empirical Formulae

The general goal of empirical and semi-empirical formulae is to represent a body of observed cross section data by a relatively simple expression containing a few parameters determined largely (but not necessarily exclusively) from those data. These formulae can then be incorporated into computer models of plasmas, etc. (see Chapter 9). Because of the great need for reasonable estimates of ionization cross section functions and the present scarcity of accurate data, a number of empirical formulas for the ionization cross section function have appeared in the literature, including those reported by e.g. Morgulis (1934), de la Ripelle (1949), Elwert (1952), Vainshtein (1957), Knorr (1958), Seaton (1959), Lorquet (1960), Drawin (1961), Post (1961), Seaton (1964), Vriens (1965), Green and Barth (1965), Percival (1966), Lotz (1967 a, b, 1968, 1970), Rudge (1968), Krinberg (1969), Green and McNeal (1971), Vainshtein *et al.* (1973), Canto and Daltabuit (1974), Khare *et al.* (1974), Jain and Khare (1975, 1976), Burgess *et al.* (1977), Franco and Daltabuit (1978), Djuric *et al.*

(1981), Casnati *et al.* (1982), Strzondola (1983 a, b), Burgess and Chidichimo (1983). Some of these formulae are given in Table 2-1. Most of the formulas apply to the partial cross section for single ionization (or ejection of at least one electron, i.e. counting cross sections), some may also be applied to the description of total ionization cross sections.

Table 2-1. Selected formulae for electron impact ionization

Thomson (1912):

$$\sigma_{in} = 4\pi a_0^2 \xi_n \left(\frac{R}{I_n}\right)^2 \frac{u-1}{u^2}$$

Elwert (1952):

$$\sigma_{in} = 2\pi a_0^2 \xi_n \left(\frac{R}{I_n}\right)^2 \frac{u-1}{u^2} [1 + 0.3(u-1)]$$

Gryzinski (1959) — Stabler (1964):

$$\sigma_{in} = 4\pi a_0^2 \xi_n \left(\frac{R}{I_n}\right)^2 \left(\frac{u}{u+1}\right)^{3/2} g(u)$$

$$g(u) = \left(\frac{5}{3u} - \frac{2}{u^2}\right) \qquad\qquad u \geq 2$$

$$\quad = \frac{4\sqrt{2}}{3u}\left(1 - \frac{1}{u}\right)^{3/2} \qquad 1 \leq u < 2$$

Post (1961):

$$\sigma_{in} = 4\pi a_0^2 \left(\frac{R}{I_n}\right)^2 a_n g(u)$$

$$g(u) = \frac{\varepsilon}{u}\ln(u), \; \varepsilon = 2.7183$$

Drawin (1961):

$$\sigma_{in} = 2.66\,\pi a_0^2 \xi_n \left(\frac{R}{I_n}\right)^2 f_1 \frac{u-1}{u^2}\ln(1.25 f_2 u)$$

f_1 and f_2 are adjustable parameters (≈ 1)

Seaton (1964):

$$\sigma_{in} = 2.2\,\xi_n\,\pi a_0^2 \left(\frac{R}{I_n}\right)^2 \frac{u-1}{u}$$

Gryzinski II (1965) (Single ionization):

$$\sigma_{in} = 4\pi a_0^2 \xi_n \left(\frac{R}{I_n}\right)^2 g(u)$$

$$g(u) = \frac{1}{u}\left(\frac{u-1}{u+1}\right)^{3/2}\left\{1 + \frac{2}{3}\left(1 - \frac{1}{2u}\right)\ln[2.7 + (u-1)^{1/2}]\right\}$$

Gryzinski II (1965) (Double ionization, McFarland 1967):

$$\sigma_{in} = \left(\frac{4\pi a_0^2 R^2}{E_1 E_2}\right)^2 \frac{\xi_e(\xi_e - 1)}{\pi r^2} g\left(\frac{T}{E_1 + E_2}\right)$$

Table 2-1 (*continued*)

Vriens (1966)-Burgess (1963):

$$\sigma_{in} = 4\pi a_0^2 \xi_n \frac{R^2}{T+I_n+E_t}\left[\left(\frac{1}{I_n}-\frac{1}{T}\right)+\frac{2}{3}E_t\left(\frac{1}{I_n^2}-\frac{1}{T^2}\right)-\phi\frac{\ln u}{T+I_n}\right]$$

$$\phi=1 \qquad I_n > R$$
$$\phi=0 \qquad I_n < R$$

Lotz (1967):

$$\sigma_{in} = a_n \xi_n \frac{\ln u}{T\cdot I_n}\left[1-b_n e^{-c_n(u-1)}\right]$$

a_n, b_n, c_n variables

Vainshtein, Sobelman, Yukov (1973):

$$\sigma_{in} = \pi a_0^2 \frac{\xi_n}{2l+1}\left(\frac{R}{I_n}\right)^2\left(\frac{u-1}{u}\right)^{3/2}\frac{c}{u-1+\Phi}$$

c, Φ variables

with

$$\sigma_i = \sum_{n=1}^{N_s} \sigma_{in}$$

σ_i	"Total" partial cross section for ion i
σ_{in}	Cross section for electron impact ionization of a target leading to ion i considering only electrons from n-th subshell
T	Energy of incident electron
I	Ionization potential of target
R	Ionization potential of H
I_n	Binding energy of electrons in n-th subshell
$u =$	T/I_n
ξ_n	Number of equivalent electrons in n-th subshell
Z	Nuclear charge
N	Total number of electrons in target
E_t	Kinetic energy of target electron
N_s	Number of subshells
E_1	Binding energy of the first ejected electron for the case of double ionization
E_2	Binding energy of second ejected electron in the field left by the removal of the first
ξ_e	Number of electrons in a specific energy state for which a transfer from the incident electron of a quantity of energy, $E_1 + E_2$, would give rise to double ionization
r	Mean radius of the atomic system having ξ_e electrons

Bell *et al.* (1982) have recently compiled a set of recommended ionization cross sections for some light atoms and ions. They fitted all cross sections to a formula with several fitting parameters ensuring the correct behavior at both low and high incident electron energies, i.e.,

$$\sigma(T) = \frac{1}{T\cdot I}\left\{A\ln(u) + \sum_{n=1}^{N_s} B_n\left(1-\frac{I_n}{T}\right)^n\right\}. \tag{2-3}$$

(For notation see Table 2-1. See also Eq. 8-16 and Table 8-2 for parameters recommended for use by Bell *et al.*)

2.2.1 The Lotz Formulae

Perhaps the most widely used empirical formulae are due to Lotz (1967 a, b, 1968, 1970), who attempted to extract from the then available experimental partial ionization cross section data a set of parameters describing the electron ionization of all atoms and ions. For atoms where experimental data is available he provides individual parameters giving a very close fit to the data for that ion. For the remainder of the periodic table he provides a set of recommended parameters depending on the principal quantum number.

The Lotz formula for the "total" electron ionization cross section for atoms and atomic ions is given in Table 2-1. The sum extends over the outermost two or three subshells, according to rules given in Lotz's papers. For ions with charge greater than or equal to three Lotz approximates the cross section by a rough fit to the Coulomb-Born cross section for a high Z hydrogenic $1s$ orbital:

$$\sigma_i = \sum_{n=1}^{N_s} \xi_n \frac{a \cdot \ln u}{T \cdot I_n} \tag{2-4}$$

with $a = 4.5 \times 10^{-14} \, \text{cm}^2 \, (\text{eV})^2$ (Lotz 1967 a, b). This formula has been recently extended to account for autoionization effects and threshold behavior and successfully applied to complex ions (Burgess et al. 1977, Burgess and Chidichimo 1983).

2.2.2 The Seaton Formula for Highly Charged Ions

Based on some theoretical work and experimental data for some neutral atoms and ions Seaton (1964) suggested (see Table 2-1) a renormalization of a cross section formula proposed by Thomson (1912). Jordan (1969) used this relation to calculate the ionization equilibrium in a hot thin plasma for the atoms C to Ni. Note that Seaton's formula does not have the correct functional form at high energies and only applies to near threshold collisions.

2.2.3 The Jain-Khare Formula for Molecules

Recently, Jain and Khare (1975, 1976) have reported a semi-empirical formula for the calculation of the energy loss cross section $d\sigma(T, E)/dE$ (with E the energy loss suffered by the incident electron in the ionizing collision (e.g., see also Khare 1969, Green and Sawada 1972, Khare et al. 1974)) and of the total ionization cross sections of molecules, using a combination of the Born Bethe approximation (describing high incident electron energy collisions with small amount of energy transfer: soft collisions) and of the Mott-Möller formula (describing collisions with fast secondary electrons: hard collisions), i.e.:

$$\frac{d\sigma}{dE} = \frac{4\pi a_0^2 R^2}{T} \left[f_1 \left(\frac{1}{E} \frac{df}{dE} \ln TC \right) + f_2 \left(\frac{1}{\varepsilon^2} - \frac{1}{\varepsilon(T-\varepsilon)} + \frac{1}{(T-\varepsilon)^2} \right) s \right] \tag{2-5}$$

with df/dE the differential oscillator strength, ε the energy of ejected secondary electron, s the number of electrons which can participate in hard collisions, C a collisional parameter as defined by Miller and Platzman (1957), ε_0 a parameter (see Jain and Khare 1976, chosen so that the theoretical results are in best agreement with the experimental data for $d\sigma/dE$ at $T = 500$ eV). R is the Rydberg constant, and

$$f_1 = \frac{1}{1 + I/T} \cdot \left[1 - \frac{\varepsilon}{T - I} \frac{\ln\left(1 - C(T - I)\right)}{\ln C \cdot T} \right] \tag{2-6}$$

$$f_2 = \varepsilon^3 (1 - I/T)/(\varepsilon^3 + \varepsilon_0^3) \tag{2-7}$$

The empirical functions f_1 and f_2 extrapolate the Born-Bethe and Mott-Möller cross sections to low incident electron energies and control the mixing of the soft and hard collisions. The results of Jain and Khare (1976) for CO, H_2O, CO_2 and CH_4 are in fair agreement with experimental data. Djuric *et al.* (1981) have recently extended these calculations introducing a better fit for the low energy regime of df/dE (e.g., see Fig. 14 in Märk 1982 b).

2.2.4 *Additivity Rule and Maximum Ionization Cross Sections*

Another semi-empirical approach to obtain quantitative information on relative atomic and molecular total ionization cross sections is based (Fano 1946) on a result by Bethe (1930) that the ionization of an atomic electron with quantum number (n, l) is approximately proportional to the mean square radius of the electron shell (n, l) (Otvos and Stevenson 1956; Batabyal *et al.* 1965; Mann 1967, 1970; and Tiwari *et al.* 1969). This leads to proposed correlations of (in some cases maximum) electron impact total ionization cross sections with polarizability, diamagnetic susceptibility, and also to the (modified) *additivity* rule for atomic ionization cross sections (Otvos and Stevenson 1956; Lampe *et al.* 1957; Stevenson and Schissler 1961; Batabyal *et al.* 1965; Schram *et al.* 1965; Harrison *et al.* 1966; Pottie 1966; Drowart and Goldfinger 1967; Beran and Kevan 1969; Grosse and Bothe 1970; Stafford 1971; Flaim and Ownby 1971; Center and Mandl 1972; Blackburn and Danielson 1972; Alberti *et al.* 1974; Arai and Hotta 1975; Rao *et al.* 1979; Märk 1982 a; Bartmess and Georgiadis 1983; Fitch and Sauter 1983). None of these correlations, however, seems to be generally valid, but each appears to be valid for certain molecular classes. For instance, the concept of additivity applies to hydrocarbons (Otvos and Stevenson 1956) and there is a single linear correlation with diamagnetic suscepti-bility for all nonfluorine-substituted compounds (Beran and Kevan 1969). Moreover, there is an excellent correlation with polarizability for aromatics, ethers high, nitroalkanes low, chlorocarbons, ketones, alcohols and thiols (Bartmess and Georgiadis 1983). In addition, Franco and Daltabuit (1978) (see also Franco 1981) pointed out a simple empirical relation between the maximum ionization cross section, its position and the ionization potential, i.e.,

$$\sigma_{\max} \cdot u_{\max} = \text{const. } \xi \left(\frac{R}{I} \right)^2 \tag{2-8}$$

which is the high energy limit of the classical Thomson formula.

2.3 Semi-Classical Formulae

2.3.1 The Binary Encounter Approximation (BEA)

Semi-classical methods are used in an attempt to circumvent the formidable theoretical problems associated with electron impact ionization by making classical approximations to the quantum scattering problem. For some semi-classical methods an attempt is made to improve their accuracy by the use of adjustable parameters determined by comparison of the basic theory to known ionization cross section data. The use of classical mechanics (with the incorporation of certain features of the quantal treatment) to describe electron impact ionization is useful, because the final state lies in the continuum (a state which is a defined both classically and quantally) with a significant number of possible angular momenta. According to Rudge (1968) three basic approximations must be used in treating electron impact ionization by classical methods:

a) A classical description must be found for the initial state of the bound electron. Several descriptions have been used assuming the electron to be at rest, to have a fixed velocity or to have some prescribed velocity distribution (e.g., see Tripathi and Rai 1972).

b) The collision has to be described in the frame of classical laws of motion.

c) Since even for the simplest case (e.g., $H + e \rightarrow H^+ + 2e$) the collision is a difficult three-body problem, in most treatments the collision process has been described as though it were a two-body one (*binary-encounter approximation*).

The earliest treatment (see also Davis 1918 and Rosseland 1923) of ionizing collisions using the classical binary-encounter approximation and assuming that the target electron is at rest, is due to Thomson (1912) (e.g., see Table 2-1). Thomson's model gives too sharp a peak in the cross section at too low an energy, and the cross section decreases too rapidly at high electron energy, where the correct behavior is $\sigma \sim (\ln T)/T$ rather than $\sigma \sim 1/T$. A useful prediction of the Thomson theory is that ionization cross sections obey a scaling law expressed by a *reduced* ionization cross section $\bar{\sigma}(u)$

$$\sigma(u) = \xi_n \left(\frac{R}{I_n} \right)^2 \bar{\sigma}. \tag{2-9}$$

Comparison of this reduced ionization cross section with experimental data have been effected by e.g., Elwert (1952), Seaton (1962, 1964), Bauer and Bartky (1965), Rudge (1968). Thomas (1927), Williams (1927) and Webster *et al.* (1933) refined the Thomson theory by partly taking into account the velocity of the target electron. Thomas (1927) also argued that the binary encounter takes place in a region small compared to the atomic dimension (hard collision) and hence the incident electron must gain energy from the atomic field (*symmetrized binary encounter*). New progress was made by Gryzinski (1959, 1965a, b) and other authors extending Gryzinski's derivations (e.g., Ochkur and Petrunkin 1963, Stabler 1964, Kingston 1964, Vriens 1964a, b, c, 1966a, Mapleton 1966, McDowell 1966, Gerjuoy 1966,

Abrines *et al.* 1966, Catlow and McDowell 1967, Garcia *et al.* 1968, Flannery 1970, 1971, Tripathi and Rai 1972). This simple approximation yields cross sections in remarkable agreement with experimental data for a broad range of target atoms (Stafford 1966, 1968; Lin and Stafford 1968; Märk 1982 a). Ton-That and Flannery (1977), Ton-That *et al.* (1977), McCann *et al.* (1979) and Flannery *et al.* (1981) have also evaluated the ionization cross sections for metastable rare gas atoms (see also Hyman 1979), rare gas dimers, N_2 and CO.

Thomas and Garcia (1969) have extended Gryzinski's formulation to include ionization of positive ions. They found that they could represent ionic cross sections reasonably well by the very simple formula

$$\sigma_i = \sum_n^{N_s} \xi_n \frac{\sigma_n}{I_n^2} \qquad (2\text{-}10)$$

where σ_n is a universal scaled cross section. Salop (1976) has applied this method to a number of highly ionized atoms, including some account of excitation-auto-ionization processes.

Bauer and Bartky (1965) and Prok *et al.* (1969) extended the Gryzinski theory to the ionization of molecules (see also Stafford 1968). This involves the use of Franck-Condon factors to permit the calculation of cross sections for the excitation of an individual vibrational level. According to these authors it is not generally necessary to take into account the Franck-Condon factors if one asks for the behavior of the "total" partial ionization cross section function for electron energies large enough so that all the vibrational levels of an electronic state can be excited.

All of these classical, improved-classical approximations and their relation to quantal approximations have been reviewed in several excellent reviews, e.g., by Williams (1945), Bohr (1948), Drawin (1961), Burgess and Percival (1968), Rudge (1968) and Vriens (1969). For this reason, we do not reproduce here these classical derivations, however, the most important formulae are given in Table 2-1 for the interested user.

2.3.2 *The Exchange Classical Impact Parameter (ECIP) Approximation*

Burgess (1963, 1964), Burgess and Summers (1967) and Vriens (1966 a, b) followed from the outset a more elegant approach, which allows for long range interaction ("symmetrical treatment"; Thomas, 1927) and including certain features of the quantal treatment (interference between direct and exchange terms) into the BEA approximation. Both, Burgess and Vriens, have given formulas for the partial ionization cross section σ_{in}, however, the results do not appear to coincide, the formula given by Vriens being much simpler than those by Burgess. The reason for this discrepancy is according to Vriens not clear, although Vriens uses essentially the same model as Burgess. The result of Vriens (1966 a, b, 1969) on carrying through the procedure suggested by Burgess is given in Table 2-1. Roy and Rai (1973 a), Kumar and Roy (1978 a, b, 1979), and Chatterjee *et al.* (1982) extended this procedure to single ionization of ions and to double ionization of atoms and ions (proceeding via two binary encounters), including exchange effects and using a Hartree-Fock velocity distribution function for the target electron (see also Roy and Rai 1973 b, 1983). Kunc (1980, 1981) has recently applied this model to targets with a one-electron outer shell in both the ground and excited state.

2.3.3 The Scaled Hydrogenic Formula

Although the scaled hydrogenic method or "infinite Z approximation" (Golden and Sampson 1977, 1980, Golden et al. 1978, Moores et al. 1980) is actually a quantum mechanical rather than a semi-classical approximation we include it here because of its general applicability to the calculation of cross sections for multiply-charged ions. In this model a cross section is computed in a Coulomb-Born-exchange approximation in which all continuum and bound orbitals are $Z=1$ Coulomb functions. The resulting cross sections scale with Z, the nuclear charge, and can be accurately represented by the analytic expression:

$$\sigma_{z=\infty} = \frac{1}{u}\left[A\ln(u) + D\left(1-\frac{1}{u}\right)^2 + \left(\frac{c}{u}+\frac{d}{u^2}\right)\left(1-\frac{1}{u}\right)\right] \qquad (2\text{-}11)$$

where A, D, c and d are parameters given in Table 2-2. The authors of this method have proposed a variety of scaling laws to relate $\sigma_{z=\infty}$ to cross sections for individual ions.

Table 2-2. *Values for the parameters in the fits to the reduced infinite Z approximation cross section* $\sigma_{z=\infty}(n\,l, u)$ *for various sublevels $n\,l$ (Moores et al. 1980)*

$n\,l$	$A''(n\,l)$	$D''(n\,l)$	$c''(n\,l)$	$d''(n\,l)$
1 s	1.13	4.41	−2.00	3.80
2 s	0.823	3.69	0.62	1.79
2 p	0.530	5.07	1.20	2.50
3 s	0.652	3.83	0.64	2.10
3 p	0.551	4.38	1.83	1.90
3 d	0.280	5.70	2.21	2.65

It should also be mentioned that McGuire (1977a, 1979) has calculated cross sections for a wide spectrum of atoms and ions using a plane wave Born approximation and has developed a scaling formula to approximate the results.

2.4 Some Comparisons and Applications

2.4.1 Atoms

Calculated and experimental ionization cross section functions for atomic hydrogen and nitrogen are shown in Figs. 5-32 and 5-42, respectively, in Chapter 5 of this book.

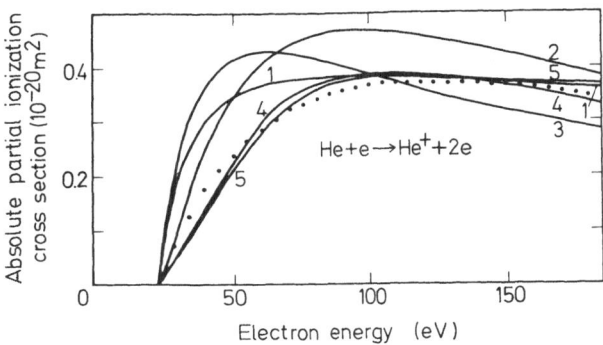

Fig. 2-1. Absolute partial ionization cross sections for $He + e \rightarrow He^+ + 2e$ after Märk (1982 a). *Experimental*: Full dots: Stephan *et al.* (1980). *Theoretical*: Curve 1: Elwert 1952 (given by Pitchford *et al.* 1980), Curve 2: Gryzinski 1965 (Pitchford *et al.* 1980), Curve 3: Burgess 1964-Vriens 1966 a, b (Pitchford *et al.* 1980), Curve 4: Gryzinski 1959 (Ochkur and Petrunkin 1963), Curve 5: Ochkur 1964 (Peach 1966)

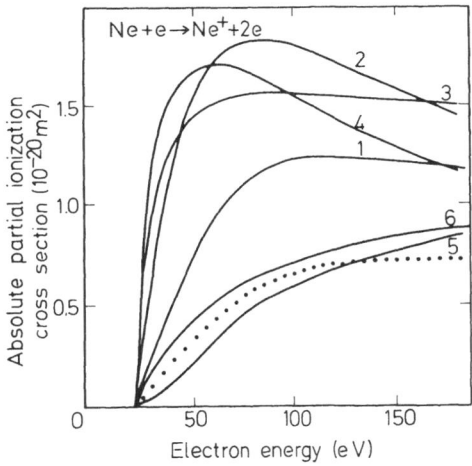

Fig. 2-2. Absolute partial ionization cross sections for $Ne + e \rightarrow Ne^+ + 2e$ after Märk (1982 a). *Experimental*: Full dots: Stephan *et al.* (1980). *Theoretical*: Curve 1: Elwert 1952 (Pitchford *et al.* 1980), Curve 2: Gryzinski 1965 (Pitchford *et al.* 1980), Curve 3: Burgess 1964-Vriens 1966 a, b (Pitchford *et al.* 1980), Curve 4: Born approximation by Wallace *et al.* (1973), Curve 5: Eikonal closure approximation by Wallace *et al.* (1973), Curve 6: Ochkur 1964 (Peach 1966)

Cross sections for rare gases have recently been reviewed in detail by Märk (1982 a, b). (See also Nagy *et al.* 1980.) For single ionization of He (Fig. 2-1) and Ne (Fig. 2-2) there is good agreement between the experimental results of Stephan *et al.* (1980) and quantal calculations by Peach (1966, 1971); see also Fig. 5-31 for Ar. The Gryzinski (1965 a) and Vriens (1966)-Burgess (1964) methods give too high values. For Kr (Fig. 2-3) the semi-classical approximations are superior to the plane wave Born calculation of McGuire (1977 b). In the case of double ionization the situation is much less encouraging. Figs. 2-4 and 2-5 compare theory and experiment for the

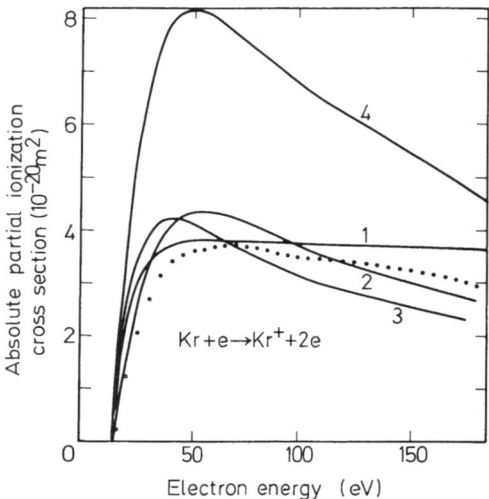

Fig. 2-3. Absolute partial ionization cross sections for $Kr + e \rightarrow Kr^+ + 2e$ after Märk (1982 a).
Experimental: Full dots: Stephan *et al.* (1980). *Theoretical*: Curve 1: Elwert 1952 (Pitchford *et al.* 1980),
Curve 2: Gryzinski 1965 (Pitchford *et al.* 1980), Curve 3: Burgess 1964-Vriens 1966 a, b (Pitchford *et al.*
1980), Curve 4: Born approximation by McGuire (1977)

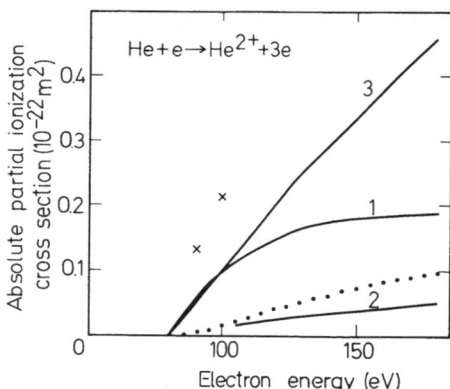

Fig. 2-4. Absolute partial ionization cross sections for $He + e \rightarrow He^{2+} + 3e$ after Märk (1982 a).
Experimental: Full dots: Stephan *et al.* (1980). *Theoretical*: Crosses: Gryzinski 1965 (Roy *et al.* 1972),
Curve 1: Burgess 1964-Vriens 1966 a, b (Roy and Rai 1973 a), Curve 2: Born no shielding approximation
(Tweed 1973), Curve 3: Born spherical average approximation (Tweed 1973)

double ionization of He and Kr, respectively, illustrating the poor agreement one
might expect from relatively simple approximations to such very complex processes.
For the ionization of certain other atoms (especially alkalis and alkaline earths)
semi-empirical, classical and semi-classical calculations are in good agreement with
the experimental data (e. g., McFarland 1967, Lin and Stafford 1968 and references

therein, Roy and Rai 1973 b, 1983, Karstensen and Schneider 1978, Dettmann and Karstensen 1982, Chatterjee *et al.* 1982 (i.e., see Figs. 5-4 and 5-6 in Chapter 5)). Still, there appears to be no generally applicable rule governing the accuracy of semi-empirical and semi-classical prescriptions for the calculation of electron ionization cross sections of atoms. Consequently the most conservative candidates would appear to be those most closely tied to actual experimental data, such as the parameterizations of Lotz.

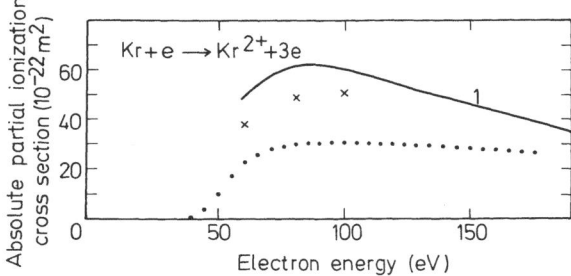

Fig. 2-5. Absolute partial ionization cross sections for $Kr + e \rightarrow Kr^{2+} + 3e$ after Märk (1982 a). *Experimental*: Full dots: Stephan *et al.* (1980). *Theoretical*: Crosses: Gryzinski 1965 (Roy *et al.* 1972), Curve 1: Burgess 1964-Vriens 1966 a, b (Kumar and Roy 1978 a)

2.4.2 Molecules

Little emphasis has been directed towards calculations of the ionization cross sections of molecules, e.g., including the studies of Lorquet (1960), Drawin (1961), Prasad and Prasad (1963), Vriens (1965), Green and Barth (1965), Bauer and Bartky (1965), Watson *et al.* (1967), Stolarski *et al.* (1967), Stafford (1968), Prok *et al.* (1969),

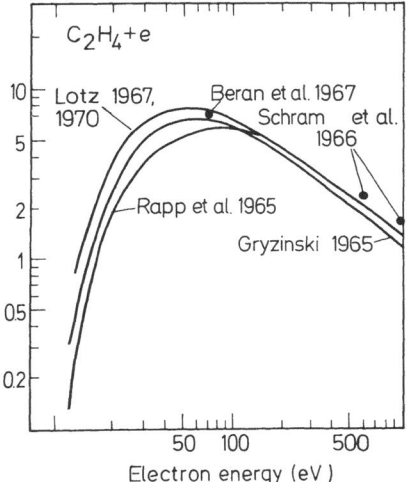

Fig. 2-6. Calculated ionization cross sections (Deutsch 1982) for C_2H_4 using the formulae of Gryzinski (1965) and Lotz (1967) as compared to the measured total ionization cross sections (Rapp and Englander-Golden 1965, Schram *et al.* 1966, Beran and Kevan 1969)

Peterson *et al.* (1969), Neckel and Sodeck (1972), Tannen (1973), Jain and Khare (1975, 1976), Ton-That and Flannery (1977), McCann *et al.* (1979), Flannery *et al.* (1981), Djuric *et al.* (1981), Deutsch and Schmidt (1982), Deutsch (1982), and Stephan *et al.* (1983 a, b). Some of these calculations have proven to be in reasonable agreement with experimental results, e.g., for H_2 (e.g., see Drawin 1961, Bauer and Bartky 1965, Prok *et al.* 1969) and hydrocarbons (e.g., see Fig. 2-6), and to a lesser degree for other molecules (i.e., N_2, O_2, CO, NO, see Fig. 5-32 in Chapter 5) and larger molecules (e.g., see Fig. 2-7 and Figs. 14 and 15 in Märk 1982 b). In addition there exist various approaches for estimation of maximum ionization cross sections (additivity rule). In Figs. 2-6 and 2-7 we compare the results of several approximations with experimental data for C_2H_4 and CF_4.

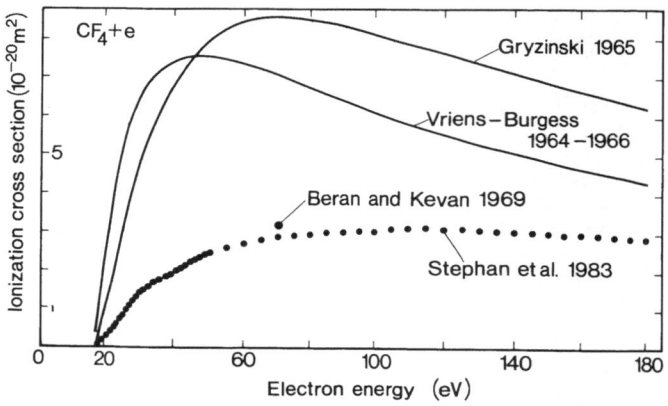

Fig. 2-7. Calculated ionization cross sections (Stephan *et al.* 1983 a) for CF_4 using the formulae of Gryzinski (1965) and Vriens (1966)-Burgess (1964) as compared to the measured total ionization cross section of Beran and Kevan (1969) and the counting ionization cross section function of Stephan *et al.* (1983 a, b)

2.4.3 Ions

The non-zero asymptotic Coulomb charge of ions makes them easier to deal with theoretically than the more complex neutral systems. Simple formulae such as the Lotz formula for ions and the infinite Z approximation compare favorably to experimental data over a wide class of ions where complex resonance structures or other special atomic structure effects are absent. Fig. 2-8 demonstrates for $He^+ + e \rightarrow He^{2+} + 2e$ the good agreement between theory and experiment typical of light ions. Fig. 2-9 illustrates a somewhat more complex example, $Si^{3+} + e \rightarrow Si^{4+} + 2e$, where inner shell ionization from the $2p$ subshell is important. Finally Fig. 2-10 demonstrates the poor performance of simple prescriptions in a moderately heavy ion ($Xe^{3+} + e \rightarrow Xe^{4+} + 2e$) where complex resonance structures dominate the cross section for $u = 1 - 4$. The high energy tail of the cross section is adequately described by most theories, but it is the threshold region that governs the calculation of ionization rate coefficients determining the charge state of such ions in a plasma.

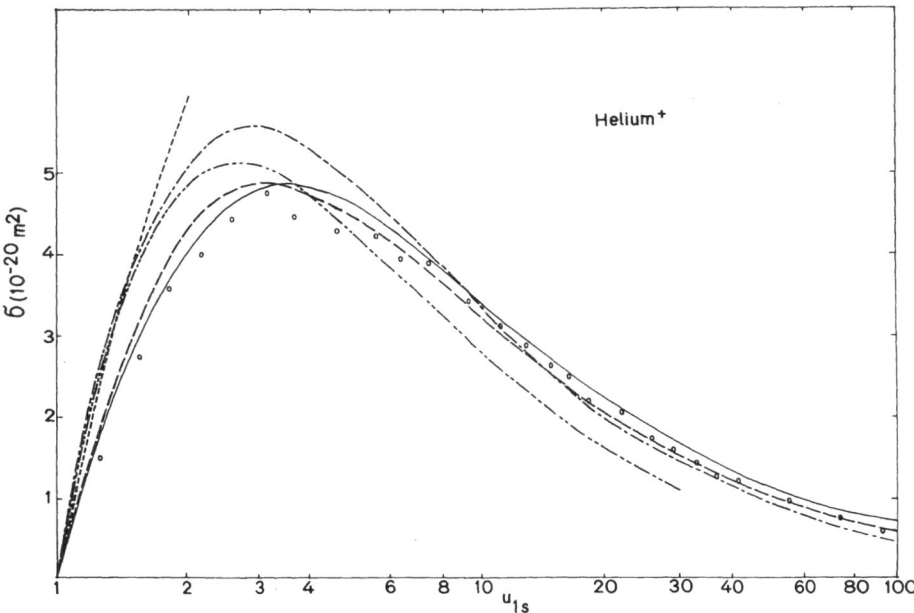

Fig. 2-8. Absolute partial ionization cross sections for $He^+ + e \rightarrow He^{2+} + 2e$. *Experimental*: Full dots: Peart *et al.* (1969). *Theoretical*: —·—·— Scaled hydrogenic formula, — — — — Seaton formula, — — — Burgess ECIP formula, ———— Lotz formula for ions, —··—··· Binary encounter approximation (Thomas and Garcia, 1969)

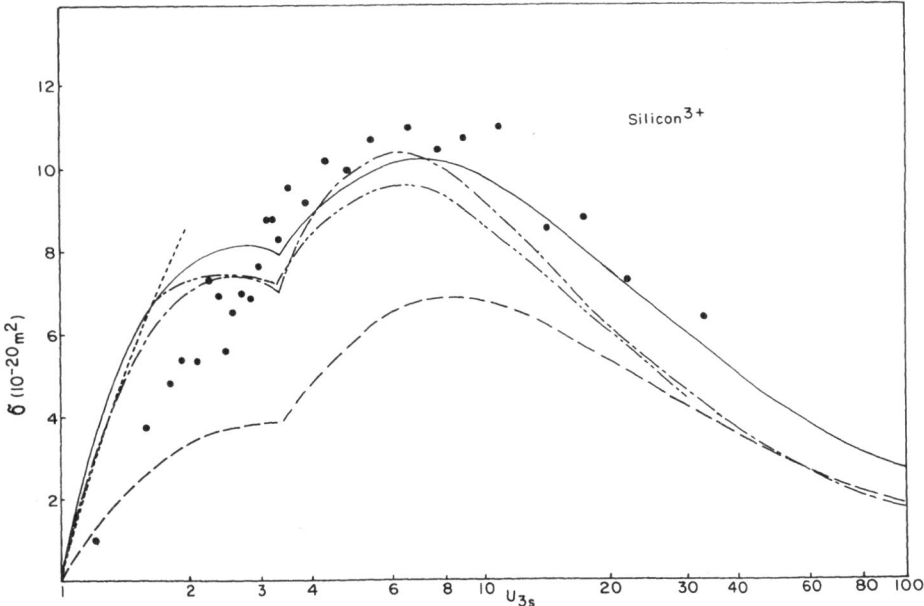

Fig. 2.9. Absolute partial ionization cross sections for $Si^{3+} + e \rightarrow Si^{4+} + 2e$. *Experimental*: Full dots: Crandall *et al.* (1982). *Theoretical*: —·—·— Scaled hydrogenic formula, — — — — Seaton formula, — — — Burgess ECIP formula, ———— Lotz formula for ions, —··—··· Binary encounter approximation (Thomas and Garcia, 1969)

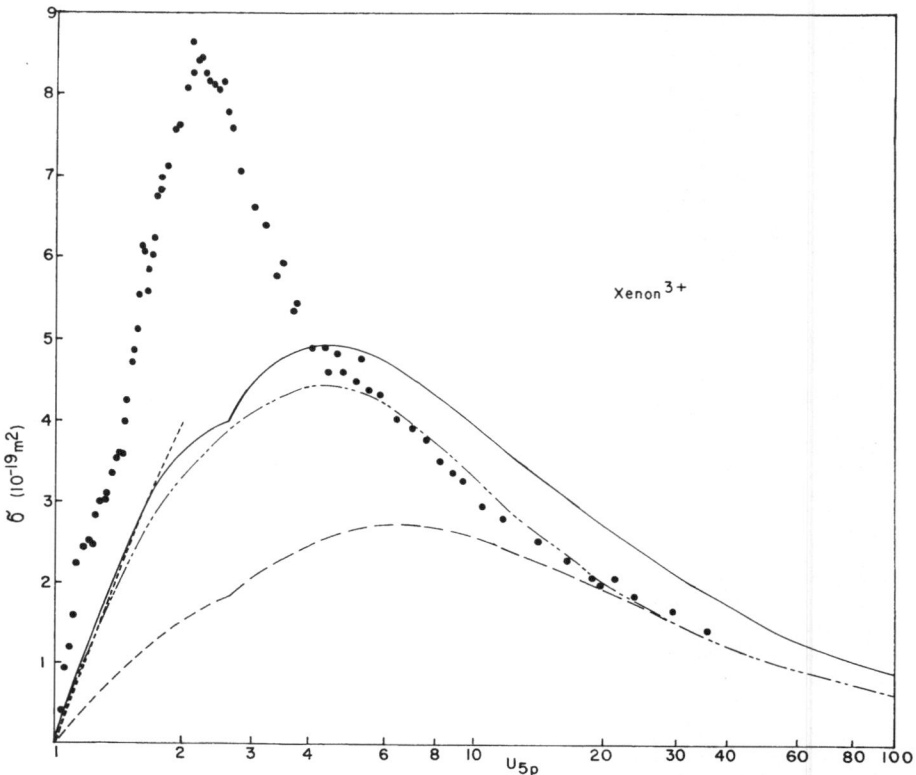

Fig 2-10. Absolute partial ionization cross sections for $Xe^{3+} + e \rightarrow Xe^{4+} + 2e$. *Experimental*: Full dots: Gregory *et al.* (1983). *Theoretical*: – – – – – Seaton formula, – – – – Burgess ECIP formula, ———
Lotz formula for ions, — · — · — Binary encounter approximation (Thomas and Garcia, 1969)

Acknowledgments

Work partially supported by the Österreichischer Fonds zur Förderung der Wissenschaftlichen Forschung (T. D. M.).

References

Abrines, R., Percival, R. I., Valentine, N. A. (1966): Proc. Phys. Soc. *89*, 515 – 523.
Alberti, R., Genoni, M. M., Pascual, C., Vogt, J. (1974): Int. J. Mass. Spectrom. Ion Phys. *14*, 89 – 98.
Arai, H., Hotta, H. (1975): Radiat. Res. *64*, 407 – 415.
Bartmess, J. E., Georgiadis, R. (1983): Vacuum TAIP *33*, 149 – 153.
Batabyal, A. K., Barna, A. K., Srivasta, B. N. (1965): Indian J. Phys. *39*, 219 – 226.
Bauer, E., Bartky, C. D. (1965): J. Chem. Phys. *43*, 2466 – 2476.
Bell, K. L., Gilbody, H. B., Hughes, J. G., Kingston, A. E., Smith, F. J. (1982): Culham Report CLM-R
     216; J. Phys. Chem. Ref. Data *12*, 891 – 916 (1983).
Beran, J. A., Kevan, L. (1969): J. Phys. Chem. *73*, 3866 – 3876.
Bethe, H. (1930): Ann. Physik *5*, 325 – 400.
Bethe, H. (1932): Z. Physik *76*, 293 – 299.

Blackburn, P. E., Danielson, P. M. (1972): J. Chem. Phys. 56, 6156−6164.

Bohr, N. (1948): Kgl. Danske Videnskab. Selskab. Mat. Fys. Medd. 18, No. 8.

Burgess, A. (1963): Proc 3rd ICPEAC, London, 237−242.

Burgess, A. (1964): Research Report, A. E. R. E., R 4818 (Proc. of Culham Symposium).

Burgess, A., Percival, I. C. (1968): Adv. Atomic Molecular Physics 4, 109−141.

Burgess, A., Summers, H. P. (1976): Mon. Not. R. astr. Soc. 174, 345−391.

Burgess, A., Summers, H. P., Cochrane, D. M., McWhirter, R. W. P. (1977): Mon. Not. R. astr. Soc. 179, 275−292.

Burgess, A., Chidichimo, M. C. (1983): Mon. Not. R. astr. Soc. 203, 1269−1280.

Canto, J., Daltabuit, E. (1974): Rev. Mexicana Astron. Astrof. 1, 5−9.

Casnati, E., Tartari, A., Baraldi, C. (1982): J. Phys. B 15, 155−167.

Catlow, G. W., McDowell, M. R. C. (1967): Proc. Phys. Soc. 92, 875−879.

Center, R. E., Mandl, A. (1972): J. Chem. Phys. 57, 4104−4106.

Chatterjee, S. N., Kumar, A., Roy, B. N. (1982): J. Phys. B 15, 1415−1419.

Crandall, D. H., Phaneuf, R. A., Falk, R. A., Belic, D. S., Dunn, G. H. (1982): Phys. Rev. A 25, 143−153.

Davis, B. (1918): Phys. Rev. 11, 433−444.

de la Ripelle, F. M. (1949): J. Phys. Radium 10, 319−329.

Dettmann, J. M., Karstensen, F. (1982): J. Phys. B 15, 287−300.

Deutsch, H. (1982): Personal Communication; see also Märk (1982 b, 1984).

Deutsch, H., Schmidt, M. (1982): Proc. 6. Tagung Physik u. Technik d. Plasmas, Leipzig, 107.

Djuric, N., Belic, D., Kurepa, M., Mack, J. U., Rothleitner, J., Märk, T. D. (1981): Proc. XIIth ICPEAC, Gatlinburg, 384−385.

Drawin, H. W. (1961): Z. Physik 164, 513−521.

Drowart, J., Goldfinger, P. (1967): Angew. Chemie 6, 581−648.

Elwert, G. (1952): Z. Naturforschg. 7 a, 432−439.

Fano, U. (1964): Phys. Rev. 70, 44−52.

Fitch, W. L., Sauter, A. D. (1983): Anal. Chem. 55, 832−835.

Flaim, T. A., Ownby, P. D. (1971): J. Vacuum Sci. Techn. 8, 661−662.

Flannery, M. R. (1970): J. Phys. B 3, 1610−1619.

Flannery, M. R. (1971): J. Phys. B 4, 892−895.

Flannery. M. R., McCann, K. J., Winter, N. W. (1981): J. Phys. B 14, 3789−3796.

Franco, J. (1981): Rev. Mex. Fisica 27, 475−485.

Franco, J., Daltabuit, E. (1978): Rev. Mex. Astron. Astrof. 2, 325−329.

Garcia, J. D., Gerjuoy, E., Welker, J. E. (1968): Phys. Rev. 165, 66−72.

Gerjuoy, E. (1966): Phys. Rev. 148, 54−59.

Golden, L. B., Sampson, D. H. (1977): J. Phys. B 10, 2229−2237.

Golden, L. B., Sampson, D. H. (1980): J. Phys. B 13, 2645−2652.

Golden, L. B., Sampson, D. H., Omidvar (1978): J. Phys. B 11, 3235−3243.

Green, A. E. S., Barth, C. A. (1965): J. Geophys. Res. 70, 1083−1092.

Green, A. E. S., McNeal, R. J. (1971): J. Geophys. Res. 76, 133−144.

Green, A. E. S., Sawada, T. (1972): J. Atm. Terr. Phys. 34, 1719−1728.

Gregory, D. C., Dittner, P. F., Crandall, D. H. (1983): Phys. Rev. A 27, 724−736.

Grosse, H. J., Bothe, H. K. (1970): Z. Naturforschg. 25 a, 1970−1976.

Gryzinski, M. (1959): Phys. Rev. 115, 374−382.

Gryzinski, M. (1965 a): Phys. Rev. 138, A305−A358.

Gryzinski, M. (1965 b): Phys. Rev. Lett. 14, 1059−1059.

Harrison, A. G., Jones, E. G., Gupta, S. K., Nagy, E. P. (1966): Can. J. Chemistry 44, 1967−1973.

Hyman, H. A. (1979): Phys. Rev. A 20, 855−859.

Inokuti, M. (1971): Revs. Mod. Phys. 43, 297−347.

Inokuti, M., Itikawa, Y., Turner, J. E. (1978): Revs. Mod. Phys. 50, 23−35.

Itikawa, Y., Kato, T. (1981): Empirical formula for ionization cross section of atomic ions for electron collisions. Report IPPJ-AM-17, Institute of Plasma Physics, Nagoya University, Japan.

Jain, D. K., Khare, S. P. (1975): Proc. 9th ICPEAC, Seattle, 484−485.

Jain, D. K., Khare, S. P. (1976): J. Phys. B 9, 1429−1438.

Jordan, C. (1969): Mon. Not. R. Astr. Soc. 142, 501−521.

Karstensen, F., Schneider, M. (1978): J. Phys. B 11, 167−172.

Kato, T. (1977): Ionization and excitation of ions by electron impact; review of empirical formulae. Report IPPJ-AM-2, Institute of Plasma Physics, Nagoya University, Japan.

Khare, S. P. (1969): Planet. Space Sci. *17*, 1257 – 1268.

Khare, S. P., Padalia, B. D., Nayak, R. M. (1974): Can. J. Phys. *52*, 1755 – 1758.

Kingston, A. E. (1964): Phys. Rev. *135*, A1537 – A1539.

Knorr, G. (1958): Z. Naturforschg. *13 a*, 941 – 950.

Krinberg, I. A. (1969): Astron. Zhurnal *46*, 993 – 997.

Kumar, A., Roy, B. N. (1978 a): Can. J. Phys. *56*, 1255 – 1260.

Kumar, A., Roy, B. N. (1978 b): Phys. Lett. *66 A*, 362 – 364.

Kumar, A., Roy, B. N. (1979): J. Phys. *B 12*, 3979 – 3986.

Kunc, J. A. (1980): J. Phys. *B 13*, 587 – 602.

Kunc, J. A. (1981): Int. J. Mass Spectrom. Ion Phys. *40*, 43 – 51.

Kurepa, M. V. (1963): Bull. Boris Kidric Inst. Nucl. Sci. *14*, 187 – 197.

Lampe, F. W., Franklin, J. L., Field, F. H. (1957): J. Am. Chem. Soc. *79*, 6129 – 6132.

Lin, S. S., Stafford, F. E. (1968): J. Chem. Phys. *48*, 3885 – 3890.

Lorquet, J. C. (1960): J. Chim. Phys. *57*, 1078 – 1084.

Lotz, W. (1967 a): Astrophys. J., Suppl. *14*, 207 – 238.

Lotz, W. (1967 b): Z. Physik *206*, 205 – 211.

Lotz, W. (1968): Z. Physik *216*, 241 – 247.

Lotz, W. (1970): Z. Physik *232*, 101 – 107.

Märk, T. D. (1982 a): Beitr. Plasmaphys. *22*, 257 – 294.

Märk, T. D. (1982 b): Int. J. Mass Spectrom. Ion Phys. *45*, 125 – 145.

Märk, T. D. (1984): Ionization of Molecules by Electron Impact. In: Electron-Molecule Interactions and Their Applications (Christophorou, L. G., ed.), Chapter 3. New York: Academic Press.

Mann, J. B. (1967): J. Chem. Phys. *46*, 1646 – 1651.

Mann, J. B. (1970): In: Recent Developments in Mass Spectroscopy (Proc. of Int. Conf. Mass Spectr., Kyoto) (Ogata, K., Hayakawa, T., eds.), 814 – 819. Baltimore: University Park Press.

Mapleton, R. A. (1966): Proc. Phys. Soc. *87*, 219 – 222.

McCann, K. J., Flannery, M. R., Hazi, A. (1979): Appl. Phys. Lett. *34*, 543 – 545.

McClure, G. W. (1953): Phys. Rev. *90*, 796 – 803.

McDowell, M. R. C. (1966): Proc. Phys. Soc. *89*, 23 – 26.

McFarland, R. H. (1967): Phys. Rev. *159*, 20 – 26.

McGuire, E. J. (1977 a): Phys. Rev. *A 16*, 73 – 79.

McGuire, E. J. (1977 b): Phys. Rev. *A 16*, 62 – 72.

McGuire, E. J. (1979): Phys. Rev. *A 20*, 445 – 456.

Miller, W. F., Platzman, R. L. (1957): Proc. Roy. Soc. (London) *A 70*, 299 – 303.

Moores, D. L., Golden, L. B., Sampson, D. H. (1980): J. Phys. *B 13*, 385 – 395.

Morgulis, N. (1934): Phys. Z. Sowjet. *5*, 407 – 417.

Nagy, P., Skutlartz, A., Schmidt, V. (1980): J. Phys. *B 13*, 1249 – 1267.

Neckel, A., Sodeck, G. (1972): Monatshefte Chem. *103*, 367 – 382.

Ochkur, V. I. (1964): Soviet Physics — JETP *18*, 503 – 508.

Ochkur, V. I., Petrunkin, A. M. (1963): Optics and Spectroscopy (USSR) *14*, 245 – 248.

Otvos, J. W., Stevenson, D. P. (1956): J. Am. Chem. Soc. *78*, 546 – 551.

Peach, G. (1966): Proc. Phys. Soc. *87*, 381 – 391.

Peach, G. (1971): J. Phys. *B 4*, 1670 – 1677.

Peart, B., Walton, D. S., Dolder, K. T. (1969): J. Phys. *B 2*, 1347 – 1352.

Percival, I. C. (1966): Nucl. Fusion *6*, 182 – 187.

Peterkop, R. (1977): Theory of Ionization of Atoms by Electron Impact (Transl. by Hummer, D. G.). Boulder, Col.: Colorado Ass. Univ. Press.

Peterson, L. R., Prasad, S. S., Green, A. E. S. (1969): Can. J. Chem. *47*, 1774 – 1780.

Pichford, L. C., Märk, T. D., Castleman, jr., A. W. (1980): Proc. X-th SPIG, Dubrovnik, 20 – 21.

Post, R. F. (1961): Plasma Physics *3*, 273 – 286.

Pottie, R. F. (1966): J. Chem. Phys. *44*, 916 – 922.

Prasad, S. S., Prasad, K. (1963): Proc. Phys. Soc. *82*, 655 – 658.

Prok, G. M., Monnin, C. F., Hettel, H. J. (1969): J. Quant. Spectrosc. Radial. Transfer *9*, 361 – 369.

Rao, B. P., Murthy, V. R., Subbaiah, D. V., Naidu, S. V. (1979): Acta Ciencia Indica *5*, 118 – 123.

Rapp, D., Englander-Golden, P. (1965): J. Chem. Phys. *43*, 1464 – 1479.

Rieke, F. F., Prepejchal, W. (1969): Proc. 6th Int. Conf. Phys. Electr. Atomic Collisions, Cambridge, 623 – 625.
Rieke, F. F., Prepejchal, W. (1972): Phys. Rev. *A6*, 1507 – 1519.
Rosseland, G. (1923): Phil. Mag. *45*, 65 – 83.
Roy, B. N., Tripathi, D. N., Rai, D. K. (1972): Can. J. Phys. *50*, 2961 – 2966.
Roy, B. N., Rai, D. K. (1973 a): J. Phys. *B6*, 816 – 822.
Roy, B. N., Rai, D. K. (1973 b): Phys. Rev. *8*, 849.
Roy, B. N., Rai, D. K. (1983): J. Phys. *B 16*, 4677 – 4685.
Rudge, M. R. H. (1968): Revs. Mod. Phys. *40*, 564 – 590.
Salop, A. (1976): Phys. Rev. *A 14*, 2095 – 2101.
Schram, B. L., van der Wiel, M. J., de Heer, F. J., Moustafa, H. R. (1966): J. Chem. Phys. *44*, 49 – 54.
Seaton, M. J. (1959): Phys. Rev. *113*, 814 – 814.
Seaton, M. J. (1962): In Atomic and Molecular Processes (Bates, D. R., ed.), 374 – 420. New York: Academic Press.
Seaton, M. J. (1964): Planet. Space Science *12*, 55 – 73.
Sobelman, I. I., Vainstein, L. A., Yukov, E. A. (1981): Excitation of Atoms and Broadening of Spectral Lines. Berlin-Heidelberg-New York: Springer.
Stabler, R. C. (1964): Phys. Rev. *133*, A 1268 – A 1273.
Stafford, F. E. (1966): J. Chem. Phys. *45*, 859 – 862.
Stafford, F. E. (1968): Adv. Chem. Ser. *72*, 115 – 126.
Stafford, F. E. (1971): High Temp. High Press. *3*, 213 – 224.
Stephan, K., Helm, H., Märk, T. D. (1980): J. Chem. Phys. *73*, 3763 – 3778.
Stephan, K., Deutsch, H., Märk, T. D. (1983 a): Proc. Ann. Conf. Mass Spectrometry and Allied Topics, Boston, 734.
Stephan, K., Leiter, K., Deutsch, H., Märk, T. D. (1983 b): Proc. XIIIth ICPEAC, Berlin, 276.
Stevenson, D. P., Schissler, D. O. (1961): In: The Chemical and Biological Action of Radiation, Vol. 5 (Harssinsky, M., ed.), 181 – 192. London: Academic Bks.
Stolarski, R. S., Dulock, V. A., Watson, C. E., Green, A. E. S. (1967): J. Geophys. Res. *72*, 3953 – 3960.
Strzondola, V. (1983 a): Scripta Fac. Sci. Nat. Univ. Purk. Brun. *13*, 279 – 282.
Strzondola, V. (1983 b): Cesk. Cas. Fyz. *33*, 251 – 258.
Tannen, P. D. (1973): Thesis, University Microfilms, Ann Arbor, Mich., Order No. 74-14940.
Thomas, B. K., Garcia, J. D. (1969): Phys. Rev. *179*, 94 – 101.
Thomas, L. H. (1927): Proc. Cambridge Philos. Soc. *23*, 829 – 831, 713 – 716.
Thomson, J. J. (1912): Phil. Mag. *23*, 449 – 457.
Tiwari, P., Rai, D. K., Rustgi, M. L. (1969): J. Chem. Phys. *50*, 3040 – 3045.
Ton-That, D., Flannery, M. R. (1977): Phys. Rev. *A 15*, 517 – 526.
Ton-That, D., Manson, S. T., Flannery, M. R. (1977): J. Phys. *B 10*, 621 – 635.
Tripathi, D. N., Rai, D. K. (1972): Ind. J. Pure Appl. Phys. *10*, 185 – 186.
Tweed, R. J. (1973): J. Phys. *B6*, 270 – 284.
Vainshtein, L. A. (1957): Opt. Spektrosk. *3*, 313.
Vainshtein, L. A., Sobelman, I. I., Yukov, E. A. (1973): Electron-Excitation Cross Sections of Atoms and Ions. Moscow: Nauka. (In Russian.)
Vriens, L. (1964 a): Physics Lett. *9*, 295 – 296.
Vriens, L. (1964 b): Physics Lett. *10*, 170 – 171.
Vriens, L. (1964 c): Physics Lett. *8*, 260 – 261.
Vriens, L. (1965): Physica *31*, 385 – 395.
Vriens, L. (1966 a): Phys. Rev. *141*, 88 – 92.
Vriens, L. (1966 b): Proc. Phys. Soc. *89*, 13 – 21.
Vriens, L. (1969): In: Case Studies in Atomic Collision Physics I (McDaniel, E. W., McDowell, M. R. C., eds.), 335 – 398. Amsterdam: North-Holland.
Vriens, L., Smeets, A. H. M. (1980): Phys. Rev. *A 22*, 940 – 951.
Wallace, S. J., Berg, R. A., Green A. E. S. (1973): Phys. Rev. *A 7*, 1616 – 1629.
Watson, C. E., Dulock, V. A., Stolarski, R. S., Green, A. E. S. (1967): J. Geophys. Res. *72*, 3961 – 3966.
Webster, D. C., Hansen, W. W., Duvaneck, F. B. (1933): Phys. Rev. *43*, 839 – 858; *43*, 384 A.
Williams, E. J. (1927): Nature *119*, 489 – 490.
Williams, E. J. (1945): Revs. Mod. Phys. *17*, 217 – 226.

3

Threshold Behaviour of Ionization Cross-Sections

F. H. Read

Schuster Laboratory, University of Manchester, Great Britain

3.1 Introduction

A relationship that is well-known in atomic physics is the Wannier law

$$\sigma_{ion} \propto E^{1.127}. \tag{3-1}$$

It gives the dependence of the cross-section for the electron-impact ionization process

$$e + A \rightarrow A^+ + e + e \tag{3-2}$$

or the photo-double-detachment process

$$h\nu + A^- \rightarrow A^+ + e + e \tag{3-3}$$

on the amount E by which the energy of the system exceeds the ionization energy of the atom or negative ion. It is a threshold law, since it applies only when E is small, and it was first derived by Wannier (1953) using a treatment based on classical mechanics. It has excited, and continues to excite, considerable interest and some controversy, and has been the subject of many theoretical and experimental studies. A recent experimental result (which will be discussed in Section 3.2.4) is shown in Fig. 3-1.

Part of the attraction of the Wannier law is that it concerns a problem that lies at the interface of classical mechanics and quantum mechanics, and so is related to a wider range of problems for which there is, at present, no generally accepted method of solution. These problems are characterised by the fact that they involve interactions between three or more particles over distances that are too large for conventional quantum-mechanical techniques to be tractable. They usually also involve motion in the vicinity of an unstable potential ridge and the existence of an exceptionally high degree of correlation between the particles involved (see for example Fano 1980 a, b, 1983 a, b, Rau 1982). The subject of near-threshold ionization is particularly important since it represents the simplest example of a problem that contains all these features. The hope is therefore that the search for a theoretical technique

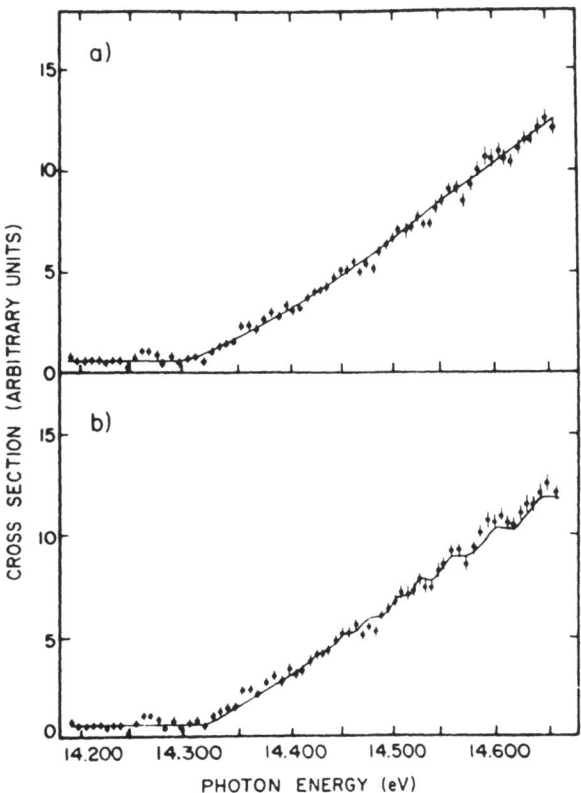

Fig. 3-1. Cross-section for the photodetachment process $hv + H^- \rightarrow H^+ + e + e$. The errors are statistical only. *a* The curve is the best fit to the power law $\sigma \propto (E + \varepsilon)^n$, where E is the excess energy above the threshold at 14.352 eV (and where ε and n are found to be 47 ± 5 meV and 1.15 ± 0.04 respectively). *b* The curve is the best fit to Temkin's modulated linear law (equation (3-101)). From Donahue *et al.* (1982, 1984)

that will give a good description of the threshold ionization process will also lead to a better understanding of the wider class of related problems. It may also lead to a better understanding of non-threshold ionization, since although the long-range correlations emphasized by the Wannier theory produce their most conspicuous effects within a few eV of the ionization threshold they may also play a significant role at higher energies. Neglect of these correlations may therefore be responsible for the marked weakness of almost all theoretical techniques in the range up to approximately 50 eV above threshold.

Yet another part of the attraction of the law and of the associated aspects of near-threshold ionization and excitation processes is of course that the experimental tests are particularly difficult and exacting, usually requiring good energy resolution, high sensitivity and the ability to handle electrons of very low energy.

The aim of this chapter is to discuss the various derivations and extensions of the Wannier law, and to compare the theoretical results with the available experimental evidence.

We start in Section 3.2 by considering when a classical approach is justified and then show in a straightforward way how this approach gives the Wannier law when the two outgoing electrons in the ionization process move collinearly. This is followed by a brief account of the general derivation of the law for non-collinear motion and a description of the experimental results. We then consider computational studies in which the ionization process is treated as a classical three-body problem and electron trajectories are integrated from their initial to their final asymptotic states, since these studies provide a useful confirmation of Wannier's analytical result and lead to various extensions of the Wannier law. The section finishes with a discussion of the effects of ion-core screening and of the form of the threshold law for multiple ionization. In Section 3.3 the quantum-mechanical and semiclassical derivations of the Wannier law are discussed and wavefunction symmetry effects are considered. In Section 3.4 we consider some of the finer details of the ionization process, such as the energy partitioning between the outgoing electrons, the distribution in angle and the distribution in angular momentum of these electrons. In Section 3.5 the discussion is broadened to cover the related process of near-threshold excitation of Rydberg states of high principal quantum number n,

$$e + A \rightarrow A^* + e. \tag{3-4}$$

The correlations that govern the behaviour of near-threshold ionization and excitation processes also give rise to the existence of the so-called "Wannier-ridge" resonances, which are discussed in this section. Finally, in Section 3.6 we draw together the various threads, to see what has been achieved so far and what remains to be done.

3.2 Classical Wannier Theory

3.2.1 Justification of the Classical Approach

For simplicity we start by considering a collinear system consisting of a stationary positive ion of charge number Z and two outgoing electrons that initially move in opposite directions. The final state of the system may be either that in which both electrons are free, corresponding to ionization, or that in which one electron is free but the other is bound to the ion and oscillates about it, corresponding to an excitation event.

We shall see below that when the total energy is small ionization occurs only when the system starts in, or near to, the symmetrical state given by $r_1 = -r_2 (=r)$ and $\dot{r}_1 = -\dot{r}_2$. Since this state plays a central role in the classical theory it is instructive to consider it in some detail.

The Coulomb energy of the system in the symmetrical state is, in atomic units,

$$V = -\frac{2Z}{r} + \frac{1}{2r}. \tag{3-5}$$

If the total energy of the system is E, then the combined kinetic energy of the two electrons is

$$K = E - V \tag{3-6}$$

and hence each electron has the wavelength

$$\lambda = \left[E + \frac{2}{r}(Z - \tfrac{1}{4}) \right]^{-1/2}. \tag{3-7}$$

We now define the "critical" radius

$$r_c = 2(Z - \tfrac{1}{4})/E \tag{3-8}$$

and use this as the unit of distance and E as the unit of energy. Then

$$\frac{V}{E} = -\frac{1}{r/r_c}, \tag{3-9}$$

$$\frac{K}{E} = 1 + \frac{1}{r/r_c}, \tag{3-10}$$

and

$$\frac{\lambda}{r_c} = \frac{E^{1/2}}{2\left(Z - \frac{1}{4}\right)\left(1 + \frac{1}{r/r_c}\right)^{1/2}}. \tag{3-11}$$

The significance of the critical radius can now be seen from equations (3-9) and (3-10): it separates the "Coulomb" zone in which K and V have similar magnitudes from the "outer" zone (also called the asymptotic or free zone) in which the magnitude of K is much larger than that of V and the two electrons are essentially free.

An electron can be treated as a classical particle only if it has a wavelength that is much smaller than the distance over which the force on the electron varies appreciably. In the present context this conditions is $\lambda \ll r$. We see from equation (3-11) that the condition is satisfied at $r = r_c$ if E is sufficiently small, and that it can also be satisfied at smaller values of r. More precisely, the condition is satisfied at values of r for which

$$r^{1/2} \gg [2(Z - \tfrac{1}{4})]^{-1/2}. \tag{3-12}$$

The Coulomb zone can therefore be defined to extend from a constant value, r_{min}, of r, for which equation (3-12) is satisfied, to the critical radius r_c. The electrons then behave approximately as classical particles in this zone and also in the outer zone. The behaviour of the system when it is in these zones can therefore be determined at small values of E, if spin-dependent effects can be ignored (see Section 3.3.3), by integrating the classical equations of motion. The division of r-space into the Coulomb and outer zones, and also into the inner zone ($r = 0$ to r_{min}) is illustrated in Fig. 3-2.

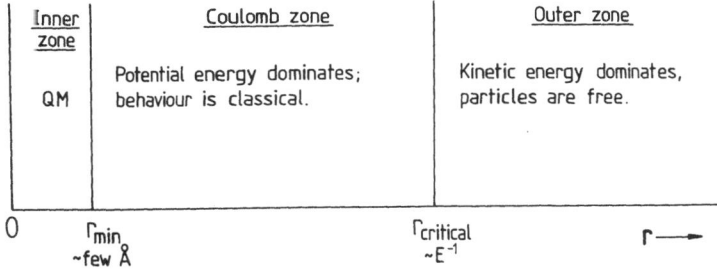

Fig. 3-2. The division of R-space into the inner, Coulomb and outer zones

The condition of the system at the inner boundary of the Coulomb zone is of course determined by the non-classical behaviour in the inner zone. However Wannier (1953) made the assumption that the phase-space distribution of the system at the inner boundary of the Coulomb zone is quasi-ergodic, in the sense that there are no strong singularities, so that the distribution in the small part of phase space that leads to ionization is finite and approximately uniform. If this assumption is valid the determination of the threshold ionization law at small E becomes a purely classical problem.

3.2.2 Collinear Motion

The Wannier law can be derived in a straightforward way by considering the states of motion in which two electrons recede in opposite directions, along the same straight line, from a central positive ion of charge number Z. The ion is assumed to be point-like and to be at rest. The classical equations of motion of the two electrons are then, in atomic units,

$$\ddot{r}_1 = -\frac{Z r_1}{|r_1|^3} + \frac{r_1 - r_2}{|r_1 - r_2|^3},$$

$$\ddot{r}_2 = -\frac{Z r_2}{|r_2|^3} + \frac{r_2 - r_1}{|r_1 - r_2|^3}, \tag{3-13}$$

and the energy of the system is given by

$$E = \frac{1}{2}(\dot{r}_1{}^2 + \dot{r}_2{}^2) - \frac{Z}{|r_1|} - \frac{Z}{|r_2|} + \frac{1}{|r_1 - r_2|}. \tag{3-14}$$

As noted by Wannier (1953) these equations can be scaled by taking

$$r \to \alpha r,$$
$$t \to \alpha^{3/2} t, \tag{3-15}$$
$$E \to \alpha^{-1} E,$$

so that if

$$r_j = f_j(t), \ j = 1 \text{ or } 2 \tag{3-16}$$

are solutions at the positive energy E, then so are

$$r_j = \alpha^{-1} f_j(\alpha^{3/2} t) \tag{3-17}$$

at the energy αE, for any positive value of α. Similarly if the corresponding velocities $v_{1,2}$ are given by

$$v_j = g_j(t) \tag{3-18}$$

at energy E, then they are given by

$$v_j = \alpha^{1/2} g_j(\alpha^{3/2} t) \tag{3-19}$$

at energy αE. Using equations (3-17) and (3-19) to eliminate t, we see that $\alpha^{-1/2} v_j$ is a function of αr_j, and therefore a set of universal trajectories can be obtained by plotting $E^{-1/2} v_j$ versus $E r_j$ for trajectories covering the initial and final conditions of interest. That is,

$$E^{-1/2}v_j = h_j(Er_j), \ j = 1, 2,$$ (3-20)

where the functions h_j depend only on the initial values of $E^{-1/2}v_j$ and Er_j.

A schematic representation of the universal plots is given in Fig. 3-3. The extent of the Coulomb and outer zones is indicated, and four trajectories are shown. The two labelled a and b correspond to ionization events in which the asymptotic energies of the outgoing electrons are 0 and E. Any trajectory that lies between these two therefore corresponds to an ionization event in which both electrons escape with non-zero final energies. On the other hand the trajectories labelled c and d correspond to excitation events, since one of the electrons escapes with a final energy greater than E while the other remains bound to the ion: this process will be discussed in Section 3.5.

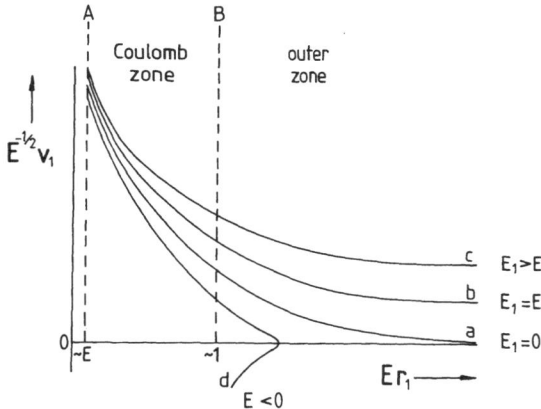

Fig. 3-3. Schematic representation of universal trajectory plots

As a first step towards finding the form of the universal trajectories we can integrate the equations of motion for the particular case $E = 0$. For the symmetric condition

$$r_1 = -r_2,$$ (3-21)

equations (3-13) and (3-14) then both reduce to

$$0 = \dot{r}^2 - \frac{2}{r}(Z - \tfrac{1}{4}).$$ (3-22)

Substituting

$$r = at^p,$$ (3-23)

we find that

$$a = \left[\frac{9}{2}(Z - \tfrac{1}{4})\right]^{1/3},$$ (3-24)

$$p = 2/3.$$

For the non-symmetric case we now put

$$r_1 = r + \Delta r,$$
$$r_2 = -r + \Delta r,$$ (3-25)

where

$$\Delta r = b t^q \tag{3-26}$$

and where we assume that

$$\Delta r \ll r. \tag{3-27}$$

Then

$$\ddot{r}_1 + \frac{Z}{r_1^2} - \frac{1}{(r_1 - r_2)^2} = b\left[q(q-1) - \frac{2Z}{a^3}\right]t^{q-2} + \dots \tag{3-28}$$

The leading term on the right-hand side of this equation is zero (and then E is also automatically zero) when

$$q = \frac{1}{2} \pm \frac{1}{6}\left[\frac{100Z - 9}{4Z - 1}\right]^{1/2}. \tag{3-29}$$

Now expressing $\dot{r}_1$ in therms of r_1 we find that

$$\dot{r}_1 = \frac{3}{2} a^{3/2} r_1^{-1/2} + b\left(q + \frac{1}{3}\right) a^{3/2(1-q)} r_1^{3/2(q-1)} + \dots \tag{3-30}$$

Defining

$$n = \frac{1}{4}\left[\left(\frac{100Z - 9}{4Z - 1}\right)^{1/2} - 1\right], \tag{3-31}$$

we find that the exponent of r_1 in the second term on the right-hand side of equation (3-30) (i.e. $^3/_2(q-1)$) either has the value $(-n-1)$, which is more negative than the exponent of the first term and is therefore of no interest in the present connection, or it has the value $n - ^1/_2$, which is positive. This positive exponent is responsible for the divergence of the trajectories shown in Fig. 3-3.

Let us consider now what happens when E is small and positive. The trajectories shown in Fig. 3-3 start at the boundary A with the same value of r_1 and with a uniform spacing in v_1. They therefore start by being uniformly distributed in phase space. As emphasized by Wannier (1953) the subsequent behaviour of the trajectories in the Coulomb zone is determined primarily by the form of the potential energy and so is almost independent of the magnitude of E when E is sufficiently small. Using equation (3-30) we see that in going from boundary A to boundary B the spacing between the trajectories increases by the factor

$$(r_B/r_A)^{n-1/2} \propto E^{-n+1/2}. \tag{3-32}$$

Therefore if the initial velocity spacing is Δv_{1i} the trajectory spacing at B is

$$\Delta(E^{-1/2} v_{1f}) \propto E^{-n} \Delta v_{1i}. \tag{3-33}$$

There is a further change in the trajectory spacing in going from the boundary B to the asymptotic region, but since the electrons are essentially free in this outer zone the change is almost independent of E. Therefore in the asymptotic region the density of trajectories, which is inversely proportional to the trajectory spacing, is given by

$$\rho \propto E^n/\Delta v_{1i}. \tag{3-34}$$

For a given initial phase-space density the probability of ionization is proportional to the density of trajectories between the limiting trajectories a and b in Fig. 3-3. This probability is therefore proportional to E^n, where n is given by equation (3-31). This is the Wannier law. For $Z = 1$, 2 and 3, n has the values 1.127, 1.056 and 1.036 respectively.

3.2.3 Total Ionization Cross-Section

To consider the general, non-collinear motion of the two outgoing electrons it is convenient to use the hyperspherical coordinates

$$R = (r_1^2 + r_2^2)^{1/2},$$
$$\theta_{12} = \cos^{-1}(\breve{r}_1 \cdot \breve{r}_2), \tag{3-35}$$
$$\alpha = \tan^{-1} r_2/r_1.$$

R gives the scale size, θ_{12} is the angle between r_1 and r_2, and α is a quasi-angle that depends on the asymmetry between the lengths r_1 and r_2. These three coordinates therefore define the size and shape of the triangle formed by the vectors r_1 and r_2. Three other coordinates, the three Euler angles, define the orientation of this triangle in space, but since the Wannier treatment is limited to systems that have a zero total angular momentum their behaviour is isotropic with respect to these angles, which therefore need not be taken into account.

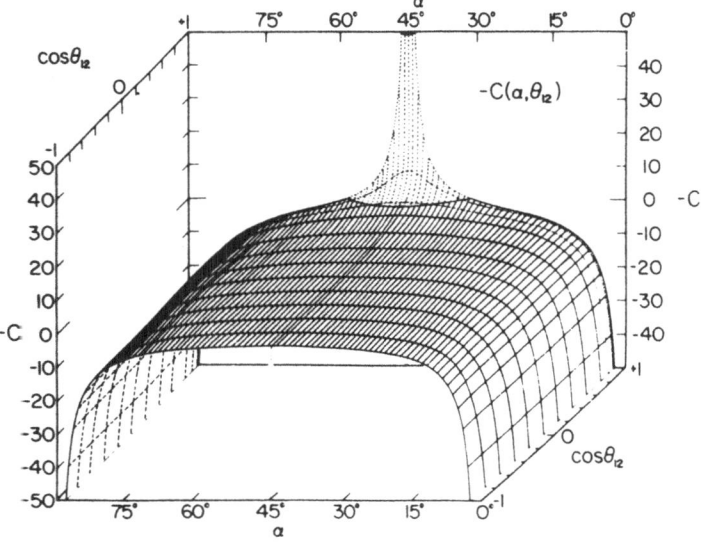

Fig. 3-4. The dependence of the potential function C (equation (3-37)) on α and $\cos\theta_{12}$. From Fano and Lin (1975). The function $-C$ plotted here is equal to twice the function defined by equation (3-37). C is symmetrical about the plane through $\theta_{12} = \pi$

The Coulomb energy expressed in terms of the hyperspherical coordinates is

$$V = \frac{C(\alpha, \theta_{12})}{R} \tag{3-36}$$

where

$$C(\alpha, \theta_{12}) = -\frac{Z}{\cos \alpha} - \frac{Z}{\sin \alpha} + \frac{1}{(1 - \sin 2\alpha \cos \theta_{12})^{1/2}}. \tag{3-37}$$

The dependence of C on α and $\cos \theta_{12}$ is shown in Fig. 3-4 (taken from Fano and Lin (1975)). The part of C that corresponds to $\theta_{12} = \pi$ to 2π is not shown. C is symmetrical about the point $\alpha = \pi/4$, $\theta_{12} = \pi$ (i.e. $r_1 = -r_2$), now often called the *Wannier point*.

In the previous section we saw that when E is small ionization occurs only when the initial state is nearly symmetric (i.e. when $r_1 \simeq -r_2$, $v_1 \simeq -v_2$). This is true also for non-collinear motion, the necessary initial conditions being $r_1 \simeq -r_2$, $v_1 \simeq -v_2$. In other words the system must start near the Wannier point. Expanding the potential in the region of this point, we find that

$$C(\alpha, \theta_{12}) = -2^{3/2}(Z - 1/4) - 2^{-3/2}(12Z - 1)\beta^2 + 2^{-7/2}\gamma^2 + \ldots \tag{3-38}$$

where

$$\beta = \pi/4 - \alpha \tag{3-39}$$

and

$$\gamma = \pi - \theta_{12}. \tag{3-40}$$

The region around the Wannier-point is therefore saddle-shaped, with C increasing in the θ_{12} direction but decreasing in the α direction. The angle θ_{12} therefore tends to converge towards π as the system evolves and R increases, but the quasi-angle α tends to diverge from $\pi/4$. If this divergence is great enough the system will "fall" into one of the valleys at $\alpha = 0$ or $\pi/2$, leaving one electron bound to the ion while the other electron goes free. Ionization (i.e. double-escape) can occur only if the system avoids this "fall" by staying in the vicinity of the Wannier point until R has increased to a value at which the two electrons are essentially independent of each other. Since it takes a long time for R to reach this value when E is small, we can now see in qualitative terms that the ionization cross-section is necessarily very small when E is small.

We now summarize the derivation given by Wannier (1953). In terms of the hyperspherical coordinates (equations (3-35), (3-39), and (3-40)) and the potential function C (equation (3-37)) the Hamiltonian of the system is

$$H = \frac{1}{2} p_R^2 + \frac{1}{2R^2} p_\beta^2 + \frac{2}{R^2 \sin^2 2\alpha} p_\gamma^2 + \frac{C}{R}, \tag{3-41}$$

where p_R, p_β and p_γ are the generalized momenta that correspond to the coordinates R, β and γ respectively, and are given by

$$p_R = \dot{R}$$

$$p_\beta = R^2 \dot{\beta}, \tag{3-42}$$

$$p_\gamma = \frac{R^2 \sin^2 2\alpha}{4} \dot{\gamma}.$$

This Hamiltonian leads to the equations of motion

$$\ddot{R} = R\,\beta^2 + \frac{1}{4}\,R\sin^2 2\alpha\cos 2\alpha\,\dot{\gamma}^2 + \frac{C}{R^2},$$

$$\frac{d}{dt}(R_2\,\beta) = \frac{1}{2}\,R^2\sin 2\alpha\cos 2\alpha\,\dot{\gamma}^2 - \frac{1}{R}\,\frac{\partial C}{\partial\beta}, \qquad (3\text{-}43)$$

$$\frac{d}{dt}\left[\frac{1}{4}\,R^2\sin^2 2\alpha\,\dot{\gamma}\right] = -\frac{1}{R}\,\frac{\partial C}{\partial\gamma}.$$

The energy is given by

$$E = \frac{1}{2}\,\dot{R}^2 + \frac{1}{2}\,R^2\,\beta^2 + \frac{1}{8}\,R^2\sin^2 2\alpha\,\dot{\gamma}^2 + C/R. \qquad (3\text{-}44)$$

It is convenient at this stage to replace the time t by the independent variable $q(t)$, defined by

$$R = b\,e^q, \qquad (3\text{-}45)$$

where b is a constant. Then for any variable x,

$$x' = (R/\dot{R})\,\dot{x}, \qquad (3\text{-}46)$$

where the prime denotes differentiation with respect to q. Putting $E=0$ and assuming that $\beta\ll 1$ and $\gamma\ll 1$, it can now be shown that

$$\beta'' - \frac{1}{2}\,\beta' + \frac{12Z-1}{2(4Z-1)}\,\beta = 0,$$

$$\gamma'' + \frac{1}{2}\,\gamma' + \frac{1}{2(4Z-1)}\,\gamma = 0. \qquad (3\text{-}47)$$

These equations therefore replace the equations of motion (3-43). An important feature of them is that β and γ are decoupled.
The general solution of equations (3-47) is

$$\beta = a_1\,e^{-(n+1/2)q} + a_2\,e^{nq},$$

$$\gamma = a_3\,e^{-1/4\,q}\cos\left(\frac{1}{2}\,\rho q + a_4\right), \qquad (3\text{-}48)$$

where n is given by equation (3-31) and

$$\rho = \frac{1}{2}\left(\frac{9-4Z}{4Z-1}\right)^{1/2}. \qquad (3\text{-}49)$$

Since

$$n \geq 1$$

the first term in β converges while the second diverges as q, and hence R, increases. Therefore if $a_2=0$ the two electrons finally have $r_1 = r_2$ and ionization occurs, but if $a_2 \neq 0$ the system falls away from the Wannier region and into one of the valleys of the potential function (see Fig. 3-4), giving rise to an excitation event.

When E is non-zero the solutions (3-48) are still approximately valid in the Coulomb zone, the outer limit of which is the critical radius

$$R_c = 2^{3/2} (Z - \tfrac{1}{4}) E^{-1}. \tag{3-50}$$

For trajectories that have a non-zero value of a_2 the value of β at the critical radius is

$$\beta_c \simeq a_2 (R_c/b)^n. \tag{3-51}$$

Ionization will occur when $|\beta_c|$ is less than some value β_{max}. Therefore the maximum value of $|a_2|$ for which ionization is possible is

$$a_{2,max} = \beta_{max} \left[\frac{b}{2^{3/2}(Z - \tfrac{1}{4})} \right]^n E^n. \tag{3-52}$$

The next step in Wannier's argument was to consider the initial volume Ω of phase space that corresponds to constant values of E and R, and to establish that the rate at which this volume is swept out by a set of representative points is given by

$$\frac{d\Omega}{dt} = \int d\beta \, dp_\beta \, d\gamma \, dp_\gamma. \tag{3-53}$$

This integral can be transformed into one over a_1, a_2, a_3 and a_4, treated now as independent variables, giving

$$\frac{d\Omega}{dt} \propto \int a_3 \, da_1 \, da_2 \, da_3 \, da_4. \tag{3-54}$$

We have seen above that the range of values of a_2 is limited, if ionization is to occur. The range of a_1 is not limited in this way, since a_1 is associated with the converging term in β, and similarly the ranges of a_3 and a_4 do not affect the ionization probability, since γ always tends towards zero according to equation (3-48). Thus if the distribution in phase space of the trajectories as they enter the Coulomb zone is uniform (i.e. if ergodicity is assumed) the ionization probability is proportional to $a_{2,max}$ and hence is proportional to E^n when E is small. This is still true if the distribution in a_2 is non-uniform, but with no zeros or infinities (i.e. the quasi-ergodic assumption).

The remaining important step in Wannier's argument was to assume that the initial phase-space distribution is indeed quasi-ergodic, in analogy with Wigner's (1948) analysis of single-particle threshold escape. This thus gives the Wannier-law (1).

3.2.4 Experimental Tests

The result of the most accurate direct text (Donahue et al. 1982, 1984) of the Wannier law has already been shown in Fig. 3-1. In this experiment two electrons are detached from H^- by photoabsorption (3-3). The threshold of the process occurs at a photon energy of 14.352 eV. The H^- ions have an energy of 800 meV and are irradiated by laser light having $h\nu = 4.65$ eV at an angle α to the direction of motion of the ions. The photon energy in the rest frame of the ions is

$$E = h\nu \left[1 - \frac{v^2}{c^2} \right]^{-1/2} \left[1 + \frac{v}{c} \cos\alpha \right], \tag{3-55}$$

where v is the velocity of the ions. This energy can be tuned from 1.63 to 15.6 eV by varying α.

The yield shown in Fig. 3-1 has two components, one from the photo-double-detachment process (3-3) and the other from the two-step process in which highly excited H atoms are produced,

$$h\nu + \mathrm{H}^- \rightarrow \mathrm{H}^* + e \qquad (3\text{-}56)$$

and are then ionized by the motional electric field which they experience when they pass through a region of magnetic field that forms part of the detection system. This field ionization occurs for states that have $n \geq 15$ and hence binding energies less than B_m, where $B_m = 60$ meV. Because of this extra contribution the yield is not proportional to E^n. However, Read (1984) has shown (see also Sections 3.4.1 and 3.5.1) that the yield should be approximately proportional to $(E + B_m)^n$. In fact Donahue et al. (1982, 1984) fitted their results to

$$\sigma \sim (E + E_0)^n \qquad (3\text{-}57)$$

and found that $E_0 = 47 \pm 5$ meV and $n = 1.15 \pm 0.04$. This value of n is consistent with that expected for the Wannier law, 1.127.

Earlier direct tests of the Wannier law (e.g. McGowan and Clarke 1968, Brion and Thomas 1967, 1968 a, b, Krige et al. 1968, Marchand et al. 1969) were made by studying the electron-impact ionization process (3-2) and all of these indicated that the power-law exponent is slightly greater than unity. A weakness of these studies was that the spread in energy (typically ≥ 70 meV) of the incident electrons caused the measured ionization yield to be rounded-off at threshold, so that even a cross-section that rises linearly from threshold would have given a measured yield that starts with a power-law exponent greater than unity. Nevertheless in the most accurate of such experiments, that of Marchand et al. (1969), it was possible to allow for this effect in the analysis of the results and to deduce that $n = 1.16 \pm 0.03$.

In a test of the Wannier law for $Z = 2$, Van der Wiel (1972) used a simulated photon technique to measure the yield of the reaction

$$h\nu + \mathrm{He} \rightarrow \mathrm{He}^{++} + e + e. \qquad (3\text{-}58)$$

He found that the exponent is approximately 1.06, which is consistent with the value 1.056 given by equation (3-31).

The most accurate (although indirect) measurement of the Wannier exponent n remains that of Cvejanović and Read (1974 a). In a preliminary experiment (described in more detail in Section 3.4.1) they established that the partitioning in energy of the two outgoing electrons is approximately uniform at low E, as had been predicted by Vinkalns and Gailitis (1967 a). In other words the probability distribution of the energy E_1 of one of these electrons is uniform between the limits 0 and E (see also Section 3.4.1). They then used the "penetrating-field technique" (see also Cvejanović and Read 1974 b) to measure a partial yield — the yield of electrons having an energy E_1 between 0 and E_m, where E_m is constant and is approximately 20 meV. Their result is shown in Fig. 3-5.

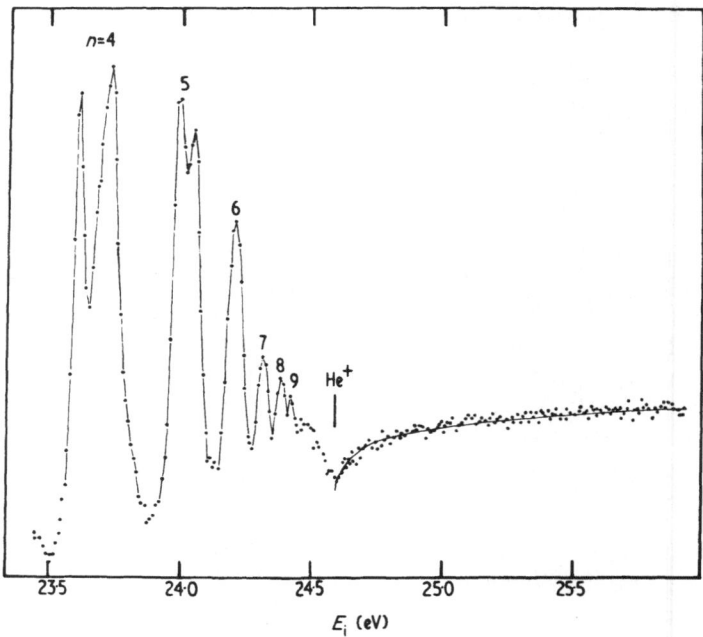

Fig. 3-5. The yield of very low energy electrons ($\leqslant 20$ meV) resulting from electron-helium impact, as a function of the incident energy. The curve drawn through the points above the ionization energy I is proportional to $E^{0.127}$, where $E = E_i - I$. From Cvejanović and Read (1974)

The ratio of the partial yield to the total ionization yield is equal to E_m/E if the energy partitioning is uniform, in which case the measured yield above the ionization energy has the energy dependence

$$\text{yield} \propto E^{n-1}. \tag{3-59}$$

The line through the data in Fig. 3-5 has $n - 1 = 0.127$. An analysis of this and other spectra gave

$$n = 1.131 \pm 0.019. \tag{3-60}$$

This dependence holds for values of E from 0 up to at least 1.7 eV. The validity of this analysis does not depend on the assumption of uniform energy partitioning (see Section 3.4.1).

The part of the spectrum of Fig. 3-5 that lies below the ionization threshold is approximately a mirror image of the part above the threshold, if an average is taken through the peaks that lie at the energies of discrete excited states. An "ionization cusp" is therefore formed. The significance of the part below the ionization threshold will be discussed in Section 3.5. Ionization cusps have been seen subsequently in many threshold electron yields for a variety of atoms and molecules (e. g. Spence 1975, Hammond et al. 1984 b). The dependence of the depth of the cusp on the energy of the detected electrons is shown for helium in Fig. 3-6 (Spence 1975). This result will be discussed further in Section 3.5.

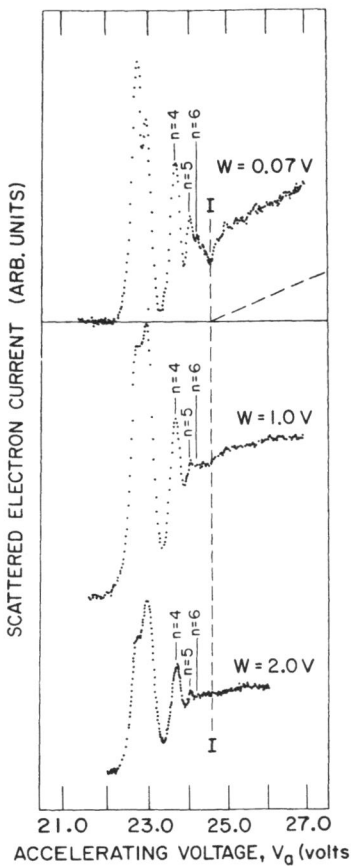

Fig. 3-6. Spectra showing the total yield of scattered electrons form He with final energies eW, as a function of the incident energy $e(V_a + W)$. The ionization threshold is marked I. From Spence (1975)

3.2.5 Trajectory Integrations

Computational studies in which the classical equations of motion are integrated numerically serve two functions: they confirm the analytical treatment outlined in Section 3.2.3. but more importantly they provide additional information about certain details of the near-threshold ionization process and also the near-threshold excitation process that it has not yet been possible to obtain analytically. These details are dealt with in Sections 3.4 and 3.5. In the present section we discuss only how the computational studies confirm and illustrate the analytical treatment.

The change in behaviour when the system of two electrons and an ion passes from the Coulomb zone to the outer zone is illustrated by the results of a numerical integration of the collinear equations of motion (3-13) (the computational details being essentially as described by Read (1984)). Rather than plot $E^{-1/2} v_1$ versus $E r_1$, as in Fig. 3-3, it is more instructive to remove the leading term in the expression (3-30) for v_1 so that the behaviour of the second and higher terms can be seen more clearly. This can be done effectively by considering the velocity difference

$$\Delta v_1 = v_1 - \left[E + \frac{2}{r_1}(Z - \tfrac{1}{4}) \right]^{1/2}.$$

(3-61)

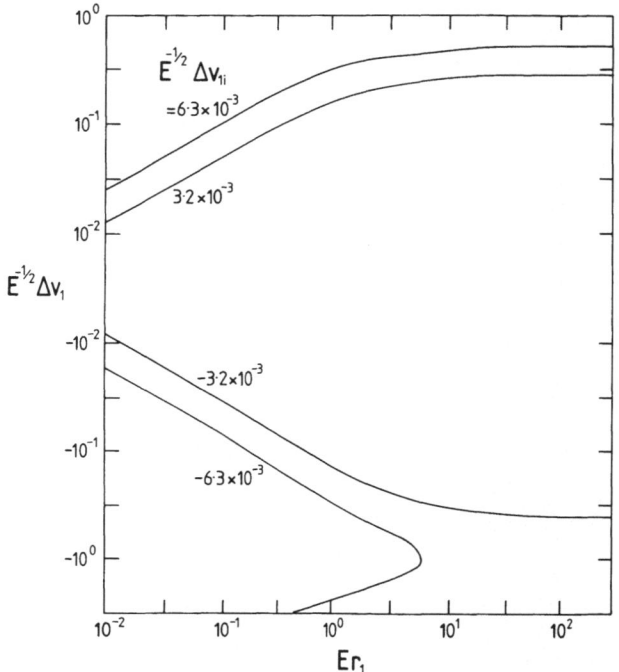

Fig. 3-7. Computed dependence of $E^{-1/2}\Delta v_1$ on Er_1 for collinear motion and for $Z=1$, where Δv_1 is defined by equation (3-61)

Fig. 3-7 shows the dependence of $E^{-1/2}\Delta v_1$ on Er_1 for $Z=1$ and for four different initial conditions. The boundary between the Coulomb and outer zones lies at $Er_1 \sim 1$. We see that for three of the trajectories this boundary does indeed mark the transition from the dependence

$$\Delta v_1 \propto r_1^{\,n-1/2} = r_1^{\,0.627}$$

(3-62)

given by equations (3-29), (3-30) and (3-31), to the dependence

$$\Delta v_1 = \text{constant}$$

(3-63)

that applies asymptotically when the electron is free. For the fourth trajectory the electron fails to escape from the field of the ion.

Approximate confirmation of the Wannier law itself has been provided, for $Z = 1$, by the numerical results of Banks et al. (1969), Peterkop and Tsukerman (1969, 1970), Grujić (1972), Cvejanović and Grujić (1975) and Boesten et al. (1976). In all these studies the initial configuration is taken to be an incident electron moving towards a classical target atom. One initial parameter, such as the impact parameter or the initial position of the target electron in its Keplerian orbit, is varied, and the range of values of this parameter that lead to ionization is found for various values of E. The ionization cross-section is taken to be proportional to the length of this range. The results are then fitted to a power law.

Higher accuracy is obtained when the initial configuration is the collision complex consisting of an ion and two nearby electrons (Read 1984). This reproduces the "half-scattering" problem that is treated analytically by Wannier (1953). The main advantage of this approach is that the ranges of initial conditions can be chosen to ensure that all the integrated trajectories lead to the final conditions of interest. By taking seven values of E in the range from 5×10^{-4} to 3.2×10^{-2} au, and by almost completely covering the initial phase-space distribution according to the ergodic hypothesis Read (1984) finds that the Wannier law is indeed reproduced. Other results of this approach will be described in detail in Section 3.4.

3.2.6 Core Screening

Klar (1981) has considered the form of the threshold ionization law when the central ion acts as a screened charge that has a magnitude $Z(r)$ at a distance r from the centre. $Z(r)$ varies from the charge of the nucleus when r is small to an asymptotic value equal to the change of the ion. Klar finds that the Wannier exponent is still given by equation (3-31), where Z is the charge of the ion, provided that a certain condition is satisfied. This condition concerns the rate of charge of $Z(r)$ at the critical radius R_c that separates the Coulomb and free zones. Taking $r = 2^{-1/2} R$ (equation (3-35)) and using equation (3-50) for the critical radius, we see that R_c is the root of

$$R_c = 2^{3/2} [Z(2^{-1/2} R_c) - {}^1/_4] E^{-1}. \tag{3-64}$$

Putting

$$r_w = 2^{-1/2} R_c, \tag{3-65}$$

Klar's condition becomes

$$\left| \frac{r}{Z(r)} \frac{dZ(r)}{dr} \right|_{r_w} \ll 1. \tag{3-66}$$

The WKB treatment that he uses to derive the threshold law is then valid.

As an experimental test Hippler et al. (1983) have measured the energy dependence of the electron-impact cross-section for ionization of the Ar K shell and the Xe L_3 shell. They find for example a constant exponent n up to 70 eV above threshold for Ar. At this energy $r_w \simeq 0.078$ nm, $Z(r_w) \simeq 2.2$, and condition (3-66) is approximately satisfied. The wavelength of the electron at $r = r_w$ is approximately 0.021 nm, which although smaller than r_w is perhaps not sufficiently small for the Wannier model to be applied with confidence.

2.2.7 Multiple Ionization

Two years after his work on single ionization Wannier (1955) went on to consider processes that result in an ion and N electrons, where $N \geq 3$, such as

$$e + A \rightarrow A^{(N-1)+} + e_1 + \ldots + e_N \tag{3-67}$$

and

$$hv + A \rightarrow A^{N+} + e_1 + \ldots + e_N. \tag{3-68}$$

He argued that in the absence of long-range electron-electron interactions the cross-section is governed by the volume of phase-space that is available to the N electrons. If the electrons have final energies $E_1, E_2, \ldots E_N$, and the total energy is E, this gives (see also Rau 1971, 1984b)

$$\sigma \propto \int \delta(E - E_1 - E_2 \ldots - E_N) \, dE_1 \, dE_2 \ldots dE_N. \tag{3-69}$$

For example when $N = 2$

$$\sigma \propto \int \delta(E - E_1 - E_2) \, dE_1 \, dE_2 = \int dE_1 = E. \tag{3-70}$$

More generally,

$$\sigma \propto E^{N-1}, \tag{3-71}$$

for $N \geq 1$. Therefore the threshold cross-section for simple photoionization is a constant, as is well-known, that for photo-double-ionization or for electron-impact ionization is proportional to E (in the absence of long-range correlations), and that for photo-triple-ionization or for electron-impact double-ionization is proportional to E^2, etc.

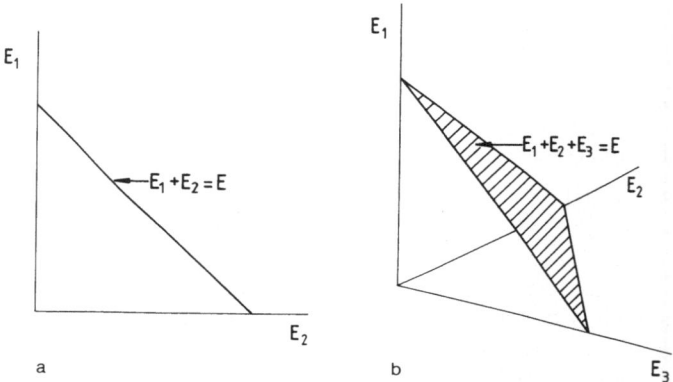

Fig. 3-8. The volume of phase-space available (in the absence of long-range electron-electron interactions) for the outgoing electrons in the reaction $e + A \rightarrow A^{(N-1)+} + e_1 + \ldots + e_N$. *a* When $N = 2$ the volume is proportional to the length of the full line that represents $E_1 + E_2 = E$. *b* When $N = 3$ it is proportional to the shaded area on the surface that represents $E_1 + E_2 + E_3 = E$

A pictorial representation of these results is given by considering the hypersurface defined by the equation $E_1 + E_2 + \ldots + E_N = E$ in the multidimensional space $E_1, E_2 \ldots E_N$. The integral in equation (3-69) is the area occupied by this surface in the energetically accessible region, defined by $E \geq E_i \geq 0$ for all i. For example when $N = 2$ the surface is the line shown in Fig. 3-8 a, and when $N = 3$ it is the two-dimensional surface shown in Fig. 3-8 b. We see that the dimensionality of the hypersurface is $N - 1$ and that its area is therefore proportional to E^{N-1}.

These results are modified when the long-range interactions and correlations between the outgoing electrons are taken into account. When $N = 2$ the work of Wannier (1953) shows that an instability in the potential function causes a reduction in the ionization probability, leading to the threshold law E^n where n is now greater than unity. The difference $(n - 1)$ decreases as Z increases, since the electron-electron interactions then become less significant compared to the electron-ion interactions. The threshold law is similarly modified in the general case of N outgoing electrons. The exponent is therefore

$$n > N - 1 \qquad (3\text{-}72)$$

and the difference $\{n - (N - 1)\}$ decreases as Z increases.

The triple-escape process (i.e. $N = 3$) has been considered in more detail by Klar and Schlecht (1976), Dimitrijević and Grujić (1981) and Grujić (1983 a). Klar and Schlecht (1976) used a WKB-like approximation and established that triple-escape is possible only when the electrons recede with approximate triangular symmetry, so that their three radius vectors approximately form an equilateral triangle. The effective (screened) charge seen by each electron in the symmetrical configuration is $(Z - 1/\sqrt{3})$. Klar and Schlecht found that the threshold law is E^n where

$$n = \frac{1}{2}\left[\frac{13\,Z\,3^{1/2} - 11 + 2\,(108\,Z^2 + 1)^{1/2}}{3^{1/2}\,Z - 1}\right]^{1/2} - \frac{1}{2}. \qquad (3\text{-}73)$$

This exponent is $2.826, 2.270$ and 2.126 for $Z = 1, 2$ and 3 respectively. The difference $(n - 2)$ is therefore quite large, especially for $Z = 1$, indicating the strong influence of the long-range electron-electron interactions. In deriving this result it was assumed that the spin-state of the electrons is such that the spatial wavefunction does not have a node at the position of triangular symmetry, and so the result applies to a ^2S state.

Other aspects of the triple-escape process, such as the distributions in energy, angle and angular momentum, have been considered by Dimitrijević and Grujić (1981) and Grujić (1983 a). The only experimental test of the threshold law appears to be that of Brion and Thomas (1968 b), who measured the yield of He^{++} ions in the reaction

$$e + He \rightarrow He^{++} + e + e + e, \qquad (3\text{-}74)$$

and found that $n \simeq 2$ in the energy region $E = 0$ to 25 eV. There is clearly a need for further experimental studies.

Grujić (1983 b) has considered the threshold law for $N = 4$. In the zero-energy limit $(E = 0)$ the four outgoing electrons recede with maximum symmetry, in the tetrahedral configuration. The effective charge that each experiences is then

$Z - (27/32)^{1/2}$. When $Z = 3$, corresponding to electron-impact triple-ionization of a neutral target, $n = 3.525$. There are no experimental tests of this. Grujić (1983 b) has defined the "fractional exponent"

$$\kappa_N = n/(N-1) \tag{3-75}$$

for reaction (3-67) on a neutral target. Its magnitude is 1.127, 1.135 and 1.175 for $N = 2$, 3 and 4 respectively. Similar trends exist for non-neutral targets.

3.3 Semiclassical and Quantum-Mechanical Treatments

In the earliest quantum-mechanical studies of the ionization process (e.g. Geltman 1965, Rudge and Seaton 1964, 1965, Rudge 1968) the special effects of the long-range electron-electron interaction were not taken into account, and a linear law was obtained. This would also follow from the general phase-space argument presented in Section 3.2.7, without a detailed treatment of the wavefunction involved (see also Rau 1971, 1984b).

In the present section we start by considering the semiclassical treatment of Peterkop (1971), in which the electron-electron correlations are taken into account by using hyperspherical coordinates. The wavefunction is expressed in a semiclassical form and the wave equation is solved using a WKB-like approximation. We then describe the work of Rau (1971), who used a similar wave function and a similar approximation. In Section 3.3.2 we describe the studies that are based on more conventional quantum-mechanical techniques and finish in Section 3.3.3 by considering the effects that the symmetries of the spatial and spin wavefunctions have on the cross-sections.

3.3.1 Semiclassical and WKB-Like Treatments

Peterkop (1971, 1977) used the WKB approximation to investigate the threshold ionization law, following an earlier one-dimensional study (Peterkop and Liepinsh 1969). The wavefunction for $L = 0$ is expressed in the semi-classical form (in atomic units)

$$\Psi_0 = P^{1/2} e^{iS} \tag{3-76}$$

where P is the classical density and S is the classical action. Substituting this in the Schrödinger equation and ignoring terms that contain differentials of P, we find that S is a solution of the Hamilton-Jacobi equation

$$(\nabla_1 S)^2 + (\nabla_2 S)^2 = 2(E - V) \tag{3-77}$$

and that P is a solution of the continuity equation

$$\nabla_1 (P \nabla_1 S) + \nabla_2 (P \nabla_2 S) = 0. \tag{3-78}$$

Peterkop took S and P to be functions of the hyperspherical coordinates (equations (3-35), (3-39) and (3-40)), and expanded each in terms of β^2 and γ^2. He established that in the asymptotic region Ψ_0 has the form

$$\Psi_0 = P_0 (R)^{1/2} \exp i \left[(2E)^{1/2} R - (2E)^{-1/2} C(\alpha, \theta_{12}) \ln R\right] \tag{3-79}$$

where C is given by equation (3-37), and that in the Coulomb zone it has a form that involves the Wannier exponent n (equation 3-31) and the quantity ρ (equation 3-49). Matching the two forms he found that the normalization condition

$$P_0(R) \to R^{-5} \quad \text{as} \quad R \to \infty \tag{3-80}$$

implies that

$$P_0(R) = R^{-5} E^{1-n} g(E, R), \tag{3-81}$$

where g tends to a constant value as E tends to zero at fixed R.

The wave function Ψ in the inner zone is not given by the WKB approximation but is assumed to be approximately independent of E when E is small. At some value R_0 of R this inner wave function matches the WKB wave function, so that

$$\Psi(R_0) = f \Psi_0(R_0), \tag{3-82}$$

where f is a matching coefficient. The differential cross-section corresponding to the final coordinates α and θ_{12} is then

$$\sigma(\alpha, \theta_{12}) = (E/I)^{1/2} |f|^2, \tag{3-83}$$

where I is the ionization energy. Since $\Psi(R_0)$ must remain finite as E tends to zero, it follows that

$$|f P_0^{1/2}| \to \text{const.}, \tag{3-84}$$

and hence

$$|f|^2 \to \propto E^{n-1}. \tag{3-85}$$

Therefore

$$\sigma(\alpha, \theta_{12}) \propto E^{n-1/2}. \tag{3-86}$$

The range of final values of θ_{12} is proportional to $E^{1/4}$ (see Section 3.4.2) and so the corresponding solid angle is proportional to $E^{1/2}$. Integration over the final values of α does not introduce any further dependence on E, and so the "total" ionization cross-section is proportional to E^n, thus reproducing the Wannier law.

In the treatment of Rau (1971) the wavefunction for $L = 0$ is written as

$$\Psi(R, \alpha, \theta) = R^{-5/2} \operatorname{cosec} 2\alpha \, \phi(R, \alpha, \theta_{12}). \tag{3-87}$$

Substituting this in the Schrödinger equation shows that ϕ satisfies

$$\begin{bmatrix} \dfrac{\partial^2}{\partial R^2} + \dfrac{1}{4R^2} + 2E + \dfrac{1}{R^2} \dfrac{\partial^2}{\partial \alpha^2} + \\ + \dfrac{4}{R^2 \sin^2 2\alpha \sin \theta} \dfrac{\partial}{\partial \theta_{12}} \left[\sin \theta_{12} \dfrac{\partial}{\partial \theta_{12}} \right] - \dfrac{2C(\alpha, \theta_{12})}{R} \end{bmatrix} \phi(R, \alpha, \theta_{12}) = 0 \tag{3-88}$$

where C is the potential function defined by equation (3-37). After expanding C in the vicinity of the Wannier point, as given by equation (3-38), an inspection of equation (3-88) shows that in the "Coulomb zone", defined here by

$$R \ll 2^{9/2}(Z - 1/4) E^{-1}, \tag{3-89}$$

ϕ has approximately the form of a single-particle Coulomb wave-function for zero energy. Substituting a modified form of this wave-function, namely

$$\phi(R, \beta, \gamma) = \exp[icR^{1/2}(1 + a\beta^2 + ib\gamma^2)] \chi(R), \tag{3-90}$$

it is found that the lowest-order terms in equation (3-88) cancel when

$$c = 2^{9/4}(Z - \tfrac{1}{4})^{1/2},$$
$$a = -\tfrac{1}{4} \pm (n + \tfrac{1}{4}), \tag{3-91}$$
$$b = \tfrac{1}{64}(i \pm 2\rho),$$

where n and ρ are the constants that appear in Wannier's classical treatment and are given by equations (3-31) and (3-49) respectively. The two values of a are

$$a_d = n, \quad a_c = -n - \tfrac{1}{2}, \tag{3-92}$$

and in analogy with the Wannier treatment it is the positive value, a_d, that causes divergent behaviour and thus governs the form of the threshold law. This is illustrated in Fig. 3-9, which shows surfaces, in the (R, β) plane, that have a constant value of the phase $cR^{1/2}(1 + a\beta^2)$ that appears in the expression for ϕ. The figure also shows the curves that are orthogonal to these surfaces, corresponding to the classical trajectories.

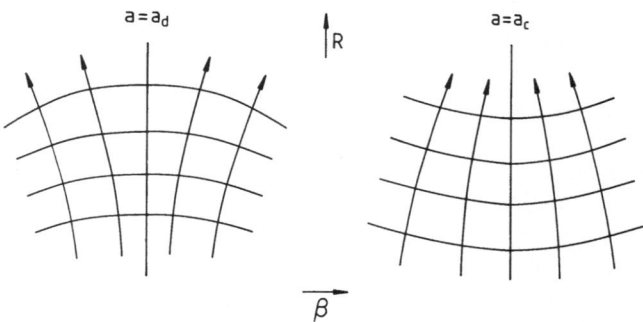

Fig. 3-9. Sketch of surface of constant phase in the (R, β) plane, with corresponding classical trajectories, for the positive (a_d) and negative (a_c) values of a (adapted from Rau (1971))

The equation for the radial function $\chi(R)$ is

$$\left[\frac{d^2}{dR^2} + 2E - \frac{1}{4R^2} + \frac{ic}{R^{1/2}} \left(\frac{d}{dR} + \frac{2a + 16ib - \tfrac{1}{4}}{R} \right) \right] \chi(R) = 0. \tag{3-93}$$

When $E = 0$ the solution of this, at large R, is

$$\chi(R) = R^{-2a - 16ib + 1/4}. \tag{3-94}$$

Substituting this into equation (3-90), ϕ is obtained as a sum of two independent terms, corresponding to the two values of a. The relative amplitude of these terms is determined by the form of the solution in the inner reaction zone and is independent of E. By considering the dependence on E of the amplitudes of these terms in the Coulomb and outer zones at finite E, and by matching the solutions across the zone boundaries, Rau established the normalization of the divergent term, and was thus able to verify that the ionization cross-section follows the Wannier law. A weakness in Rau's normalization procedure has been pointed out and corrected by Peterkop (1983).

Rau (1971) laid particular emphasis on the role played by what he termed the "dynamic screening" of the two electrons. This refers to the fact that the effective core screening that each electron provides for the other changes with time. This dynamic screening had been considered in the work of Wannier (1953) but it had not been considered in conventional quantum-mechanical treatments since in these the effective charges Z_1 and Z_2 seen by the two electrons had been assumed to be independent of time or position.

To see the effect of dynamic screening let us consider two electrons that start at the same distance from the central ion, so that they both experience the same effective charge, but with velocities that are slightly different. Because the slower electron moves less far from the ion it becomes less well screened from the ion than the faster electron, which results in an enhancement of the velocity difference. Thus the slower electron will be progressively slowed down and will perhaps be prevented from escaping. Dynamic screening thus causes the probability for excitation to be increased, at the expense of the probability for ionization. Dynamic screening is of course directly related to the instability in α, as given by equation (3-38) and illustrated in Fig. 3-4.

Yet another WKB-like approach is that of Klar and Schlecht (1976). They expressed the wavefunction in the form

$$\Psi = R^m \exp\left[i(8R)^{1/2} Q(\alpha, \theta_{12})\right]. \tag{3-95}$$

Substituting this in the Schrödinger equation and neglecting terms in R^{-2} and higher-order terms they found that Q is a solution of the WKB-like equation

$$Q^2 + 4(\nabla Q)^2 + C(\alpha, \theta_{12}) = 0 \tag{3-96}$$

and that

$$m = -\frac{9}{4} + \frac{\Lambda^2 Q}{Q}\bigg|_{\alpha = \frac{\pi}{4}, \theta_{12} = \pi}. \tag{3-97}$$

Here Λ^2 is the grand angular momentum: it is a function of α and θ_{12} only and is proportional to the terms in equation (3-88) that involve differentiation with respect to α and θ_{12}. They verified the Wannier law and were able to extend their technique to cover the triple-escape process (see Section 3.2.7). More recently Feagin (1984) has used a WKB-like method based on relative Jacobi coordinates to investigate the energy dependence of the Coulomb break-up of the general 3-particle system with arbitrary angular momentum. He verifies the Wannier law for $L=0$. Other aspects of Feagin's treatment will be dealt with in sections 3.3.3 and 3.4.2.

The classical and WKB-like treatments discussed so far have all (except that of Feagin 1984) been concerned only with ionization processes for which the angular momentum L is zero. Roth (1972) has shown however, using an extension of Rau's technique, that the Schrödinger equation for $L=1$ is equivalent to that for $L=0$ for threshold purposes, so that the threshold law is unchanged. Peterkop and Tsukerman (1969) have shown similarly that the threshold behaviour of the classical ionization cross-section is the same for $L=0$ and $L=1$. A discussion of later more detailed studies of the effect of non-zero values of L will be given in Section 3.3.3.

3.3.2 Quantum-Mechanical Treatments

A treatment that does not give the Wannier law is that of Temkin (1966, 1969, 1974, 1982, 1984) and Temkin and Hahn (1974). For example, Temkin (1982) treats the two-dimensional model, with $Z = 1$. The potential energy is (in atomic units)

$$V = -\frac{1}{r_1} - \frac{1}{r_2} + \frac{1}{r_1 + r_2}. \tag{3-98}$$

When $r_1 > r_2$ this can be expressed as

$$V = -\frac{1}{r_2} - \frac{r_2}{r_1^2} + \frac{1}{r_1} \sum_{n=2}^{\infty} \left[\frac{-r_2}{r_1} \right]^n. \tag{3-99}$$

Temkin (1982) suggests that only the first two terms need be retained when $r_1 \geq 2 r_2$, and that these two terms can be interpreted as the sum of a Coulomb and a dipole potential, seen by the inner and outer electrons respectively. It is argued that threshold ionization takes place predominantly in the part of space for which $r_1 \geq 2 r_2$ (or $r_2 \geq 2 r_1$), and so the description "Coulomb-dipole theory" is given to this approach. The condition $r_1 \geq 2 r_2$ is assumed to be maintained for all values of r_1 and r_2, and so the process of dynamic screening (which would allow the electrons to start with $r_1 \simeq r_2$ but finish with $r_1 \geq 2 r_2$) is not taken fully into account. The wave function is expressed as a product of a Coulomb wavefunction for the inner electron and a dipole wavefunction for the outer electron, and the near-threshold ionization cross-section is found to be

$$\sigma \propto E/(\ln E)^2 \left[1 + \sum_L C_L \sin (A_L \ln E + B_L) \right] \tag{3-100}$$

where A_L, B_L and C_L are adjustable parameters. A similar form,

$$\sigma \propto E \left[1 + C \sin (A \ln E + B) \right] \tag{3-101}$$

has also been proposed (Temkin 1974, Temkin and Hahn 1974). We see that in this latter form a linear dependence is modulated by the sine term. Fig. 3-1 b shows a fit of Temkin's modulated linear law (101) to the results of Donahue et al. (1982, 1984). Although the oscillations in the solid curve appear to follow the data more closely than the power-law fit of Figure 1 (a), Donahue et al. (1982, 1984) point out that there are more parameters in the fit of Figure 1 (b) (six versus four), and that the confidence level for the second fit is only marginally higher (25% versus 19%).
Temkin (1974, 1984) points out that although in his opinion the modulated linear form applies to electrons that originate from a limited region of phase-space, the Wannier law still applies to the remainder of the ionization yield. He emphasises that the ratio of the linear to Wannier contributions is proportional to $E^{-0.127}$ and that the linear contribution, if it exists, must therefore dominate at sufficiently small E. The possibility of the existence of a linear contribution is also indicated by the work of Bottcher (1982), who studied the evolution in time of a wave-packet representing an electron incident on an atom.

3.3.3 Wavefunction Symmetry Effects

The two main restrictions in the analyses of Wannier (1953), Peterkop (1971) and Rau (1971) are that the total angular momentum is zero and that the symmetry of the spin state of the two electrons does not introduce any antisymmetry in the spatial wavefunction at the Wannier point. The total spin is therefore $S = 0$ and the system is in the $^1S^e$ state. In this section we consider the form of the threshold ionization law for other states of the system. We start by outlining the theoretical work of Klar and Schlecht (1976), Greene and Rau (1982, 1983), Stauffer (1982) and Feagin (1984) and then consider the results of experimental investigations.

The starting point in the treatment of Greene and Rau (1983) is to express the wavefunction of the state having quantum numbers L, S and Π (representing the total orbital and spin angular momenta and the parity respectively) in the form used by Klar and Schlecht (1976),

$$\Psi_{LS\Pi} \varpropto f(\omega) R^m \exp\left[i(8 R)^{1/2} Q(\omega)\right]. \tag{3-102}$$

This is a generalization of equation (3-95). Here ω denotes the set of five angular coordinates consisting of the two hypershperical coordinates $\beta(=\pi/4-\alpha)$ and $\gamma(=\pi-\theta_{12})$ and the three Euler angles that define the orientation in space of the triangle formed by the vectors r_1 and r_2 (see Section 3.2.3). The Euler angles do not enter the problem when $L = 0$, because the wavefunction is then isotropic in these angles, but they must be considered when $L \neq 0$.

The symmetry of the spin state determines the symmetry of the spatial wavefunction with respect to exchange of the two electrons,

$$\Psi_{LS\Pi}(r_2, r_1) = (-1)^S \Psi_{LS\Pi}(r_1, r_2). \tag{3-103}$$

Similarly the parity of the state imposes the condition

$$\Psi_{LS\Pi}(-r_1, -r_2) = \Pi \, \Psi_{LS\Pi}(r_1, r_2). \tag{3-104}$$

A wavefunction that satisfies these conditions can be constructed by making use of the coupled spherical harmonics $Y_{l_1 l_2 LM}(\theta_1 \phi_1, \theta_2 \phi_2)$. These represent the angular wavefunctions of states having the quantum numbers L and M, formed from electrons having angular momenta l_1 and l_2, and have the symmetry property (Edmonds 1974, Stauffer 1982)

$$Y_{l_1 l_2 LM}(\theta_2 \phi_2, \theta_1 \phi_1) = (-1)^{l_1+l_2-L} Y_{l_2 l_1 LM}(\theta_1 \phi_1, \theta_2 \phi_2). \tag{3-105}$$

The angular function $f(\omega)$ is now expressed in the form (Greene and Rau 1983)

$$f(\omega) = \sum_{l_1 \leq l_2} \left[g_{l_1 l_2}(R; \beta) Y_{l_2 l_1 LM}(\theta_1 \phi_1, \theta_2 \phi_2) + \right.$$
$$\left. + (-1)^{l_1+l_2+L+S} g_{l_1 l_2}(R; -\beta) Y_{l_2 l_1 LM}(\theta_1 \phi_1, \theta_2 \phi_2) \right]. \tag{3-106}$$

Since $Q(\omega)$ is symmetric with respect to the operation $r_1 \leftrightarrow r_2$, we see from equation (3-105) that condition (3-103) is satisfied with this form of $f(\omega)$. Condition (3-104) is also satisfied, provided that the summation is limited to l values that satisfy

$$(-1)^{l_1+l_2} = \Pi. \tag{3-107}$$

The dependence of $f(\omega)$ on the coordinate β is particularly important in determining the threshold law. If $f(\omega)$ is even under $\beta \to -\beta$ the threshold exponent n is still as given by equation (3-31), but if $f(\omega)$ is odd under $\beta \to -\beta$ there is a node in the wavefunction at the Wannier point. Peterkop (1983) and Feagin (1984) show that the threshold exponent is then

$$m = 3n. \tag{3-108}$$

This has the value 3.381 when $Z = 1$. The exponent is not affected by the symmetry of the wavefunction under the operation $\gamma \to -\gamma$, although this symmetry can affect the magnitude of the cross-section, as we shall discuss later.

The symmetry of $f(\omega)$ with respect to the operation $\beta \to -\beta$ can easily be established for states that have $\Pi = +1$ and $L = 0$ or 1 (i.e. for the states $^1S^e$, $^3S^e$, $^1P^e$ and $^3P^e$), because then only terms that have $l_1 = l_2$ can appear in equation (3-106), which thus becomes

$$f(\omega) = \sum_l [g_{ll}(R;\beta) + (-1)^{L+S}g_{ll}(R;-\beta)]\, Y_{llLM}(\theta_1\,\phi_1, \theta_2\,\phi_2). \tag{3-109}$$

The operation $\beta \to -\beta$ then gives

$$f(\omega) \to (-1)^{L+S} f(\omega). \tag{3-110}$$

Therefore the $^1S^e$ and $^3P^e$ states are symmetric in β and give rise to the threshold exponent 1.127, whereas the $^3S^e$ and $^1P^e$ states are antisymmetric and give the exponent 3.381. For all other states the function $f(\omega)$ is neither even nor odd with respect to the operation $\beta \to -\beta$, but can be expressed as a linear combination of even and odd components. For example the even component is

$$f_+(\omega) = \frac{1}{2} \sum_{l_1 \leq l_2} [g_{l_1 l_2}(R;\beta) + g_{l_1 l_2}(R;-\beta)] \times$$
$$\times [Y_{l_1 l_2 LM}(\theta_1\,\phi_1, \theta_2\,\phi_2) + (-)^{l_1+l_2+L+S}\, Y_{l_2 l_1}(\theta_1\,\phi_1, \theta_2\,\phi_2)]. \tag{3-111}$$

This component gives the exponent $m = 1.127$ while the corresponding odd component gives $m = 3.381$. The lower value dominates of course. Therefore the exponent 3.381 would only be observed for the $^3S^e$ and $^1P^e$ states.

The symmetry in β determines the value of the exponent but the symmetry in γ can influence the magnitude of the cross-section. This happens because for certain combinations of L, S and Π the even component $f_+(\omega)$ (which is the only one of interest, except for the $^3S^e$ and $^1P^e$ states) in odd with respect to the operation $\gamma \to -\gamma$, and hence is zero at the Wannier point. In fact it follows from equation (3-111) and the properties of the functions Y (Edmonds 1974) that $f_+(\omega)$ is non-zero at the Wannier point only when L, S and Π are either all even or all odd (i.e. when the state symmetry is $^1S^e$, $^3P^0$, $^1D^e$, etc). In all other cases there is a node in γ at the Wannier point, and as a consequence the cross-section is expected to be reduced. Therefore the threshold ionization process, and also double excitation of atoms to highly excited states (see Fano 1976 and Section 3.5.3), are dominated by the $^1S^e$, $^3P^0$, $^1D^e$, etc. states of the two outgoing electrons. All these results have also been obtained, using similar reasoning, by Stauffer (1982).

The simplest forms of the angular functions of the dominant part of $f(\omega)$ are shown in Table 3-1 for all states having $L \leq 2$. These forms are derived from those given by Greene and Rau (1983). Also shown in the table are the corresponding values of the exponent m and the lowest-order behaviour of the wavefunction near the Wannier point. The lowest-order behaviour illustrates all the points made above: only $f_-(\omega)$ exists for the $^3S^e$ and $^1P^e$ states, which thus have a node in β and a higher value of m, there is no node in γ in the $f_+(\omega)$ of the $^1S^e$, $^3P^0$ and $^1D^e$ states, and there is a node in γ for all the remaining states.

Table 3-1. *The simplest forms of the angular functions of the dominant part of $f(\omega)$ for each type of state (up to $L=2$), together with the corresponding exponent (for $Z=1$) and the lowest-order behaviour near the Wannier point*

State	Angular function of dominant part	Exponent	Behaviour near Wannier point
$^1S^e$	$f_+ = 1$	1.127	const.
$^3S^e$	$f_- = \cos^2\alpha - \sin^2\alpha$	3.381	β
$^1P^e$	$f_- = \sin 2\alpha \cos 2\alpha \sin\theta_1 \sin\theta_2 \sin(\phi_2 - \phi_1)$	3.381	$\beta\gamma$
$^3P^e$	$f_+ = \sin\theta_1 \sin\theta_2 \sin(\phi_2 - \phi_1)$	1.127	γ
$^1P^0$	$f_+ = \cos\theta_1 + \cos\theta_2$	1.127	γ
$^3P^0$	$f_+ = \cos\theta_1 - \cos\theta_2$	1.127	const.
$^1D^e$	$f_+ = (3\cos^2\theta_1 - 1) + (3\cos^2\theta_2 - 1)$	1.127	const.
$^3D^e$	$f_+ = (3\cos^2\theta_1 - 1) - (3\cos^2\theta_2 - 1)$	1.127	γ
$^1D^0$	$f_+ = \sin 2\theta_1 \sin\theta_2 \sin(\phi_2 - \phi_1) + \sin 2\theta_2 \sin\theta_1 \sin(\phi_1 - \phi_2)$	1.127	γ
$^3D^0$	$f_+ = \sin 2\theta_1 \sin\theta_2 \sin(\phi_2 - \phi_1) - \sin 2\theta_2 \sin\theta_1 \sin(\phi_1 - \phi_2)$	1.127	γ

The $^1P^0$ state is particularly important since it is the only state formed in photo-double-detachment of H$^-$ or He. The exponent is 1.127 but the final state has a node in γ and hence the cross-section is reduced.

Before the work of Greene and Rau (1982, 1983) and Stauffer (1982), Klar and Schlecht (1976) had erroneously predicted that an exponent 3.881 would apply to all triplet final states of the two electrons. This prediction gave rise to several experiments (Alguard *et al.* 1977, Hils and Kleinpoppen 1978, Hils *et al.* 1980, Baum *et al.* 1981 and Gay *et al.* 1982) in which spin-polarized electrons are used to ionize spin-polarized atoms that have only one valence electron. The total spin state of the initial electron-atom system becomes the spin state of the two outgoing electrons. A measurement of the asymmetry parameter

$$A = (\sigma^{\downarrow\uparrow} - \sigma^{\uparrow\uparrow})/(\sigma^{\downarrow\uparrow} + \sigma^{\uparrow\uparrow}) \tag{3-112}$$

thus allows the ratio of triplet to singlet cross-sections to be deduced. According to the prediction of Klar and Schlecht A was expected to be unity near the ionization threshold. The experimental results of Alguard *et al.* (1977) on electron-impact ionization of H atoms are shown in Fig. 3-10, together with the results of various non-threshold theoretical calculations. At 2 eV above threshold the value of A is far from unity. The experiments of Hils and Kleinpoppen (1978), Hils *et al.* (1980), Baum *et al.* (1981) and Gay *et al.* (1982), on K, Na, Li and H respectively, have also shown that A converges to a threshold value that is not unity. In fact this is the expected result since the dominant components ($^1S^e$, $^1D^e$ etc. for the singlet cross-section, and $^3P^0$, $^3F^0$ etc. for the triplet cross-section) all have the exponent 1.127. In the most recent experiment of this type Kelley et al. (1983) measured A and also the unnormalized total cross-section for electron-impact ionization of sodium from the

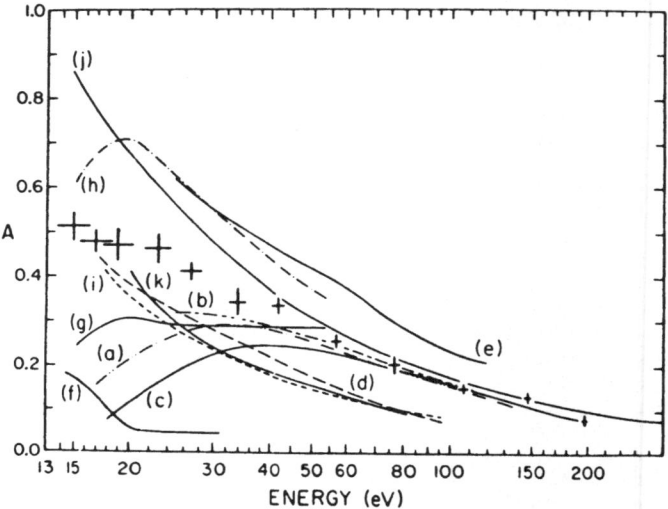

Fig. 3-10. Values of the asymmetry parameter A for electron-impact ionization of H atoms, measured by Alguard et al. (1977), together with the results of various non-threshold theoretical calculations (see Alguard et al. 1977 for details)

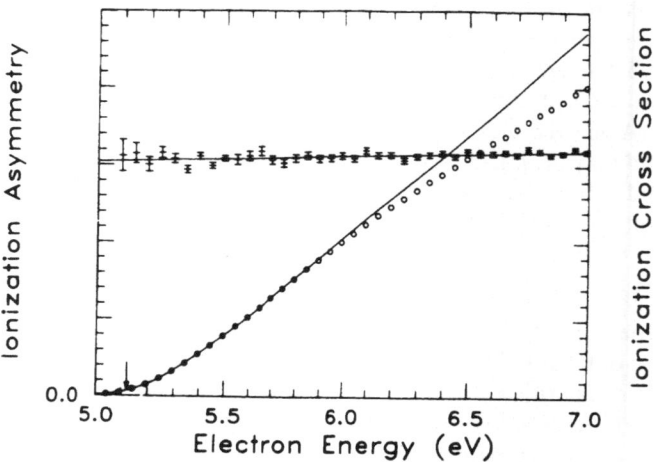

Fig. 3-11. Unnormalized values of the spin-state asymmetry parameter A (equation (3-111)) and the cross-section (circles) for electron-impact ionization of sodium. From Kelley et al. (1983)

threshold to 2 eV above. Their results are shown in Fig. 3-11. It can be seen that A is almost constant.

The experiment of Kelley *et al.* (1983) also shows that the total cross-section follows a power law up to about 0.8 eV above threshold, and that the exponent is 1.097 ± 0.17, consistent with that expected for the Wannier law. The range over which the power law holds for sodium is somewhat less than the range 1.7 eV used in the experiment of Cvejanović and Read (1974) on helium. There are as yet no indications (see for example Marchand *et al.* 1969, Donahue *et al.* 1982) that the range of validity of the Wannier law is less than approximately 1 eV, except when autoionizing levels are present (as in the experiments of Hammond *et al.* (1984 b), on Ar, Kr and Xe). This is in stark contrast to the range of validity of the Wigner threshold law for processes in which the final state consists of a neutral atom and a free electron, which can be as small as 6 μeV (Mead *et al.* 1984). The reason for the large range of validity of the Wannier law is presumably that the law is governed by the Coulomb force and that this force dominates any others that may be present (such as the dipole, centrifugal, polarisation and quadrupole forces), except at very short distances.

3.4 Differential Ionization Cross-Sections

So far we have considered only the total (integrated, see page 5) cross-section for near-threshold ionization. We now turn out attention to more detailed aspects of the near-threshold ionization process, starting with the way in which the two outgoing electrons share the available energy and then going on to consider the relative directions and individual angular momenta of the two electrons.

3.4.1 Energy Partitioning

In this section we consider the differential cross-section $\sigma_E(E_1) dE_1$ for producing two free electrons at the excess energy E, one having an energy in the range E_1 to $E_1 + dE_1$, the other having an energy in the range $(E - E_1 - dE_1)$ to $(E - E_1)$. The discussion is restricted to systems in which (i) the total angular momentum is zero, (ii) the electrons are in a singlet spin state, since no account is taken of the antisymmetries discussed in Section 3.3.3, and (iii) the ion charge number is unity. We start by describing the results of a recent computational study (Read 1984) and then consider the results of other computational and experimental studies.

The main feature of the differential cross-section $\sigma_E(E_1)$ can be explained by referring again to Fig. 3-3. It was shown in Section 3.2.2 that the asymptotic density of the trajectories that lie between those labelled a and b (which mark the limits of the ionization process) is proportional to E^n, where n is given by equation (3-31). There is no *a priori* reason to suppose however that this density is constant in the interval between a and b, although it will be a constant function of the ordinate

$$y = E^{-1/2} v_{1f} = (2 E_1 / E)^{1/2}. \tag{3-113}$$

The differential cross-section must therefore have the form

$$\sigma_E(E_1) = E^{n-1} f(E_1/E), \tag{3-114}$$

since the integral of this over the interval $E_1 = 0 \rightarrow E$ is proportional to E^n.

The form of the function f has been determined (Read 1984) by accurate trajectory integrations of the classical equations of motion

$$\ddot{r}_i = -\frac{Z}{|r_i|^3}r_i + \frac{r_i - r_j}{|r_i - r_j|^3}, \quad i = 1, 2, \ j \neq i. \tag{3-115}$$

The result is shown in Fig. 3-12. The figure also shows the form of $f(x)$ when $x > 1$, and the form of an analogous function $f_<$ that applies when $E < 0$: this extra information is relevant to near-threshold excitation processes, and will be discussed in Section 3.5. At present we are concerned only with $f_>(x)$ in the range $x = 0 \rightarrow 1$.

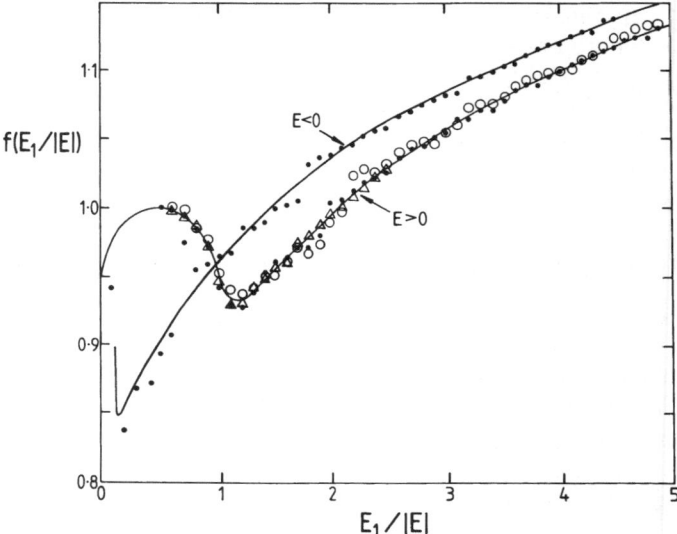

Fig. 3.12. The points are computed values of the universal functions $f_<$ and $f_>$ from which the classical cross-sections for near-threshold excitation and ionization process can be derived, for negative and positive values respectively of the total energy E. E_1 is the asymptotic energy of the more energetic of the two final electrons. The curves are the empirical fits given by equations (3-117), (3-145) and (3-146), except and where these fits are inappropriate arbitrary curves have been drawn through the points. The classical cross-sections are proportional to $|E|^{0.127} f(E_1/|E|)$. From Read (1984)

To compute the form of f it is necessary to select a set of initial conditions that represents a uniform distribution in phase space. This is achieved by using hyperspherical coordinates (equations (3-35), (3-39) and (3-40)) and using the fact that the volume element of phase space for fixed values of R and E is (Wannier 1953)

$$d\tau = d\beta \, dp_\beta \, d\gamma \, dp_\gamma \tag{3-116}$$

(see also equation (3-53)). Here p_β and p_γ are the generalized momenta defined by equations (3-42). In practice the initial value of β is set equal to zero (i.e. $r_{1i} = r_{2i}$) and uniform intervals are taken in the remaining three parameters γ, p_γ and p_β, or p_β is set equal to zero (i.e. $v_{1i} = v_{2i}$) and uniform intervals are taken in β, γ and p_γ. The initial conditions are then expressed in terms of cartesian coordinates and the equations of motion integrated to obtain the asymptotic energy E_1, using a library code based on the Runge-Kutta-Merson method. Three different values of E have been used

$(5 \times 10^{-4}, 10^{-3}$ and 2×10^{-3} au), to confirm that the dependence on E is as given by equation (3-114).

As can be seen in Fig. 3-12, $f_>(E_1/E)$ is not uniform in the interval $E_1 = 0 \to E$, as had been previously supposed (see for example Vinkalns and Gailitis 1967 a, b, Cvejanović and Grujić 1975), but is approximately 5% higher at the centre of the interval than at the ends. In other words the energy partitioning between the two electrons in the threshold ionization process is not uniform: instead the probability decreases as the electron energies become more dissimilar. Another feature is that at the ends of the interval, where the process of excitation is approached, the slope of f is large: in a detailed study of this region using closely spaced initial parameters for the trajectories it has not been possible to place an upper limit on this slope, which may therefore be infinitely large. An empirical formula that fits the computed results well in the interval $E_1 = 0$ to E is

$$f_>(E_1/E) \propto 0.95 + 0.1 \left[\frac{E_1}{E} - \left(\frac{E_1}{E} \right)^2 \right]^{1/2}. \tag{3-117}$$

A possible cause of the non-uniformity of the energy partitioning lies in the definition of the critical radius R_c, equation (3-50). Although this definition is appropriate over most of the interval $E_1 = 0$ to E, it may not be appropriate near the ends of the interval. Consider for example the situation in which $E_1 \simeq E$, and hence $E_2 \simeq 0$. In the outer zone r_i ($i = 1$ or 2) is approximately proportional to r_i, which in turn is proportional to $E_i^{1/2}$. Hence

$$(r_2/r_1) \simeq (E_2/E_1)^{1/2}. \tag{3-118}$$

Assuming that this is also true at the critical radius, equation (3-35) implies that

$$r_{2,c} \simeq R_c (E_2/E)^{1/2}. \tag{3-119}$$

Therefore when E_2/E is very small $r_{2,c}$ may be small enough to invalidate the assumption that $\lambda_{2,c} \ll r_{2,c}$. This problem can perhaps be overcome by taking R_c to be larger when E_2 is small. The range of the parameter a_2 defined by equations (3-48), (3-51) and (3-52) would then be reduced, causing a reduction in the ionization yield.

The earlier numerical and theoretical studies of energy partitioning in the threshold ionization process (Vinkalns and Gailitis 1976 a, b, Cvejanović and Grujić 1975, Yurev 1977, Peterkop and Liepinsh 1981) were not accurate enough (see Read 1984) to have revealed the existence of a small non-uniformity. This is also true for the experimental studies, of which there have been three. In the first of these Cvejanović and Read (1974) investigated the partitioning function by a time-of-flight technique. They measured the difference Δt in the time taken for the two ionization electrons to travel from the electron-helium collision centre along field-free paths of length l ($= 10$ mm) in opposite directions. This time difference is related to the velocities of the electrons by

$$\Delta t = l \left[\frac{1}{v_1} - \frac{1}{v_2} \right]. \tag{3-120}$$

Since the excess energy E was known in these experiments, each measurement of Δt gave the velocities, and hence the energies E_1 and $(E - E_1)$, of the two electrons. Analysis of the time-of-flight spectra then gave the probability distribution for E_1.

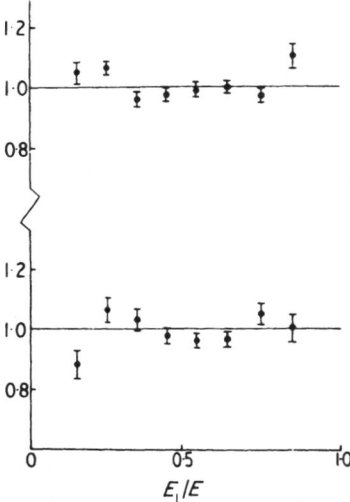

Fig. 3-13. Energy-partitioning probability deduced by Cvejanović and Read (1974) from time-of-flight measurements. The excess energy E is 0.37 eV and 0.60 eV for the upper and lower results respectively.

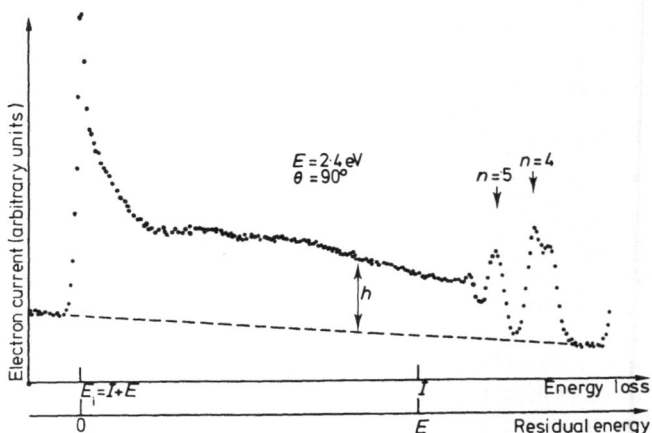

Fig. 3-14. The yield of electrons resulting from $e - $ He impact, as a function of the energy (labelled residual energy) of the detected electrons. The incident electron energy is 2.4 eV above the ionization threshold and the angle of scattering is 90°. When the residual energy is less than 2.4 eV the detected electrons arise from the ionization process $e + $ He $\rightarrow$ He$^+$ $+ e + e$, and when the residual energy is greater than 2.4 eV they arise from the excitation process $e + $ He $\rightarrow$ He* $+ e$. From Pichou *et al.* (1978)

Fig. 3-13 shows the results obtained at two values of E. The energy partitioning is seen to be uniform to within approximately 5%. The form of the partitioning near the ends of the range (i.e. at $E_1 \simeq 0$ and $E_1 \simeq E$) could not be tested however since there one of the electrons has a very low energy and hence is susceptible to small stray fields of unknown strength. The second experiment is that of Pichou *et al.* (1978), who looked at the energy spectrum of the ionization or scattered electrons resulting from electron impact on helium at a constant value of E. Their spectra, one of which is shown in Fig. 3-14, should therefore give the function $f(E_1/E)$ directly, but in practice the sensitivity is limited and also the behaviour at small E_1 is masked by a strong dependence on E_1 of the efficiency of the electron detection system. They found that f is approximately uniform at small values of E but that it has a minimum in the centre of the range when $E \gtrsim 3.6$ eV. On the other hand in the third, similar experiment Keenan et al. (1982) found that f is approximately uniform up to $E = 5.5$ eV, but again did not have the sensitivity and accuracy to be able to see a 5% non-uniformity at smaller values of E. A definitive experiment is therefore awaited.

3.4.2 Angular Distributions

The two electrons in the near-threshold ionization process emerge asymptotically with an angle θ_{12} between their directions of motion. In this section we consider the form, and dependence on E, of the probability distribution function for θ_{12}.
The dependence of the relative angle θ_{12} on the excess energy E was first established by Vinkalns and Gailitis (1967 a). Rather than use their line of reasoning we can make use of the equations already presented in Section 3.2.3. From equations (3-40), (3-45) and (3-48) we see that when $E = 0$,

$$\gamma = \pi - \theta_{12} = a_3 (R/b)^{-1/4} \cos\left(\tfrac{1}{2} \rho \ln(R/b) + a_4\right),$$ (3-121)

where b is an arbitrary constant, a_3 and a_4 are constants that are determined by the initial conditions, and ρ is the constant given by equation (3-49). Assuming that equation (3-121) remains valid, in the Coulomb zone, when E is finite but small, it follows from the energy-dependence of R_c (equation (3-50)) that the value of γ at the outer limit R_c of the Coulomb zone is

$$\gamma_c \propto a_3 E^{1/4} \cos\left(\tfrac{1}{2} \rho \ln(R_c/b) + a_4\right).$$ (3-122)

Although γ may change slightly in the outer zone, no further dependence on E is expected to be introduced by the motion in this zone. The width $\gamma_{1/2} (FWHM)$ of the asymptotic probability distribution of γ is therefore proportional to $E^{1/4}$,

$$\gamma_{1/2} = a E^{1/4}.$$ (3-123)

This is the result derived by Vinkalns and Gailitis (1967 a).
Read (1984) has shown that it is possible to go further and to use equation (3-122) to deduce the form of the classical probability distribution $P(\gamma)$. It is also necessary to use equation (3-54), which shows that the probability of finding the parameters a_3 and a_4 in the intervals da_3 and da_4 is proportional to $a_3 da_3 da_4$. The result, expressed in terms of

$$\chi = \gamma E^{-1/4},$$ (3-124)

is

$$P(\chi) = \frac{2}{\pi \chi_m^2} (\chi_m^2 - \chi^2)^{1/2},$$ (3-125)

where χ_m is the maximum value of $|\chi|$ and is determined by the initial range of values of γ. This probability distribution has a maximum at $\gamma = 0$ and has sharp cut-offs at $\gamma = \pm \chi_m E^{1/4}$. It is shown as the full curve in Fig. 3-15, where it has been used to fit the results obtained from trajectory integrations. It is found that

$$\chi_m = 2.79$$ (3-126)

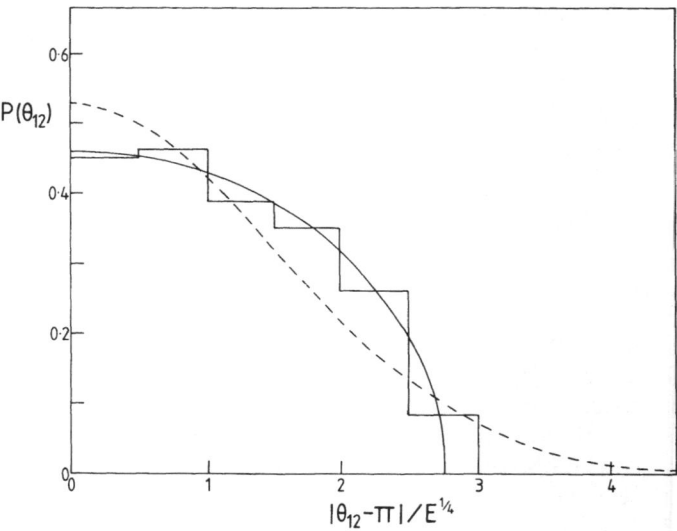

Fig. 3-15. The histogram gives the result of a computational study to find the probability distribution of the asymptotic angle θ between the directions of the two outgoing electrons. The full and broken lines are fits to the functions $[(\theta_m - \pi)^2 - (\theta - \pi)^2]^{1\,2}$ and $\exp[-\alpha(\theta - \pi)^2]$ respectively. From Read (1984)

and that the corresponding width is

$$\chi_{1/2} = 4.83. \tag{3-127}$$

The results given by equations (3-125), (3-126) and (3-127) should be treated with some caution however since although the classical approach is valid in the sense that the electron wavelengths are much smaller than the inter-particle distances at sufficiently small E, the motion in the θ_{12} direction may nevertheless be non-classical. This would arise if the wavelength for the motion in this direction is not much smaller than the excursion in the same direction. For example if we take as an approximation that the motion in the θ_{12} direction is governed solely by the final potential energy term on the right-hand side of equation (3-38), giving simple harmonic motion, and that the maximum excursion is $\chi_m E^{1/4}$, we find that the ratio of the classical energy to the quantal zero-point energy is of the order of E, and is therefore smaller than unity when E is small. This type of failure in the classical picture would not however affect the deductions concerning the total ionization cross-section or the forms of $f_<$ and $f_>$ (see equations 3-143 and 3-144).

The energy dependence given by equation (3-123) has also been obtained by the WKB-like and quantum-mechanical treatments of Rau (1971, 1976), Altick (1984) and Feagin (1984), but each of these treatments also gives the result that the distribution function has a Gaussian shape,

$$P(\chi) \propto \exp\left[-\alpha\left(\frac{\gamma}{E^{1/4}}\right)^2\right]. \tag{3-128}$$

Rau (1971, 1976) (see also Section 3.3.1) finds that the width (FWHM) of the distribution is

$$\left[\frac{64\ln 2}{15^{1/2}}\right]^{1/2} E^{1/4} = 3.38\, E^{1/4} \;\; (=85°\,(E/\text{eV})^{1/4}). \tag{3-129}$$

Altick (1984) suggests that the asymptotic part of the wavefunction is such that

$$|\psi(r_1,r_2,\theta)|^2 \propto e^{-2\pi k_{12}}, \tag{3-130}$$

where k_{12} ($=k_1-k_2$) is the difference between the asymptotic wave-numbers of the two electrons. When E is small the resulting width of the angular distribution is

$$\left[\frac{32\ln 2}{\pi}\right]^{1/2} E^{1/4} = 2.66\, E^{1/4}. \tag{3-131}$$

Feagin (1984) has used relative (Jacobi) coordinates to represent the correlated three-particle motion, and has found, amongst other things, that the width of the angular distribution is $2.71\, E^{1/4}$.

Fitting the computational results shown in Fig. 3-15 to a Gaussian distribution (the broken curve in the figure) gives the width $3.55\, E^{1/4}$. This result and the three theoretical results mentioned above are therefore within 30% of each other.

These predictions can be tested against the experimental evidence of Cvejanović and Read (1974), who measured the ratio of the yields at two values of θ_{12} (180° and 150°, with $\theta_1 = \theta_2 = 90°$ and an out-of-plane geometry) for five values of E, from 0.2 to 3.0 eV. If a Gaussian form is assumed then the measured ratios imply that the constant a in equation (3-123) has the value 2.1 ± 0.1 at 0.2 eV and 1.82 ± 0.02 at 3 eV, and is therefore significantly smaller than the results quoted above. On the other hand if the form given by equation (3-125) is assumed then a has the value 3.12 ± 0.01 at 0.2 eV but 1.82 ± 0.04 at 3 eV. As well as being significantly different from each other, these values are also significantly smaller than that given in equation (127). More recently Fournier-Lagarde et al. (1984) have measured the dependence on θ_1 and θ_2 of the electron impact ionization cross-section of helium at $E = 1$ to 6 eV. Their in-plane geometry does not allow ϕ_1 and ϕ_2 to be varied. The results show that values of L up to at least 2 contribute at these energies, and a preliminary analysis shows that the dependence on θ_{12} is consistent with the width given by Rau (1971, 1976), equation (3-129). Further experimental studies, and also theoretical and computational studies on the effects of higher values of L, are clearly needed.

3.4.3 Angular Momenta

When the total angular momentum $L\hbar$ of the system is zero the two electrons emerge in the threshold ionization process with the angular momenta $l\hbar$ and $-l\hbar$ respectively. In this section we consider the form and the energy dependence of the probability distribution $P(l)$ for l.

As with the form of $P(\gamma)$ discussed in the previous section, it is possible to extend the Wannier treatment to find the form of $P(l)$. For example when $\alpha = \pi/4$ (i.e. $r_1 = r_2$, $E_1 = E_2$) the magnitude of l at the critical radius is

$$l = \frac{1}{4} R_c^2\, \dot{\gamma}_c \tag{3-132}$$

where γ_c is given by equation (3-122). Repeating the reasoning used to obtain equation (3-125) it follows (Read 1984) that

$$P(l) = \frac{4}{\pi l_m^2} (l_m^2 - l^2)^{1/2},$$ (3-133)

where

$$l_m = a E^{-1/4}$$ (3-134)

and a is a constant. Equation (3-133) fits the probability distribution resulting from trajectory integrations very well, as can be seen in Fig. 3-16. The resulting value of a is 4.53. The root-mean-square value of l is

$$l_{RMS} = {}^1\!/_2 \, l_m = 2.27 \, E^{-1/4}.$$ (3-135)

To find the form of $P(l)$ for $\alpha \neq \pi/4$ (i.e. $E_1 \neq E_2$) Read (1984) has made certain approximations and has found that equation (3-133) is still valid, but with l_m being replaced by

$$l_n = \frac{4 E_1 (E - E_1)}{E^2} \, l_m.$$ (3-136)

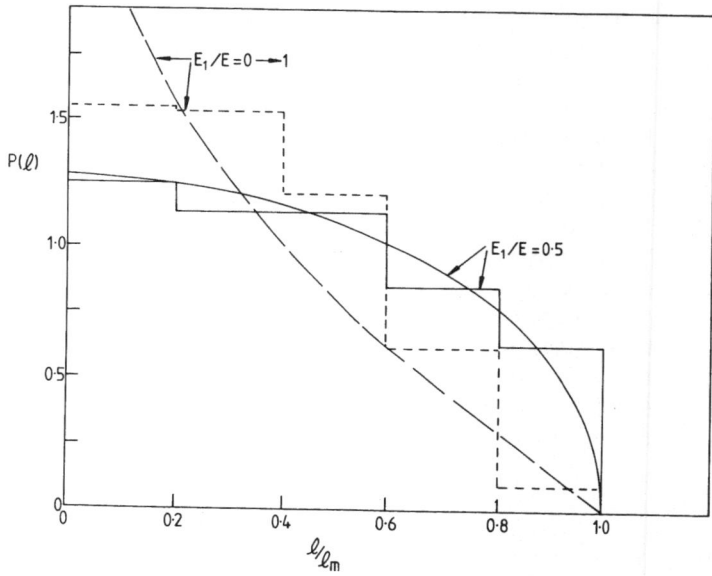

Fig. 3-16. The two histograms show the results of a computational study to find the probability distribution of the angular momentum l of each of the outgoing electrons. The full and broken histograms correspond to $E_1 = 0.5 E$ and $E = 0$ to E respectively, and the full and broken curves correspond to the functions given by equation (3-133) and by equation (3-136) averaged over the range of values of E_1, respectively. From Read (1984)

Averaging the resulting probability distribution over all values of E_1 in the range from 0 to E gives the broken curve shown in Fig. 3-16. This distribution is singular at $l = 0$ (it is proportional to $\ln(2/l)$ at small l) and so does not give a good fit to the histogram resulting from the trajectory integrations. The cause of this discrepancy can be traced to the value of l when either E_1 or E_2 is approximately zero. We see from equation (3-136) that l_n, and hence l, is then proportional to E_1 or E_2, whichever is the smaller. This proportionality follows from the classical behaviour of l at $R = R_c$, namely (Read 1984)

$$l_c \simeq \frac{E_1 E_2}{E^2} R_c^2 \theta_c, \tag{3-137}$$

and from the assumption that l does not change appreciably in the outer zone, and it gives rise to the singularity. As mentioned in Section 3.4.1 in connection with the non-uniform energy partitioning, the definition of R_c given by equation (3-50) may not be appropriate when either E_1 or E_2 is small, and it may then be necessary to use a larger value of R_c. This would cause l_c to be increased, which would presumably remove the singularity.

The computational results show that the root-mean-square value of l, averaged over the interval $E_1 = 0 \rightarrow E$, is

$$l_{RMS} = 1.82 \, E^{-1/4}. \tag{3-138}$$

Previous computational studies of the energy-dependence of l in threshold ionization are those of Cvejanović and Grujić (1975), Boesten et al. (1976) and van der Water et al. (1978). Cvejanović and Grujić (1975) found the dependence

$$l_{max} \simeq 2.25 \, E^{-1.4}. \tag{3-139}$$

The exponent is that given by equation (3-134) but the constant of proportionality is approximated halved. On the other hand Boesten et al. (1976) found

$$l \simeq 0.36 \, E^{-0.5} \tag{3-140}$$

and van der Water (1978) found the similar result

$$l \simeq 0.30 \, E^{-0.55}. \tag{3-141}$$

All these authors have integrated trajectories from an initial condition that represents an electron moving towards a classical target atom, and have varied only one initial parameter, and so have been able to obtain l-values for only a small number of trajectories leading to the ionization condition.

As mentioned above in connection with the classical calculation of the probability distribution for θ_{12}, the motion in θ_{12} may be non-classical even at low values of E, in which case the classical treatment of the angular momenta would also be inaccurate.

3.5 Related Excitation Cross-Sections

Information has alrady been presented in Figs. 3-5, 3-12, and 3-14, that relates to the near-threshold excitation process

$$e + A \rightarrow A^* \, (\text{high-}n) + e. \tag{3-142}$$

Here one electron escapes with a small positive energy E_1 while the other is left bound to the A^+ ion, in a high-n Rydberg state, with a small but negative total energy E_2. The excitation process can be viewed, in theoretical terms, as an analytic continuation of the ionization process, to which it is therefore closely related. This close relationship also exists in many experimental measurements, the yield of the excitation process appearing as a continuation of that of the ionization process. Of particular relevance to the present Chapter is the fact that the high degree of electron-electron correlation that is present in near-threshold ionization must also be present in near-threshold excitation of high-n states. Fano (1974) was the first to point out this connection and to argue that strong evidence for it is given by the existence of the nearly-symmetric "ionization cusp" that can be seen in the spectrum of Fig. 3-12.

In this section we consider firstly the energy-dependence of the near-threshold excitation process and then the distribution of l-values (of the free and bound electrons) to which it gives rise.

3.5.1 Differential Excitation Cross-Section

We see from equation (3-114) that when $E > 0$ and when the ion core has $Z = 1$, the differential cross-section for ionization processes in which E_1 is in the range E_1 to $E_1 + dE_1$, where $0 \leq E_1 \leq E$, is

$$\sigma_E(E_1) dE_1 \propto E^{0.127} f_>(E_1/E) dE_1. \tag{3-143}$$

The function $f_>(E_1/E)$ is shown in Fig. 3-12. Also shown in the figure is the form of the function for $E_1 > E$ (and hence $E_2 < 0$). This part of the function gives the yield of the process (3-142) when the atom is excited to states that have binding energies in the range $|E_2|$ to $|E_2| + dE_1$. In other words the high-n excited states are regarded as a continuum, and the function $f_>(E_1/E)$ gives, through equation (3-143), the cross-section for exciting a band of width dE_1 in this continuum.

High-n states of binding energy $|E_2|$ can also be excited at negative values of E, provided that $E \geq -|E_2|$. Read (1984) has shown that the yield has a form analogous to that given by equation (3-143), namely

$$\sigma_E(E_1) dE_1 \propto |E|^{0.127} f_<(E_1/|E|) dE_1, \quad E < 0 \tag{3-144}$$

where $f_<(x)$ is different from $f_>(x)$. The form of $f_<(E_1/|E|)$ has been derived from trajectory integrations and is shown in Fig. 3-12. Earlier numerical studies of the form of $f_<(E_1/|E|)$ (Vinkalns and Gailitis 1967 b, Cvejanović and Grujić 1975) are significantly less accurate (see Read 1984).

The spike in $f_<(E_1/|E|)$ at $E_1 = 0$ is caused by trajectories that initially evolve towards the "forbidden" condition $E_1 < 0$, $E_2 < 0$. If such a final condition could exist the function $f_<(E_1/|E|)$ would exist in the region $-\frac{1}{2}|E| \leq E_1 \leq 0$ (remembering that we take $E_1 > E_2$), but because the condition is unstable the yield corresponding to it is redistributed to positive values of E_1. In fact the trajectory integrations show that when the two electrons initially evolve towards the forbidden condition they usually both perform many orbits about the ion before one of them eventually emerges with a value of E_1 that is typically much smaller than $|E|$. Hence the redistribution is concentrated at the smallest positive value of E_1, as shown in Fig. 3-12. The states of oscillation of the quasi-forbidden trajectories presumably correspond to the "Wannier-ridge" resonances, to be discussed in Section 3.5.3.

The broken curves in Fig. 3-12 show the empirical fits

$$F_>(x) \propto (x - 0.3)^{0.127}, \quad x \geq 1.3, \tag{3-145}$$

$$F_<(x) \propto (x + 0.3)^{0.127}, \quad x \geq 0.2. \tag{3-146}$$

The corresponding differential cross-sections are then

$$\sigma \propto (E_1 - 0.3\,E)^{0.127}.$$ (3.147)

This is valid, with the same proportionality constant, for negative and positive values of E and for excitation and ionization processes. It is therefore an expression that is useful for interpreting a wide range of experimental measurements. The constant 0.3 is empirical, and equally good fits are obtained if it lies in the range 0.29 to 0.31.

Expressions (3-146) and (3-147) are not valid when $E < 0$ and $E_1 < 0.2\,|E|$, but the average value of the cross-section in this region is approximately twice that given by equation (3-147). Similarly expressions (3-145) and (3-147) are not valid when $E > 0$ and $E_1 < 1.3\,E$, but for $E_1 < E$ it is possible to use equation (3-117) or to take the simpler, less accurate approximation

$$\sigma_> \propto E^{0.127}, \quad 0 \le E_1 \le E.$$ (3-148)

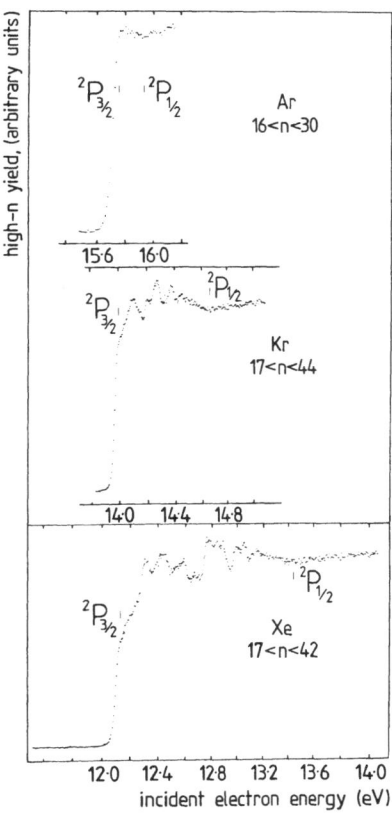

Fig. 3-17. Near-threshold excitation functions of Rydberg states having the indicated bands of n-values. The experimental energy resolution is 30 meV. The threshold onsets are consistent with the cross-sections being finite at threshold. From Hammond $et\ al.$ (1984 a)

The threshold electron spectrum of Cvejanović and Read (1974), shown in Fig. 3-5, provides an experimental test of the validity of equation (3-144) for the excitation process, as well as of equation (3-143) for the ionization process. In this experiment the yield of electrons that have E_1 in the range 0 to E_m, where E_m is constant and is approximately 20 meV, is measured as a function of E, as E is swept through negative and positive values. A dip, or "ionization cusp" is seen at $E = 0$. The yield above the cusp should be proportional to $E^{0.127} f_>$, according to equation (3-143), where the average of $f_>$ is taken over the appropriate range of E_1/E. Since this average varies by less than 3% over the range of E studied, the yield is expected to be nearly proportional to $E^{0.127}$. This is confirmed by the excellent fit of the data points to the curve, shown in Fig. 3-5, that has the form $E^{0.127}$. This evidence has already been discussed in Section 3.2.4. Below the ionization cusp the yield is proportional to $|E|^{0.127} \bar{f}_<$, according to equation (3-144). In fact $\bar{f}_>$ and $\bar{f}_<$ differ by less than 15% (see Read 1984), which is consistent with the observation that the cusp is approximately symmetric.

The experiment of Spence (1975) also provides confirmation of the validity of equations (3-143) and (3-144). In this experiment the yield of electrons having a fixed but comparatively large value of E_1 is measured as a function of E. The ionization cusp is again seen, but with a reduced amplitude, and the reduction is consistent with that expected from equations (3-143) and (3-144) and from the forms of $f_<$ and $f_>$.

Another relevant experimental result is that of Hammond et al. (1984 a), who have investigated the near-threshold cross-section for excitation of Rydberg states of high n (in the band from $n = 16$ to 43) and have found that these cross-sections appear to be finite at threshold. This indicates that $f_<(0)$ is non-zero, as found in the numerical studies. Some of the excitation functions measured by Hammond et al. (1984 a) are shown in Fig. 3-17. A detailed analysis of the excitation functions in terms of the empirical functions $F_<$ and $F_>$, equations (3-145) and (3-146), has been given by Hammond et al. (1984 a).

3.5.2 Angular Momentum Distributions

Fano (1974) was the first to suggest that in near-threshold excitation of high-n states the presence of electron-electron correlations causes the angular momentum quantum number l of the excited states to be high. This suggestion is supported qualitatively by the experimental results of Hammond et al. (1984 b) and also by those of Hammond et al. (1984 a), mentioned above, and by the results of the earlier similar experiments of Tarr et al. (1980, 1981) carried out at a lower energy resolution. The even earlier experiments of Heideman and co-workers (see for example Heideman et al. 1976 and the review by Heideman 1984) on electron-impact optical excitation functions of helium also indicated that the correlations affect the form of the excitation functions of the higher-l states and the degree of polarization of the decay radiation, when the incident energy is near the ionization threshold.

Quantitative information on the distribution of l-values and of the energy-dependence is provided by the theoretical studies of Drukarev (1982) and Rau (1984a) and the numerical studies of Read (1984). Drukarev (1982) considered the form of the wavefunction of a state of principal quantum number n when it is excited at threshold by electron impact. He used a parabolic coordinate frame for the description of the atomic electron and took the correlation between the excited and scattered electron to have its strongest form, so that the wavefunction of the excited electron is directed as far away as possible from the direction of the outgoing scattered e.ectron. Decomposing this wavefunction into those that are characterized by the angular momentum quantum number l, he found that the distribution in l is given by

$$W_l = \frac{(2l+1)\,[(n-1)!]^2}{(n+l)!\,(n-l-1)!}.\tag{3-149}$$

When $n \gg 1$ and $l \gg 1$ this can be approximated by (Drukarev 1982)

$$W_l \simeq \frac{2l+1}{n}\,e^{-l(l+1)/n},\tag{3-150}$$

from which we find that

$$\bar{l} \simeq \frac{(n\pi)^{1/2}}{2} - \frac{1}{2} + \ldots,\tag{3-151}$$

$$l_{\max} = \left[\frac{n}{2}\right]^{1/2} - \frac{1}{2},\tag{3-152}$$

$$l_{RMS} \simeq n^{1/2} - \frac{\pi^{1/2}}{4} + \ldots\tag{3-153}$$

Using a technique that involves reduction of the four-dimensional rotation group that has been shown to give a good description of angular correlations in two-electron states (see Section 3.5.3 and Wulfman 1973, Herrick and Sinanoglu 1975, Crance and Armstrong 1982), Rau (1984a) has found that

$$W_l = \frac{(2n-1)(2l+1)\,[(n-1)!]^4}{[(n+l)!\,(n-l-1)!]^2}.\tag{3-154}$$

This can be approximated by

$$W_l \simeq \frac{2(2l+1)}{n}\,e^{-2l(l+1)/n}\tag{3-155}$$

giving

$$\bar{l} \simeq \left[\frac{n\pi}{8}\right]^{1/2} - \frac{1}{2} + \ldots,\tag{3-156}$$

$$l_{\max} = \frac{1}{2}\,n^{1/2} - \frac{1}{2},\tag{3-157}$$

$$l_{RMS} \simeq \left[\frac{n}{2}\right]^{1/2} - \frac{\pi^{1/2}}{4} + \ldots\tag{3-158}$$

The higher terms in equations (3-151), (3-153), (3-156) and (3-158) are significant, at the 10% level, if $n \leq 20$.
Using classical trajectory integrations Read (1984) has found that for the range of

n-values from approximately 10 to 22,

$$l_{RMS} \simeq 0.49 \, n^{0.82}. \tag{3-159}$$

Using Drukarev's distribution (3-149) to find the n-dependence of l_{RMS} over the same range and then fitting to a power-law in n gives

$$l_{RMS} \simeq 0.33 \, n^{0.82}, \tag{3-160}$$

while Rau's distribution gives

$$l_{RMS} \simeq 0.43 \, n^{0.61}. \tag{3-161}$$

Clearly the values of n in this range are too small for the $n^{1/2}$ dependence of the first terms in approximations (3-153) and (3-158) to hold.

3.5.3 Wannier-Ridge Resonances

The high degree of electron-electron correlation that exists in the processes of near-threshold ionization and near-threshold excitation of high-n states can also exist in systems in which two electrons have a small energy but are bound, namely in doubly-excited atomic states of high excitation energy. The electrons in such atomic states will reside in the vicinity of the Wannier point (i.e. with $r_1 \simeq -r_2$) and will be at

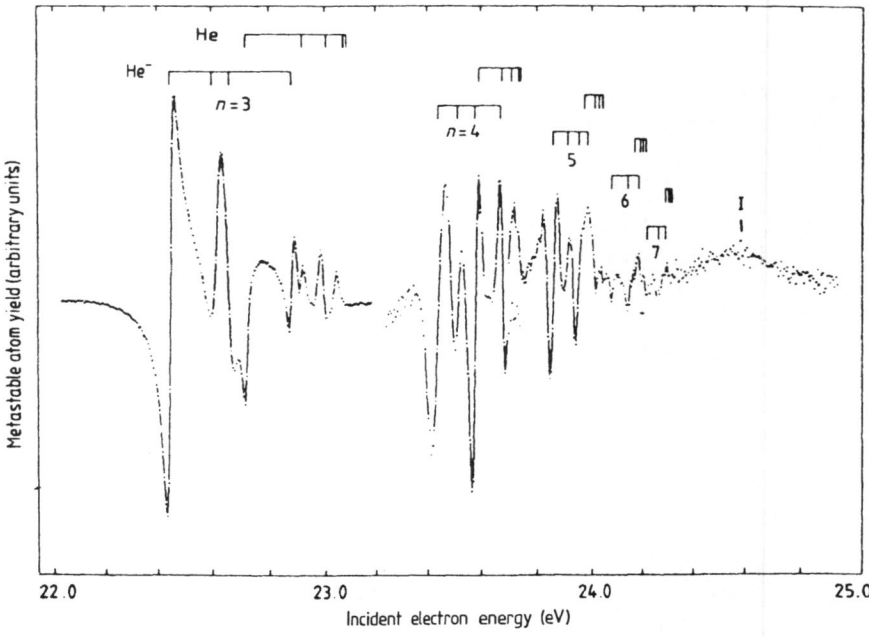

Fig. 3-18. Yield of metastable Helium atoms resulting from electron impact on Helium, taken over three ranges of the incident electron energy. In each region a sloping background has been subtracted and the vertical scale has been expanded by an arbitrary factor. The energies of the more prominent He⁻ resonances are indicated, as well as those of the 1sns, 1snp and 1snd states of He, for $n = 3$ to 7. From Buckman et al. (1983)

large distances from the central ion core ($r_{1,2} \gg 0.1$ nm). If the distances become sufficiently large the motion of the electron will become approximately classical. Clear evidence for the existence of highly-correlated doubly-excited states is given by the spectrum shown in Fig. 3-18 (Buckman *et al.* 1983). The structures in the spectrum correspond to He$^-$ resonances formed as short-lived intermediate complexes in the reaction

$$e + \text{He} \rightarrow \text{He}^- \rightarrow \text{He}^* + e. \tag{3-162}$$

In these resonances one electron is in the 1s state but both the other two are in Rydberg states. For the lower member of each group of resonances the two Rydberg electrons have the same value of n, while for the higher members (for example the highest of those identified in Fig. 3-18 and those lying still higher) one Rydberg electron will have a higher value of n than the other. Earlier, less precise observations of these resonances had been made by Heddle, Keesing and co-workers (see Heddle 1976, 1977) and Brunt *et al.* (1977).

Three properties of the He$^-$ resonances show clearly that they have a structure that is dominated by strong electron-electron correlations. The first property concerns the energies of the resonances. In particular we consider the energy of the lowest member of each group. These energies are well approximated (Buckman *et al.* 1983) by the modified Rydberg formula (Read 1977)

$$E_n = I - \frac{2R(Z_{\text{core}} - \sigma)^2}{(n - \delta_n)^2}, \tag{3-163}$$

where I is the ionization energy (i.e. the energy required to remove both electrons, 24.586 eV for He$^-$), R is the Rydberg energy (13.602 eV for He), Z_{core} is the charge number of the ion core (1 for He$^+$), σ is an effective screening constant and δ_n is an effective quantum defect. Formula (3-163) is based on the grandparent model (see also Read (1982)) in which the two outer electrons, treated as being equivalent, couple to form a pair state that interacts with the central ion core only through the effective Coulomb potential provided by the core. Although σ and δ_n can be treated as adjustable parameters a good fit to the observed He$^-$ energies is obtained if σ is given the value 0.25 that corresponds to two electrons at the Wannier point (see equation (3-38)) and if δ_n is put equal to the spin-averaged quantum defect of the ns electron in the 1sns configuration of the He atom. Equation (3-163) then contains no adjustable parameters. The resonances thus form a Rydberg series, but with a structure different from that of singly-excited atomic Rydberg series since the effective charge is non-integral and is equal to that expected for an electron pair at the Wannier point. This is therefore strong evidence that the wavefunctions of these resonances are peaked at the Wannier ridge, on or near the Wannier point. Buckman *et al.* (1983) have therefore described them as "Wannier-ridge resonances". The associated theoretical evidence will be discussed separately below.

In passing it is relevant to note that the modified Rydberg formula (3-163) accurately reproduces the energies of a wide range of atomic states that have the configuration [core] (ns)^2, including negative ions, highly stripped positive ions, autoionizing states and resonances, if σ is taken to be 0.25 and δ_n is put equal to the spin-averaged δ_{ns} of the atom that has the configuration [core] ns (Read 1977). The significance of this choice of δ_n is however not yet explained.

The second property of the He⁻ resonances that underlines their correlated structure is the fact that their intensities are observed to be approximately proportional to n^{-6} (Buckman et al. 1983). This is markedly different from the $(n^*)^{-3}$ proportionality obtained for excitation cross-sections of unperturbed singly-excited states (Fano and Cooper 1968), where $n^*(=n-\delta)$ is the effective principal quantum number. Rau (1983) has pointed out that the grandparent model (mentioned above) can account for an $(n^*)^{-6}$ dependence, since this can be viewed as the product of two $(n^*)^{-3}$ factors, one for each of the electrons. More rigorously, Feagin and Macek (1984) have considered the n-dependence of the normalization of the resonance wavefunctions, and have found that the wavefunction amplitude at small R is proportional to $n^{-3/2-m}$, where $m(=1.127)$ is the Wannier index. The intensities (and also the widths) of the resonances are proportional to the square of this amplitude, namely $n^{-3-2m}=n^{-5.254}$, which is consistent with the experimental observations. Feagin and Macek (1984) also arrive at this n-dependence by arguing that the average intensity of the resonances should be given very near threshold by the Wannier law for two-electron escape just above threshold. They refute an argument used by Rau (1983) based on the dimensionality of the Coulomb function that describes the resonances.

Yet another property of the He⁻ resonances that is different from the corresponding property of singly-excited states is the behaviour of the energy spacings within the groups of resonances. These spacings vary approximately as $(n^*)^{-2}$, and thus do not have the $(n^*)^{-3}$ proportionality that would be expected for conventional groups of unperturbed singly-excited Rydberg series (Fano and Cooper 1968). This difference also manifests itself in the fact that neighbouring groups of resonances begin to overlap at high n ($n \geq 6$), a behaviour that is impossible for conventional groups of Rydberg states. The reasons for this behaviour, and the possible relevance to the correlated structure of the resonances, have not yet been explained. By contrast, Wulfman (1973) and Herrick, Sinanoglu, Kellman and co-workers (e. g. Herrick and Sinanoglu 1975, Kellman and Herrick 1978, Herrick and Kellman 1980, Herrick et al. 1980) have had some success in accounting for the energies and symmetries of many of the high-lying doubly-excited states of H⁻ and He in terms of a new supermultiplet classification scheme, but these studies have not yet been extended to cover atoms such as He⁻, in which the central core is not a bare nucleus.

Finally we discuss the relevant theoretical work that has not already been mentioned. Rau (1983) has considered the wavefunction of the He⁻ resonances in terms of hyperspherical coordinates and has proposed the formula

$$E_n = I - \frac{8\,R\,(Z_{\text{core}} - \sigma - {}^1\!/_4)^2}{\left(n + \dfrac{3}{2} - \delta\right)^2} \tag{3-164}$$

for the resonance energies, where I, R and Z_{core} are defined as in the modified Rydberg formula (3-161), and σ and δ are adjustable parameters. He obtains a good fit to the energies of the lowest member in each group by taking $\sigma = 0.354$ and $\delta = 1.67$. More recently (Rau 1984 c) the energies of the doubly excited states of H⁻ and He with $n \leq 5$ have been fitted by equation (3-164), but with

$$\sigma = \frac{1}{2} Z - 0.164,$$

$$\delta = \frac{3}{2} - \frac{3}{20\,Z}.$$

(3-165)

Equation (3-164) then becomes

$$E_n = I - \frac{2\,R\,(Z - 0.172)^2}{\left(n + \dfrac{3}{20\,Z}\right)^2},$$

(3-166)

which is close to the modified Rydberg formula (3-163). Other recent theoretical work that is relevant to the understanding of the Wannier-ridge resonances and of analogous high-n doubly-excited states is that of Watanabe et al. (1982), who have used the hyperspherical formalism (Macek 1968) to study the $n = 3$ multiplet of He^-, and that of Crance and Armstrong (1982), who have developed the approach of Herrick and co-workers to obtain an approximate operator expression for $|r_1 - r_2|^{-1}$, from which they obtain an expression for the doubly-excited states of He. There is also a considerable amount of work on the lower-lying doubly-excited states of H^- and He (see Lin 1982, 1983 and references therein), using the hyperspherical formalism. The nature of the correlations between the two excited electrons in these atoms is shown clearly for example by the surface plots of charge-density distributions that Lin (see for example Lin 1982) has presented. Berry and co-workers (see for example Yuh et al. 1981, Ezra and Berry 1983) have presented similar plots of wavefunctions calculated by more conventional techniques.

3.6 Conclusions

Threshold ionization is one of those rare subjects in physics for which classical, semiclassical and quantum-mechanical treatments all have some validity. The value of the interplay of these treatments is a theme that has run throughout the present Chapter, and there is now little dissent from the view that convergence has been reached on all the major aspects of the subject. Some differences in detail remain and clearly these need to be investigated further, although the necessary experimental and theoretical work will almost certainly present considerable difficulties. Doubtless new facets of the subject will emerge from these further studies. But there remains also a wider problem not yet solved — the development of a comprehensive theoretical technique that can embrace and explain all the phenomena associated with near-threshold ionization and excitation processes, a technique in which the classical, semi-classical and quantum-mechanical treatments merge, rather than overlap. Perhaps this aim is unrealistic, but should such a technique be developed it would surely find ready application in the wider class of problems concerning long-range correlations between three or more particles.

Acknowledgments

The author is grateful to Drs. P. Grujić, G. C. King, R. Peterkop and A. R. P. Rau for their critical comments on the manuscript and to the authors whose figures have been reproduced, for their permission to do so.

References

Alguard, M. J., Hughes, V. W., Lubell, M. S., Wainwright, P. R. (1977): Phys. Rev. Lett. *39*, 334 – 338.
Altick, P. L., 1984 (in press).
Banks, D., Percival. I. C., Valentine, N. A. (1969): Abstracts of the Sixth International Conference on the Physics of Electronic and Atomic Collisions, 215 – 216. Cambridge, Mass.: MIT Press.
Baum, G., Kisker, E., Raith, W., Schroeder, W., Sillmen, U., Zenzes, D. (1981): J. Phys. B: Atom. Molec. Phys. *14*, 4377 – 4388.
Boesten, L. G. J., Banks, D., Heideman, H. G. M. (1976): J. Phys. B: Atom. Molec. Phys. *9*, L97 – 100.
Bottcher, C. (1982): J. Phys. B: Atom. Molec. Phys. *15*, L463 – 469.
Brion, C. E., Thomas, G. E. (1967). Vth ICPEAC, 53 – 55. Leningrad: Nauka.
Brion, C. E., Thomas, G. E. (1968 a): Phys. Rev. Lett. *20*, 241 – 242.
Brion, C. E., Thomas, G. E. (1968 b): J. Mass. Spectrom. Ion. Phys. *1*, 25 – 39.
Brunt, J. N. H., King, G. C., Read, F. H. (1977): J. Phys. B: Atom. Molec. Phys. *10*, 433 – 448.
Buckman, S. J., Hammond, P., Read, F. H., King, G. C. (1983): J. Phys. B: Atom. Molec. Phys. *16*, 4039 – 4047.
Crance, M., Armstrong, L. (1982); Phys. Rev. *A 26*, 694 – 696.
Cvejanović, S., Grujić, P. (1975): J. Phys. B: Atom. Molec. Phys. *8*, L305 – 309.
Cvejanović, S., Read, F. H. (1974 a): J. Phys. B: Atom. Molec. Phys. *7*, 1841 – 1852.
Cvejanović, S., Read, F. H. (1974 b): J. Phys. B: Atom. Molec. Phys. *7*, 1179 – 1193.
Dimitrijević, M. S., Grujić, P. (1981): J. Phys. B: Atom. Molec. Phys. *14*, 1663 – 1674.
Donahue, J. B., Gram, P. A. M., Hynes, M. V., Hamm, R. W., Frost, C. A., Bryant, H. C., Butterfield, K. B., Clark, D. A., Smith, W. W. (1982): Phys. Rev. Lett. *48*, 1538 – 1541.
Donahue, J. B., Gram, P. A. M., Hynes, M. V., Hamm, R. W., Frost, C. A., Bryant, T. C., Butterfield, K. B., Clark, D. A., Smith, W. W. (1984): Phys. Rev. Lett. *52*, 164.
Drukarev, G. F. (1982): Sov. Phys. JETP *56*, 532.
Edmonds, A. R. (1974): Angular Momentum in Quantum Mechanics. Princeton, N. J.: Princeton University Press.
Ezra, G. S., Berry, R. S. (1983): Phys. Rev. *A 25*, 1974 – 1988.
Fano, U. (1974): J. Phys. B: Atom. Molec. Phys. *7*, L401 – 404.
Fano, U. (1976): Physics Today *29*, 32 – 41.
Fano, U. (1980 a): Phys. Rev. *A 22*, 2660 – 2671.
Fano, U. (1980 b): J. Phys. B: Atom. Molec. Phys. *13*, L519 – 523.
Fano, U. (1983 a): Atomic Physics *8* (Lindgren, I., Rosen, A., Svanberg, S., eds.), 5 – 22. New York: Plenum.
Fano, U. (1983 b): Repts. on Prog. Phys. *46*, 97 – 165.
Fano, U., Cooper, J. W. (1968): Rev. Mod. Phys. *40*, 441 – 507.
Fano, U., Lin, C. D. (1975): Atomic Physics *4*, 47 – 70.
Feagin, J. M. (1984): J. Phys. B: Atom. Molec. Phys. *17*, 2433 – 2451.
Feagin, J. M., Macek, J. (1984): J. Phys. B: Atom. Molec. Phys. *17*, L245 – 247.
Fournier-Lagarde, P., Mazeau, J., Huetz, A. (1984): J. Phys. B: Atom. Molec. Phys. *17* (in press).
Gay, T. J., Fletcher, M. J., Alguard, M. J., Hughes, V. W., Wainwright, P. F., Lubell, M. S. (1982): Phys. Rev. *A 26*, 3664 – 3667.
Geltman, S. (1956): Phys. Rev. *102*, 171 – 179.
Geltman, S. (1983): J. Phys. B.: Atom. Molec. Phys. *16*, L525 – 528.
Greene, C. H., Rau, A. R. P. (1983): J. Phys. B: Atom. Molec. Phys. *16*, 99 – 106.
Greene, C. H., Rau, A. R. P. (1982): Phys. Rev. Lett. *48*, 533 – 537.
Grujić, P. (1972): J. Phys. B: Atom. Molec. Phys. *5*, L137 – 139.

Grujić, P. (1982): J. Phys. B: Atom. Molec. Phys. *15*, 1913 – 1928.
Grujić, P. (1983 a): J. Phys. B: Atom. Molec. Phys. *16*, 2567 – 2576.
Grujić, P. (1983 b): Phys. Lett. *96 A*, 233 – 235.
Hammond, P., Jureta, J. J., Read, F. H., King, G. C. (1984 b): J. Elect. Spect. (in press).
Hammond, P., Read, F. H., King, G. C. (1984 a): J. Phys. B: Atom. Molec. Phys. *17*, 2925 – 2941.
Hammond, P., Read, F. H., King, G. C. (1984 b). J. Phys. B: Atom. Molec. Phys. (in press).
Heddle, D. W. O. (1976): Contemp. Phys. *17*, 443 – 460.
Heddle, D. W. O. (1977): Proc. Roy. Soc. *A 352*, 441 – 449.
Heideman, H. G. M., van der Water, W., Nienhuis, G., Peeters, P. H. (1976): J. Phys. B: Atom. Molec. Phys. *9*, L 523 – 526.
Heideman, H. G. M. (1984): Electronic and Atomic Collisions (Eichler, J., Hertel, I. V., Stolterfoht, N., eds.), 743 – 754. Amsterdam: Elsevier.
Herrick, D. R , Kellman, M. E. (1980): Phys. Rev. *A 21*, 418 – 425.
Herrick, D. R., Kellman, M. E., Poliak, R. D. (1980): Phys. Rev. *A 22*, 1517 – 1535.
Herrick, D. R., Sinanoglu, O. (1975): Phys. Rev. *A 11*, 97 – 110.
Hils, D., Jitschin, W., Kleinpoppen, H. (1982): J. Phys. B: Atom. Molec. Phys. *15*, 3347 – 3357.
Hils, D., Kleinpoppen, H. (1978): J. Phys. B: Atom. Molec. Phys. *11*, L 283 – 287.
Hils, D., Rubin, K. Kleinpoppen, H. (1980): Coherence and Correlation in Atomic Collision (Kleinpoppen, H., Williams, J. F., eds.), 689 – 696. New York: Plenum.
Hippler, R., Klar, H., Saeed, K., McGregor, I., Duncan, A. J., Kleinpoppen, H. (1983): J. Phys. B: Atom. Molec. Phys. *16*, L 617 – 621.
Kelley, M. H., Rogers, W. T., Celotta, R. J., Mielczarek, S. R. (1983); Phys. Rev. Lett. *51*, 2191 – 2193.
Kellman, M. E., Herrick, D. R. (1978): J. Phys. B: Atom. Molec. Phys. *11*, L 755 – 759.
Klar, H. (1981): J. Phys. B.: Atom. Molec. Phys. *14*, 3255 – 3265.
Klar, H., Schlecht, W. (1976): J. Phys. B: Atom. Molec. Phys. *9*, 1699 – 1711.
Krige, G. J., Gordon, S. M., Haarhoff, P. C. (1968): Z. Naturforschung *23 a*, 1383 – 1385.
Lin, C. D. (1982): Phys. Rev. *A 25*, 76 – 87.
Lin, C. D. (1983): Phys. Rev. Lett. *51*, 1348 – 1351.
Macek, J. H. (1968): J. Phys. B: Atom. Molec. Phys. *1*, 831 – 843.
Marchand, P., Paquet, C., Marmet, P. (1969): Phys. Rev. *180*, 123 – 132.
McGowan, J. W., Clarke, E. (1968): Phys. Rev. *167*, 43 – 51.
Mead, R. D., Lykke, K. R., Lineberger, W. C. (1984): Electronic and Atomic Collisions (Eichler, J., Hertel, I. V., Stolterfoht, N., eds.), 721 – 730. Amsterdam: Elsevier.
Peterkop, R. (1971): J. Phys. B: Atom. Molec. Phys. *4*, 513 – 521.
Peterkop, R. (1977): Theory of Ionization of Atoms by Electron Impact. (English Edition.) Boulder, Colorado: Associated University Press.
Peterkop, R. (1983): J. Phys. B: Atom. Molec. Phys. *16*, L 587 – 593.
Peterkop, R., Liepinsh, A. (1969): Abstracts of the Sixth International Conference of the Physics of Electronic and Atomic Collisions, 212 – 214. Cambridge, Mass.: MIT Press.
Peterkop, R., Liepinsh, A. (1981): J. Phys. B: Atom. Molec. Phys. *14*, 4125 – 4135.
Peterkop, R., Tsukerman, P. (1969): Abstracts of the Sixth International Conference on the Physics of Electronic and Atomic Collisions, 209 – 211. Cambridge, Mass.: MIT Press.
Peterkop, R., Tsukerman, P. (1970): Sov. Phys. JETP *31*, 374 – 377.
Pichou, F., Huetz, A., Joyez, G., Landau, M. (1978): J. Phys. B: Atom. Molec. Phys. *11*, 3683 – 3692.
Rau, A. R. P. (1971): Phys. Rev. *A 4*, 207 – 220.
Rau, A. R. P. (1976): J. Phys. B.: Atom. Molec. Phys. *9*, L 283 – 288.
Rau, A. R. P. (1982): J. de Physique *43*, C 2, 211 – 221.
Rau, A. R. P. (1983): J. Phys. B: Atom. Molec. Phys. *16*, L 699 – 705.
Rau, A. R. P. (1984 a): J. Phys. B: Atom. Molec. Phys. *17*, L 75 – 78.
Rau, A. R. P. (1984 b): Electronic and Atomic Collisions (Eichler, J., Hertel, I. V., Stolterfoht, N., eds.), 711 – 719. Amsterdam: Elsevier.
Rau, A. R. P. (1984 c): PRAMANA (in press).
Read, F. H. (1977): J. Phys. B: Atom. Molec. Phys. *12*, 449 – 458.
Read, F. H. (1982): Aust. J. Phys. *35*, 475 – 499.
Read, F. H. (1984): J. Phys. B: Atom. Molec. Phys. (in press).
Roth, T. A. (1972): Phys. Rev. *A 5*, 476 – 478.
Rudge, M. R. H (1968): Rev. Mod. Phys. *40*, 564 – 590.

Rudge, M. R. H., Seaton, M. J. (1964): Proc. Phys. Soc. *83*, 680 – 682.
Rudge, M. R. H., Seaton, M. J. (1965). Proc. Roy. Soc. *A 283*, 262 – 290.
Spence, D. (1975): Phys. Rev. *A 11*, 1539 – 1542.
Stauffer, A. D. (1982): Phys. Lett. *91 A*, 114 – 116.
Tarr, S. M., Schiavone, J. A. Freund, R. S. (1980): Phys. Rev. Lett *44*, 1660 – 1663.
Tarr, S. M., Schiavone, J. A., Freund, R. S. (1981), J. Chem. Phys. *74*, 2869 – 2878.
Temkin, A. (1966): Phys. Rev. Lett. *16*, 835 – 839.
Temkin, A. (1969): Physics of the One and Two-Electron Atoms (Bopp, F., Kleinpoppen, H., eds.),
    655 – 668. Amsterdam: North-Holland.
Temkin, A (1974): J. Phys. B: Atom. Molec. Phys. *7*, L450 – 453.
Temkin, A. (1982): Phys. Rev. Lett. *49*, 365 – 368.
Temkin, A. (1984): Electronic and Atomic Collisions (Eichler, J., Hertel, I. V., Stolterfoht, N., eds.),
    755 – 765. Amsterdam: Elsevier.
Temkin, A., Hahn, Y. (1974): Phys. Rev. *A 9*, 708 – 724.
van der Water, W., Kets, F. B., Boesten, L. G. J., Heideman, H. G. M. (1978): J. Phys. B: Atom. Molec.
    Phys. *11*, L465 – 468.
van der Wiel, M. J. (1972): Phys. Lett. *41 A*, 389 – 390.
Vinkalns, I., Gailitis, M. (1967 a): Abstracts of the Fifth International Conference on the Physics of
    Electronic and Atomic Collisions, 648 – 650. Leningrad: Nauka.
Vinkalns, I., Gailitis, M. (1967 b): Latvian Academy of Science Report no. 4, 17 – 34. Riga: Zinatne.
Wannier, G. H. (1953): Phys. Rev. *90*, 817 – 825.
Wannier, G. H. (1955): Phys. Rev. *100*, 1180.
Watanabe, S., Le Dourneuf, M., Pelamourgues, L. (1982): J. Physique *43*, C2 223 – 241.
Wigner, E. P. (1948): Phys. Rev. *73*, 1002 – 1009.
Wulfman, C. E. (1973): Chem. Phys. Lett. *23*, 370.
Yuh, H. J., Ezra, G. S., Rehmus, D., Berry, R. S. (1981): Phys. Rev. Lett. *47*, 497 – 500.
Yurev, M. S. (1977): Opt. Spectrosc. *42*, 594 – 596.

4

Differential Ionization Cross Sections

P. J. O. Teubner

Institute for Atomic Studies, School of Physical Sciences, The Flinders University of South Australia, Bedford Park, South Australia

4.1 Introduction

Ionization cross sections play such an important role in so many branches of physical science that a considerable amount of effort has been devoted both to the theoretical and experimental determinations of these cross sections.

The theoretical problem can be summarised by the observation that the final state of the system consists of at least three bodies all of which interact through the long range Coulomb force. Thus a description of the ionization mechanism requires a solution to the many body problem. It is not surprising that there has been only limited success in providing even total cross sections at energies less than those at which the Born approximation is valid.

The most complete description of an ionization event is provided by determining the energy and momentum of all particles involved in the collision. The triple differential cross section thus defined is given by (Ehrhardt *et al.* (1972 a))

$$\frac{d^3\sigma}{dE\,d\Omega_A\,d\Omega_B} = f_3(E_0, E_A, \theta_A, \theta_B, \phi_B). \qquad (4\text{-}1)$$

Where the scattering angles are defined in Fig. 4-1, E_0 is the incident energy and E_A the energy of one of the electrons. The energy of the other electron E_B is determined by the conservation of energy which also allows for the separation energy or ionization potential of the electron in the target. This cross section depends on five variables and therefore is sometimes referred to as the five-fold differential cross section. More recently the term $(e, 2e)$ cross section has been used to describe these processes. The measurement of this cross section requires the coincident detection of both post collision electrons.

The cross section depends not only on a description of the ionization mechanism but also on the structure of the target and the ion. In certain regions of phase space the ionization mechanism can be described accurately, thus the $(e, 2e)$ cross section can be used to provide structural information. Such is the case for certain symmetric collision geometries and the differential cross section has been used to probe the momentum space wavefunctions of valence electrons in atoms and molecules (McCarthy and Weigold (1976)). The other limiting case concerns ionizing events in which the ion core plays a significant role in the collision. This class of events provides a very sensitive test of the details of the ionization mechanism and it is here that most theoretical studies have been singularly unsuccessful. Triple differential or $(e, 2e)$ cross sections will be discussed in Section 4.2.

The double differential cross section is obtained by measuring the intensity distribution of one of the electrons as a function of energy and angle. This is equivalent to integrating the $(e, 2e)$ cross section over the solid angle of the unobserved electron with a consequent loss of information about the ionization process. These cross sections will be discussed in Section 4.3.

Information about electron impact ionization can also be gained in principle by observing the energy and angular distribution of the ions arising from the collision. McConkey et al. (1972) have studied the angular distribution of singly charged ions arising from electron collisons with helium and argon. These distributions showed similar features to the triple differential cross section measurements in helium by Ehrhardt et al. (1972). In electron molecule collisons which lead to dissociative ionization the situation is more complicated than in the atomic case because of the potentially large number of intermediate states which can participate in the collision. Dunn's selection rules (Dunn (1962)) have been used as a basis to predict energy and angular distribution of the fragment ions emitted in the dissociative ionization. These rules have stimulated a significant experimental effort in the area which is discussed in Section 4.4.

Inner shell processes near the K shell edge in molecules which lead to the formation of shape resonances and to ionization are discussed in Section 4.5.

4.2 Triple Differential or (e, 2e) Cross Sections

Ionizing collisions between electrons and neutral targets in which the kinematics of both outgoing electrons are completely determined, yield triple differential or $(e, 2e)$ cross sections. Such measurements have provided an increased understanding of the ionization mechanism and of momentum distributions of target electrons over the past decade. If the energy of the incident electron is known and the kinematics of the outgoing electrons is determined completely by coincidence techniques, then the following general relationships apply.

$$E_0 - \varepsilon = E_A + E_B \tag{4-2}$$

where E_0, E_A and E_B are the energies of the incident and two outgoing electrons respectively; ε is the binding energy of the electron in the target atom or molecule. The momentum transfer q in this case is given by

$$q = (\underline{k}_A + \underline{k}_B) - \underline{k}_0 \tag{4-3}$$

where $\underline{k}_0$, k_A and k_B are the momenta of the incoming and outgoing electrons defined as above. Conservation of momentum requires that q is equal to the recoil momentum of the ion.

Thus one is faced with all of phase space in which experiments can be done. However three particular regions have emerged as being particularly tractable to measurement over the past decade or so. These regimes are characterized by the choice of the kinematic variables in equations (4-2) and (4-3) and can be further divided into symmetric geometries ($E_A = E_B$) $\theta_A = \theta_B = 45°$ and asymmetric geometries where E_A is chosen to be very much greater than E_B and consequently $\theta_A \cong 0°$.

Some insight into the relative importance of collision dynamics in these regions can be gained from the following intuitive description (Weigold 1982). Three limiting types of ionizing events can be described; namely those in which the electron interacts predominantly with the core, those in which the electron makes a distant collision with the target and those in which the electron interacts directly with the target electron.

In general, interactions with the core give rise to one class of electrons of low energy and momentum which will be emitted in the backward directions. It is conventional when discussing the electrons involved in these collisions to ignore the principle of indistinguishability and to describe the slow electron as the "*ejected*" electron whilst describing the fast electron as the "*scattered*" electron. Such ionizing collisions contribute predominantly to the "total" cross section (Ehrhardt *et al.* (1982 a)). These events give rise to asymmetric kinematics in energy obviously but also in geometry since the fast electron is scattered in the forward direction ($\theta < 10°$).

Those events involving distant collisions also give rise to fast electrons scattered at very forward angles ($\theta_A \cong 0$). The slow electron can then be observed in coincidence at $\theta = 90°$. Such measurements have been shown (Brion (1975)) to be equivalent to the absorption of a photon of energy $E_0 - E_A$; where E_A is the energy of the electron in the forward direction. Binding energy spectra obtained from these asymmetric experiments compare favourably with photoelectron spectra (Brion (1975)). To date it has not been possible to deduce momentum distributions.

In collisions involving direct interaction between the electrons, the ion core plays the role of a spectator. At incident energies $E_0 \gg \varepsilon$, the electron-electron interaction can be considered elastic, and thus the angle between the two outgoing electrons would be $90°$. Such collisions were initially described by Glassgold and Ialongo (1968) as quasi-elastic scattering. Clearly the theoretical description of such collisions requires an adequate description of both the electron-electron interaction and the momentum distribution of the electron in the target. We shall see that for a certain geometry — the symmetric non coplanar geometry — such a description exists and has proven very successful as a probe for atomic and molecular wavefunctions.

Weigold (1982) has given a semi-classical argument which illustrates the way in which this third class of collisions can yield information on the momentum distribution of the struck electron. In the case of atomic hydrogen it is assumed that the proton enters the interaction only via the conservation of momentum. Then the differential cross section is given by the Rutherford cross section multiplied by the probability that the electron has momentum q, defined in equation (4-2) which is required by momentum conservation.

Thus the $(e, 2\epsilon)$ cross section in this semi-classical argument is given by

$$\sigma(q) \propto \frac{4}{K^4} |\phi_{1s}(q)|^2 \tag{4-4}$$

where the Rutherford cross section αK^{-4} and the square of the momentum space wavefunction is $|\phi_{1s}(q)|^2$. Such an experiment has been performed (Lohmann and Weigold (1981)) and will be described in more detail in Section (4.2.1.4).

$(e, 2e)$ experiments which yield momentum space information then demand symmetric conditions i.e. $E_A = E_B$ and $\theta_A = \theta_B$.

4.2.1 Symmetric Kinematics

In this case the indistinguishability of the electron leads to an advantageous experimental situation in which $\theta_A = \theta_B = \theta$ and $E_A = E_B$.

Here the magnitude of the recoil momentum can be shown from equation (4-3) to be

$$q = \left[(2 k_A \cos \theta - k_0)^2 + 4 k_A^2 \sin^2 \theta \sin^2 (\phi/2) \right]^{\frac{1}{2}} \qquad (4\text{-}5)$$

where the angles θ, ϕ and the momenta are defined in Fig. 4-1.

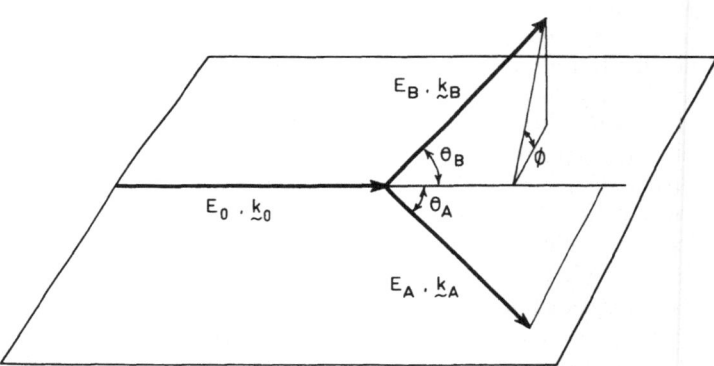

Fig. 4-1. Schematic diagram of the kinematics which apply to $(e, 2e)$ measurements. The momentum of the incident electron is $\underline{k}_0$ whilst $\underline{k}_A$, $\underline{k}_B$ are the momenta of the scattered electrons. The angles θ_A, θ_B measure deviation from the incident beam direction whilst ϕ measures the out of plane angle

Two symmetrical experimental situations have been used to obtain differential cross sections. The *coplanar symmetric case* (see e. g. Weigold *et al.* (1979), Camilloni *et al.* (1977), van Wingerden *et al.* (1979)). Here $E_A = E_B$, $k_A = k_B$, $\phi = 0$. For a given binding energy $\varepsilon = E_0 - E_A - E_B$ different values of q are chosen by varying the angles θ about the quasi free scattering angle $\theta = 45°$.

The first experiment in the coplanar symmetric geometry was carried out by Amaldi *et al.* (1969). This followed early suggestions of Baker *et al.* (1960), Smirnov and Neudatchin (1966) and Glassgold (1967). The early experiment was improved by Camilloni *et al.* (1972) who measured momentum distributions from the 1 s state in carbon.

More recent experiments on various other targets have demonstrated serious difficulties in the theoretical description of the ionization mechanism in this geometry. Thus quantitative descriptions of the momentum distributions of electrons in target atoms are clouded by uncertainties in the electron-electron interaction, that is the Mott scattering t matrix, which depends very strongly on the polar angle θ.

On the other hand the Mott scattering t matrix is only weakly dependent on the azimuthal angle ϕ. This leads to the other symmetric geometry which has been employed with much success in $(e, 2e)$ experiments on a wide range of atoms and molecules (see for example Weigold and McCarthy (1978)). This geometry is described as the *non coplanar symmetric geometry*. Here $k_A = k_B$, $\theta_A = \theta_B$ and the values of q are chosen by varying the azimuthal angle ϕ.

The potential of this method for the determination of the momentum distributions of the valence states in atoms was first demonstrated by Weigold *et al.* (1973) in an $(e, 2e)$ experiment on argon. In an experiment on methane, Hood *et al.* (1973) showed that the technique could also be applied to molecules.

4.2.1.1 Theory

The theory of the $(e, 2e)$ reaction for symmetric geometries has been developed most fully by McCarthy and collaborators and has been reviewed extensively by McCarthy and Weigold (1976), Weigold and McCarthy (1978) to which one should refer for more than the brief summary which is presented here.

Early calculations based on the Born approximation by Glassgold and Ialongo (1968) and by Smirnov and Neudatchin (1966) were sufficient to show that momentum distributions could be determined from symmetric $(e, 2e)$ geometry. These theories have been modified most significantly by the incorporation of distorted waves in the theory.

The differential ionization cross section for an $(e, 2e)$ reaction can be written, in atomic units. McCarthy and Weigold (1976),

$$\sigma(\underline{k}_0, \underline{k}_A, \underline{k}_B) = (2\pi)^4 \frac{k_A k_B}{k_0} \sum_{av} | M_f(\underline{k}_0, \underline{k}_A, \underline{k}_B) |^2 \qquad (4\text{-}6)$$

where $\sum_{av}$ denotes the sum and average over degeneracies.

$$M_f = \langle \chi_A^{(-)}(\underline{k}_A) \chi_B^{(-)}(\underline{k}_B) \psi_f^{N-1} | T(E) | \psi_0^N \chi_0^{(+)}(\underline{k}_0) \rangle. \qquad (4\text{-}7)$$

Antisymmetry is implicit throughout.

$T(E)$ is the antisymmetrized two electron Mott scattering T matrix. ψ_0^N is the eigenfunction describing the ground state of the target and χ_f^{N-1} describes the final state of the ion. The $\chi^{(\pm)}$ describe the incoming and outgoing electron waves respectively.

For symmetric collisions the T matrix depends only on the incident electron (a) and the initially bound electron (b). This assumption, which is called the binary encounter approximation, assumes that the state vector commutes with $T(E)$ since it contains neither electron (a) nor (b). Therefore

$$M_f = \langle \chi_A^{(-)} \chi_B^{(-)} | T(E) | (\psi_f^{N-1} | \psi_0^N) \chi_0^{(+)} \rangle. \qquad (4\text{-}8)$$

Two major approximations have been used to proceed from this point. The plane wave impulse approximation (PWIA) arises by replacing electron waves with plane waves. The two electron t matrix replaces T and factorizes giving

$$M_f = \langle \underline{k}^1 \mid t(E) \mid \underline{k} \rangle \langle \underline{q} \psi_f^{N-1} \mid \psi_0^N \rangle \qquad (4\text{-}9)$$

where

$$\underline{k}^1 = \frac{1}{2}(\underline{k}_A - \underline{k}_B), \qquad (4\text{-}10)$$

$$\underline{k} = \frac{1}{2}(\underline{k}_0 - \underline{q}) \qquad (4\text{-}11)$$

and q is given by equation (4-3).
The cross section in atomic units is

$$\sigma_f = (2\pi)^4 \frac{k_A k_B}{k_0} f_{ee} \sum_{av} \mid \langle \underline{q} \psi_f^{N-1} \mid \psi_0^N \rangle \mid^2. \qquad (4\text{-}12)$$

The Mott scattering factor

$$f_{ee} = A \mid \langle \underline{k}^1 \mid t(E) \mid \underline{k} \rangle \mid^2 \qquad (4\text{-}13)$$

where

$$f_{ee} = \left(\frac{1}{2\pi}\right)^2 \frac{2\pi\eta}{\exp(2\pi\eta) - 1} \left[\frac{1}{\mid \underline{k}^1 - k \mid^4} + \frac{1}{\mid \underline{k}^1 + \underline{k} \mid^4} - \frac{1}{\mid \underline{k} - \underline{k}^1 \mid^2} \frac{1}{\mid \underline{k} + \underline{k}^1 \mid^2} \right.$$

$$\left. \cos\eta \ln\left(\frac{\mid \underline{k} + \underline{k}^1 \mid^2}{\mid \underline{k} - \underline{k}^1 \mid^2}\right) \right] \qquad (4\text{-}14)$$

where

$$\eta = \frac{1}{2 k^1}. \qquad (4\text{-}15)$$

The Mott factor is essentially independent of the azimuthal scattering angle ϕ. Thus in the non coplanar case, where θ is fixed, f_{ee} is a constant and the differential ionization cross section is proportional to the square of the momentum representation of the overlap amplitude. On the other hand f_{ee} varies rapidly with θ thus differential cross sections measured in the coplanar symmetric case test the validity of the description of the interaction mechanism as well as structural information on the initial and final states. Indeed inadequacies in the PWIA have been demonstrated in coplanar symmetric experiments on atomic hydrogen by Weigold et al. (1979). Thus information on the wavefunctions has in general been restricted to experiments in the non coplanar symmetric geometry.

An alternative description of the interaction is provided by the Distorted Wave Impulse Approximation (DWIA) (McCarthy and Weigold (1976)). In this approximation, the electron-electron t matrix is again used for T in equation (6) but the distorted waves $\chi_i(\underline{k}_i)$ are computed in relevant optical model potentials. f_{ee} is factorized out and the differential ionization cross section becomes

$$\sigma_f = (2\pi)^4 \frac{k_A k_B}{k_0} f_{ee} \sum_m |\langle \chi_A^{(-)} \chi_B^{(-)} | (\psi_f^{N-1} | \psi_0^N) \chi_0^+ \rangle|^2 . \tag{4-16}$$

The major improvement which is seen by this approximation in the non coplanar symmetric geometry is that it provides a more accurate description of the cross section at regions of large momentum transfer q. These regions are equivalent to the inner regions of the atom where the role of distortion is expected to be more significant. Although some improvement over the PWIA in the coplanar case is observed by Weigold et al. (1979) such improvement is not sufficient to supplant the non coplanar geometry as a test of the properties of the initial and final state wavefunctions.

For molecules it is usual (McCarthy and Weigold (1976)) to make the Born-Oppenheimer approximation. In this approximation the molecular wave function is the product of separate electronic vibrational and rotational functions. The experiments do not resolve the final vibrational and rotational states thus closure is used in the analysis. Under these conditions the PWIA reduces to (Weigold (1981))

$$\sigma_f = (2\pi)^4 f_{ee} \frac{k_A k_B}{k_0} \int d\Omega \int dv \, |\langle q \psi_f^{N-1} | \psi_0^N \rangle|^2 \tag{4-17}$$

where the spherical averaging over the molecular orientations Ω is done by invoking closure. The complications of the multicentre scattering problem have so far prevented the applications of the DWIA to the study of molecules (McCarthy (1984)).

For cases in which the ground state wavefunction can be represented by a Hartree Fock configuration, McCarthy and Weigold (1976) show that the differential cross section is proportional to the spectroscopic factor $S_c^{(f)}$. This is the probability of finding the configuration involving a single hole in the orbital c in the ion state f.

$$S_c^{(f)} = [t_c^{(f)}]^2 \tag{4-18}$$

where the t_c^f are the CI coefficients for the ion. These authors also show that the spectroscopic factor satisfies a sum rule derived from the normalization and closure properties of the wavefunction i.e.

$$\sum_f S_c^{(f)} = 1. \tag{4-19}$$

4.2.1.2 The Experiments

The majority of $(e, 2e)$ experiments in symmetric geometry which have been performed have measured relative differential cross sections. Such experiments can however accurately measure ratios, thus tests with the theory involve only one normalization. Many descriptions of the general experimental detail have been given by Weigold and collaborators (McCarthy and Weigold (1976)). A schematic view of typical apparatus is shown in Fig. 4-2. A beam of electrons is directed at the neutral beam of particles under study. Two electron spectrometers A and B view the interaction region and are set to pass electrons with energies E_A and E_B. In symmetric geometries $E_A = E_B = E/2$ where $E = E_0 - \varepsilon$ from equation (4-2). Various types of electrostatic energy analysers have been used in $(e, 2e)$ experiments. These include cylindrical mirror, hemispherical and $127°$ analysers.

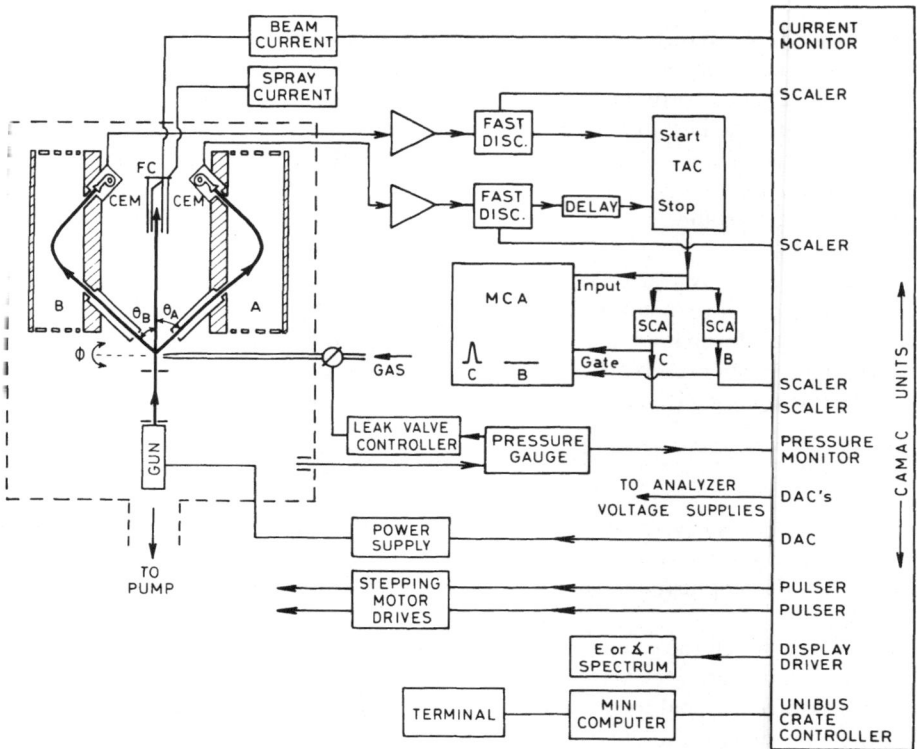

Fig. 4-2. Schematic diagram of the apparatus used in a typical $(e, 2e)$ experiment

In the non coplanar case $\theta_A = \theta_B = 45°$ and the momentum transfer q is varied by changing the azimuthal angle ϕ in accordance with equation (4-5). This can be achieved by rotating one spectrometer out of the scattering plane defined by the momentum of the incident beam and the momentum of the other electron. Other techniques have been used to vary ϕ most notably that by the group at the University of Maryland (Moore et al. (1978)) who have used the focussing properties of a cylindrical mirror electron spectrometer to advantage.

In the coplanar case the angle ϕ is held constant to $\phi = \pi$ and the momentum transfer q is varied by rotating both spectrometers together about the incident beam direction such that $\theta_A = \theta_B$.

Electrons arising from ionizing collisions in the interaction region pass through the spectrometers and are detected with electron multipliers. In the original experiments channel electron multipliers were used, however more recently micro-channel plates have been employed with some advantage. In the usual $(e, 2e)$ experiments these latter devices have proven to be able to cope with greater rates and have much faster rise times than the channel multipliers. Pulses from the detectors are amplified, delayed, processed by standard fast timing electronics and act as start and stop pulses for a time to amplitude converter (TAC). An alternative timing network has been described by Van der Wiel and co-workers (1971) and has some advantages in situations where low signal to noise ratios are encountered.

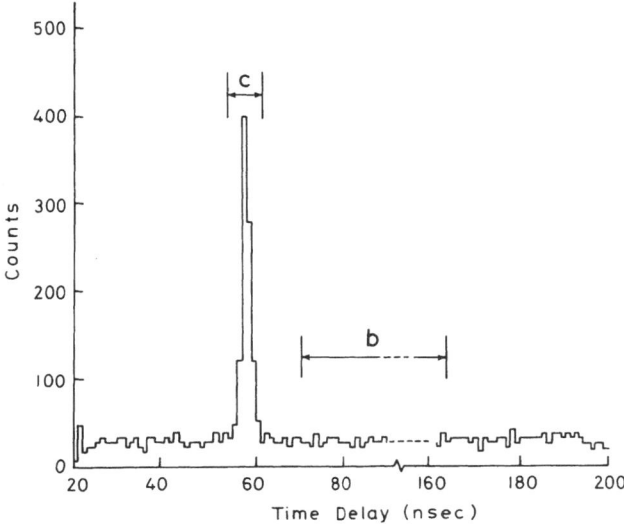

Fig. 4-3. A typical timing spectrum recorded in a conventional $(e, 2e)$ apparatus using channel electron multipliers as detectors. The signal is contained in the region "c" whilst the background is determined from the region "b"

The output of the time to amplitude converter can be monitored with a multichannel analyser. A typical timing spectrum is shown in Fig. 4-3, where the coincidence peak can be seen on top of a background of accidental coincidence counts. The experiments then involve the measurement of the number of counts in the coincidence peak as a function of q in experiments which measure momentum distribution or as a function of E_0 for experiments concerned with the measurement of binding energy spectra. In practice the true coincidence count, N_t, rate is established with two single channel analysers which monitor the output of the time to amplitude converter. One single channel analyser is set to scan the channels which hold the coincidence peak, i.e. N_c and the other to scan the random background, N_b. Usually the background is monitored over R times more channels than that containing the peak. That is $R = \dfrac{t_c}{t_b}$ the ratio of the window widths

$$N_t = N_c - \frac{N_b}{R} \tag{4-20}$$

and the statistical error in N_t i.e. one standard deviation is

$$\sigma_t = \left[N_c + \frac{N_b}{R^2} \right]^{\frac{1}{2}} \tag{4-21}$$

if the relative error in R is ignored. In practice R can be determined to arbitrary accuracy by measuring the number of counts in the two single channel analysers arising from two uncorrelated electron signals. The use of equation (4-20) assumes that there is no slope in the background that is that the background rate is independent of the time between start and stop. This is essentially a feature of the TAC but it can lead to systematic errors when large singles count rates are encountered in situations with poor signal to noise ratios.

The accidental coincidence count rate N_r in a window of width t_i is given by

$$N_r = N_A N_B t_i \qquad\qquad (4\text{-}22)$$

where N_A, N_B are the single count rates in each spectrometer. In the region containing the coincidence peak $t_i \cong$ twice the FWHM of the coincidence peak. The ratio of true coincidence counts to randoms ultimately governs the signal to noise ratio in the experiment. Clearly benefits in the signal to noise ratio can be gained by minimizing t_i. The advantage of microchannel plates in this regard is seen by comparing t_i for these devices at ~ 1 nsec wheras t_i for channel electron multipliers are typically ~ 5 nsec.

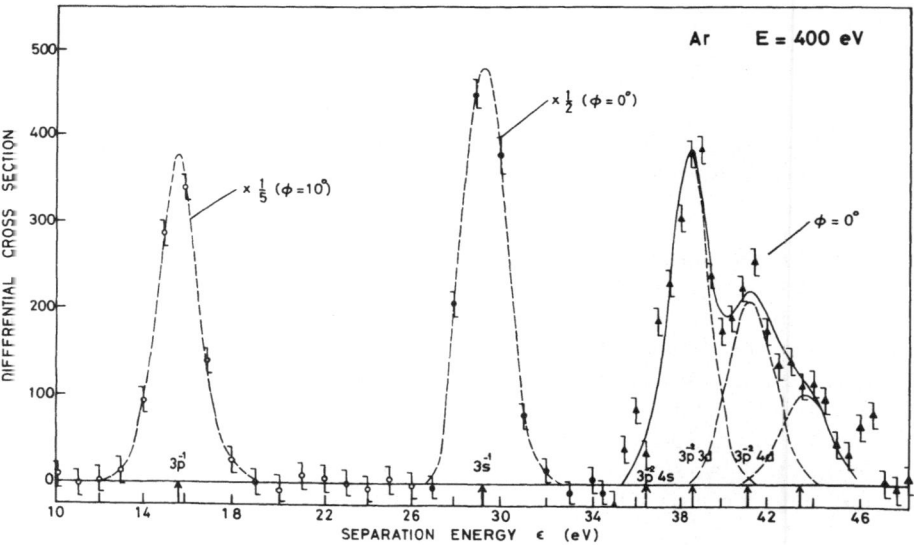

Fig. 4-4. Differential cross sections for the electron impact ionization of argon at a total energy E of 400 eV (McCarthy and Weigold 1976). The peak at a separation energy of 15.76 eV was taken at $\phi = 10$ and corresponds to the ejection of electrons from the $3p$ shell. The arrows at increasing separation energies indicate the positions of the $3s\,3p^5$, $3s^2\,3p^4\,3d$ and $3s^2\,3p^4\,4d$ ion states respectively: the cross sections were measured at $\phi = 0$.

The use of the technique in the determination of binding energies of valence electrons in an atomic target can be seen in Fig. 4-4. Here the number of true coincidence counts arising from the ionization of argon at a total energy $E = 400$ eV is shown as a function of the incident energy which is varied from 400 eV to 450 eV

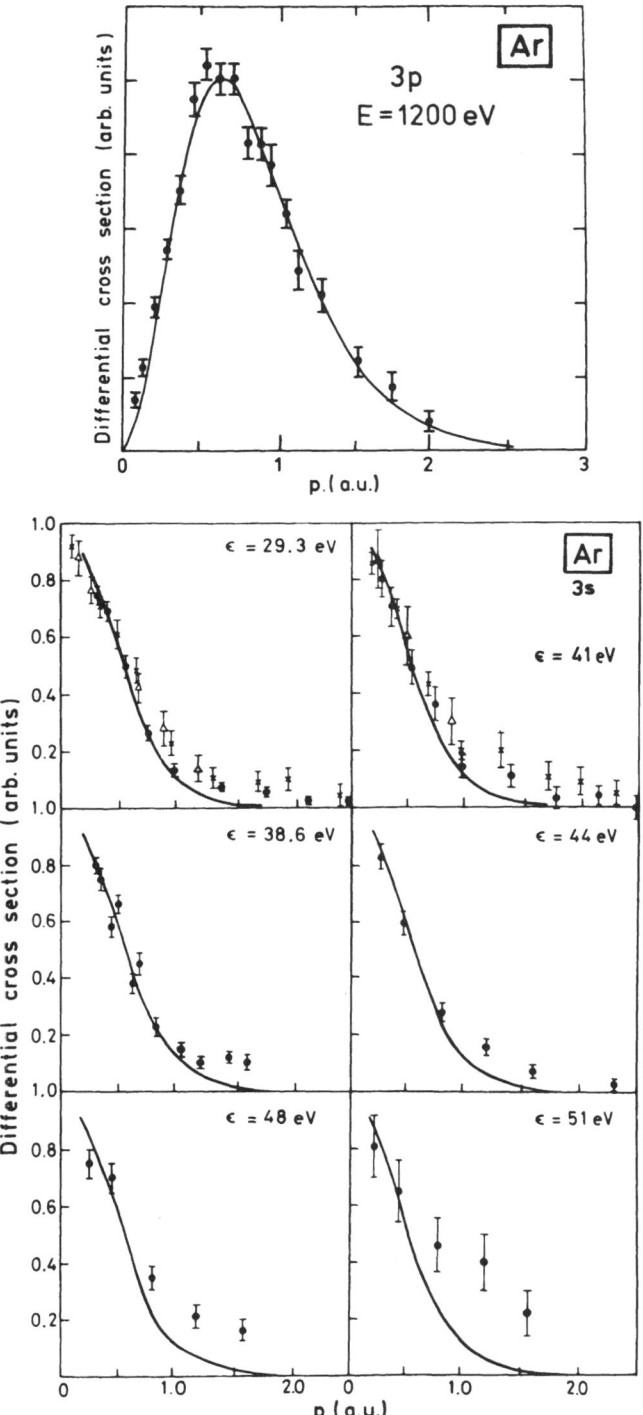

Fig. 4-5. (e, 2 e) angular correlations for the valence shells in argon taken in the symmetric non coplanar geometry at separation energies corresponding to the spectral features in Fig. 4-4 (Weigold (1981)). The momentum transfer of the ejected electron is p. The solid lines are predictions of a plane wave theory using Hartree-Fock wavefunctions

(McCarthy and Weigold (1976)). Dramatic increases in the coincidence count rate are observed when the incident energy is such that equation (4-2) is satisfied for various binding energies ε_i. The data in Fig. 4-4 were taken in the symmetric non coplanar geometry at two polar angles $\phi = 0$ and $10°$. Peaks corresponding to the ejection of electrons from the $3p$ and $3s$ single particle states are seen at binding energies of 15.75 and 29.3 eV respectively. The structure at higher binding energies corresponds to the presence of correlation in the final ion state. Such spectra can be compared at least superficially with binding energy spectra which can be obtained from x-ray photoelectron spectrometry (XPS). However the advantage of $(e, 2e)$ spectroscopy over XPS is demonstrated by the observation that it is possible to identify the peaks in the binding energy spectrum by the shape of the differential cross section as a function of the momentum transfer q. Equation (4-12) shows that the differential cross section $\sigma(q)$ is proportional to the square of the momentum space wavefunction of the struck electron. This can be seen in Fig. 4-4 where the coincidence count rate is plotted as a function of q for the $3p$ hole state and the $3s$ hole state. The difference in shape of the two peaks is obvious and characteristic of the general predictions of equation (4-12).

Such a feature is perhaps trivial in the case of the main features of Fig. 4-4 as it could be argued that the states are clearly represented by their binding energies. However the major strength of the technique can be seen in the identification of the satellite structure in Fig. 4-5. In this case the angular correlations clearly identify these structures as arising from configuration interaction in the $3s$ hole state. These correlations will be described in more detail in Section 4.2.2.

Comparison with the theory for such experiments is complicated by the fact that the data are relative. However although the absolute value of the components of the experiment which go to make up the cross section measurement are not known explicitly, the way in which the data are taken allows a comparison with the theory after only one normalization. Specifically the energies at which the outgoing electrons are analyzed are constant, thus the transmission and detection efficiencies are constant for any incident energy E_0. Therefore meaningful comparison with the theory can be made either in the determination of the relative differential cross sections for different valence shells of the target or in the measurement of spectroscopic factors.

4.2.1.3 Absolute $(e, 2e)$ Cross Sections

The problems associated with the determination of the absolute value of the $(e, 2e)$ cross section are similar to those encountered in the measurement of any differential cross section using a crossed beam configuration. In particular the problems associated with the neutral beam density, with the definition of the scattering geometry and with the transmission of the spectrometers have in general led to very large uncertainties in the designation of the cross sections.

The first attempt at an absolute $(e, 2e)$ cross section was reported by Beaty et al. (1977) on an experiment in helium. Shortly after Stefani et al. (1978) reported cross sections for the ionization of helium at incident energies from 200 to 4000 eV. Experimentally the $(e, 2e)$ cross section can be defined (van Wingerden et al. (1979))

$$\sigma = \frac{S_e}{I_0 \, N \, (l \, d\Omega_A \, d\Omega_B)_{\text{eff}} \, \tau_A \, \tau_B \, \Delta E_{\frac{1}{2}}} \tag{4-23}$$

where S is the true coincidence count rate, e the electronic charge, N the target density, I_0 the incident electron beam current, $(l \, d\Omega_A \, d\Omega_B)_{\text{eff}}$ the scattering length times solid angles as seen by the two analysers in coincidence, $\tau_A \, \tau_B$ the transmission of the two spectrometers and $\Delta E_{\frac{1}{2}}$ the half width of the energy resolution function for coincident detection of both outgoing electrons.

The experiments of Beaty et al. (1977) and Stefani et al. (1978) were carried out in a crossed beam configuration and had great difficulty in determining N. In the latter case an attempt was made to minimize the problem by using gas flow techniques. The problem of the effective solid angles of both detectors was approached by scanning the electron beam across the target beam. Beaty et al. used a time of flight analyser as one of the spectrometers and were therefore able to calculate the solid angle $d\Omega_A$, they also assumed that its transmission and efficiency η_A was unity. Both groups estimate a factor of 2 in their absolute values which unfortunately cannot be compared because they span different kinematic regions.

The far more accurate experiments of van Wingerden et al. (1979) produced an overall uncertainty of 20% in the $(e, 2e)$ cross section in helium. The data were taken in the coplanar symmetric geometry at $\theta_A = 45° = \theta_B$ at various total energies E from 200 eV to 2800 eV. The target density N was determined very accurately by the use of a static gas cell. Collimating systems determined an effective value of the scattering length times the solid angle for each analyser which was calculated numerically; these values were consistent with an analytic expression given by Kuyatt (1968). Similarly a numerical value was calculated for the term $(l \, d\Omega_A \, d\Omega_B)_{\text{eff}}$ which appears in equation (4-23).

The overall efficiencies τ_A and τ_B of each analyser at an energy $\frac{1}{2}E$ were determined by comparison with the known elastic differential cross sections at an energy of $\frac{1}{2}E$. The half width of the energy resolution function for the coincident detection of both outgoing electrons was derived by folding the separate energy resolution functions of both analysers which had been determined experimentally.

The inelastic cross section at $\frac{1}{2}E$ or the double differential cross section was measured by counting the singles count rate in one analyser and by using the previously determined factors pertinent to the spectrometer. The double differential cross sections measured in each analyser were equal to within 2% which gives confidence in the technique considering that the efficiencies, $(l \, d\Omega)_{\text{eff}}$ and energy resolution functions for both analysers were different.

The major contribution to the uncertainty in the $(e, 2e)$ cross sections determined in these experiments arises from the errors in the elastic scattering cross sections which count twice.

It is possible to compare these data with those of Stefani et al. (1978). The latter data appear to be too small at most energies by about a factor of two which can be interpreted as agreement between the two sets of experiments given the large uncertainties in the Frascati data. In a later series of experiments Stefani et al. were able to improve the error to about 50%. Nevertheless van Wingerden et al. (1981) point out that there is still significant scatter in the Frascati cross sections.

The comparison between the FOM data and various theories allow a sensitive comparison of these theories. It is found that the Coulomb projected Born approximation with exchange of Geltmann (1974) is in excellent agreement with the data over the whole energy range. The plane wave impulse approximation of McCarthy and Weigold (1976) and the plane Coulomb wave impulse approximation of Stefani et al. (1978) agree with the experimental results below 1000 eV.

4.2.1.4 Experiments on Atomic Hydrogen

Atomic hydrogen is unique as a target in that its ground state wavefunction is known analytically and in an ionizing event the final state is simply a proton. In a recent experiment on atomic hydrogen Lohmann and Weigold (1981) have demonstrated conclusively that the $(e, 2e)$ cross section is a direct measure of the momentum distribution of the electrons in the ground state and that the PWIA can be used with confidence in non coplanar symmetric geometry.

For atomic hydrogen, equation (4-12) becomes

$$\sigma = (2\pi)^4 \frac{k_A k_B}{k_0} |f_{ee}|^2 |\phi_{1s}(q)|^2 \tag{4-24}$$

where $\phi_{1s}(q)$ is the momentum space wavefunction for the ground state of atomic hydrogen. In atomic units

$$\phi_{1s} = \left(\frac{2}{\pi}\right)^{\frac{3}{2}} \frac{1}{(1+q^2)^{+2}}. \tag{4-25}$$

Thus

$$\sigma = 2^7 \pi^2 \frac{k_A^2}{k_0} |f_{ee}|^2 \frac{1}{(1+q^2)^4}. \tag{4-26}$$

In symmetric geometry the Mott scattering T matrix

$$|f_{ee}|^2 = \frac{1}{4\pi^4} \frac{2\pi\eta}{(e^{2\pi\eta}-1)} \frac{1}{K^4} \tag{4-27}$$

where

$$K = |\underline{k}_0 - \underline{k}_A| = |\underline{k}_0 - \underline{k}_B| \tag{4-28}$$

and

$$\eta = \frac{1}{|\underline{k}_A - \underline{k}_B|}. \tag{4-29}$$

In the non coplanar symmetric geometry K is constant and $|f_{ee}|^2$ is essentially independent of q for $q < 2$ a.u. Thus the $(e, 2e)$ cross section should be directly proportional to the square of the momentum space wavefunction.

Very few experiments on atomic hydrogen have been carried out because of the difficulties associated with the production of beams of hydrogen atoms. In the experiments of Lohmann and Weigold (1981) a d.c. discharge tube was used to produce the beam of hydrogen atoms. Two cylindrical mirror electron energy analysers viewed the interaction region at polar angles of 45° to the incident electron beam direction. The azimuthal angle ϕ could be varied by rotating one of the spectrometers in both directions about the zero degree position. It was confirmed that the data was symmetric about $\phi = 0$ and that the singles count rates were independent of the azimuthal angle.

The overall energy resolution in these experiments was $\sim 2\,eV$ which was sufficient to discriminate against any undissociated molecular hydrogen in the beam. The data were corrected for the small ($\Delta\theta = 1.5°$ and $\Delta\theta = 2°$) angular resolution of the spectrometers.

The coincidence count rate as a function of the momentum transfer q is shown in Fig. 4-6 for total energies $E = 400$, 800 and 1200 eV. The data are relative thus at each energy they have been normalized to the three data points of lowest q in the curve described by $(1 + q^2)^{-4}$. It was found that the inclusion of the Mott scattering term had a negligible effect on the fit to the data. The agreement between the shape of the curve predicted by equation (4-26) and the data is excellent which demonstrates conclusively that the $(e, 2e)$ cross section is proportional to the square of the momentum space wavefunction in this geometry.

Experiments in the coplanar symmetric geometry have not been so successful and illustrate the problems associated with the correct evaluation of the Mott scattering t matrix (Weigold et al. (1979)).

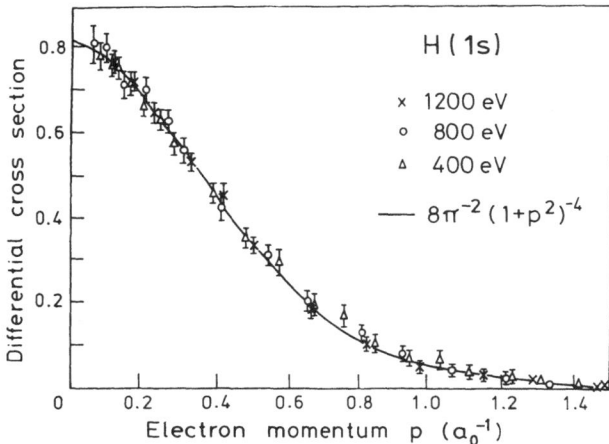

Fig. 4-6. Differential cross sections for the $(e, 2e)$ reaction in atomic hydrogen taken in the non coplanar symmetric geometry (Lohmann and Weigold (1981)). The experimental points were taken at total energies $E = 400\,eV$, 800 eV and 1200 eV and the solid line represents the calculated momentum distribution of the hydrogen atom

4.2.1.5 Correlation Effects

In favourable collision geometry and kinematics, the $(e, 2e)$ reaction probes the overlap between the ground state wavefunction of the atom and the final state wavefunction of the ion. The differential cross section is then sensitive to the effects of configuration interaction in these wavefunctions. Such effects have been studied by Weigold and collaborators in both atoms and molecules and two examples are discussed below.

(a) Ground State Correlations

In collisions involving the helium atom at energies up to about 2 keV the dominant process is that which leaves the ion in the ground 1 s state. He^+ is of course a one electron ion consequently correlation effects can only arise from the atom. There is a finite probability of exciting higher states in the ion in the ionization process. The two reactions under consideration then are

$$e + He(1\,s) \rightarrow He^+ (1\,s) + e + e \tag{4-30}$$

and

$$e + He(1\,s) \rightarrow He^+ (n=2) + e + e. \tag{4-31}$$

The relative cross sections for these two processes can be measured by the ratio of the peak heights in a binding energy spectrum. This ratio depends strongly on the description of the ground state of the atom. For example in the Hartree Fock basis, the excitation of the $n=2$ states in the ion arises because in this set the 1 s one electron wavefunctions of the atom are not orthogonal to the ns wavefunctions of the ion; they are of course orthogonal to the nl wavefunctions for $l \neq 0$.

On the other hand in a basis which employs configuration interaction the 2 p state of the ion is no longer orthogonal to the atomic wavefunction. Thus a CI basis will predict a different cross section ratio.

The ratio of the cross sections for $n=2$ to $n=1$ excitation in helium was first measured in the symmetric coplanar geometry by McCarthy et al. (1974). They found that the ratio varied from 0.72% at $\theta_A = \theta_B = \theta = 45°$ to 1.54% at $\theta = 53°$ which was in excellent agreement with the ratios calculated from the correlated wavefunction of Joachain and van der Poorten (1970). They also found that ratios calculated from the Hartree Fock basis wavefunction of Froese Fisher varied from 2.75% at $\theta = 45°$ to 1.94% at $\theta = 53°$. Thus this technique was demonstrated to be a sensitive test of the details of the ground state wavefunction. It is clear however that in this geometry the agreement between theory and experiment relied on the fact that the Mott scattering factors for the two states were essentially equal and therefore cancel. Although the experimental results were sufficiently accurate to distinguish between the Hartree Fock and CI calculations, the errors were such as to mark the weak dependence of the ratio on the details of the interaction mechanism.

A far more accurate set of measurements of the ratio in the non coplanar symmetric geometry has recently been reported by Cook et al. (1984). The experiments were performed at a total energy of 1200 eV and covered a range of q from 0.3 to 2.5 a.u. A pair of position sensitive detectors were used to detect the electrons and perform the energy analysis. These detectors resulted in a spectacular increase in the data taking capabilities of the apparatus. The choice of the non coplanar geometry removed any potential problems with the t matrix. The aim of the experiments was to provide data which could distinguish between various quality correlated wavefunctions. Calculations of the ratio were made using the Hartree Fock basis of Clementi and Roetti (1974) which they found to be independent of q and equal to 2.17%. The ratios were computed using the correlated wavefunctions of Joachain and van der Poorten (1970) (JV), of Taylor and Parr (1952) (TP) and of Nesbet and Watson (1958) (NW). These wavefunctions predict 98%, 85% and 97.7% of the correlation energies respectively.

The experimental data are compared with the computed ratios for the various wavefunctions in Fig. 4-7. The Hartree Fock computation is clearly inadequate whereas that predicted by JV is in excellent agreement for $q < 1$ a.u. and shows acceptable agreement for larger values of q. Taken over the whole range of momentum transfer, the experimental values clearly favour the correlated wavefunction which predicts the majority of the correlation energy. Clearly this technique is an extremely sensitive probe to the details of the wavefunction. Indeed Cook *et al.* (1984) stress the considerable advantages which this technique has over Compton profile measurements as a test of wavefunctions.

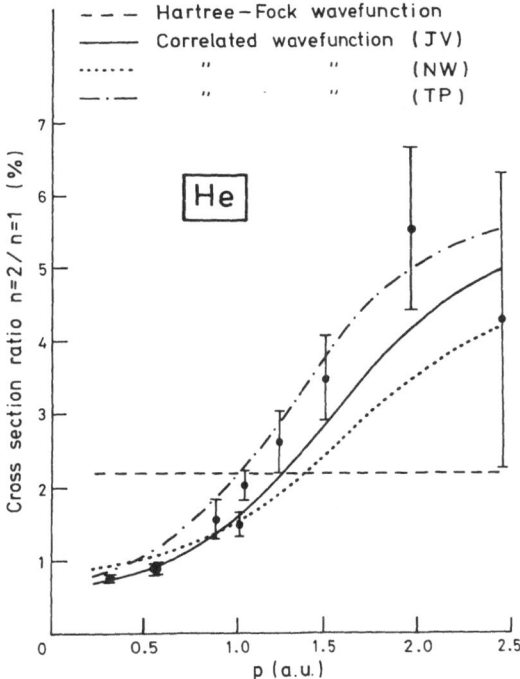

Fig. 4-7. Ratio of the cross sections for the excitation of the $n = 2$ to $n = 1$ states in the helium ion in the non coplanar symmetric $(e, 2e)$ reaction. The total energy used to measure the experimental cross sections was 1200 eV (Cook *et al.* (1984)). The theoretical calculations are described in the text

(b) Final State Correlations

In the heavy rare gases such as argon, krypton and xenon, structure is observed in the separation energy spectra which indicates the influence of correlation effects in the final ion state.

These correlation effects have also been observed for molecules such as the hydrogen halides (see for example Brion et al. (1980)) and in the halo methanes (Minchinton et al. (1982)).

The $(e, 2e)$ spectroscopy of argon offers a good example of the way in which spectroscopic information can be derived from the separation energy spectra. Similar information has been derived on the valence electron spectroscopy of krypton (Fuss et al. (1982 a)) and on xenon (Weigold and McCarthy (1978)). The separation energy spectrum of argon, taken from McCarthy and Weigold (1976), is shown in Fig. 4-4. The peaks at 15.75 eV and 29.3 eV arise from the ejection of electrons from the $3p$ and $3s$ shells respectively. The additional structures at 38.6 eV and 41.2 eV correspond to configurations with the same spin and parity as the $3s$ hole state namely the valence configurations $3s^2 3p^4 3d$ and $3s^2 3p^4 4d$. Angular correlations taken in the symmetric non coplanar geometry at the higher separation energies clearly identify these states with the $3s$ hole state; that is the cross section peak at $q \approx 0$. Indeed almost half of the spectroscopic strength of this state is distributed amongst the higher configurations. This effect was first observed by Weigold et al. (1973) and was subsequently confirmed at higher energies by Giardini-Guidoni et al. (1976). Williams (1978) carried out a set of experiments at 300 eV total energy with an order of magnitude better energy resolution than the previous experiments. Apart from observing the spin-orbit splitting in the $3p$ hole state, Williams detected the weak influence from the configuration $3s^2 3p^4 4s$ which was not observed in the earlier experiments. No evidence of contributions to the $3p$ hole state from additional configurations was seen by any of these groups.

All of these experiments were relative but it was possible to derive the spectroscopic factors (equation (4-18)) from the relative cross sections in the separation energy spectrum. Two techniques have been employed to deduce the spectroscopic factors from the spectra as in Fig. 4-4. The one reported by McCarthy and Weigold (1976) involved taking the areas of the peaks corresponding to the $3s$ hole state relative to that of the $3p$ hole state. These ratios were then multiplied by the ratio of the theoretical cross sections to account for the difference in the momentum distributions of the $3p$ and $3s$ states. This procedure rests on the observed fact that all of the spectroscopic strength of the $3p$ state lies in the peak at 15.75 eV and therefore $S_{3p}^f = 1$. It also depends on the assumption that the ratio of the theoretical cross sections is known accurately. Whilst the plane wave impulse approximation accurately described the shape of the cross sections at small momentum transfer, the possibility of different absorption for the $3s$ and $3p$ hole states could not be dismissed "a priori". A reliable distorted wave theory was required to explore this point.

Thus an alternative approach has been used to determine the spectroscopic factors which involves the use of equation (4-19). Here the strength of a particular configuration is just the ratio of that peak to the total $3s$ hole strength. This technique relies on properly accounting for all of the strength in the continuum.

The influence of inadequacies in the theory and hence the first method would be apparent if energy dependence in the spectroscopic factors, was observed. Such a comparison between the results of the three groups mentioned above is difficult for the $3s 3p^6$ configuration because Giardini-Guidoni et al. give no error limits. Nevertheless the results of Williams (1978) at 300 eV and of McCarthy and Weigold at 1200 eV are in good agreement. In respect of the configurations at higher separation energies, error limits are given by each group and favourable agreement is observed over the whole energy range.

There is marked disagreement between the spectroscopic factors derived from $(e, 2e)$ spectroscopy in argon to those determined from x-ray photo electron spectroscopy (XPS) (Spears *et al.* (1974)). The XPS data show 82% of the strength in the $3s\,3p^6$ hole state compared to 53% in $(e, 2e)$ spectroscopy. This discrepancy is usually attributed to the fact that the two methods probe different regions of the atom and therefore of the wavefunction (Weigold and McCarthy (1978)). In terms of momentum transfer, XPS gives spectroscopic strengths for high q regions which are very close to the nucleus whereas the $(e, 2e)$ experiments probe regions where $q \simeq 0$ and therefore the electron is more likely to be found.

There is also marked disagreement between both $(e, 2e)$ and XPS data and various sophisticated theoretical calculations employing electron correlation effects (e.g. Smid and Hansen (1983), Williams (1979)).

4.2.1.6 Molecular $(e, 2e)$ Spectroscopy

The success of the $(e, 2e)$ reaction in the non-coplanar symmetric geometry in accurately describing the momentum distributions of electrons in atomic orbitals has prompted extensive studies into its use to probe the details of molecular wavefunctions. The great strength of the technique rests in the ability to provide not only separation energy spectra but also accurate details of momentum distributions for regions of low momentum transfer q. These regions correspond to the outermost regions of the target and are therefore very relevant in chemical descriptions of bonding. The emphasis on the outer areas of the molecule overcomes the severe computational difficulties associated with a complete distorted wave description of the reaction mechanism since here the distortion of the incoming and outgoing electron waves is minimal. In addition if the energies of the electrons involved in the event are high enough, the plane wave impulse approximation is expected to provide an adequate approximation to the reaction mechanism (Weigold and McCarthy (1978)). This premise has been tested on various molecular targets and two examples will be given below.

The extension of the $(e, 2e)$ technique to molecules has however, raised several experimental problems which do not necessarily apply to atomic targets. Perhaps the most significant of these is a consequence of the fact that the energy levels in molecular ions are relatively more closely spaced. This feature places rather stringent restrictions on the overall energy resolution which can be tolerated. Typically the energy resolution which has been used in molecular $(e, 2e)$ experiments is 1.5 eV, although Coplan *et al.* (1978) have worked with $\Delta E = 1.0$ eV.

Such resolution is sufficient to identify the major features of simple molecules such as N_2 and H_2 but it is only barely adequate to resolve the detail in a more complicated molecule such as CH_3Cl (Minchinton *et al.* (1982)). Clearly the study of vibrationally resolved structure is beyond the realm of conventional experiments.

The reliance on the plane wave impulse approximation for the description of the interaction mechanism places a further restriction on the experimental conditions. The assumption of the use of plane waves in the ionizing event demands that both the incident and outgoing electron energies must be high i.e. $E_A = E_B \geq 600\,\text{eV}$. Thus molecular $(e, 2e)$ spectroscopy requires good energy resolution at high energies. For electrostatic energy analysers the resolution is directly proportional to the analysing energy (Kuyatt (1968)) thus these two features demand that the electrons be retarded in energy before analysis. This solution to the problem is however achieved at the expense of transmission; for example in a cylindrical mirror spectrometer the transmission is proportional $R^{1/3}$ where R is the resolution of the device (Sar-el (1967)). But the true coincidence count rate is proportional to the square of the transmission, thus improvement in the resolution implies much longer data collection times. Inevitably then, some compromise must be struck between signal and data collecting rates; the practical limit in a conventional apparatus of $\sim 1.5\,\text{eV}$ reflects this compromise. Dramatic improvements in the energy resolution to less than $0.5\,\text{eV}$ would require the use of an electron monochromoter which does not seem to be a feasable approach in the symmetric configuration. Williams (1978) has achieved resolutions better than this in experiments at 300 eV total energy in argon. Even at this impact energy, in excess of one day was required to accumulate data of sufficient accuracy for each point in both separation energy spectra and angular correlations.

Two developments which show significant improvements in the signal rate are the novel design introduced by Moore et al. (1978) and the use of position sensitive detectors (Cook et al. (1984)). The electron spectrometer developed by Moore et al. uses the focal properties of a spherical electrostatic analyser. Electrons, which leave the interaction region, travel along the surface of a cone in the analyser. After analysis they pass through an exit aperture and diverge along the surface of another cone. Detectors are located around the surface of the exit cone; the relative position of a pair of detectors determines the azimuthal angle ϕ in the experiment. Up to ten detectors can be used, each pair representing a different azimuthal angle. The resulting data are processed by a computer. Signal rates of up to twenty five times that in a conventional apparatus have been reported by devices of this type (Coplan et al. (1982)).

The other major new development lies in the use of position sensitive detectors. Although these devices have not yet been used in the study of molecular $(e, 2e)$ spectra, Cook et al. (1984) have demonstrated the considerable advantages of such devices in an $(e, 2e)$ experiment on helium. In this case the position sensitive detectors were used in conjunction with a pair of hemispherical electron analysers. For these spectrometers the radial exit position of an electron arriving normal to an entrance aperture is directly proportional to the electron energy. Thus the position of detection is directly proportional to the energy of electrons entering the analyser.

Notwithstanding these significant experimental problems there is a growing compendium of data on various molecules by several experimental groups. A survey of the work at Flinders University has been given by Minchinton (1982). This work has focussed on three major areas of $(e, 2e)$ spectroscopy namely the development and testing of various impulse approximations, the assignment of outer valence states and on the study of the effects of configuration interaction in the splitting of inner valence hole states. To a greater or lesser extent these areas have also formed the basis of studies by Giardini-Guidoni and collaborators in Frascati (Giardini-Guidoni *et al.* (1980), by Brion and co-workers at the University of British Columbia (Cook *et al.* (1980)) and by Coplan, Moore and collaborators at the University of Maryland (Tossell *et al.* (1982)).

The experimental procedures which are followed with molecular targets are similar to those which have been described for $(e, 2e)$ collisions on atomic targets. Thus both separation energy spectra are obtained and angular correlations taken on peaks in the energy spectra. The angular correlation data are then interpreted as a measure of the molecular overlap function and therefore as a test of the basis functions used to describe the initial and final states. This interpretation rests on the validity of the PWIA in most cases, to describe the ionization mechanism. This approximation has been tested in the symmetric non coplanar experiments on H_2 by Dey *et al.* (1975). Here the molecular wavefunctions are so well known that any lack of agreement between theory and experiment can be attributed to the PWIA. In this work it was found that the calculated momentum distribution for the $1\sigma_g$ molecular orbital was in excellent agreement with the shape of the measured differential cross section out to $q \simeq 1.5$ a.u. for several total energies E between 300 eV and 1200 eV. The calculations involved the use of the plane wave impulse approximation and the wave functions of Snyder and Basch (1972). Absolute cross sections were not measured in the experiments. Subsequently van Wingerden *et al.* (1981) measured absolute $(e, 2e)$ cross sections in H_2 at fixed scattering geometry ($\theta = 45°$, $\phi = 0$) at several energies E between 200 eV and 2800 eV. These measurements, which corresponded to small q, demonstrated acceptable agreement between the PWIA and experiment provided E was greater than ~ 800 eV.

Coupled with the observations of Dey *et al.* regarding the shape of the momentum distributions one can conclude that the PWIA accurately describes the differential cross section at these energies. The results of van Wingerden *et al.*, however, reinforce the view that some caution must be exercised for $E < 800$ eV because significant differences are observed between theory and experiment in this energy range.

The validity of the PWIA has also been tested in a series of experiments on N_2 (Weigold *et al.* (1977 a)). Here it has been demonstrated that the measured angular correlation for the $3\sigma_g$ orbital was in excellent agreement with theory over an extensive range of q even at a total energy of 400 eV. In addition the relative magnitudes of the $1\pi_u$ and $2\sigma_g$ orbitals are very well described for small q i.e. less than ~ 1 a.u. At larger q there are significant discrepancies between the computed cross section for the $1\pi_u$ orbital. These regions correspond to areas close to the nucleus where distortion is expected to play a major role. The influence of distortion can be clearly seen when the total energy E is increased to 1200 eV (Dey *et al.* (1977)). In this case better agreement is found between the computed and measured cross-sections.

From these experiments and others on molecules, it is clear that the PWIA can be used to provide an accurate description of the shapes of differential cross sections for small q even at low energies; the range of validity is increased as the energy increases. Thus for small q less than ~ 1 a.u. any discrepancies between theory and experiment can, with some confidence, be attributed to inadequacies in the basis wavefunctions.

Such a conclusion has been drawn by Minchinton *et al.* (1983) on the calculated momentum distributions of the outer valence regions of CH_3CN.

The group at the University of Maryland have proposed a generalized technique to test molecular wavefunctions (Tossell *et al.* (1982)). They compute auto correlation functions, $B(r)$, from measured momentum distributions, $\rho(q)$, by taking the Fourier transform of $\rho(q)$. The derived functions $B(r)$ for the anti bounding orbitals of NO and O_2 are compared with those generated from calculations involving a restricted Hartree Fock basis. Although these authors conclude that the calculations in general overestimate polarisation and underestimate diffuseness, it should be noted that the $(e, 2e)$ data were taken at an incident energy of 400 eV. Thus the validity of the theoretical description of the ionization mechanism is open to some debate.

The positive identification of orbital assignment in complex molecules is, to a large extent, free from uncertainties surrounding the validity of the theory. In this case momentum distributions are taken at features in the separation energy spectrum. The shapes of these distributions removes any ambiguity surrounding the orbital assignment. This feature of $(e, 2e)$ spectroscopy was first demonstrated by Dey *et al.* (1976) in the resolution of the controversy surrounding the photoelectron spectroscopy identification of the ground state of $C_2H_6^+$. Angular correlations of the $3\sigma_{1g}$ and $1e_g$ orbitals clearly identified each. Similar demonstrations have been given in chloromethane by Minchinton *et al.* (1982) and on the identification of the $3l_{2g}$ orbital of benzene by Fuss *et al.* (1981 b).

As in the case in the rare gases, the $(e, 2e)$ cross section offers a sensitive test of electron-electron correlations in the initial state. For example Weigold *et al.* (1977 b) measured the relative intensities of exciting the $1s\sigma_g H_2^+$ ground and the $2p\sigma_u, 2p\pi_u$ and $2s\sigma_g$ ion states. Similar effects were seen in these experiments to those which have been described in the ionization of helium. In the case of H_2 the ratio of the cross sections for the excited state to the ground state was found to depend strongly on q rising from a value of 2% at small q. Angular correlations were taken on the excited state data and compared with the shapes predicted from the H_2 ground state configuration interaction wavefunction of McLean *et al.* (1960). It was found that the shape of the $2p\sigma_u$ transition was in good agreement with the calculated distribution but the calculated cross sections for the $2s\sigma_g$ and $2p\pi_u$ transactions did not agree with the experiments.

Final state correlations have been studied in a series of experiments in the hydrogen halides (Brion *et al.* (1980, 1982)). In general these experiments have shown similarity between HCl, HBr and HI with their isoelectronic equivalent inert gases; that is the outer valence states remain unsplit whereas the inner $ns\sigma$ transition is significantly split among many ion states. Brion *et al.* compared the observed spectroscopic factors for these molecules with those predicted from several one particle Greens function calculations using the two particle hole Tamm-Dancoff approximation. It was found that the agreement between the measurements and calculations was not good even though the approximation allows for correlation in the initial and final states as well as relaxation.

4.2.2 Asymmetric Collisions

1969 was a good year for coincidence experiments for it was in that year that the two pioneering papers which have come to define the field of triple differential cross sections were published. The first experiment with symmetric geometry was described by Amaldi *et al.* (1969). Such experiments we have seen have led to the use of this geometry to gain insight into momentum distributions of electrons in atoms and molecules. The first experiments with asymmetric geometry were described by Ehrhardt *et al.* (1969). Such experiments have concentrated on providing sensitive data to test the ionization mechanism. The work on asymmetric collisions can be conveniently divided into two energy regimes. Low energies i.e. $\varepsilon \leq E_0 \leq 500\,\text{eV}$ and high energies $E_0 > 3\,\text{keV}$. In the low energy regime the primary aim of the experiments is to provide data to test the ionization mechanism; whereas experiments in the high energy regime have provided information on binding energies of atoms and molecules. These compare favourably with those produced in photoelectron spectroscopy using synchrotron radiation.

4.2.2.1 Low Energies

A study of typical single and double differential cross section measurements shows that for ionizing collisions where the incident electron energy is greater than $\sim 200\,\text{eV}$, the two electrons leave the interaction region in a very asymmetric configuration. That is one of these electrons is fast and is scattered in a narrow cone in the forward direction whilst the other is slow and is ejected into a much larger angle. Ehrhardt *et al.* (1982 a) have made semi-quantitative estimates of the contributions of such asymmetric collisions to the "total" cross section in helium. Using the data of Burnett *et al.* (1976), they find that $\sim 85\%$ of the ionization cross section comes from these collisions. Thus it is not surprising that results from these asymmetric configurations have presented the most severe test of the theory.

The first $(e, 2e)$ cross sections in this kinematic range were measured by Ehrhardt *et al.* (1969). The target was the ground state of the helium atom and the incident electron beam energies which were used were 114 eV and 50 eV.

As is the case in the symmetric experiments conservation of energy requires, if thermal energies are ignored,

$$E = E_0 - \varepsilon = E_A + E_B. \tag{4-32}$$

In the ionizing event the two outgoing electrons are undistinguishable, usually however, one of the electrons is referred to as the "*scattered*" electron and the other is the "*ejected*" electron. For the purposes of this discussion the scattered electron will be assigned an energy E_A which is considerably greater than the ejected electron energy E_B.

The apparatus which is used in these experiments is similar to that which has been described previously in the symmetric collisions. Two electron spectrometers view the interaction region which is defined by the intersection of an electron beam with the neutral target beam. The work which has been reported by Ehrhardt and collaborators used 127° electrostatic analyzers whereas an experiment carried out by Beaty *et al.* (1977) used a time of flight analyzer for the "*ejected*" electron and a hemispherical analyzer for the scattered electron. The energy resolution in the incident electron beam reported by Ehrhardt *et al.* (1972b) was between 100 and 200 meV which was derived by energy selection in the incident beam. This feature reflects the significantly larger cross sections here than in the symmetric case. It is usual in these experiments to fix spectrometer A at an angle, θ_A, close to the incident beam direction i.e. $< \sim 10°$. The small angle limit being set by the angle at which interference with the electron beam is experienced. The energies of the "*ejected*" electrons are then set at values between 5 and 10 eV which then fixes the energy at which the scattered electrons are analysed in accordance with equation (4-32).

Fig. 4-8 shows a polar plot of a typical set of data from all such experiments. Two peaks in the coincidence count rate are observed; one at forward angles which is called the *binary* peak and another at backwards angles − the *recoil* peak. No absolute cross sections have yet been measured in this configuration but various features of the distribution are quite sensitive to the approximations used in the theory. These include the ratio of the intensity of the binary peak to the recoil peak; and the angles at which the binary and recoil peaks occur.

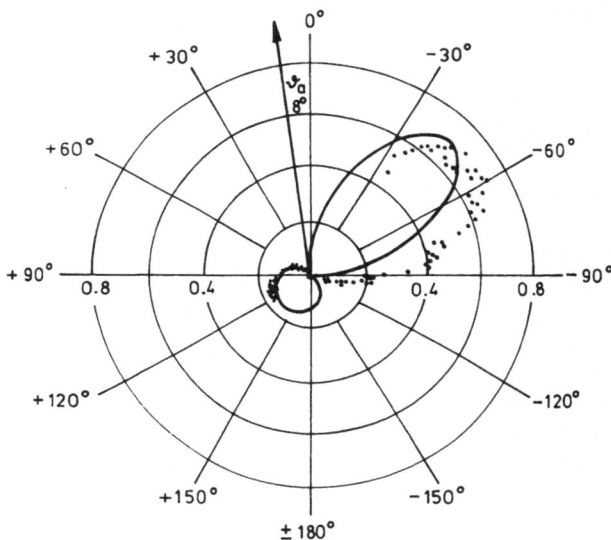

Fig. 4-8. Examples of a polar plot of typical triple differential cross sections for the ionization of helium taken in an asymmetric collision geometry (Ehrhardt (1982)). The dots represent the measurements whilst the full line symbolizes a theoretical result. The fast electron is detected at a scattering angle of $\theta_A = 8°$

Ehrhardt *et al.* (1972c) have given a qualitative description of the origin of the binary peak which is based on classical arguments and the first Born approximation. They define the momentum transfer $\underline{K}$ to the atomic electron as

$$\underline{K} = \underline{k}_0 - \underline{k}_A \tag{4-33}$$

which is fixed for each experiment. The momentum of the ejected electron, k_B, can be determined if the momentum of the electron in the target is considered. Such considerations lead to the prediction that the binary encounter peak should be symmetric about the K direction. The actual shape of this peak depends sensitively on the momentum distribution of the target electron. This feature has been demonstrated by Ehrhardt et al. (1980) in experiments on neon. For the ejection of a p electron from the $2p$ shell in neon a node is observed in the binary peak whereas no node is observed in the ejection of an electron from the $2s$ shell. Although qualitative differences in the binary peak can be observed in these two cases in neon, the derivation of momentum densities from such asymmetric collisions is not possible at present because of the difficulties associated with the description of the ionization mechanism.

The symmetry axis of the binary peak does not in general coincide with the K direction. This observation is consistent with post collisional correlation between the two outgoing electrons (Ehrhardt 1982).

Ehrhardt (1982) has recently summarized the following general points about various features of the binary peak.

(a) For high and intermediate energies i. e. $E_0 > \sim 3\varepsilon$ the binary peak is the dominant feature of the cross section. This is generally true for all angles of the "*scattered*" electron θ_A.

(b) For high energies E_B (i. e. $E_B \geq 20\,\mathrm{eV}$) the binary peak represents a very high portion of the cross section. At these energies the recoil peak practically vanishes.

(c) At low impact energies (i. e. $E_0 \leq 3\varepsilon$) the binary peak gradually vanishes.

The case of symmetric $(e, 2e)$ collisions can then be seen as a special case of conditions (a) and (b) above. Here the recoil peak has vanished, the impulse approximation gives a good description of the ionization mechanism and thus the momentum distribution of the struck electron can be obtained from the $(e, 2e)$ cross section.

Unlike the binary peak, the recoil peak bears no relationship to the momentum distribution of the struck electron. This has been demonstrated by Ehrhardt and co-workers over a wide range of energies and angles. This peak then requires for its description consideration of both the outgoing electrons and the ion. It can be considered to have its origins in the interaction of several strong effects in the collision complex. These have been listed by Ehrhardt et al. (1982 a) as electron-electron interactions between the target and the fast incoming electron; the interaction of the ejected electron with the ion core resulting in large momentum transfer to the ion; and final state interactions of the slow electron with the ion. It is not surprising therefore that various first order theoretical models have been singularly unsuccessful in finding agreement with the position and relative magnitudes of the recoil peak. These calculations include the work based on the first Born approximation by Schultz (1973), Jacobs (1974), Robb et al. (1975); the Coulomb-Born calculations of Schultz (1973), of Geltman and Hidalgo (1974) and of Geltman (1974). Various distorted wave models have also been proposed by Baluja and Taylor (1976), by Madison et al. (1977), by Bransden et al. (1978) and by Tweed (1980). Byron et al. (1980) point out that although the distorted wave and Coulomb-

Born calculations represent improvements on the first Born they fail to account properly for the electron-electron repulsion in the final state. These authors have accounted for this interaction in a second Born calculation in hydrogen (Byron *et al.* (1980)) and in helium (Byron *et al.* (1982)).

In this treatment Byron *et al.* (1982) have calculated the differential cross section according to the formula

$$\sigma = \frac{k_A k_B}{k_0} (f_{B1} + f_{B2})^2 \tag{4-34}$$

where f_{B1} is the first Born approximation to the direct scattering amplitude and f_{B2} the second Born term calculated in the closure approximation. This prescription is justified in the asymmetric geometry because the effects of exchange are small and because the third Born term can be neglected (Ehrhardt *et al.* (1982 b)). In this calculation the ground state wavefunction for the helium atom has been described in an analytical fit to the Hartree Fock wavefunction; the final state wavefunctions of the ion are of course known exactly.

A comparison of the results of this calculation with recent data on helium taken in asymmetric geometry shows a marked improvement in the first Born calculations (Ehrhardt *et al.* (1982 b)). Although the experimental results are not absolute the improvement in the calculations can be seen by the fact that the second Born term results exhibit shifts of both the binary peak and recoil peak to larger angles than the direction $\underline{K}$ predicted by the first Born approximation. In addition the magnitude of the recoil peak is significantly greater in this calculation than in the first order theory.

In spite of these dramatic improvements, the theory still over-estimates the value of the ratio of the intensity at 500 eV. Ehrhardt *et al.* (1982 b) consider that the second order theory accounts for all of the relevant dynamical collision effects in the calculations and that the remaining discrepancies could be removed by using more elaborate wavefunctions in the description of the helium ground state and also in the description of the slowly ejected electron in the field of the ion.

This assertion has recently been tested in the case of asymmetric $(e, 2e)$ collisions on atomic hydrogen (Lohmann *et al.* (1984)). Here the uncertainties concerning the adequacy of the ground state wavefunction of the target are removed because they are known exactly.

Lohmann *et al.* (1984) produced a beam of atomic hydrogen in an RF source which was based on the designs of Toennies *et al.* (1979) and of Slevin and Stirling (1981). This source had the advantage over the DC source used in the non-coplanar experiments on atomic hydrogen by the same authors in that it produced less noise during operation. The experiments were performed at an incident energy of 250 eV and the slow electron was detected at energies of 5, 10 and 14 eV. The energies of the electrons after the collision were analyzed with two cylindrical mirror electron spectrometers which were calibrated at the energy of the $L_3 M_{23} M_{23} {}^1D_2$ Auger line in argon i. e. at 203.26 eV. Calibration of the energy of the slow electrons was complicated by the fact that the electron gun would not work below 30 eV. However it was demonstrated that the analyzer constant for the slow electron analyzer was independent of energy over an energy range from 30 eV to 300 eV which is consistent with such devices (Risley (1972)). This constant therefore was used to define the energy of the slow electron.

Although the experiments were not absolute, relative normalizations were measured between the data for different energies E_A and E_B and for different values of the scattering angles θ_A. The accuracy of the relative normlizations was estimated to be better than 10%. Care was taken to ensure that the efficiency of the low energy electron analyzer was independent of the energy. This was demonstrated by measuring double differential cross sections in helium at 250 eV and 70°, which were compared with the recommended distributions of Kim (1983). Similar double differential cross section measurements at 200 eV and energies, E_B, of 10 eV confirmed that no correction was necessary for the possibility that the slow electron analyzer did not view the full collision volume over the angular range of interest.

A polar diagram of data at an incident energy of 250 eV is shown in Fig. 4-9; here $\theta_A = 3°$ and $E_B = 5$ eV. The data have been normalized to the theory of Byron et al. (1983). It is clear that the experimental data show similar characteristics to those observed in helium by Ehrhardt and co-workers namely the presence of a binary peak and a clearly defined recoil peak. In this case both the second Born and the eikonal Born series calculations of Byron, Joachain and Piraux (1983) fit the binary peak very well but both predict a recoil peak which is significantly smaller than that which is observed. This feature is common to the data at each energy and scattering angle.

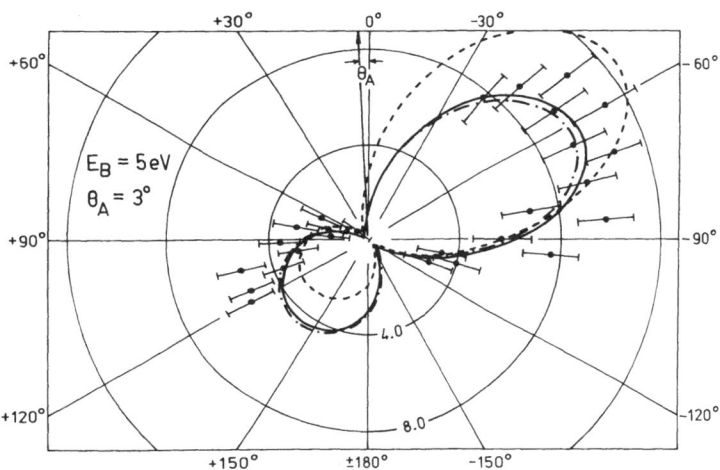

Fig. 4-9. Polar diagrams of the $(e, 2\,e)$ cross sections for atomic hydrogen measured in the coplanar asymmetric geometry. The incident electron energy was 250 eV, $\theta_A = 3°$ and $E_B = 5$ eV. The data have been normalized to the second order Born calculation of Byron et al. (1983). The calculated cross sections are: $------$ first order Born, $\underline{\hspace{1.5cm}}$ second order Born, $-\cdot-\cdot-$ eikonal Born series (Byron et al. (1983)). The cross sections are in atomic units (Lohmann et al. (1984))

The data have also been compared with various distorted wave impulse approximations Weigold et al. (1979), McCarthy et al. (1977) each of which made different assumptions concerning the distortion of the outgoing electron wave functions; the incident electron being represented by a distorted wave calculated in the full optical model potential. It is found that these simpler approximations give a surprisingly good fit to the data although the relative magnitudes are not always predicted accurately. The best description of the data being obtained from the approximation which represents the fast outgoing electron as a distorted wave calculated in the static interaction potential for the ground state of the hydrogen atom.

4.2.2.2 High Energies

At sufficiently high energies that is, those at which the first Born approximation can be used, coincidence experiments in which both electrons are detected in an ionizing event are equivalent to photo-ionization experiments. The justification for this statement rests on the work of Bethe (1930) who used the Born approximation to describe the interaction of an electron with a target atom. Under conditions of small momentum transfer, defined here as $\underline{K} = \underline{k}_0 - \underline{k}_A$, the cross section for the scattering event is proportional to the optical oscillator strength. This feature is often used in inelastic scattering experiments to provide absolute values for both total and differential cross sections (e. g. Dolder and Peart (1976), Lassettre *et al.* (1964)). The Bethe-Born approximation can be generalized to ionizing events in which at least two of the post collision products are detected. Van der Wiel, Brion and co-workers have shown that in this case binding energy spectra and partial oscillator strengths so obtained are equivalent to those which are derived from photo-electron spectroscopy. Clearly the experiments must be performed at high energies and at small momentum transfer K. Thus asymmetric kinematics are required.

The Bethe-Born approximation gives for the differential cross section in atomic units

$$\sigma = \frac{4}{E} \frac{k_A}{k_0} \frac{1}{k^2} f(k, E) \tag{4-35}$$

where E is the energy loss in the collision and is $E_0 - E_A$. The generalized oscillator strength

$$f(k, E) = E \left| \frac{1}{K} < \psi_f \left| \sum_{j=1}^{N} \exp(i \underline{K} \cdot \underline{r}_1) \right| \psi_i > \right|^2 \tag{4-36}$$

where ψ_f, ψ_i are the final and initial state wavefunctions respectively and r_j is the instantaneous position of the j-th electron. For small K the exponential may be expanded as a power series in K and the generalized oscillator strength becomes

$$f(K, E) = f(o, E) + A K^2 + B K^4 + \ldots \tag{4-37}$$

where $f(o, E)$ is the optical oscillator strength and is proportional to the square of the dipole matrix element for the collision. As $K^2 \to o$ the generalized oscillator strength tends to the optical oscillator strength $f(o, E)$ and in the limit of zero momentum transfer

$$\sigma = \frac{2}{E} \frac{k_A}{k_0} \frac{1}{k^2} \frac{mC}{he^2 \pi} \sigma_{\text{photo}}. \tag{4-38}$$

At high energies and small $\theta_A \simeq o$ the cross section is then equivalent to that produced by the absorption of a photon of energy $E = E_0 - E_A$. Thus this method can be used to simulate photoionizing events. The advantages which electron impact techniques have over conventional photoelectron spectroscopy have been explored by Brion (1975) and include the facts that the energy of the probe can be varied easily and that the resolution available is superior to the optical case.

A typical experimental configuration has been described by van der Wiel and Brion (1972/73) and has many similarities to that which has been described in the low energy case. The major differences are that the incident energy is high enough such that the first Born approximation can be invoked with confidence. The "scattered electron" is observed at very small angles ($< 1°$) so that the momentum transfer K tends to the optical limit. The target region consists of a collision chamber which is filled to a pressure of order 10^{-3} torr. The "ejected" electron is observed at either $90°$ or $54°$ to the incident beam direction and the energy analyzed with a $127°$ analyzer. The measuring procedure consists of setting the energy loss of the incident electrons at a fixed value as detected in spectrometer A — this defines the pseudo photon energy and then recording the true coincidence count rate as a function of the ejected electron energy. The data were taken with a constant field in the $127°$ analyzer; the energy was varied by accelerating or retarding the ejected electrons.

This procedure yields energy loss spectra for a given electron energy loss which can be converted into partial oscillator strengths for various molecular ionic states. Relative populations of ionic states derived in this way can be compared with those derived from conventional photoelectron spectroscopy data. In general the agreement between the data from both techniques is very good.

Lahmam-Bennani et al. (1983) have carried out asymmetric $(e, 2e)$ experiments on argon at incident energies of around 8.1 keV. These are similar in concept to the Ehrhardt type experiments at low energies except in this case the forward scattered electron has an energy of 8 keV and the slow electron an energy of 100 eV. They observe that the cross sections are symmetric about the momentum transfer direction $\hat{K}$ which is consistent with the predictions of the first Born approximation. The recoil peak is quite evident at small scattering angles θ_A but for $\theta_A \geq 4°$, this peak essentially disappears so that the collision can be considered to be dominated by binary processes. Absolute values have been assigned to the cross sections by integrating them over the solid angle of the slow electron. These double differential cross sections are then normalized to the measured Compton profile $J(q)$ (equation 4-43) for the $3p$ state in argon. The integration over solid angle relies on the assumption that the $(e, 2e)$ cross section is cylindrically symmetric about $\underline{K}$. This is a consequence of the validity of the impulse approximation. Momentum distributions have been derived from the data and for larger scattering angles agree both in shape and magnitude with those predicted by taking the Fourier transform of a Hartree-Fock wavefunction. Significant departures are observed both in magnitude and shape for smaller scattering angles where the binary model clearly breaks down.

4.3 Double Differential Cross Sections

The first measurements of ionization cross sections which were differential in both energy and angle were carried out by Mohr and Nicoll (1934). These experiments did little more than identify the general features of the cross sections. Much later Opal, Beaty and Peterson (1972) carried out an extensive series of measurements on a wide range of target gases at incident energies from 50 to 200 eV. The rationale for this study was to provide data on the rate of production of secondary electrons which would assist a quantitative evaluation of the total energy deposition by fast particles in gases. These data stimulated theoretical interest in the problem and several attempts were made to apply the Born and related approximations to the ionization of helium. Kim and Inokuti (1973), Bell and Kingston (1975) and Manson et al. (1975) calculated double differential cross sections for incident electron energies between 50 and 2000 eV. Bell and Kingston concluded that it was difficult to define a range of energies in which the Born approximation was valid. The theory was clearly inadequate below 200 eV and at 2 keV discrepancies were observed for slow electrons in the forward direction and fast electrons in the backward direction. These shortcomings in the theory clearly reflected the absence of polarization, exchange and distortion in the calculations. The application of more sophisticated theories to the problem was to a large extent, pre-empted by the growth, at about that time, of a large body of data on triple differential cross sections.

In the high energy regime, i.e. at incident energies from 10 to 60 keV Bonham, Wellenstein and co-workers (e.g. Bonham and Wellenstein 1977) have demonstrated that double differential cross sections can be used to determine electron momentum distributions of light atoms and molecules which compare favorable with those obtained from x-ray Compton profiles.

4.3.1 Low Energy Double Differential Cross Sections

Apart from the work of Beaty and collaborators on He, N_2 and O_2 mentioned above, Oda and co-workers have measured cross section on He, CH_4, H_2O and Kr (Oda (1975)), Rudd and duBois (1976 on He, Ar and N_2), Ehrhardt et al. (1971) on He and Shyn and Sharp on He (1979a), on CO_2 (1979b) and H_2 (Shyn et al. (1981)). In general it is difficult to compare cross sections measured by each group because there is very little overlap in either the energy range of the secondaries or even target gases. Nevertheless Shyn and Sharp have been able to make a limited comparison between some of their data on He and with the cross sections of Opal et al. (1972) and with those of Rudd and duBois (1976). As a prelude to measuring triple differential cross sections in He and H_2, van Wingerden et al. (1981) measured double differential cross sections for these targets at a fixed scattering angle (45°) and at fixed secondary electron energies $\frac{1}{2}(E_0 - \varepsilon)$ for primary electron energies between 200 eV and 2.8 keV.

Similar experimental techniques were used in each of the above studies, the major difference in approach was in the preparation of the target. Opal et al. (1972), Shyn and Sharp (1979) used a beam, duBois and Rudd (1976) scattered electrons from a bulk sample of gas and van Wingerden et al. (1981) used a collision chamber filled with the target gas. All experiments used an electrostatic energy analyzer to view the interaction region. The procedure involved setting the electric field in the electron spectrometer such that electrons of a particular energy were transmitted. The spectrometer was then rotated through a large range of angles (θ) about an axis normal to the electron beam direction. Thus angular distributions of secondary electrons of a particular energy could be determined. Each group payed careful attention to the inhibition of stray secondary electrons from surfaces bounding the interaction region and consideration was given in each case to the possibility of multiple scattering processes. Helmholtz coils or magnetic shielding materials or both were used to reduce the magnetic field in the interaction region to less than 20 milligauss. There were however, significant differences in the methods used by each group to assign absolute values to the cross-sections.

The double differential cross section is given by (van Wingerden *et al.* (1981))

$$\frac{d^2 \sigma}{d\Omega dE} = \frac{S_{in} e}{N I (l d \Omega)_{\text{eff}} \tau \Delta E_{1/2}}. \tag{4-39}$$

Where S_{in} is the inelastic signal at a particular angle, e the electronic charge, N the target density, I the incident electron beam current, $(l d \Omega)_{\text{eff}}$ the effective scattering length times the solid angle, τ the efficiency of analyzer and detection system and $\Delta E_{1/2}$ the half width of the analyzer.

The groups which prepared their targets in the form of beams had significant difficulties in determining N in the above equation. Consequently they normalized the data to known cross sections. For example Shyn and Sharp (1979) integrated their distributions in helium over energy and angle to yield a total ionization cross section at 100 eV which was normalized to that of Smith (1930). They estimated a 5% uncertainty in the transmission of electrons through the detection system and an overall uncertainty of 17% in the cross sections. Opal *et al.* (1972) used the differential cross section for elastic scattering of 100 eV electrons from helium at 90° to calibrate the product $N (l d \Omega)_{\text{eff}} \tau$ in their experiments. The elastic cross section which they used was that of Williams (1969) which in turn had been obtained by normalizing to the elastic cross section measured by Vriens *et al.* (1968) at 20°. Comparison of these cross sections with those of Jansen *et al.* (1976) for elastic scattering at 100 eV, shows that the calibration cross section used by Opal *et al.* was too high by about 50%. This potential for renormalization was recognized by this group however.

The cross sections of Rudd and duBois (1976) in helium were derived from measured quantities thus they did not rely on normalizing procedures. Few details of the measurements were given but they estimate an overall uncertainty of 20% in the cross sections for secondary energies above 10 eV.

Shyn and Sharp (1979) have compared their cross sections with those of Opal *et al.* (1972) and with Rudd and duBois (1976) at an incident electron energy of 200 eV. They find that the Opal *et al.* cross sections for secondary electron energies of 87 eV are too low at forward and backward angles by about a factor of two. There is, however, reasonable agreement between each set of data at intermediate angles at this energy. At lower secondary energies there are serious discrepancies between the data from all groups. Kim (1983) has pointed out that such discrepancies make it difficult to apply the measured differential cross sections to specific problems such as energy degradation. Kim has therefore assembled a set of recommended cross sections in helium for scattering angles between 0 and 180°, secondary energies between 0 and 40 eV and incident electron energies between 100 and 2000 eV. The recommended set is based on experimental and theoretical cross sections for double differential and photoionization cross sections which are consistent with the integrated cross sections and expected asymptotic behaviour.

The helium data of van Wingerden et al. (1981) were taken at a fixed scattering angle of 45° and the differential cross section for elastic scattering from helium used to determine the efficiency τ in equation (4-39). The factor $(l\,d\Omega)_{\text{eff}}$ was computed numerically and found to be consistent with that derived from the analytical expression given by Kuyatt (1968). The number density in the target was found to an accuracy of 3% by measuring the pressure with a capacitance manometer. These cross sections had an overall uncertainty of between 6 and 8%, more than a factor of two less than the error limits in each of the other experiments. Thus a case could be made for using these cross sections to normalize future double differential cross section experiments in helium. By interpolating these cross sections at $E_0 - \varepsilon = 275.5$ eV one can compare with the helium cross section of Shyn and Sharp at an incident energy of 300 eV and $E = 137.7$ eV, $\theta = 48°$. Such a comparison shows that the Shyn and Sharp cross section is higher than that of van Wingerden et al. by about 18%. This is still within the combined errors in the two experiments. The cross section of Opal et al. at $E_0 = 300$ eV and a secondary energy of 135 eV is 17% lower than the interpolated value but again this is within the error cited by Opal et al.

Van Wingerden et al. (1981) compared their experimental cross-sections with those predicted by a plane wave Born calculation with the ground state helium represented by Hartree-Fock wavefunctions. They found acceptable agreement above 1000 eV but noted that this reflected the amount of information which was lost by integrating over the solid angle of the unobserved electron; the same calculation for the triple differential cross section yielded considerably higher cross sections than the measured values.

4.3.2 Electron Energy Loss Spectroscopy

At initial electron energies in excess of 10 keV, the first Born approximation can be used to describe the ionization process and the double differential cross section can provide structural information about the target. This proposal was first made by Hughes et al. (1938) almost fifty years ago but for various reasons was discarded until it was revived by the elegant experiments of Wellenstein and Bonham (1973). The technique has been reviewed by Bonham and Wellenstein (1977).

In the Born approximation the relationship between the double differential cross section and the continuum generalized oscillator strength $f(K, E)$ is, atomic units,

$$\frac{d^2\sigma}{dE\,d\Omega} = \frac{k_A}{k_0}\left(\frac{d\sigma}{d\Omega}\right)_R \frac{K^2}{E}\, f(K, E). \tag{4-40}$$

$f(K, E)$ is related to the structural properties of the target by equation (4-36), E is the energy loss in the collision, K is the momentum transfer between the initial momentum $\underline{k}_0$ and the final momentum $\underline{k}_A$ and $\left(\dfrac{d\sigma}{d\Omega}\right)_R$ the Rutherford cross section.

Bonham and Wellenstein (1977) show that the ratio of the differential cross sections for electron scattering to those for photon scattering is greater than 10^4 so long as the momentum transfer for electron scattering does not exceed 14 a.u. In the energy range of these experiments, this limitation on K restricts the observations to scattering angles of less than $12°$. Consideration of exchange effects in the collision also demonstrates the necessity of small angle scattering. Bonham and Wellenstein account for exchange by replacing the Rutherford cross section in equation (4-40) by the Mott cross section $\left(\dfrac{d\sigma}{d\Omega}\right)_M$ and deduce

$$\left(\frac{d\sigma}{d\Omega}\right)_M = \left(\frac{d\sigma}{d\Omega}\right)_R F_{ex}. \tag{4-41}$$

The exchange term, F_{ex}, is almost a pure function of scattering angle only and gives a correction of generally less than 10% if the momentum transfer is restricted to the range $0 < k < 14$ a.u. Relativistic corrections to equation (4-40) must also be made but these are small if the incident energy is less than 60 keV.

In regions where the Born approximation is valid, the generalized oscillator strength can be converted to an absolute scale by using the Bethe sum rule (Inokuti (1971))

$$\text{i.e.} \quad \int f(K, E)\, dE = n \tag{4-42}$$

where n is the number of target electrons and the integral is taken out over the entire energy loss spectrum. The validity of the Born approximation can be checked by observing the independence of the generalized oscillator strength for fixed K and E. The absolute generalized oscillator strengths are then related to the x-ray Compton profile, $J(q)$, by using the binary encounter approximation. That is

$$J(q) = \lim_{K \to \infty} \frac{2k^3}{E} f(K, E) \tag{4-43}$$

where in this case

$$q = \frac{E - K^2}{2K} \tag{4-44}$$

and $J(q)$ is given by

$$J(q) = 2\pi \int_{|q|}^{\infty} \rho(p)\, p\, dp \tag{4-45}$$

where $\rho(p)$ is the atomic or molecular momentum density.

For sufficiently large momentum transfer, the right hand side of equation (4-43) becomes only a factor of q. This limit is reached only for those collisions in which K^2 is large compared to the separation energy. Thus the limiting value of K increases with atomic number and restricts the technique to light atoms or molecules; that is atoms smaller than neon or molecules whose heaviest element is smaller than neon.

The experimental procedures are similar in principle to those which are used in the low energy experiments namely energy loss measurements are taken for various scattering angles. In this case, however, the incident energies are between 10 and 60 keV, energy loss is measured up to 3 keV and the range of scattering angles between 2° and 20°. There are nevertheless several important differences in detail which are unique to the high energy experiments. In particular the cross sections are extremely forward peaked thus the scattering angle must be determined to great accuracy that is typically a few seconds of arc. This in turn implies small and extremely well defined electron and target beams and magnetic fields inside the apparatus of less than 5 milligauss. The criterion which determines acceptable energy resolution in these experiments is approximately 1% of the width of the Compton profile; thus electron spectrometers with resolutions $\frac{\Delta E}{E} \sim 10^{-4}$ must be used. The major development in this field has come from the use of Mollenstadt energy analysers which have the required energy resolution and acceptance angles (Bonham and Wellenstein (1977)).

The derived momentum distributions for light targets such as He and H_2 compare favorably with those obtained from other Compton profile techniques. For example the momentum distribution in helium measured by Lee (1977) is in good agreement with both Hartree Fock and correlated ground state wavefunctions but cannot distinguish between them. Such a technique cannot observe the influence of other correlation effects which have been discussed in Section 4.2.1.5.

As the precision of the experiments has been increased to the order of 0.1%, some doubt has been cast on the validity of the Born approximation at this level of accuracy. Wellenstein and co-workers have demonstrated the presence of non impulse effects which appear as the departure of the profile maximum away from $q = 0$, profile asymmetry $J(q) \neq J(-q)$ and the appearance of secondary maximum (Barlas et al. (1978)). Modifications have been made to the Born calculations by incorporating distorted waves to describe the ejected electrons (Mendelsohn (1982)). Generally encouraging agreement is found between theory and experiment in this case (Wong et al. (1982)).

4.4 Dissociative Ionization

In the collision between an electron and a molecule which leads to dissociative ionization, it has been known for many years, Condon (1930), that the kinetic energy of the fragment ions depends on the details of the potential energy curves of the states involved in the collision. In addition Dunn (1962) observed that the cross sections depended on the symmetries of the states involved, thus at least to first order the angular distribution of the ions would provide more information on the ionization process. In the absence of a detailed solution to the scattering problem, Dunn's selection rules have provided a basis for a qualitative description of the ionization process.

4.4.1 Selection Rules

Dunn (1962) first deduced relevant symmetry principles for the case of dissociative attachment which he then generalized to dissociative excitation and ionization. For dissociative attachment he observed that the interaction potential in the scattering problem could be represented by the scalar sum of Coulomb terms. This ensured the preservation of symmetries which existed prior to the collision. If the incident electron beam was represented by the plane wave, $e^{i k_0 \cdot r}$, then the plane wave was symmetric with respect to all rotations about k_0, and with respect to reflections in planes containing k_0. Dunn then considered two limiting cases; that for which the target molecule had its internuclear axis parallel to k_0 at the time of the collision and that in which the internuclear axis was perpendicular to k_0. He then showed that, in some cases, zero transition probabilities in directions parallel or perpendicular to k_0 could be deduced depending on the symmetries of the initial and final state wavefunctions. These selection rules were presented for both homonuclear and heteronuclear molecules for the two limiting cases described above. For the case of attachment the symmetry axis was the direction of the incident electron beam k_0. As an example of these rules transitions from the Σ_g^+ state in a homonuclear molecule to the final states Σ_g^+ or Π_u or Δ_g would yield dissociation products normal to the symmetry axis, whereas dissociation products parallel to the symmetry axis would result only from the transitions $\Sigma_g^+ \to \Sigma_g^+$ and $\Sigma_g^+ \to \Sigma_u^+$.

The situation is more complicated in the case of dissociative excitation because here the scattered electron is a spherical outgoing wave and no definite symmetries are seen. Dunn invokes the Born approximation and notes that in this limit the scattering theory yields a term $e^{i K \cdot r}$ where $K = k_0 - k_n$ is the momentum transfer. The symmetry axis for dissociative excitation then becomes the momentum transfer direction $\hat{K}$.

Dunn recognized the additional complications associated with dissociative ionization. Here the presence of the "ejected" electron made the symmetries even less well defined than was the case for excitation. By assuming, however, that the nucleus played no part in the collision, momentum conservation showed that the electron would be emitted predominantly along the momentum transfer direction.

The assumption of lack of involvement of the core is consistent with the initial assumption of plane waves and the Born approximation. The asymmetric $(e, 2e)$ studies of Ehrhardt and co-workers on helium reinforces the view that the ejected electron is emitted along the momentum transfer direction at energies where the first Born approximation approaches validity. However as the electron energy is reduced to near threshold values, significant departures of the binary encounter peak from the momentum transfer direction are observed (Ehrhardt et al. (1980)). Indeed von Brunt and Kieffer (1970) conclude from their molecular recoil measurements that so much of the incident electrons momentum is taken up by the heavy particles that the symmetry axis is no longer along K. At these energies then it is not clear how the selection rules can be used with confidence.

The situation at threshold is improved somewhat. Here both outgoing electrons have zero energy and can be represented by s waves. Thus the wavefunctions describing each electron are spherically symmetrical so the selection rules should hold exactly. Various measurements of dissociative ionization in H_2 have been carried out to test the validity of these assumptions. These experiments will be described in detail below. The rules show that the transition probability is zero for transitions from Σ_g^+ states to Σ_u^+ for dissociation products perpendicular to the axis of symmetry. At threshold the momentum transfer $\underline{K}$ is parallel to the incident beam direction. Thus if the dominant source of protons in the dissociative ionization of H_2 is from the reaction

$$e + H_s(\Sigma_g^+) \rightarrow H_2^+ (\Sigma_u^+) + 2e \rightarrow H + H^+ + 2e \qquad (4\text{-}46)$$

then the differential cross section for proton formation at a scattering angle of $90°$ should be zero at threshold. Van Brunt and Kieffer (1970) observed significant proton production at threshold at $\theta = 90°$, which on balance they attributed to the effects of repulsive high lying Rydberg states of H_2. These states had initially been proposed by Kieffer and Dunn (1967) to explain their observed kinetic energy distributions of H^+ and appearance potential measurements.

Subsequently Burrows et al. (1980) have fitted their measured kinetic energy distributions, for incident electron energies between 30 eV and 50 eV, with calculated distributions. The symmetry of the states of H_2 and H_2^+ which were used in these calculations was chosen so that Dunn's selection rules were satisfied. That is they used states which would give non zero proton yields at observation angles of $90°$ where the experiments were performed. The states which they considered were the $Q_1 \Sigma_g^+$ and $Q_1 \Pi_u$ autoionizing states of H_2 and the $2p\pi_u\,{}^2\Pi_u$ state of H_2^+. They found good agreement between calculations and measurements near threshold and reasonable agreement even at 50 eV. They do, however, sound a note of caution about the uniqueness of the calculations.

4.4.2 Dissociative Ionization in H_2

The study of the production of protons by the ionization of the hydrogen molecule has dominated the field for many years. This reflects the relative simplicity of the molecule but also reflects the considerable disagreement between the measurements of various groups concerning the shapes of the proton spectra and the angular distributions of the fragment ions. Indeed Dunn and Kieffer (1963) cited Bleakney (1930) who introduced his paper describing the first observation of protons in the ionization of H_2 with "The ions produced by electron impact in hydrogen has been studied by the method of positive ray analysis so many times by so many investigators that it might, at first sight, seem useless to try to make much more progress in this direction."

This section will concentrate on the hydrogen molecule because it demonstrates most of the general features of the dissociative process. Nevertheless a considerable amount of effort has been devoted to other molecules most of which amplifies the problems which are apparent in the H_2 study.

Dunn and Kieffer (1963) have reviewed the field prior to 1963 and have described the origins of the general feature of the proton spectrum. Electrons which collide with H_2 in the ground $^1\Sigma_g^+$ state may ionize the H_2, leaving the H_2^+ in the $^2\Sigma_g^+$ bound state, the $^2\Sigma_u^-$ dissociative state or higher dissociative states. Hence two groups of protons will in general be seen, one group with thermal energy which arises from excitations above the dissociation limit of the $^2\Sigma_g^+$ state and a group of higher energy protons corresponding to excitation to the $^2\Sigma_u^+$ state of H_2^+. Measurement of the proton energy spectrum then in principle provides information on the role of the states participating in the ionization process or on the Franck-Condon principle. In addition the angular distribution of the fragment ions, in conjunction with Dunn's selection rules, has been used to infer possible reaction mechanisms.

Thus the broad features of this description shows that the protons will be emitted in an energy range from thermal energies up to 14 eV. The experiments are difficult to perform because the cross sections are small and in addition most groups have incorporated some form of mass filter as well as an energy analyser. The low energy ions can be seriously affected by surface charges and contact potentials whilst the use of ion optics to increase the effective solid angle can seriously discriminate against the high energy ions. The discrepancies in the observed energy spectra between the many experiments no doubt reflect these difficulties and to put them into perspective it is necessary to examine the methods and techniques used in some detail.

4.4.3 Methods and Techniques

4.4.3.1 Energy Distributions

The general description given above shows that the ions involved in these experiments on H_2 are thermal H_2^+, thermal H^+ and more energetic protons. Consequently most groups have incorporated some form of mass filter as well as an energy analyser in the apparatus to measure the kinetic energy of the protons emitted in the dissociative ionization process. The types of mass filters which have been used include magnetic sector, quadrupole and time of flight devices.

Dunn and Kieffer (1963), Kieffer and Dunn (1967), van Brunt and Kieffer (1970) and van Brunt (1977) used the first order focussing condition of a 60° sector magnet which related the energy E of an ion mass M by

$$E = \frac{B^2}{M\alpha^2} \qquad\qquad (4\text{-}47)$$

where B was the magnetic field strength. The geometric factor α was determined by accelerating H_2^+ ions through known electric fields and measuring the field strength required to focus the ions. Care was taken to measure contact potentials and the energy scale was estimated to be accurate to within 2% against systematic drifts. The spectrometer was operated in a mode which discriminated against thermal energy ions by allowing the ions to drift from the interaction region without the use of accelerating potentials or ion lenses. It was demonstrated that lenses could increase the proton signal but with a consequent distortion of the measured energy distribution. Kieffer and Dunn (1967) showed that the spectra were reproducable above 1.0 eV, however discrimination against low energy ions by surface charges subsequently forced van Brunt (1977) to limit his spectra to energies greater than 3.0 eV.

The energy resolution $\left(\dfrac{\Delta E}{E}\right)$ in each of these experiments was 10%. Typically the earth's magnetic field was reduced to 35 milligauss by the use of shielding material. The magnet was designed so that its field dropped off rapidly at short distances beyond the exit and entrance apertures. They found that the field inside the scattering chamber changed by 20 milligauss for magnet field changes from 0 to 1 kilogauss.

The original experiments of Dunn and Kieffer (1963) found significant discrepancy between the observed spectra and those predicted by the Franck-Condon principle. Improvements in the calculations and experiments by Kieffer and Dunn (1967) showed that the improved data could not be explained by the assumption of the sole participation of the $^2\Sigma_u$ state. It was observed that the peak of the energy distribution of the ions was shifted to higher energies than the calculations predicted, the experimental peak was much broader than the calculated peak and there was a shoulder on the lower energy side. These general features were confirmed by van Brunt and Kieffer (1970). van Brunt (1977) demonstrated that not only the shape of the shoulder depended on the angle at which the observations were made but also the peak of the distribution.

Köllmann (1975) also used a 60° sector magnetic mass spectrometer to analyze the mass of the ions but preceded it with a 90° sector electrostatic energy analyzer to measure the kinetic energy of the protons. In this case the ions were extracted from the interaction region by applying an electric field between two spherical grids and then accelerated to the fixed transmission energy of the energy analyzer. The voltage between the grids was scanned in the region of interest and the kinetic energy of the protons thus determined. Köllmann gives details of the ray tracing technique which he used to demonstrate that the overall transmission of the ion lens was constant to within 20% over the energy range between 2 and 12 eV. His data confirmed those of van Brunt and Kieffer (1970) for an incident electron energy of 150 eV.

Crowe and McConkey (1973a) analyzed the energy of the ions with a cylindrical mirror energy analyzer and analyzed the mass with a quadrupole mass filter. The ion spectrometer was equipped with an input lens system which focussed ions from the interaction region on to the entrance of the spectrometer and an output lens system which focussed the analyzed ions on to the entrance aperture of the mass filter. The properties of the spectrometer depended critically on the fact that the ions of different energies were all focussed at the same point on the axis of the mirror, all with the same angular spread as no collimating slits were used. Crowe and McConkey acknowledged the difficulty of this problem in ion optics but gave no detailed information as to its solution. Cylindrical lenses were used throughout. In addition the lenses on the output relied on calculated parameters to ensure that the launch angle into the mass filter remained constant with ion energy. The procedure which was employed in these experiments was that the ion lenses were adjusted for optimum transmission at one energy and the energy measured by varying the field in the cylindrical mirror analyzer. The transmission of the system was not measured and the energy resolution was between 2% and 3%. Considerable care was taken to remove stray electric fields, surface charges and contact potentials. The earth's magnetic field was reduced to less than 5 milligauss in the chamber by shielding with magnetic shielding materials.

Crowe and McConkey found pronounced peaks in the proton spectrum at energies of 8, 4 and 2 eV for impact energies between 50 and 300 eV and observation angles of

$23°$. These peaks had not been seen in the experiments of Dunn and Kieffer (1963), Kieffer and Dunn (1967) and van Brunt and Kieffer (1970) although as previously noted a shoulder was observed at about 4 eV in some of the earlier work. The difference in the energy resolution in the earlier experiments was not sufficient to explain the discrepancy. Köllmann (1975, 1978) was unable to reproduce the structure observed by Crowe and McConkey even though he made a thorough search for possible sources of systematic errors in his apparatus.

The prominent peaks in the proton spectrum stimulated theoretical interest, for example Bottcher and Docken (1974), Hazi (1974, 1975), Bottcher (1974), as they appeared to indicate the existence of dissociation channels from resonance intermediate states which would autoionize into the dissociation continuum of the ground state of H_2^+.

Stockdale et al. (1975) and Burrows et al. (1980) have used time of flight techniques to measure energy distributions. The former group were forced to work with D^+ from D_2 because of various technical problems and found a 10% variation in the transmission of the quadrupole mass filter with different pole bias conditions. This coupled with their acknowledged lack of reliability at lower energies and the smoothing technique used to reduce statistical errors makes it difficult to compare their data with that of van Brunt and Kieffer (1970) on D_2. They did however see a shoulder in this distribution at ~ 4 eV, and reported similar distributions to those observed by van Brunt and Kieffer (1970).

The procedure employed by Burrows et al. (1980) was based on time of flight to measure the kinetic energy and a novel retarding device to discriminate between the masses. This mass filter was not essential in their experiments on H_2 and D_2 and data taken with the filter in and out were indistinguishable. They found acceptable agreement with the data of van Brunt (1977) even though the peaks of the distributions differed by about 1 eV. This difference was partially attributed to uncertainties in their energy scale. No structure as reported by Crowe and McConkey (1973 a) was found at incident electron energies E_0 of 50 or 100 eV. As has been outlined above Burrows et al. (1980) were able to account for the shapes of the proton energy distributions for energies E_0 from threshold to 50 eV in calculations using dissociative channels predicted by Dunn's selection rules. These states in turn had been suggested by Bottcher and Docken (1974), Hazi (1975) and by Bottcher (1974) in attempts to explain the structure observed by Crowe and McConkey.

Johnson and Franklin (1980) used a retarding potential device for energy analysis and a quadrupole mass filter. In addition they used a series of lenses to extract the ions from the collision region and accelerated them to the mass filter. No corrections were made for transmission effects and although they observed considerable structure in the proton spectrum for $E_0 = 50$ eV it does not agree with any of the energy distributions from any other group. It is clear from such comparisons that their apparatus discriminated seriously against the higher energy ions in the spectrum.

Landau et al. (1981) made use of the fact that the H_2^+ ions were all formed with thermal energies and dispensed with the mass filter. Thus the apparatus could not be used to study low energy ions and proton distributions are only given for energies in excess of 0.8 eV. The transmission of the $127°$ electrostatic energy analyzer which was used to measure proton energies was calibrated by studying the threshold behaviour of electrons in the ionization of helium. The transmission function, which was measured by Pichou et al. (1978), then enabled them to calibrate the spectrometer over the whole energy range. A dramatic illustration of the way in which the energy distributions can be affected by variations in transmission in the analyzer can be seen in Fig. 7 of Landau et al. (1981). The raw data show a prominent peak at 4 eV which is substantially removed when corrections are made for the high energy transmission function. At lower energies, $E_0 < 50$ eV, Landau et al. compare their data with that of Crowe and McConkey (1973 a) and find that the structure observed by the latter group was only weakly discernible, although the data were taken at vastly different scattering angles.

It is clear that the use of measured proton spectra to derive detailed information about the ionization mechanism is limited by the uncertainties surrounding the shapes. There is generally good agreement in the shapes reported by Kieffer and Dunn (1967), van Brunt and Kieffer (1970), Köllmann (1975) and by van Brunt (1977) but it should be remembered that the small differences between each of these profiles make detailed fits of questionable reliability. There is now a growing weight of evidence which strongly suggests that the structure observed by Crowe and McConkey (1973 a) was in large part a reflection of an artifice in their apparatus. Nevertheless this work did stimulate theoretical interest in the role of autoionizing transitions in the ionization process so that there now appears to be consensus that at least the $Q_1\,{}^1\Sigma_g^+$ and $Q\,{}^1\Pi_u$ autoionizing states of H_2 play a significant part in the interaction. Additional information in this regard has been provided by the electron-ion coincidence experiments of van der Wiel and co-workers (Backx et al. (1976)). In this case a beam of 8 keV electrons ionized the hydrogen molecules, electrons which were scattered at very forward angles ($0-2 \times 10^{-2}$ rad.) were energy analyzed and measured in coincidence with the ionic fragments. These coincidence measurements yielded oscillator strengths as a function of the energy loss of the forward scattered electron. The protons arising from the collisions were extracted with an electrostatic lens system which was designed and demonstrated to have 100% transmission efficiency for ions in the energy range from 0 to 20 eV (Backx et al. (1975)) Inspection of the oscillator strengths showed that there was a contribution from autoionizing states, presumably Rydberg states converging on the $2p\sigma_u$ state of H_2^+. In addition, the presence of a shoulder in the oscillator strength at energy losses around 37 eV indicated the influence of a partially non autoionizing channel which was designated as a $^1\Pi_u$ state.

4.4.3.2 Angular Distributions

Sasaki and Nakao (1941) observed that the angular distribution of protons emitted in the dissociative ionization of molecular hydrogen was strongly peaked in the direction of the incident electron beam. Dunn (1962) demonstrated that this phenomenon was consistent with that expected from the application of symmetry rules to the collision process. Subsequently Dunn and Kieffer (1963) carried out a comprehensive series of experiments which measured the angular distribution of protons of a particular energy for collisions in the incident energy range

$50 \, \mathrm{eV} \leq E_0 \leq 1500 \, \mathrm{eV}$. The results of these experiments showed qualitative agreement with those predicted from a dipole Born model. Namely at high incident energies the distributions became more isotropic whilst near threshold a $\cos^2 \theta$ distribution was to be expected, where θ was observing angle.

van Brunt and Kieffer (1970) repeated the measurements for an incident electron energy of 33 eV where Dunn's selection rules were expected to hold exactly. They found that the angular distribution consisted of two parts, an anisotropic component and an isotropic component which could not be fitted with a function of the form $a + b \cos^2 \theta$. Zare (1967) had used the Born approximation to predict angular distributions in the dissociative ionization of H_2^+ and, encouraged by the similarity in the shapes of the cross sections for the molecular case, van Brunt and Kieffer applied Zare's relationship to their data. This relationship depended on the parameter $K \cdot r_0$ where K was the momentum transfer and r_0 the internuclear separation. They used a value of K which was derived from molecular recoil experiments. They found excellent agreement between the measurements and Zare's relationship and concluded that the theoretical basis of the anisotropic component in the production of protons of energy 7.6 eV was very similar to that which described the dissociative ionization of H_2^+, namely a single repulsive potential curve; the $(2p\sigma_u) \, ^2\Sigma_u^+$ state of H_2^+. van Brunt and Kieffer offered no quantitative explanation for the isotropic component but found that the angular distributions were insensitive to the energy down to $E_0 = 29 \, \mathrm{eV}$. They attributed this behaviour to the presence of the autoionizing states which had been proposed earlier by Kieffer and Dunn (1967).

Köllmann (1978) extended these measurements to $E_0 = 25 \, \mathrm{eV}$ and observed protons with energies between 1 eV and 4 eV. He found significant forward backward asymmetry in the distribution which has been observed at higher incident energies for protons of energy 8.6 eV by Dunn and Kieffer (1963). Köllmann found that the asymmetry was more pronounced at $E_0 = 27 \, \mathrm{eV}$ than at 25 eV and used the theory of van Brunt (1974) to conclude that this asymmetry was due to the excitation of the resonant autoionizing state of H_2, $Q_1 \, \Sigma_g^+$. van Brunt (1974) had considered the influence of higher order multipole corrections to the simple dipole approximation and found that these could lead to significant asymmetries which in turn depended on the momentum transfer in the collision and on the symmetries of the states involved in the collision. The influence of this state has subsequently been recognized in the energy distributions data of Burrows *et al.* (1980).

van Brunt (1977) has recorded proton energy spectra for energies greater than 3 eV at $E_0 = 75 \, \mathrm{eV}$ for scattering angles of θ of $23°$, $42°$ and $90°$. The shape of these spectra changes as a function of θ as does the position of the peak. van Brunt concludes that this demonstrates the influence of other states in the dissociation mechanism which is consistent with the qualitative predictions of Dunn's selection rules. He also emphasizes that care must be exercised when comparing data taken at different angles, at different incident energies and for different proton energies. Consequently it is not possible to compare the angular distributions of Crowe and McConkey (1973 b) with any other work because they did not analyze the proton energies.

Angular distributions alone cannot uniquely specify the states involved in the dissociative ionization of H_2. Dunn's selection rules have had some success in enabling the role of some states to be designated and, taken with the energy distribution data, they can be used to discuss the problem qualitatively. This may be somewhat surprising given the assumptions which are involved in the application of these rules to the ionization process even at threshold. Nevertheless they have formed the basis of discussion of theoretical aspects of the problem for many years and will continue to do so until a more complete theoretical description is developed.

4.5 Inner Shell Processes

The gross features in the absorption spectra of molecules near inner shell edges have been classified by Van der Wiel (1980). These are

(i) a single intense peak, which does not show normal Rydberg structure, can be seen at energies below the edge;
(ii) features which stand out prominently above a smooth continuum above the threshold;
(iii) a weak fine structure which extends over hundreds of electron volts into the continuum.

The excitation of these inner shell states involves the escape of an inner electron through the atomic and molecular fields, thus their study can provide information about these fields. Two complementary theoretical approaches have been used to describe the general features of the absorption spectra. One is a multiple scattering approach (Dehmer and Dill (1975)) and the other a molecular approach (Gianturco et al. (1972), Rescigno and Langhoff (1977)). In the multiple scattering treatment, it is assumed that the atomic electron escapes with low initial angular momentum from a highly localized source through the anisotropic molecular field. Interactions with this field raise the angular momentum of the ejected electron to higher states which then add a centrifugal barrier to the electrostatic potential. The penetration of barriers thus formed is responsible for the strong resonances both above and below the threshold. At higher angular momentum values no barrier is formed and the result is a much more extensive and weak oscillatory structure i.e. X-ray absorption fine structure (XAFS).

Inner shell processes can be studied either by electron impact or by synchrotron radiation (e.g. Nakamura et al. (1969)). The electron impact technique offers considerable advantages over photo absorption. As has been pointed out elsewhere in this chapter, the generalized oscillator strength for electron scattering from valence shells at high energies tends to the optical oscillator strength as the momentum transfer tends to zero. The application of this principle to inner shell excitation in atoms has been demonstrated by King et al. (1977 a) and in molecules by for example Van der Wiel et al. (1970). The most striking advantage of electron energy loss spectroscopy lies in the resolution which the technique offers. For example, Tronc et al. (1976) have obtained an energy resolution, which is independent of the incident energy, of 70 meV at an excitation energy of 290 eV. This is equivalent to a photon source with a band pass of 1×10^{-3} at 4.3 nm.

Apart from the experimental advantages, the excitation of inner shells with electrons offers the possibility of studying states which are forbidden by electric dipole selection rules. For example Shaw *et al.* (1982) have observed the parity forbidden transition to the $(2\,p_{3/2})^{-1}\,4\,p$ state in Ar.

The major problem with the technique stems from the fact that the cross section for inner shell processes contains a factor of $(\Delta E)^{-3}$. This means that long (in excess of 24 hours) data collecting periods are required and effectively limits the incident electron energy to a maximum of 2 keV (van der Wiel (1980)).

Two techniques have been used to study inner shell processes by electron impact. These are electron energy loss spectroscopy and electron ion coincidence techniques.

4.5.1 *Electron Energy Loss Spectroscopy*

In this technique an electron beam is passed through a cell containing the target gas and the electrons which are scattered at very forward angles are observed with an electron spectrometer as a function of energy loss (Wight *et al.* (1972/73)). Experiments with much higher energy resolution have been carried out by the group at the University of Manchester (King *et al.* (1977 a)). Here an electron monochromator is used to produce an electron beam with an energy spread of 50 meV at an energy of 1.5 keV. Electrons which are inelastically scattered from the target through a small angular range ($\sim 0.3°$) are retarded to an energy of ~ 6 eV and analyzed with a 180° hemispherical analyzer. The energy loss spectrum is obtained by providing a ramp between the target region and the local earth of the electron energy analyzer. Spectra are calibrated against well resolved valence shell states which are known accurately from spectroscopic data.

The results obtained from such experiments have mainly concentrated on the spectroscopy of inner shell excitation. For example in the study of N_2 and CO by Wight *et al.* (1972/73) the emphasis was on the designation of the structure of the states involved in the inner shell excitation rather than on the measurement of the cross sections. This study however did reinforce the concept of an equivalent core model which had been used previously in photo absorption studies on N_2 by Nakamura *et al.* (1969). Wight *et al.* (1972/73) showed that the K shell energy loss spectrum of N_2 was indeed very similar to the energy spectrum of NO as was the carbon K shell energy loss spectrum of CO. This model has led Wight and Brion (1974) to predict the energy levels and ionization potentials of the unstable radicals NH_4, H_3O and H_2F from the K shell energy loss spectra of CH_4, CH_3 and H_2O respectively.

Shaw *et al.* (1982) have observed vibrational structure in the $(1\,s)^{-1}\,(\pi\,2\,p)^{1}\pi$ state of N_2 at 401 eV at high incident energies (1.3 keV). This vibrational structure was first seen by King *et al.* (1977 b), however Shaw *et al.* observed that as they reduced the incident energy, additional structure appeared at lower energy loss values. This structure was the dominant feature at an incident energy of 480 eV. They attributed this structure to the $(1\,s)^{-1}\,\pi\,2\,p^{3}\pi$ state of N_2, the excitation of which is forbidden by dipole selection rules.

Core excited negative ion resonances have been observed by detecting the number of positive ions released from N_2 and from CO as a function of the incident energy by King *et al.* (1977 c). These ions are formed by the capture of the incident electron in the $\pi 2p$ orbital in a way which is analogous to the formation of negative ion resonances in low energy electron molecule scattering (King *et al.* (1980)). The shapes of the resonances $[N_2^k]^-$ and $[C^kO]^-$ are observed to be similar which is consistent with the predictions of the equivalent core model. Such similarities have led King *et al.* (1980) to suggest that the technique could be used to study resonances in equivalent core molecules which are difficult to study by other means. For example the structure of CF^- should correspond to that observed in oxygen K shell excited CO^-.

Accurate oscillator strengths for the absorption spectra of N_2 and CO have been measured by Kay *et al.* (1977a) from zero degree scattering of 8 keV electrons. In N_2 they find agreement between their integrated oscillator strength for the intense peak and that predicted by Dehmer and Dill (1976) using the multiple scattering model. The moment theory MO calculation of Rescigno and Langhoff (1977) is in good agreement with the experiments in the continuum.

X-ray absorption fine structure has been seen in electron impact of the chloromethanes $CHCl_3$ and CCl_4 by Hitchcock and Brion (1978). The oscillations were observed in the carbon and chlorine continua.

4.5.2 Electron Ion Coincidence Experiments

The technique used in these experiments is the same as has been outlined in Section 4.4.3 concerning the coincidence experiments on dissociative ionization in H_2 (Backx *et al.* (1976)). The success of the technique rests on the ability of the apparatus to collect even the most energetic ions with 100% efficiency. Two procedures can be used. First the electron energy is set at a fixed energy loss and all the ions are measured in coincidence with electrons of this energy. This yields an ion fraction spectrum from which ion oscillator strengths can be derived. The second procedure involves the measurement of coincidences between a particular ion and electron as a function of the energy loss. Here a relative ion oscillator strength for a particular ion is obtained.

Van der Wiel (1980) has described the three main areas in which these techniques have been used.

Inner-shell ionized molecules normally produce a doubly charged molecular ion in the final state which will decay into two ionic fragments. This double dissociative ionization has been shown to be the dominant process for K ionized CO (Kay *et al.* (1977b)) and for $5\,2p$ ionized SF_6 (Hitchcock *et al.* (1978) where in excess of 85% of all events lead to double dissociative ionization.

Another use of the technique has been described by Kay *et al.* (1977b) who searched for vibrational structure in the continuum resonances by measuring the individual fragment oscillator strengths over the energy range of these resonances. No evidence for this phenomenon was found either in carbon K ionized CO or in SF_6 (Hitchcock *et al.* (1978)). It was observed that the shapes of the individual ion oscillator strengths in both targets closely followed the shapes of the total absorption oscillator strength.

The increased sensitivity of ion studies over photo-electron spectroscopy in the observation of post collision interaction effects has been demonstrated by Kay *et al.*

(1977 b). They note that the oscillator strengths for CO^{2+} and $(C^+ + O^+)$ production in CO are suppressed near the carbon K threshold which is evidence for the effects of post collision interaction for the ionization event.

References

Amaldi, V., Egidi, A., Marconero, R., Pizella, G. (1969): Rev. Sci. Inst. *40*, 1001 – 1004.

Backx, C., Wight, G. R., Tol, R. R., Van der Wiel, M. J. (1975): J. Phys. *B 8*, 3007 – 3019.

Backx, C., Wight, G. R., Van der Wiel, M. J. (1976): J. Phys. *B 9*, 315 – 331.

Baker, G. A., jr., McCarthy, I. E., Porter, C. E. (1960): Phys. Rev. *120*, 254 – 264.

Baluja, K. L., Taylor, H. S. (1976): J. Phys. *B 9*, 829 – 835.

Barlas, A. D., Rueckner, W. H. E., Wellenstein, H. F. (1978): J. Phys. *B 11*, 3381 – 3400.

Beaty, E. C., Hesselbacher, K. H., Hong, S. P., Moore, J. H. (1977): J. Phys. *B 10*, 611 – 620.

Bell, K. L., Kingston, A. E. (1975): J. Phys. *B 8*, 2666 – 2678.

Bleakney, W. (1930): Phys. Rev. *35*, 1180 – 1186.

Bonham, R. A., Wellenstein, H. F. (1977): Compton Scattering (Williams, B., ed.). New York: McGraw-Hill.

Bottcher, C. (1974): J. Phys. *B 7*, L 352 – 357.

Bottcher, C., Docken, K. (1974): J. Phys. *B 7*, L 5 – 8.

Bransden, B. H., Smith, J. J., Winters, K. H. (1978): J. Phys. *B 11*, 3095 – 3114.

Brion, C. E. (1975): Rad. Res. *64*, 37 – 52.

Brion, C. E., Hood, S. T., Suzuki, I. H., Weigold, E., Williams, G. R. J. (1980): J. Elect. Spect. *21*, 71 – 90.

Brion, C. E., McCarthy, I. E., Suzuki, I. H., Weigold, E., Williams, G. R. J., Bedford, K. L., Kunz, A. B., Weidman, R. (1982): J. Elect. Spect. *27*, 83 – 107.

Burnett, T., Rountree, S. P., Doolen, G. (1976): Phys. Rev. *A 13*, 626 – 631.

Burrows, M. D., McIntyre, L. C., jr., Ryan, S. R., Lamb, W. E., jr. (1980): Phys. Rev. *A 21*, 1841 – 1846.

Byron, F. W., jr., Joachain, C. J., Piraux, B. (1980): J. Phys. *B 13*, L 673 – 676.

Byron, F. W., jr., Joachain, C. J., Piraux, B. (1982): J. Phys. *B 15*, L 293 – 296.

Byron, F. W., jr., Joachain, C. J., Piraux, B. (1983): Phys. Lett. *99 A*, 427 – 431.

Camilloni, R., Giardini-Guidoni, A., Tirribelli, R., Stefani, G. (1972): Phys. Rev. Lett. *29*, 618 – 621.

Camilloni, R., Giardini-Guidoni, A., McCarthy, I. E., Stefani, G. (1978): Phys. Rev. *A 17*, 1634 – 1641.

Clementi, E., Roetti, C. (1974): At. Data and Nuclear Data Tbl. *14*, 177 – 478.

Condon, E. U. (1930): Phys. Rev. *35*, 658.

Cook, J. P. D., Brion, C. E., Hamnett, A. (1980): Chem. Phys. *45*, 1 – 13.

Cook, J. P. D., McCarthy, I. E., Stelbovics, A. T., Weigold, E. (1984): J. Phys. *B 17*, 2339 – 2352.

Coplan, M. A., Migdall, A. L., Moore, J. H., Tossell, J. A. (1978): J. Am. Chem. Soc. *100*, 5008 – 5011.

Coplan, M. A., Tossell, J. A., Moore, J. H. (1982): A.I.P. Conf. Proc. *86*, 82 – 89.

Crowe, A., McConkey, J. W. (1973 a): Phys. Rev. Lett. *31*, 192 – 196.

Crowe, A., McConkey, J. W. (1973 b): J. Phys. *B 6*, 2088 – 2117.

Dehmer, J. L., Dill, D. (1975): Phys. Rev. Lett. *35*, 213 – 215.

Dehmer, J. L., Dill, D. (1976): J. Chem. Phys. *65*, 5327 – 5334.

Dey, S., McCarthy, I. E., Weigold, E., Teubner, P. J. O. (1975): Phys. Rev. Lett. *34*, 782 – 785.

Dey, S., Dixon, A. J., McCarthy, I. E., Weigold, E. (1976): J. Elect. Spect. *9*, 397 – 412.

Dey, S., Dixon, A. J., Lassey, K. R., McCarthy, I. E., Teubner, P. J. O., Weigold, E., Bagus, P. S., Viinikka, E. K. (1977): Phys. Rev. *A 15*, 102 – 111.

Dolder, K. T., Peart, R. (1976): Rep. Prog. Phys. *39*, 693 – 749.

Dunn, G. H. (1962): Phys. Rev. Lett. *8*, 62 – 64.

Dunn, G. H., Kieffer, L. J. (1963): Phys. Rev. *132*, 2109 – 2117.

Ehrhardt, H., Schulz, M. Tekaat, T., Willmann, K. (1969): Phys. Rev. Lett. *22*, 89 – 92.

Ehrhardt, H., Hesselbacher, K. H., Jung, K., Schulz, M., Tekaat, T., Willmann, K. (1971): Z. Physik *244*, 254 – 267.

Ehrhardt, H., Hesselbacher, K. H., Jung, K., Willmann, K. (1972 a): In: Case Studies in Atomic Collision Physics II (McDaniel, E. W., McDowell, M. R. C., eds.), 161 – 210. Amsterdam-London: North-Holland.

Ehrhardt, H., Hesselbacher, K. H., Jung, K., Willmann, K. (1972b): J. Phys. *B 5*, 1559 – 1571.

Ehrhardt, H., Hesselbacher, K. H., Jung, K., Schulz, M., Willmann, K. (1972c): J. Phys. *B 5*, 2107 – 2116.

Ehrhardt, H., Jung, K., Schubert, E. (1980): In: Coherence and Correlation in Atomic Collisions (Kleinpoppen, H., Williams, J. F., eds.), 13 – 40. New York-London: Plenum.

Ehrhardt, H. (1982): A. I. P. Conf. Proc. *86*, 183 – 198.

Ehrhardt, H., Fischer, M., Jung, K. (1982a): Z. Phys. *A 304*, 119 – 124.

Ehrhardt, H., Fischer, M., Jung, K., Byron, F. W., jr., Joachain, C. J., Piraux, B. (1982b): Phys. Rev. Lett. *26*, 1807 – 1810.

Fuss, I., McCarthy, I. E., Minchinton, A., Weigold, E. (1981a): J. Phys. *B 14*, 3277 – 3287.

Fuss, I., McCarthy, I. E., Minchinton, A., Weigold, E., Larkins, F. P. (1981b): Chem. Phys. *63*, 19 – 30.

Geltman, S. (1974): J. Phys. *B 7*, 1994 – 2002.

Geltman, S., Hidalgo, M. B. (1974): J. Phys. *B 7*, 831 – 839.

Gianturco, F. A., Guidotti, C., Lamanna, U. (1972): J. Chem. Phys. *57*, 840 – 846.

Giardini-Guidoni, A., Missoni, G., Camilloni, R. (1976): In: Electron and Photon Interactions with Atoms (Kleinpoppen, H., McDowell, M. R. C., eds.), 149 – 160. New York-London: Plenum.

Giardini-Guidoni, A., Camilloni, R., Stefani, G. (1980): In: Coherence and Correlation in Atomic Collisions (Kleinpoppen, H., Williams, J. F., eds.), 13 – 40. New York-London: Plenum.

Glassgold, A. E. (1967): Abstracts of Papers VICPEAC 646 – 648.

Glassgold, A. E., Ialongo, G. (1968): Phys. Rev. *175*, 151 – 159.

Hazi, A. (1974): Chem. Phys. Lett. *25*, 259 – 262.

Hazi, A. V. (1975): J. Phys. *B 8*, L 262 – 264.

Hitchcock, A. P., Brion, C. E. (1978): J. Elect. Spect. *14*, 417 – 441.

Hitchcock, A. P., Brion, C. E., Van der Wiel, M. J. (1978): J. Phys. *B 11*, 3245 – 3261.

Hood, S. T., Weigold, E., McCarthy, I. E., Teubner, P. J. O. (1973): Nature (Phys. Sci.) *245*, 65 – 68.

Hughes, A. L., Mann, M. M. (1938): Phys. Rev. *53*, 50 – 63.

Inokuti, M. (1971): Rev. Mod. Phys. *43*, 297 – 347.

Jacobs, V. L. (1974): Phys. Rev. *A 10*, 499 – 507.

Jensen, R. H. J., de Heer, F. J., van Wingerden, B., Blaauw, H. J. (1976): J. Phys. *B 9*, 185 – 212.

Joachain, C. J., van der Poorten, R. (1970): Physica *46*, 333 – 343.

Johnson, J. P., Franklin, J. L. (1980): Int. J. Mass Spectrom. Ion Phys. *33*, 393 – 407.

Kay, R. B., van der Leeuw, Ph. E., van der Wiel, M. J. (1977a): J. Phys. *B 10*, 2513 – 2519.

Kay, R. B., van der Leeuw, Ph. E., van der Wiel, M. J. (1977b): J. Phys. *B 10*, 2521 – 2528.

Kieffer, L. J., Dunn, G. H. (1967): Phys. Rev. *158*, 61 – 65.

Kim, Y.-K. (1983): Phys. Rev. *A 28*, 656 – 666.

Kim, Y.-K., Inokuti, M. (1973): Phys. Rev. *A 7*, 1257 – 1260.

King, G. C., Tronc, M., Read, F. H., Bradford, R. C. (1977a): J. Phys. *B 10*, 2479 – 2495.

King, G. C., Read, F. H., Tronc, M. (1977b): Chem. Phys. Lett. *52*, 50 – 54.

King, G. C., McConkey, J. W., Read, F. H. (1977c): J. Phys. *B 10*, L 541 – 543.

King, G. C., McConkey, J. W., Read, F. H., Dobson, B. (1980): J. Phys. *B 13*, 4315 – 4323.

Köllmann, K. (1975): Int. J. Mass Spectrom. Ion Phys. *17*, 261 – 285.

Köllmann, K. (1978): J. Phys. *B 11*, 339 – 355.

Kuyatt, C. E. (1968): In: Methods of Experimental Physics (Marton, J., ed.), Vol. 7, Part a, 1 – 41. New York: Academic Press.

Lahmam-Bennani, A., Wellenstein, H. F., Duguet, A., Rouault, M. (1983): J. Phys. *B 16*, 121 – 130.

Landau, M., Hall, R. I., Pichou, F. (1981): J. Phys. *B 14*, 1509 – 1524.

Lassettre, E. N., Krasnow, M. E., Silverman, S. (1964): J. Chem. Phys. *40*, 1242 – 1248.

Lee, J. S. (1977): J. Chem. Phys. *66*, 4906 – 4914.

Lohmann, B., Weigold, E. (1981): Phys. Lett. *86 A*, 139 – 141.

Lohmann, B., McCarthy, I. E., Stelbovics, A. T., Weigold, E. (1984): Phys. Rev. *A 30*, 758 – 767.

Madison, D. H., Calhoun, T. C., Shelton, W. N. (1977): Phys. Rev. *A 16*, 552 – 562.

Manson, S. T., Toburen, L. H., Madison, D. H., Stolterfoht, N. (1975): Phys. Rev. *A 12*, 60 – 79.

McCarthy, I. E., Ugbabe, A., Weigold, E., Teubner, P. J. O. (1974): Phys. Rev. Lett. *33*, 459 – 462.

McCarthy, I. E., Weigold, E. (1976): Phys. Rep. *27 C*, 275 – 371.

McCarthy, I. E., Noble, C. J., Phillips, B. A., Turnbull, A. D. (1977): Phys. Rev. *A 15*, 2173 – 2185.

McCarthy, I. E. (1984): Personal Communication.

McConkey, J. W., Crowe, A., Hender, M. A. (1972): Phys. Rev. Lett. *29*, 1 – 4.

McLean, A. D., Weiss, A., Yoshimine, S. (1960): Rev. Mod. Phys. *32*, 211 – 218.

Mendelsohn, L. B. (1982): A.I.P. Conf. Proc. *86*, 211 – 222.

Minchinton, A. (1982): A.I.P. Conf. Proc. *86*, 115 – 143.

Minchinton, A., Brion, C. E., Cook, J. P. D., Weigold, E. (1983): Chem. Phys. *76*, 89 – 101.

Minchinton, A., Giardini-Guidoni, A., Weigold, E., Larkins, F. P., Wilson, R. M. (1982): J. Elect. Spect. *27*, 191 – 203.

Mohr, C. B. O., Nicoll, R. H. (1934): Proc. Roy. Soc. (Lond.) *A 144*, 596 – 608.

Moore, J. H., Coplan, M. A., Skillman, T. L., jr., Brooks, E. D. (1978): Rev. Sci. Inst. *49*, 463 – 468.

Nakamura, M., Sasanuma, M., Sato, S., Watanabe, M., Yamashita, H., Iguchi, Y., Ejira, A., Nakai, S., Yamaguchi, S., Sagawa, T., Nakai, Y., Oshio, T. (1969): Phys. Rev. *178*, 80 – 82.

Nesbet, R. K., Watson, R. E. (1958): Phys. Rev. *110*, 1073 – 1076.

Oda, N. (1975): Radiat. Res. *64*, 80 – 95.

Opal, C. B., Beaty, E. C., Peterson, W. K. (1972): At. Data *4*, 209 – 253.

Pichou, F., Huetz, A., Joyez, G., Landau, M. (1978): J. Phys. *B 11*, 3683 – 3692.

Rescigno, T. N., Langhoff, P. W. (1977): Chem. Phys. Lett. *51*, 65 – 70.

Risley, J. S. (1972): Rev. Sci. Inst. *43*, 95 – 103.

Robb, W. D., Rountree, S. P., Burnett, T. (1975): Phys. Rev. *A 11*, 1193 – 1199.

Rudd, M. E., Dubois, R. D. (1977): Phys. Rev. *A 16*, 26 – 32.

Sar-el, H. Z. (1967): Rev. Sci. Inst. *38*, 1210 – 1216.

Sasaki, N., Nakao, T. (1935): Proc. Imp. Acad. (Tokyo) *11*, 138 – 140.

Schultz, M. (1973): J. Phys. *B 6*, 2580 – 2599.

Shaw, D. A., King, G. C., Read, F. H., Cvejanovic, D. (1982): J. Phys. *B 15*, 1785 – 1793.

Shyn, T. W., Sharp, W. E. (1979a): Phys. Rev. *A 19*, 557 – 567.

Shyn, T. W., Sharp, W. E. (1979b): Phys. Rev. *A 20*, 2332 – 2339.

Shyn, T. W., Sharp, W. E., Kim, Y.-K. (1981): Phys. Rev. *A 24*, 79 – 88.

Slevin, J., Stirling, W. (1981): Rev. Sci. Inst. *52*, 1780 – 1782.

Smid, H., Hansen, J. E. (1983): J. Phys. *B 16*, 3339 – 3370.

Smirnov, Yu. F., Neudatchin, V. G. (1966): J.E.T.P. Lett. *3*, 192 – 193.

Smith, P. T. (1930): Phys. Rev. *36*, 1293 – 1302.

Snyder, L. C., Basch, H. (1972): Molecular Wavefunctions and Properties. New York: Wiley.

Spears, D. P., Fishbeck, H. J., Carlson, T. A. (1974): Phys. Rev. *A 9*, 1603 – 1611.

Stefani, G., Camilloni, R., Giardini-Guidoni, A. (1978): Phys. Lett. *64 A*, 364 – 366.

Stockdale, J. A. D., Anderson, V. E., Carter, A. E., Deleanu, L. (1975): J. Chem. Phys. *63*, 3886 – 3897.

Taylor, G. R., Parr, R. G. (1952): Proc. Nat. Acad. Sci. U.S.A. *38*, 154 – 160.

Toennies, J. P., Welz, W., Wolf, G. (1979): J. Chem. Phys. *71*, 614 – 641.

Tossell, J. A., Moore, J. H., Coplan, M. A., Stefani, G., Camilloni, R. (1982): J. Am. Chem. Soc. *104*, 7416 – 7423.

Tronc, M., King, G. C., Read, F. H. (1979): J. Phys. *B 12*, 137 – 157.

Tweed, R. J. (1980): J. Phys. *B 13*, 4467 – 4479.

van Brunt, R. J. (1974): J. Chem. Phys. *60*, 3064 – 3070.

van Brunt, R. J., Kieffer, L. J. (1970): Phys. Rev. *A 2*, 1293 – 1304.

van Brunt, R. J. (1977): Phys. Rev. *A 16*, 1309 – 1311.

van der Wiel, M. J., El-Sherbini, Th. M., Brion, C. E. (1970): Chem. Phys. Lett. *7*, 161 – 164.

van der Wiel, M. J., Brion, C. E. (1972/73): J. Electron Spectrosc. *1*, 309 – 318.

van der Wiel, M. J., Wiebes, G. (1971): Physica *53*, 225 – 255.

van der Wiel, M. J. (1980): In: Electronic and Atomic Collisions (Oda, N., Takayanagi, K., eds.), 209 – 218. Amsterdam: North-Holland.

van Wingerden, B., Kimman, J. T., van Tilburg, M., Weigold, E., Joachain, C. J., Piraux, B., de Heer, F. J. (1979): J. Phys. *B 12*, L627 – 631.

van Wingerden, B., Kimman, J. T. N., van Tilburg, M., de Heer, F. J. (1981): J. Phys. *B 14*, 2475 – 2498.

Vriens, L., Kuyatt, C. E., Mielczarek, S. R. (1968): Phys. Rev. *170*, 163 – 169.

Weigold, E., Hood, S. T., Teubner, P. J. O. (1973): Phys. Rev. Lett. *30*, 475 – 478.

Weigold, E., Dey, S., Dixon, A. J., Lassey, K. R., Teubner, P. J. O. (1977a): J. Elect. Spect. *10*, 177 – 191.

Weigold, E., McCarthy, I. E., Dixon, A. J., Dey, S. (1977b): Chem. Phys. Lett. *47*, 209 – 212.

Weigold, E., McCarthy, I. E. (1978): Adv. At. Mol. Phys. *14*, 127 – 179.

Weigold, E., Noble, C. J., Hood, S. T., Fuss, I. (1979): J. Phys. *B 12*, 291 – 313.

Weigold, E. (1981): Nuclear Physics *A 353*, 327c – 340c.

Weigold, E. (1982): A.I.P. Conf. Proc. *86*, 1−4.
Wellenstein, H. F., Bonham, R. A. (1973): Phys. Rev. *A 7*, 1568−1572.
Wight, G. R., Brion, C. E., van der Wiel, M. J. (1972/73): J. Elect. Spect. *1*, 457−469.
Wight, G. R., Brion, C. E. (1974): Chem. Phys. Lett. *26*, 607−609.
Williams, G. R. J. (1979): J. Elect. Spect. *15*, 247−252.
Williams, J. F. (1978): J. Phys. *B 11*, 2015−2020.
Williams, K. G. (1969): Abstracts of Papers VI ICPEAC 735−737.
Wong, T. C., Mendelsohn, L. B., Grossman, H., Wellenstein, H. F. (1982): Phys. Rev. *A 26*, 181−185.
Zare, R. N. (1967): J. Chem. Phys. *47*, 204−215.

5

Partial Ionization Cross Sections

T. D. Märk

Institut für Experimentalphysik, Leopold-Franzens-Universität,
Innsbruck, Austria

5.1 Introduction

Electron impact ionization involves the collision of an electron with a target particle and the subsequent production of an ion (and a neutral) and the respective ejected electron(s). A measure of the probability for this reaction is the total ionization cross section (see Chapter 7), i.e., the cross section for producing a single positive charge in the exit channel of the ionization reaction regardless of the details of the reaction products (mass to charge ratio, electronic states, neutral fragments etc.). Another (more detailed) measure is the partial ionization cross section, which describes the probability for an individual ionization process, including dissociative ionization, multiple ionization, and more generally, ionization to a particular final state of the produced ion.

The determination of partial ionization cross sections is not only important in its own right but also because of the many applications, i.e., mass spectrometry, plasma physics and chemistry, atmospheric and interstellar physics etc. (e.g. see Chapter 9). The measurement (and the theoretical calculation) of partial ionization cross sections is more difficult than the determination of total ionization cross sections, because it is necessary to make a detailed analysis of the reaction products. In this chapter the various methods, their results and the associated problems and errors will be discussed with particular regard to the most salient experimental pitfalls and the most recent developments to overcome their difficulties.

In addition, another purpose of this review is to summarize available data, to discuss critically their reliability and to provide a guide-line (to select recommended data) for use of partial ionization cross sections. Another aim of this chapter will be to elucidate the physics of the electron impact ionization process in the framework of a state to state (partial) ionization reaction.

5.1.1 Ionization Processes and Mechanism

a) Ionization Processes

If the energy of an electron beam colliding with a gas phase target (atom, molecule or cluster) is greater than a critical value (ionization or appearance energy), some target species (depending on the corresponding cross section) will be ionized. As the electron energy is increased, the variety and abundance of the ions produced will increase, because the electron impact ionization process may proceed via different reaction channels. For the simple case of a diatomic molecule AB these reaction channels are (not including negative ion production):

$$AB + e \rightarrow AB^+ + e_s + e_e \qquad \text{single ionization} \qquad (5\text{-}1)$$
$$\rightarrow AB^{2+} + e_s + 2\,e_e \qquad \text{double ionization} \qquad (5\text{-}2)$$
$$\rightarrow AB^{z+} + e_s + z \cdot e_e \qquad \text{multiple ionization} \qquad (5\text{-}3)$$
$$\rightarrow AB^{K+} + e_s + e_e \qquad K \text{ shell (inner) ionization} \qquad (5\text{-}4)$$
$$\rightarrow AB^{**} + e_s \rightarrow AB^+ + e_s + e_e \qquad \text{autoionization} \qquad (5\text{-}5)$$
$$\rightarrow AB^{+*} + e_s + e_e \rightarrow A^+ + B + e_s + e_e \qquad \text{fragmentation} \qquad (5\text{-}6)$$
$$\rightarrow AB^{2+} + e_s + 2\,e_e \qquad \text{autoionization} \qquad (5\text{-}7)$$
$$\rightarrow AB^+ + e_s + e_e + h\nu \qquad \text{radiative ionization} \qquad (5\text{-}8)$$
$$\rightarrow A^+ + B + e_s + e_e \qquad \text{dissociative ionization} \qquad (5\text{-}9)$$
$$\rightarrow A^+ + B^- + e_s \qquad \text{ion pair formation} \qquad (5\text{-}10)$$

b) Franck-Condon Principle

Electron impact ionization of an *atom* involves the transition between two well defined (electronic) states, whereas in case of a *molecule* also vibrational and rotational excitation has to be considered. The energy transferred into vibrational or rotational excitation is, however, small compared to the energy of the electronic transition. The changes in vibrational levels during the ionization event can be described for *diatomic* molecules with help of the well known *Franck-Condon principle*, whereas rotational excitation depends on the validity (above 800 eV electron energy) or breakdown (below 800 eV) of the electric dipole selection rule (Hernandez *et al.* 1982).

The Franck-Condon principle may be summarized qualitatively as follows: No (or only negligible) changes occur in the nuclear separation and velocity of relative nuclear motion during the ionization event. This is due to the great ratio of nuclear to electronic mass and the short interaction time. Hence, the arrival point on the final potential energy curve lies directly above the starting point on the initial potential energy curve (*vertical transition*). Depending on the relative shape of the potential energy curves in a specific system, different reaction channels may be possible, i.e. the accessible final level lies (1) within the region of discrete vibrational states, (2) not only within this region, but includes some part of the continuum and (3) entirely within the continuum of a repulsive state and all transitions lead to dissociative ionization (see Fig. 5-1).

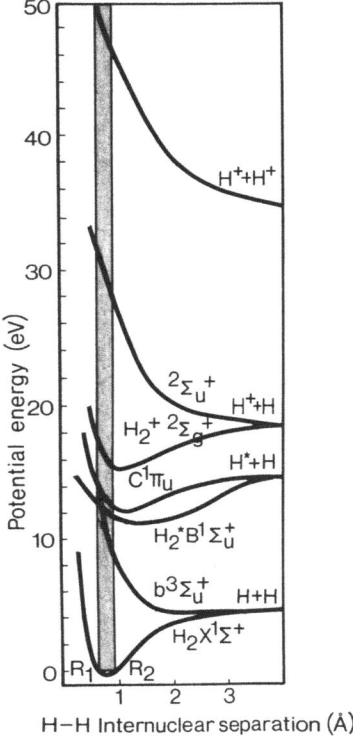

Fig. 5-1. Schematic potential energy diagram for the ground state of H_2 and some states of H_2^*, H_2^+ and H_2^{2+} (after Märk 1984 a). The nuclear separation in the ground vibrational state lies between the limits R_1 and R_2 and according to the Franck-Condon principle it must still lie between R_1 and R_2 after any transition to higher states (*effective Franck-Condon region*: shaded area)

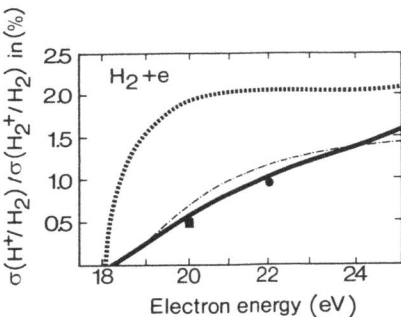

Fig. 5-2. Partial ionization cross section ratio $\sigma(H^+/H_2)/\sigma(H_2^+/H_2)$ as a function of incident electron energy: ● Hipple (see Bleakney *et al.* 1937), ■ Adamczyk *et al.* 1966, ——— electron impact results by Crowe and McConkey 1973 a, —·—·—· electron impact curve predicted by Browning and Fryar 1973, ■ ■ ■ ■ photon impact results by Browning and Fryar 1973. Note: Photoionization (the so called soft ionization technique) leads in this energy regime to more fragmentation than electron impact ionization!

In addition, the Franck-Condon principle can be used to treat quantitatively electron impact ionization (fragmentation) of diatomic or pseudo-diatomic molecules at low electron energy (Stevenson 1947, Schaeffer 1955, Dunn 1966, Browning and Fryar 1973, Crowe and McConkey 1973a). Results have been compared with observed atomic to diatomic ion ratios also including isotope effects (Stevenson 1947, Bauer and Beach 1947, Schaeffer and Hastings 1950, Schaeffer 1955, Stevenson 1960, Adamczyk *et al.* 1966, Crowe and McConkey 1973a). Fig. 5-2 shows results for the H^+/H_2^+ ratio up to 25 eV, i.e. including only H^+ from the dissociation of the H_2^+ ($^2\Sigma_g^+$) state (see Fig. 5-1).

In case of polyatomic molecules (with the exception of small polyatomics of high symmetry which can be treated in a similar way as diatomics using correlation rules to predict electronic states and reaction paths involved; see e.g. Pople 1975, Momigny *et al.* 1980, Lorquet 1980, Märk 1984a) the two-dimensional potential energy curves (Fig. 5-1) have to be replaced by n-dimensional potential energy hypersurfaces (with n the number of atoms in the molecule). Although electron impact ionization proceeds without nuclear displacements (Franck-Condon principle), the resulting (excited) polyatomic ion can undergo further internal transitions leading to subsequent unimolecular decomposition. Therefore, it is impossible to interpret fully the ionization process in terms of a detailed knowledge of the hypersurface. It is necessary to use statistical methods, i.e. the quasi equilibrium theory (QET) or RRKM theory (Glasstone *et al.* 1941, Marcus and Rice 1951, Rosenstock *et al.* 1952).

c) Ionization Mechanism

Most of the ionization reactions summarized in the previous sections (e.g. process (5-1) through (5-4), (5-9) and 5-10)), can be classified as *direct ionization* processes, where the ejected and the scattered electron leave the ion within 10^{-16} s of each other (Berry 1974). Conversely, there exists an alternate ionization channel (competing with direct ionization), where the electrons are ejected one after another. This autoionization event (e.g. process (5-5) and (5-7)) can be described as a two-step reaction. First, a neutral molecule (or atom) is raised into a superexcited state which can exist for some finite time interval. Then, radiationless transition into the continuum occurs. For molecules the upper autoionization rate (and hence the ionization cross section) is limited by the characteristic energy storage mode frequency. In addition, if predissociation (into two neutrals) is faster than autoionization the latter will not occur at an appreciable rate. Moreover, autoionization is a resonance process and this will complicate (structure) the respective ionization cross section function at low electron energy (see e.g. Figs. 7 to 9 in Johnson *et al.* 1978 (threshold ionization curves of Kr^+, Xe^+ and N_2^+, respectively), a review on autoionization in heavy alkali metals by Nygaard (1975), or results of Morrison 1964, Winters *et al.* 1966, Collins *et al.* 1968, Maeda *et al.* 1968, Marchand *et al.* 1969, Marmet *et al.* 1972, Carboneau and Marmet 1972, Morrison and Traeger 1973, Lefaivre and Marmet 1978, Locht *et al.* 1979, Selim 1980, Miletic *et al.* 1980, Hubin-Franskin *et al.* 1980, Hashizume and Wasada 1980, Mathur 1981a, b, Mathur and Frost, 1981 and Pichou *et al.* 1983), but also at higher

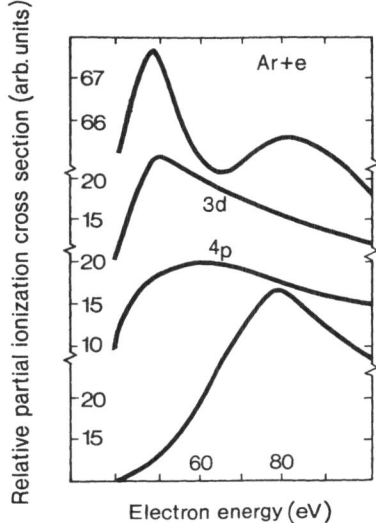

Fig. 5-3. Partial ionization cross section curves for Ar$^+$ after Crowe *et al.* 1972. Top curve shows the variation of the cross section function summing over all possible ionization mechanisms. (This curve is in good agreement with recent measurements of Stephan *et al.* 1980 a, b.) Middle curves illustrate the variation of the strengths of the 3 d and 4 p autoionization processes. The bottom curve shows the behavior of the cross section function for Ar$^+$ production by direct ionization. These results were obtained by taking energy loss spectra at different incident electron energies and scattering angles

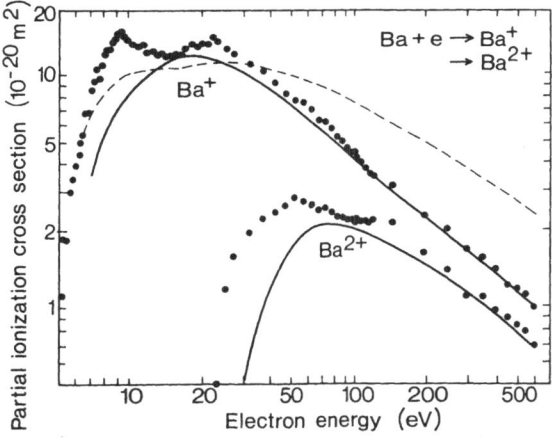

Fig. 5-4. Partial ionization cross section curves for Ba$^+$ and Ba^{2+}: ● Ba $+ e \to$ Ba$^+$ measured by Dettmann and Karstensen 1982, ● Ba $+ e \to$ Ba^{2+} measured by Dettmann and Karstensen 1982, ——— scaled Born approximation by Mc Guire (1979), – – – – Lotz (1970). Relative cross section curves measured by Okudaira (1970) and Ziesel and Abouaf (1971) are not shown for the sake of clarity. Dettmann and Karstensen (1982) interpret the maximum of the Ba$^+$ at ~ 9 eV by direct ionization of a 6 s electron, whereas the maximum at ~ 22 eV is caused, according to these authors, by the autoionization process Ba $(5 p^6 6 s^2) + e \to$ Ba $(5 p^5 6 s^2 5 d) + e \to$ Ba $(5 p^6 6 s) + 2 e$. Ziesel and Abouaf (1971) find the threshold for this autoionization process to be ~ 12.5 eV. The maximum of the Ba^{2+} curve at 55 and 110 eV is interpreted by Dettmann and Karstensen by contribution from direct inner shell ionization, whereas a break at about 20 eV (not shown in this figure) is ascribed due to an Auger process following the removal of one 5 p electron resulting in the removal of two 6 s electrons

energies. Figs. 5-3 and 5-4 show as an example the ionization cross section curves of Ar^+ and Ba^+, clearly demonstrating the strong influence of autoionization on the general shape of the ionization curve (see also discussion of other influences, i.e. excited states (Kerwin *et al.* 1969, Rosenstock 1976) or competitive ionization (Lifshitz and Long 1964, Cantone *et al.* 1966, Rapp 1971, Lifshitz 1971, Stephan *et al.* 1980 b)).

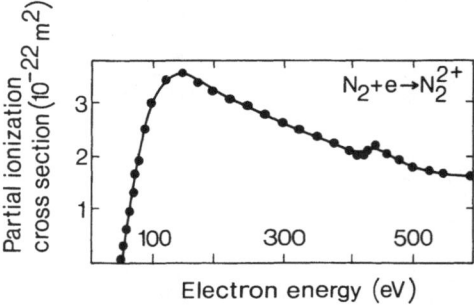

Fig. 5-5. Partial ionization cross section as a function of electron energy for the process $N_2 + e \rightarrow N_2^{2+} + 3e$ after Halas and Adamczyk (1972). At an electron energy of 427 eV (Nesbet 1964) an Auger process is possible and can be seen to influence the shape of the cross section function

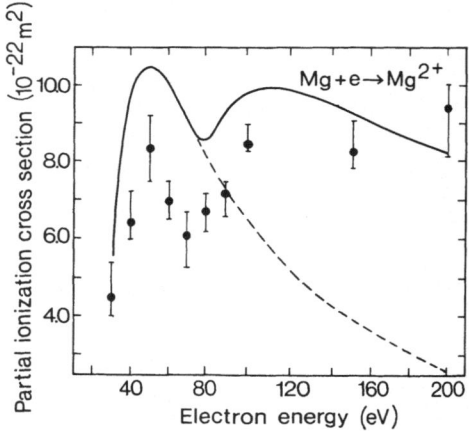

Fig. 5-6. Partial ionization cross section as a function of electron energy for the process $Mg + e \rightarrow Mg^{2+} + 3e$. ● : experimental results by Karstensen and Schneider (1978), dashed line: double binary encounter calculation (valence shell contribution) by Chatterjee *et al.* 1982, and full line: double binary encounter calculation (with inner shell contribution) by Chatterjee *et al.* 1982

Quite similarly, multiply charged ions can be formed in a two-step autoionization process. First, a singly charged ion is produced by the ejection of an electron from an innershell ((process (5-4); see Chapter 6); this internally ionized atom (molecule) is then transformed into a multiply charged ion by a series of radiationless transitions (Auger effect). Fig. 5-5 shows as an example the ionization cross section curve of

N_2^{2+}, demonstrating the occurrence and influence of this Auger autoionization process. Auger ionization and its electron energy dependence has also been studied in detail by Fiquet-Fayard (1962), Fiquet-Fayard and Lahmani (1962), Fiquet-Fayard and Ziesel (1963 a, b), Ziesel (1965, 1967 a, b), Ziesel and Abouaf (1967), Fiquet-Fayard *et al.* (1968), Karstensen and Schneider (1978) (see Fig. 5-6), Dettmann and Karstensen (1982) (see Fig. 5-4) and Chatterjee *et al.* (1982) (see Fig. 5-6).

5.1.2 Types of Individual Ions Produced

Electron impact ionization of atoms will result in the production of singly and multiply charged ions. On the other hand ionization of molecules gives rise to a number of different ions, whose properties will be discussed in the following (see also Märk 1984 a):

a) Parent Ions

Parent ions are positively charged ions as produced by reaction (5-1) through removal of one electron from the neutral precursor molecule. The production of these parent ions (with lifetimes of $\geq 10^{-5}$ s) relative to other ions originating from the same neutral precursor is depending on the electron energy and on the properties of the neutral molecule. At, and just above the ionization potential only singly

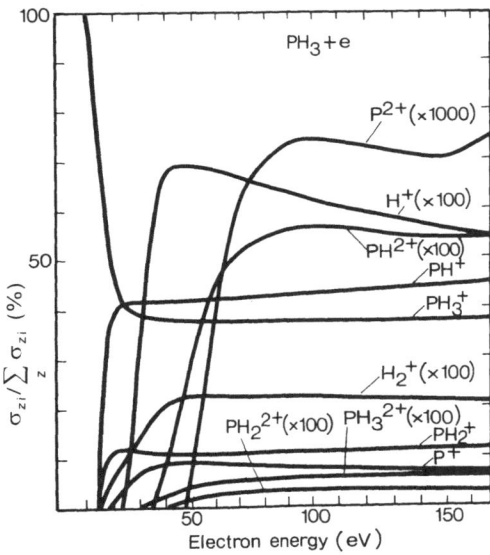

Fig. 5-7. Clastogram for phosphine after Märk and Egger 1977. Plotted are the respective partial ionization cross sections σ_{zi} (see page 151) divided by the sum of all partial ionization cross sections $\sum_z \sigma_{zi}$ as a function of electron energy

charged parent ions are produced, but at higher electron energies other ions (see below) are observed. The relative importance of the various ions as a function of electron energy can be demonstrated with help of clastograms (see, e.g. clastogram of PH_3 in Fig. 5-7). In general, for small molecules the parent ion is the dominant ion at all energies. Conversely, for large molecules the parent ion intensity usually decreases with increasing molecular weight and with higher electron energies (see e.g. Fig. 8 and 9 in Märk 1984 a, and discussion in Brunnee and Voshage 1964, Spiteller 1966, Field and Franklin 1971, Litzow and Spalding 1973).

Note. In this conjunction it is interesting to point out that the reason for running mass spectra at relatively high electron energy of 50 to 100 eV (Meisels 1982) is the fact that fragmentation patterns do not vary very much with electron energy in this energy range (e.g. Fig. 5-7) and that the partial ionization cross section functions (and thus the mass spectrometer detection efficiency) have their respective maximum in this energy range (e.g. Fig. 5-8). On the other hand a superfluity of peaks is present in this case (which makes it sometimes hard to detect the parent peak, Beynon, 1960), because so much energy can be transferred to the molecular ion in this energy range leading to heavy fragmentation (see below). Because of this fact electron impact ionization is considered a "non-soft" ionization technique as compared to photoionization. However, if similar energies are used for photo-ionization and electron impact ionization very similar mass spectra have been observed (e.g. see Fig. 5-2 present work and Figs. 7 and 8 im Maccoll 1982).

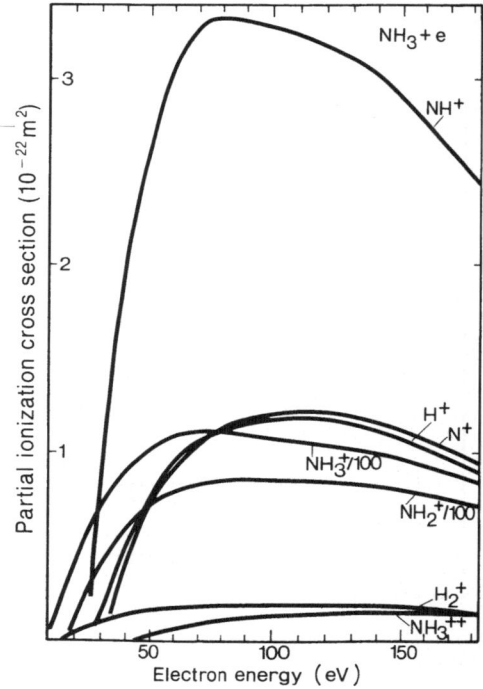

Fig. 5-8. Partial ionization cross section functions for ammonia after Märk *et al.* 1977 a

The parent ion intensity is also depending on the temperature of the molecular gas target due to the fact that the molecular kinetic energy is a function of temperature, i.e. an increase of translational energy (due to higher temperature) under constant ion source gas pressure leads to a decrease of all specific ion intensities (Stevenson 1949), whereas an increase in vibrational energies of the molecular ion leads to an appreciable decrease of the relative parent ion intensity due to a higher fragmentation rate (see, e.g. Fig. 5-9). For polyatomic molecules this latter effect can be explained in terms of QET (Vestal 1968). Moreover, well defined relations for the temperature dependence of relative parent ion intensities (pattern ratios) have been reported (Field and Franklin, 1970). The temperature dependence of small molecules will depend on both the electronic transition moment and the Franck Condon factor (Jackson *et al.* 1974).

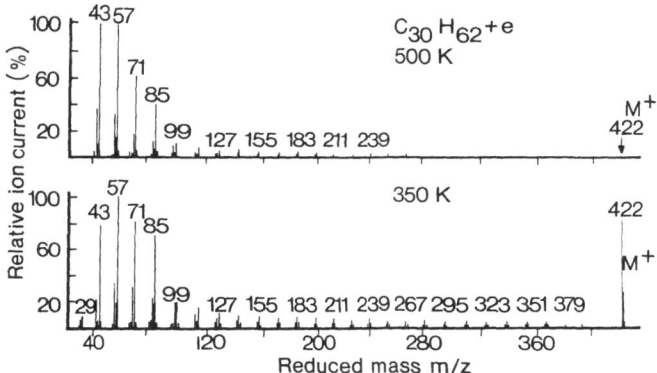

Fig. 5-9. Schematic mass spectrum of triacontane at 70 eV electron energy after Remberg *et al.* 1968 for two different gas target temperatures

b) Fragment Ions

If the energy of the incident electron is increased above the ionization potential of a molecule, fragment ions appear, e.g. for diatomics as produced by reaction (5-9). In case of polyatomic molecules a wide range of fragment channels are available, i.e. the molecular ion can immediately decay into an even or odd electron fragment ion (primary fragment ion). These primary fragment ions may be produced in excited states and immediately decay into further fragments, and so on.

The number of fragments ions and their relative cross section functions are characteristic of the corresponding parent molecule. In case of diatomic parent molecules, the fragmentation can be treated quantitatively in terms of the Franck-Condon principle (see 5.1.1.b.). For the discussion of small polyatomic molecules an approach can be used where spectroscopic and quantum-theoretical ideas are utilized to determine the dissociation path. To deal with large polyatomic molecules the QET theory has to be utilized. This theory (recently reviewed by Rosenstock and Krauss 1963, Rosenstock 1968, Vestal 1968, Wahrhaftig 1972, Cooks *et al.* 1973, Lifshitz 1978) provides the most successful theoretical approach for the discussion of dissociative ionization of large molecules, especially if combined with quantum mechanical calculations (Lorquet 1980, 1981). Levsen (1978) has shown that at least qualitative, in some cases quantitative, agreement between QET results and experimental observation of breakdown graphs (which are directly related to partial ionization cross section functions) has been observed.

In general, the relative abundance of any fragment ion is related to its rate of formation and its rate of dissociation by unimolecular decomposition. Hence, a measured mass spectrum is a record in time of the position of this "quasi-equilibrium" of those rates. In other words, because the mass spectrum is a cut in the three-dimensional plot of ion current as a function of electron energy and mass to charge ratio, also the respective partial ionization cross section functions will depend on the time after formation of the primary ion. Lifshitz and Gefen (1980) have used the trapped ion mass spectrometry technique (see Section 5.2.3.f.) to investigate this phenomenon for the electron impact ionization of 1.5 hexadiyne. Typical results, for short and long delay times, are shown for $C_6H_6^+$, $C_6H_5^+$ and $C_6H_4^+$ in Fig. 5-10. The increased fragmentation for $C_6H_6^+$ is obvious at long delay times.

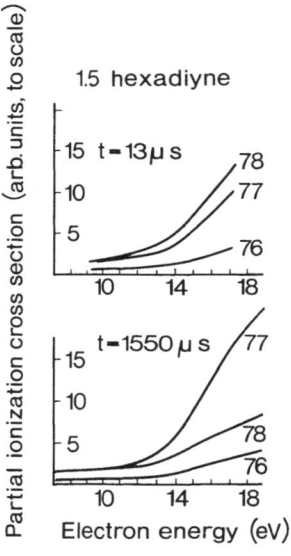

Fig. 5-10. Partial ionization cross section functions for ions of 1.5-hexadiyne (a linear isomer of benzene) with $m/ze = 76$, 77 and 78 at two different times following electron impact ionization after Lifshitz and Gefen (1980)

c) Metastable Ions

Ions produced by electron impact with sufficient internal energy to dissociate before reaching the detector of a mass spectrometer (i.e. $\gtrsim 10^{-5}$ s) are called *metastable ions*. Conversely, *unstable ions* are those which decompose immediately into fragments in the ion source before leaving the ionization region (i.e. $\gtrsim 10^{-6}$ s). More generally, however, metastable ions with lifetimes between 10^{-3} and 10^{-11} s have been reported (Lifshitz 1978, Brenton *et al.* 1979).

The existence of metastable ions can be explained by different mechanisms depending on the size and property of the precursor ion, i.e. forbidden predissociation, tunneling through a barrier, vibrational (statistical) predissociation and rearrangement transitions (e.g. Herzberg 1967, Kovacs 1969, Beynon *et al.* 1971, Cooks *et al.* 1973, Beynon and Cooks 1976, Brenton *et al.* 1979, Illies *et al.* 1982, Stephan *et al.* 1982a, Beynon *et al.* 1982, Märk 1982a, c, 1984a).

The intensity of *metastable peaks* (see Hipple and Condon 1945, Cooks *et al.* 1973) in a mass spectrum is usually less than 1% of the base peak. Conversely, for certain metastable transitions, e.g. $CH_4^{+*} \rightarrow CH_3^+$; $CCl_4^{+*} \rightarrow CCl_3^+$; $CF_4^{+*} \rightarrow CF_3^+$ (Brehm *et al.* 1974, Stephan *et al.* 1983 d, e, Leiter *et al.* 1984 b, Deutsch *et al.* 1984), it is not possible to detect any precursor ion signal at all. In addition, the temperature dependence of the metastable decomposition reaction rate has been studied (Stephan and Märk 1982 a, Griffiths *et al.* 1982, Illies *et al.* 1982) and it has been found that for certain ions the ratio between metastable and precursor ion is a strong function of the temperature of the neutral precursor molecule (see e.g., $Ar_3^{+*} \rightarrow Ar_2^+$ in Fig. 5-11 (Märk 1982a, Stephan and Märk 1983)).

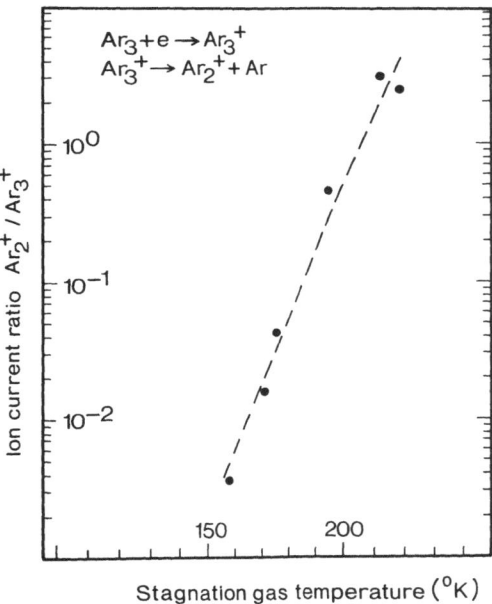

Fig. 5-11. A plot of the relative metastable to stable precursor ion intensity ratio against stagnation chamber temperature after Stephan and Märk 1983. Note that the stagnation chamber temperature determines the final temperature of the neutral Ar_3 cluster (produced in the free jet expansion) before being ionized by electrons

c) Multiply Charged Ions

Multiply charged atomic and molecular ions were observed and identified very early (Thomson 1912, 1921, Conrad 1930). Subsequent observation of numerous doubly charged ions followed (e.g. see reviews by Mohler *et al.* 1949, Ast 1980, Mathur *et al.* 1980, Teleshefsky *et al.* 1982, Märk 1984 a, Brehm *et al.* 1984). In addition, triply charged molecular ions have been detected (Märk 1984 b) in low abundance for various organic and anorganic compounds (Dibeler *et al.* 1953, Quinn and Mohler 1959, Dorman and Morrison 1961, Meyerson and Van der Haar 1962, Van Brunt and Wacks 1964, Newton 1964, Brion and Paddock 1968, Fiquet-Fayard *et al.* 1968; see also Kellogg 1981, 1982, Becker and Dietze 1983). Moreover, electron impact ionization of free Van der Waals clusters can lead to multiply charged ions with up to five elementary charges (Henkes and Isenberg 1970, Gspann and Körting 1973, Echt *et al.* 1982; see also Dole *et al.* 1968).

Most of the doubly charged diatomic ions AB^{2+} observed (a summary of those ions is given in Märk 1984 a) satisfy

$$IP(A^+) + IP(B^+) < IP(A^{2+}) \le IP(B^{2+}) \tag{5-11}$$

with IP the respective single and double ionization potentials for the single atoms A and B. In this case the repulsive Coulomb state arising from $A^+ + B^+$ lies energetically below weakly bound states arising from $A^{2+} + B$, at all internuclear distances. Metastable AB^{2+} exist when short range chemical forces impose a sufficiently strong attractive well in the repulsive Coulomb interaction (see also Sattler *et al.* 1981). Fig. 5-12 gives as an example the potential curves of He_2^{2+}.

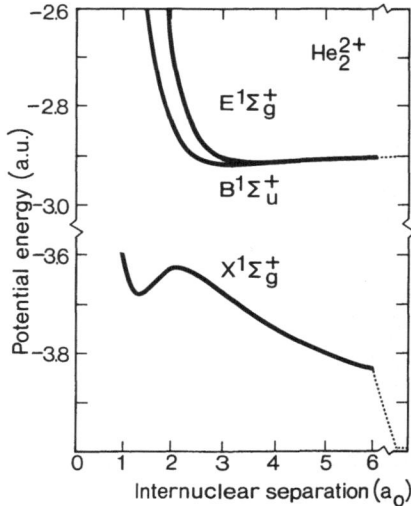

Fig. 5-12. Potential energy curves calculated by Cohen and Bardsley (1978) (see Nikulin and Samoylov (1982) and Castro *et al.* (1982)) for $He^+\,(^2S) + He\,(^2S)$ and $He^{2+} + He\,(^1S)$ interactions. It can be seen that the lowest doubly charged molecular state $He_2^{2+}\,(X^1\Sigma_g^+)$ has a local minimum around $1.4\,a_0$. Moreover, the attractive $B^1\Sigma_u^+$ state arising from $He^{2+} + He$ is short lived since the $B \rightarrow X$ transition is dipole allowed (Helm *et al.* 1981)

Conversely, as has been recently pointed out by Helm *et al.* (1981), there is a second category of doubly charged diatomic ions, satisfying

$$IP(A^+) + IP(B^+) > IP(AB^{2+}).$$ (5-12)

Here, the repulsive Coulomb state arising from A^+ and B^+ lies above the weakly bound state arising from $A^{2+} + B$ (see Fig. 5-13) allowing the formation of stable AB^{2+} ions. Three rare gas dimers (namely, $HeXe^{2+}$, $HeKr^{2+}$ and $NeXe^{2+}$) fall into this group. One of these ions, $NeXe^{2+}$, has recently been observed experimentally (Johnson and Biondi 1979, Helm *et al.* 1981), and Helm *et al.* (1981) and Stephan *et al.* (1982 b) reported semi-quantitative potential energy curves consistent with the measured appearance potential of $NeXe^{2+}$, $ArXe^{2+}$ and $NeKr^{2+}$.

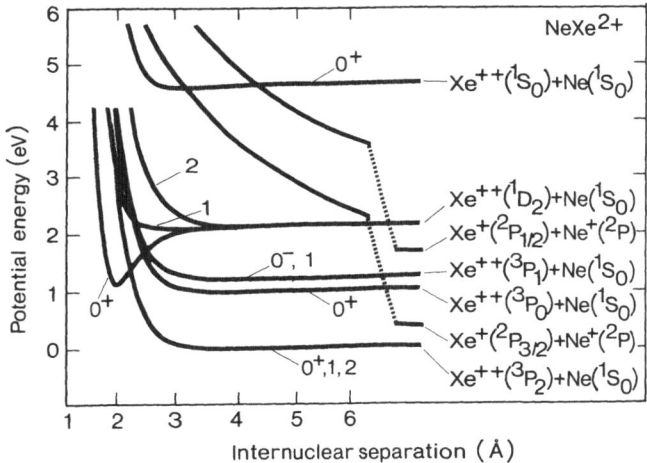

Fig. 5-13. Potential energy curves reported by Helm *et al.* 1981 for $Xe^{2+} + Ne(^1S_0)$ and $Xe^+(^2P) + Ne^+(^2P)$ interactions. The well depths of the lowest states arising from $Xe^{2+}(^3P_2) + Ne(^1S_0)$ are 0.019, 0.020, and 0.023 eV for $\Omega = 0^+$, 1 and 2, respectively

Unstable or metastable doubly charged molecules can decompose in two different ways, depending on how the two charges will be distributed in the reaction, i.e.

$$AB^{2+} \rightarrow A^+ + B^+,$$ (5-13)

$$AB^{2+} \rightarrow A^{2+} + B.$$ (5-14)

In general, most of the electron impact multiple ionization processes of molecules lead to singly charged fragment ions via process (5-13) (e.g. Brehm and De Frenes 1978, 1980), rather than to one multiply charged fragment ion as shown in reaction (5-14) (e.g. Beynon *et al.* 1959, Rabrenovic *et al.* 1983). The charge separation process (5-13) (sometimes termed Coulomb explosion) has recently been reviewed by Ast (1980) (see also Sattler *et al.* 1981, Jentsch *et al.* 1982 and Gemmell and Kanter 1984).

The partial ionization cross sections for the production of multiply charged atomic and molecular ions rarely exceed $1 - 5\%$ of that of the dominant singly charged ions

(Beynon 1960, Biemann 1962, Kieffer and Dunn 1966, Märk 1982 b, 1984 a). Certain atoms (e.g. see review by Kieffer and Dunn 1966) and certain compounds, however, have been found to possess, increased ability to sustain two positive charges, i.e. aromatic, heteroaromatic and polyfluor compounds (Kienitz 1968), aminoboranes (Dewar and Rona 1965), and organometallic compounds (King 1969). In some cases doubly charged parent ions were found to be more abundant than the corresponding singly charged ion (Waight 1969, Hellwinkel and Wünsche 1969, Solomon and Mandelbaum 1969).

It is interesting to mention in this section, that there exist also other cases of multiple ionization (see also Chapter 8), the most simplest case being the reaction $H^- + e \rightarrow H^+ + 2 e$. Cross sections for these reactions have been recently studied by Feart et al. (1971) and Defrance et al. (1982).

5.1.3 Principle of Ionization Cross Section Experiment and Definitions

Experimental determination of electron impact ionization cross section functions usually involves an experimental arrangement of simple conceptual design (Fig. 5-14). A parallel, homogeneous and monoenergetic beam of electrons crosses a semi-infinite medium containing N_t target particles per cm^3 at rest. If $n(O)_e$ represents the initial intensity of the incident electrons per cm^2 s, the density of the electron beam at depth x is given by the exponential absorption law

$$n(x)_e = n(O)_e \cdot \exp(-N_t \cdot \sigma \cdot x).$$ (5-15)

If $N_t \cdot \sigma \cdot x \ll 1$ (single collision condition), the number of ions generated per s along the collision interaction path $x = L$ (over which the ions are collected and analyzed) is

$$n(L)_i = n(O)_e \cdot N_t \cdot \sigma_c \cdot L$$ (5-16)

with σ_c the counting ionization cross section in cm^2. The total positive ion current i_t produced in this interaction volume is given by

$$i_t = n(O)_e \cdot e \cdot N_t \cdot \sigma_t \cdot L$$ (5-17)

with σ_t the total ionization cross section. If the produced ions are analyzed with respect to their mass m and charge $z \cdot e$, the respective individual ion current is given by

$$i_{ms} = n(O)_e \cdot e \cdot N_t \cdot \sigma_{zi} \cdot L$$ (5-18)

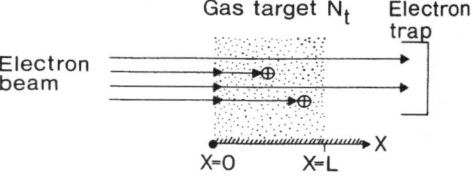

Fig. 5-14. Schematic view of an electron impact ionization experiment (analysis see text). L interaction length

with σ_{zi} partial ionization cross section for the production of a specific ion i with charge $z \cdot e$. Total and counting ionization cross sections of a specific target system are the weighted and simple sum of the various single and multiple partial cross sections, respectively

$$\sigma_t = \Sigma \, \sigma_{zi} \cdot z \quad \text{and} \quad \sigma_c = \Sigma \, \sigma_{zi}. \tag{5-19}$$

According to a previous review (Kieffer and Dunn 1966) there exist severe difficulties in obtaining accurate measurements of all the quantities in the above equations necessary to determine accurate ionization cross sections. Kieffer and Dunn have summarized the most important experimental prerequisites to be met in order to obtain high precision cross sections (see Table 1 in Kieffer and Dunn 1966). In case of the determination of *partial* ionization cross sections the most serious problem is the collection of a known fraction of the individual ions under study, a problem which arises from *discrimination* effects along the ion path from the region of production through the mass analyzer to the collector. In addition, other problems involve the absolute calibration of measured relative partial cross sections and the determination and definition the incident electron energy.

5.2 Experimental Considerations

5.2.1 Discrimination Effects in Ion Source-Mass Spectrometer Systems

In order to measure partial ionization cross sections it is necessary to analyze the individual ions produced in the ion source with help of a mass analyzing system. It is not possible to give, in the space of this section, a comprehensive review of the many methods used in mass spectrometry. Standard techniques and apparatus are extensively discussed in a number of review books, e.g. Beynon 1960, Mc Lafferty 1963, Mc Dowell 1963, Brunnee and Voshage 1964, Blauth 1965, Mc Lafferty 1966, Spiteller 1966, Kienitz 1968, White 1968, Field and Franklin 1970, Melton 1970, Budzikiewicz 1972, Litzow and Spalding 1973, Maccoll 1975, Dawson 1976, Beynon and Morgan 1978, Johnstone 1980, Budzikiewicz 1981, Rose and Johnstone 1982. Instead, the performance and inherent discrimination effects of a representative and commonly used ion source mass spectrometer system will be discussed in the following, i.e. the three electrode, Nier type, ion source (Nier 1940, 1947; Hipple 1948, Brunnee and Voshage 1964) in combination with sector type mass spectrometers.

It was pointed out by a number of investigators that discrimination effects occur at the slits of this ion source and mass spectrometer system. The extraction of ions from the ion source (see Fig. 5-15) into the mass spectrometer depends on various experimental parameters, i.e. the initial energy of the ions produced, their mass to charge ratio, and the extraction field (shape and strength). The *extraction efficiency* and its dependence on experimental parameters directly determines the accuracy of measured ion currents as a function of electron energy (partial ionization cross section function) or the accuracy of measured ion current ratios (partial ionization cross section ratios). Therefore, it is imperative that the experimental parameters which can influence the *extraction efficiency* (or *collection efficiency*, if the efficiency of the overall system from the production region to the ion collector at the end of the mass spectrometer is considered) be investigated for each ion source geometry and subsequent sector field mass spectrometer.

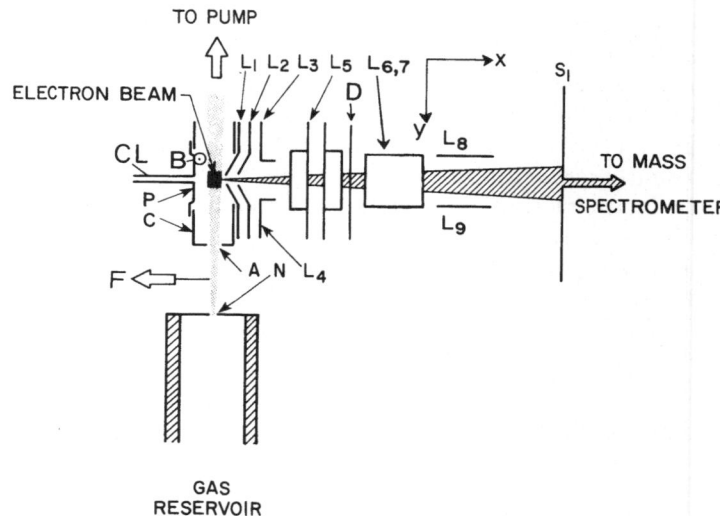

Fig. 5-15. Schematic view of three-electrode, Nier type, ion source after Märk (1984 a). *CL* capillary leak for stagnant gas target, *N* nozzle (10 to 50 μm) for molecular beam gas target, *A* aperture, *F* molecular beam flag, *C* collision chamber, *P* pusher (repeller), L_1 extraction electrodes (direct extraction), L_2 extraction electrodes (penetrating field extraction), L_3, L_4 beam centering half-plates, L_5 earth slit, *D* defining aperture, $L_{6,7}$ and L_8, L_9 deflection plates, S_1 adjustable mass spectrometer entrance slit, *B* guiding magnetic field

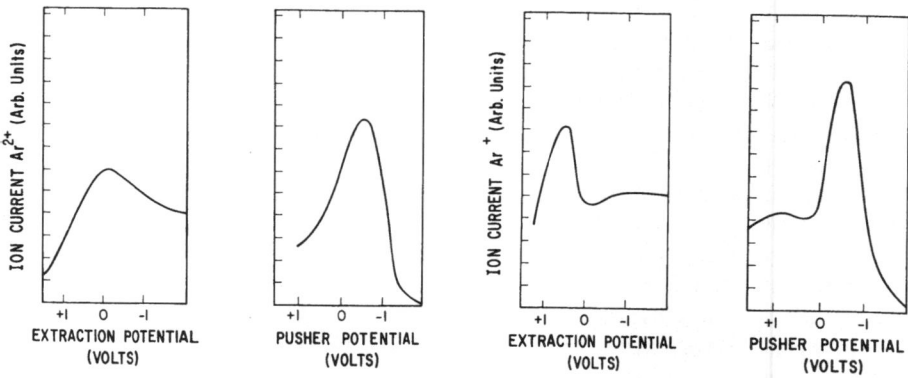

Fig. 5-16. Ion current Ar^+ and Ar^{2+} versus repeller (pusher) potential (repeller curve) and versus extraction electrode potential (extraction curve) after Hille and Märk (1979). Electron energy: 95 eV

Information about the extraction efficiency of an ion source may be obtained by either studying theoretically the focal properties of an ion source or by measuring the extraction characteristic. For the three-electrode, Nier type, ion source (Fig. 5-15) the extraction field in the collision chamber of the ion source is usually provided by either applying a positive potential to the pusher P with respect to the collision chamber C and the extraction electrode L_1, or a negative potential to the extraction electrode L_1 with respect to the collision chamber and pusher (repeller) or both. If the applied potential is varied and the mass-analyzed ion current is measured the resulting ion current characteristic is called *repeller curve* (ion current versus repeller potential, see Barnard 1953) or *extraction curve* (ion current versus extraction electrode potential). Examples of such curves are shown in Fig. 5-16. It can be seen that the extraction efficiency shows a similar overall dependence on the repeller (pusher) and on the extraction electrode potential for the same experimental conditions in the ion source.

Some of the first to note discrimination effects at the collision chamber exit slit were Hagstrum and Tate (1941). They calculated the efficiency of collection of ions as a function of initial kinetic energy, showing that ions of very low or zero kinetic energy are much more efficiently collected than those of higher energy. Their investigation, however, has been simplified (neglecting space charge and guiding magnetic field) and has been used to study the effect of initial kinetic energy on the ion peak shape, thus obtaining information on the dissociative ionization process.

An early review of Jordan and Coggeshall (1942) is devoted to the measurements of and possible sources of errors in relative isotope abundances. Because in most cases ions are drawn from a region in the ion source in which there is a crossed electric and magnetic field (see e.g. Märk 1982 b), Jordan and Coggeshall (1942) have quantitatively studied the influence of the electric and magnetic fields in the ion source on the shape of the electron beam. They also discussed (see also Coggeshall 1944, Nier 1950, Schaeffer 1950) the dependence of measured isotope abundances on the ion path (see also Bleakney 1936) through the extraction electrode and a second slit (e.g. mass spectrometer entrance slit) and on space charge conditions. Bainbridge (1931) pointed out, when space charge conditions exist, a number of discriminative factors enter in, even if all the ions come from the same point in the ion source. Jordan and Coggeshall (1942) reported also that for a 180° sector field mass spectrometer where the ion source is immersed in the analyzing magnetic field, mass discrimination at the extraction electrode and a second slit (at a fixed electron energy) should be a minimum if magnetic scanning is used for mass selection, because (thermal) ions which are collected, although they have different masses, originate at the same position. Moreover, Coggeshall (1944) showed in his analysis of a magnetic field free ion source (Coggeshall and Jordan 1943) that already thermal velocities are great enough to cause appreciable discrimination at the second slit (e. g. earth slit or mass spectrometer entrance slit) for ions which have passed the extraction electrodes (see also Berry 1950). Coggeshall also calculated the ion current as a function of the sum of the electric fields in the ion source and between the extraction electrode and second slit. This dependence has been studied experimentally by Careri and Nencini (1950), giving results in contradiction to Coggeshall's calculations, and also by Washburn and Berry (1946), yielding information on dissociation energies. Moreover, Washburn and Berry were the first to point out (1) that discrimination at the mass spectrometer exit slit could be of the same order of magnitude as that at the mass spectrometer entrance slit (except for thermal ions at high accelerating voltage, Berry 1950), and (2) that the effects of motion of the electron beam due to changing either the electric or the magnetic field are negligible compared with the effects of initial energies. This study was expanded later by Berry (1950) exploring theoretically and experimentally the effects of discrimination by the entrance and exit slit of a mass spectrometer and showing that measured relative abundances of energetic ions depend strongly on the geometry of the system and on the accelerating voltage between the extraction electrode and the second slit.

Reese and Hipple (1949) and Berry (1950) used the discrimination at the mass spectrometer exit slit due to initial velocity components parallel to the magnetic field to determine the initial velocity distribution with help of a pair of deflecting electrodes placed immediately behind the mass spectrometer entrance slit. This deflection method has been used by several authors to measure initial energy distribution functions (Märk 1984 a) of fragment ions (Reese and Hipple 1949, Berry 1950, Osberghaus and Taubert 1951, Dibeler et al. 1956, Taubert 1959, Bracher et al. 1953, Durup and Heitz 1964, Appell et al. 1964, Taubert 1964, Erhardt and Tekaat 1964, Fuchs and Taubert 1965, Erhardt and Kresling 1967, Rowland et al. 1969, Rowland 1971, Appell and Durup 1972, Sen Sherma and Franklin 1974, Drewitz and Taubert 1976, Drewitz 1976, Futrell et al. 1981).

In order to obtain information on the ion source extraction efficiency and to determine mechanisms limiting the ion source efficiency, however, it is necessary to study the focal properties of the ion source as a function of the extraction field within the ion source. In a first study Bertein 1950 and Vauthier 1950 estimated the focal properties and aberrations of such an ion source. Using a ray tracing technique, Vauthier (1955) showed that the number of cross-overs of the ion trajectories within the ion source changes from one to two if the repeller potential is changed from a large positive value to zero. Extending this study, Coggeshall (1962), Naidu and Westphal (1966) and Fock (1969) studied in detail the influence of the extraction field on the focal properties. Coggeshall (1962) calculated initial kinetic energy discrimination effects in a crossed field ion source (Coggeshall 1946) by investigating the characteristics of the ion orbits. Moreover, Coggeshall appears to be the first to report experimentally determined ion beam shapes for different repeller voltages by pushing the ion beam across the mass spectrometer entrance slit by changing the voltage applied to a pair of deflector half-plates. It is interesting to note that the integrated ion flux obtained by this procedure shows no saturation as a function of repeller voltage (Figs. 6 and 7 in Coggeshall 1962; see also Stephan et al. 1980 b and below). Moreover, the width and maximum intensity of the ion beam were found to depend on ion source operating conditions and it has been mentioned by Coggeshall that misleading results may be obtained depending upon how the mass spectrometer entrance slit samples the ion beam coming from the ion source. By calculating the ion withdrawal space Naidu and Westphal (1966) demonstrated that the number of ions extracted from the ion source depends strongly on the repeller voltage. This study was extended by Fock (1969) to ions with initial kinetic energy. The slits considered by both, however, allowed only for infinitely thin electrodes.

These studies were extended by Wallington (1970, 1971) to slit systems with finite thickness. Moreover, Wallington also calculated the dependence of the ion source

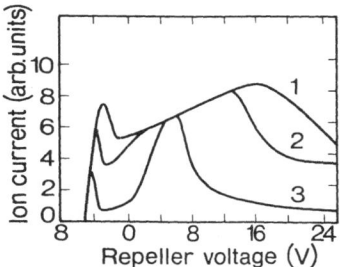

Fig. 5-17. Repeller curves calculated by Wallington (1971) for three different sizes (1 : 0.9 mm, 2 : 0.5 mm, 3 : 0.1 mm) of the infinitesimally thin electron beam. The slit width is 0.05 mm and the accelerating voltage 8 kV. It can be seen that each curve has two maxima

sensitivity (extracted ion current) on the repeller voltage (Fig. 5-17), demonstrating that the structure of the repeller curve is related to the multiplicity of real images that can be formed within the ion source. Additional computations showed that an electron beam of finite thickness results in broader repeller peaks than given in Fig. 5-17. Similarly, a spread in initial kinetic energies has the effect of diffusing the paraxial images in a way that is caused by the finite thickness of the electron beam (see also experimental evidence (repeller curve dependence on the initial kinetic energy and on m/ze) reported by Taubert, 1959; see also Schaefer 1954). It is interesting to note that Wallington interprets the two maxima of an experimental repeller curve obtained with an electron current of 20 μA in terms of these theoretical findings. This has been questioned later (see below). In addition, Wallington has shown that the source sensitivity is approximately independent from the source geometry and the voltage on the beam-centering half plates, provided the optimum repeller voltage is used. This is confirmed by the results of Barnard (1956) and Schulz et al. (1968).

Werner (1974) extended the study of Wallington (1970, 1971) using a computer program by Weber (1967), which considers a two-dimensional model taking into account both, the influence of the side walls and the finite extraction electrode thickness. Werner showed (Fig. 5-18) by means of calculated ion trajectories that the number of (thermal) ions extracted from the ion source at constant repeller voltage depends strongly on the accelerating voltage (see also Chantreau and Vauthier 1970) and on the mass to charge ratio. Werner (1974), Werner and Linssen (1974) and Werner et al. (1972) also investigated the effect of positive and negative space charge (see also Brubaker 1955), scattering at the residual gas (see also Rüdenauer 1970, 1972) and the influence of initial thermal energy.

Recently, Märk and co-workers (Märk et al. 1977 b, Egger and Märk 1978 a, Hille et al. 1978, Stephan et al. 1978, Märk and Castleman, jr. 1980, Stephan et al. 1980 a) extended the studies of Wallington (1971), Schulz et al. (1968) and Werner and co-workers by investigating systematically the source sensitivity (repeller or extraction curves) as a function of the (1) gas density and temperature in the ion source, (2) the extraction field, (3) the ion beam focussing conditions, (4) the electron current and energy (this is especially important when measuring cross section functions), (5) the guiding magnetic field, and (6) the mass and charge of the ions under study. They found that the shape of the extraction curves depends strongly on the electron

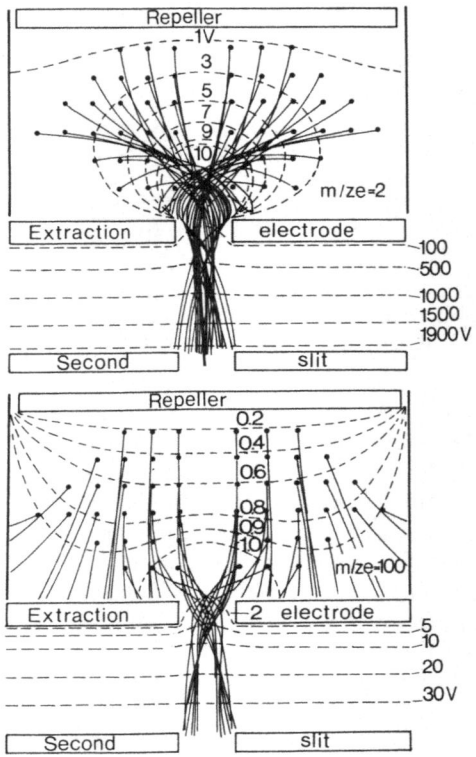

Fig. 5-18. Computer-calculated ion trajectories (full lines) and equipotentials (dashed lines) after Werner (1974). Upper part: Repeller voltage (between repeller and extraction electrode (and collision chamber)) 1 V, accelerating voltage (between extraction electrode and second slit) 2 kV, $m/ze=2$; Lower part: Repeller voltage 1 V, accelerating voltage 40 V, $m/ze=100$

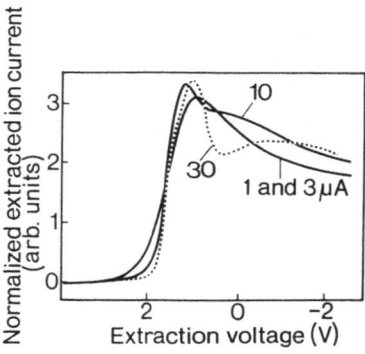

Fig. 5-19. Normalized extraction curves for Ar^+ (ion current divided by electron current) for different electron currents after Stephan (1979)

current (above 3 μA; see Fig. 5-19), on the ion charge (Fig. 5-20) and on the electron energy (Fig. 5-21, 5-22). A comparison of repeller (or extraction) curves measured by different authors at different electron currents reveals an interesting discrepancy, i.e. the curves measured by Schulz et al. (1968) at 10 μA and by Schaefer (1954) at 5 μA are in agreement (one maximum) with those of Märk and co-workers at electron currents < 10 μA (see Fig. 5-19), whereas curves at electron currents > 10 μA in Fig. 5-19 are similar (two maxima) to those reported by Wallington (1971) for 20 μA. From this it was concluded recently (Märk and Castleman, jr., 1980) that in general the predicted properties of the repeller curves (as calculated by Wallington) are not observed in the experimental results, i.e. the two maxima occurring at higher electron currents and for multiply charged ions are likely to be due to space charge effects, whereas the one maximum appearing at low electron currents results by an averaging process of many theoretical repeller curves, because assumptions made for the calculation are not met under normal experimental conditions (Wallington 1971, Naidu and Westphal 1966).

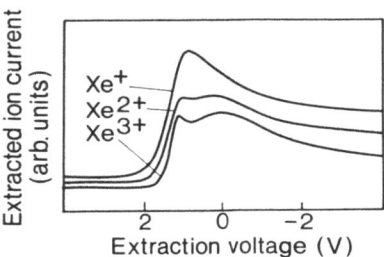

Fig. 5-20. Relative extraction curves for singly and multiply charged xenon ions for an electron current of 3 μA after Stephan (1979)

Table 5-1. *Measured partial ionization section ratio at an electron energy of 100 eV for the production of Ar$^+$ and Ar^{2+}*

$\sigma(Ar^{2+}/Ar)/\sigma(Ar^+/Ar)$	Authors
0.10	Bleakney (1930 a)
0.10	Stevenson and Hipple (1942)
0.10	Fox (1960)
0.087	Fiquet-Fayard (1962) and Fiquet-Fayard and Lahmani (1962)
0.071	Peterson (1963)
0.079	Melton and Rudolph (1967)
0.050	Gaudin and Hagemann (1967)
0.050	Morrison and Traeger (1970)
0.070	Crowe et al. (1972)
0.070	Drewitz (1976)
0.084	Egger and Märk (1978 b)
0.068	Stephan et al. (1980 a, b)

Fig. 5-20 shows as an example extraction curves for Xe^+, Xe^{2+} and Xe^{3+} at low electron current. It can be seen that the shape of the extraction curve changes with the charge of the ion and, in addition, it was found that the maximum of the curve shifts with electron energy. These facts are the reason why measured partial ionization cross section ratios between singly and multiply charged ions depend strongly on the chosen extraction voltage (see e.g. Fig. 5-23) and can be used to explain, at least partly, the large difference in reported cross section ratios (see e.g. Table 5-1). Moreover, some of these findings (Figs. 5-21 and 5-22) have lead to the conclusion (Märk 1982 b) that it is not possible to measure accurately the shape of the partial ionization cross section function (at least in the low energy regime <200 eV) for these operating conditions (i.e. extraction of ions from the Nier type ion source by a repeller and/or extraction voltage). This is illustrated in Fig. 5-21 and 5-22, where extraction curves are shown as a function of electron energy and apparent partial ionization cross sections are shown as a function of extraction voltage. It can be seen that the relative shape of the apparent partial ionization cross section functions depends strongly on the applied extraction voltage. Again, this fact can be invoked to explain, at least partly, the large difference in reported partial ionization cross section functions (see e.g. Fig. 5-24); especially in view of the fact that a large number of measurements have used this type of ion extraction (Märk 1982 b) with an unknown amount of a superimposed penetrating field from lenses beyond the extraction electrodes (e.g. Elliott 1963, Werner 1974, Kingston et al. 1982).

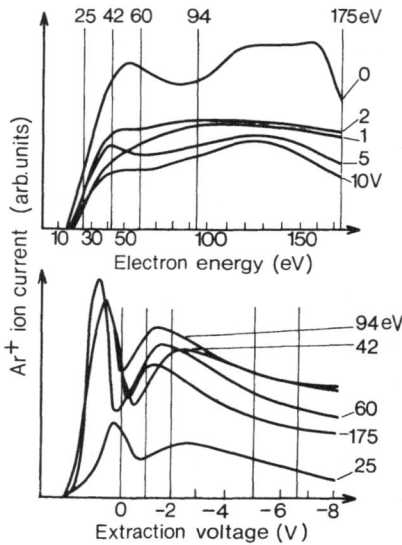

Fig. 5-21. Ar^+ ion current as a function of electron energy and extraction voltage for an electron current of $50 \mu A$ after Hille et al. (1978)

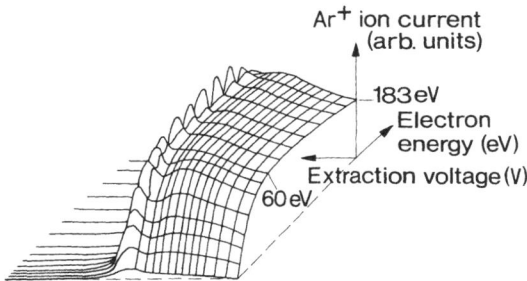

Fig. 5-22. Ar⁺ on current as a function of electron energy and extraction voltage for an electron current of
10 pA (compare with Fig. 5-21, i.e. *50 μA* electron current) after Stephan (1979)

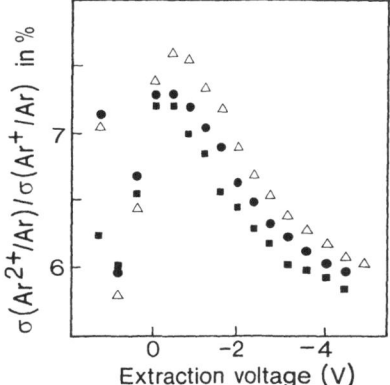

Fig. 5-23. Measured partial ionization cross section ratios for the production of Ar^+ and Ar^{2+} as a
function of extraction voltage for three different electron currents (■ ... 1 μA, ● ... 3 μA and △ ... 10 μA)
after Stephan *et al.* 1978

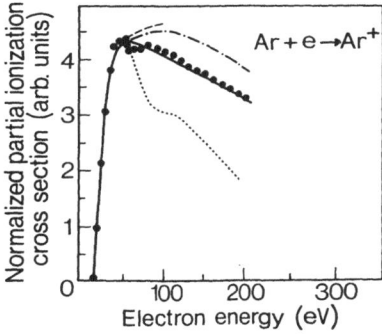

Fig. 5-24. Partial ionization cross section curves for the process $Ar + e \rightarrow Ar^+ + 2e$ as measured by
Bleakney (1930 a) ——, Stevenson and Hipple (1942) –·–·, Fox (1960) ....., Morrison and Traeger
(1970) –––, and Crowe *et al.* (1972) ●. All curves are normalized to one point at 50 eV electron energy.
The structure of the Ar⁺ curve reported by Crowe *et al.* (1972) has recently been confirmed by Stephan
et al. (1980 a, b) (not shown in this figure for the sake of clarity) and Mathur and Badrinathan (1984). Also
not shown the results of Fiquet-Fayard (1962) and Peterson (1963), which both are in disagreement with
the data of Crowe *et al.* (1972) and Stephan *et al.* (1980 a, b)

An alternate way to extract ions is to use only a penetrating field. In this case all electrodes confining the collision chamber are kept at the same potential and ions are extracted only by means of a penetrating field produced by electrodes (e.g. L_2 in Fig. 5-15) placed beyond the extraction electrodes (collision chamber exit slit) (Nier 1947, Paul 1948, Hagstrum 1951, 1953, Fiquet-Fayard and Lahmani 1962, Stanton and Monahan 1964, Locht and Momigny 1969, Drewitz 1974). There exist some general studies of this extraction mode, i.e. on the influence of the repeller voltage on ion trajectories including the limit of zero repeller voltage, and on the ion source extraction efficiency as a function of accelerating voltage (Vauthier 1955, Fock 1969, Chantreau and Vauthier 1970, Wallington 1971, Chantreau 1972, Werner 1974). This type of ion extraction has been recently studied in detail by Stephan (1979) and Stephan et al. (1980 a, b) and it has been found that this mode effectively avoids discrimination at the collision chamber exit slit (at least for ions which are produced without initial kinetic energy, see also Stephan et al. 1983 d, e, Leiter et al. 1984 a, b). Stephan et al. (1980 a, b) demonstrated that it is possible to extract all ions in this mode (obtaining saturation conditions, see e.g. Fig. 5-25) in accordance with theoretically determined ion trajectories under similar field configurations (e.g. Fig. 5-18). This penetrating field changes the energy of the electron beam, but, Stephan et al. reported that the change in electron energy is linear (e.g. 100 mV per 30 V at L_2, see Fig. 5-15) and is less severe than in cases where the electric extraction field is applied between electrodes confining the collision chamber (Honig 1948, Waldron and Wood 1952).

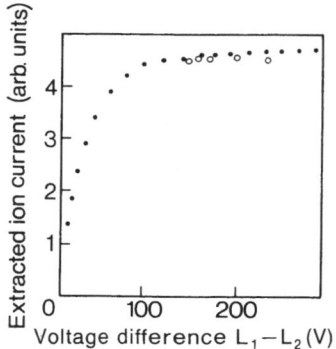

Fig. 5-25. Measured ion current (after Stephan et al. 1980 b) as a function of voltage difference between L_1 and L_2 (see Fig. 5-15) with the following electrode potentials (measured with respect to ground): $R = C = L_1 = 3\,\mathrm{kV}$, $L_3 = 2.85\,\mathrm{kV}$, $L_4 = 2.77\,\mathrm{kV}$, $L_6 = 0.06\,\mathrm{kV}$, $L_7 = 0.20\,\mathrm{kV}$, $L_5 = L_8 = L_9 = S_1 = 0\,\mathrm{kV}$. Full dots: total helium ion current measured on a collector in the plane of the mass spectrometer entrance slit; open circles: He$^+$ ion current obtained by integrating over the mass analyzed beam profile (see e.g. Fig. 5-26)

In addition, Stephan et al. (1980 a, b) improved also the transmission of the extracted ion current through the subsequent slits of the mass analyzer, because under usual conditions the transmission is only of the order of a few percent and also depends on the mass to charge ratio and the experimental conditions of the ion source. They introduced a so called *deflection method*, i.e. sweeping the extracted ion current across the mass spectrometer entrance slit (Fig. 5-15) with help of the

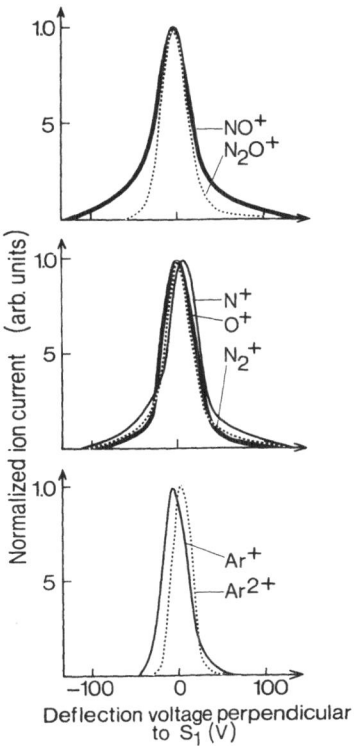

Fig. 5-26. Normalized ion beam profiles in y-direction (perpendicular to S_1) of parent and fragment ions of N_2O and of Ar^+ and Ar^{2+} after Märk *et al.* 1981 a

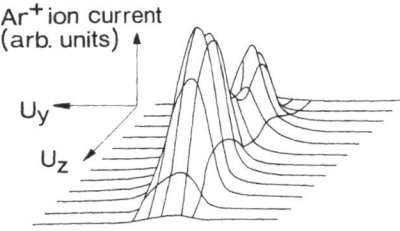

Fig. 5-27. Extracted and mass-analyzed Ar^+ current as a function of beam deflection voltage in the y-direction (perpendicular to the slits S_1), U_y, and z-direction (parallel to the slit S_1), U_z, after Stephan (1979)

deflection plates L_6 and L_7 (parallel to S_1; z-direction) and L_8 and L_9 (perpendicular to S_1; y-direction) and integrating over the ion current as a function of beam deflection voltage (e. g. see as an example Fig. 5-26 and 5-27). This allows to obtain a representative measure of the total extracted mass-analyzed ion flux (see also Coggeshall 1962) and hence accurate values of the partial ionization cross section ratios, despite the fact (which under normal operating conditions leads to erroneous ratios) that the ion beam shape and position is different for ions with different m/ze

(e. g., Fig. 5-26). If certain experimental conditions are fulfilled (Stephan *et al.* 1980 b), the accurate measurements of partial ionization cross section functions can be made, however, without sweeping the ion beam. It is interesting to note that this deflection method is a useful technique for distinguishing ions produced by electron impact ionization of the static target gas (capillary leak gas inlet in Fig. 5-15) and ions produced of neutral beam particles (molecular beam or molecular jet gas inlet via the nozzle N in Fig. 5-15, see also 5.2.3 e) (e. g. see Helm *et al.* 1979, 1981; Stephan *et al.* 1981, 1982 a — d, 1983 b — c, 1984, Stephan and Märk 1982 a, b, Futrell *et al.* 1981, 1982).

In contrast, some authors improved the transmission through the mass spectrometer entrance slit with help of additional focussing optics (e. g. quadrupole lenses etc.), thus succeeding to focus the fanned out ion beam onto S_1 and avoiding discrimination at S_1 (e. g. Hagstrum 1951, 1953, Drewitz 1976, Drewitz and Taubert 1976, Schmidt *et al.* 1976, Nagy *et al.* 1980).

5.2.2 Other Experimental Considerations

a) Electron Beam Properties

For the accurate determination of partial ionization cross section functions via equation (5-18) it is not only essential to fulfill the conditions mentioned by Kieffer and Dunn (1966) for the measurement of $n(0)_e$, but also of great importance that the electron energy be well defined and known. The electron guns usually used for the determination of partial ionization cross section functions employ metal (oxide) filaments providing electrons principally by thermionic emission, use a set of lenses and collimators to obtain a narrow beam, and apply a voltage difference ΔV between the filament and the collision chamber to accelerate the electrons to a nominal energy of $\Delta V \cdot e$. Due to contact and surface potential differences of the electrodes and due to a potential drop across the filament, the nominal energy can be quite different from the actual electron energy, the difference amounting sometimes to more than 1 eV. This is of no concern at high electron energies (because of the slow change of the cross section), but close to the threshold, where the cross section rises steeply, this fact may make a large difference. This is especially crucial if a comparison of different experimental results is to be made (e. g. see as an example σ (He$^+$/He) close to threshold in Fig. 5-28 (Märk and de Heer 1979)). Hence, calibration of the electron energy scale is necessary, via the determination of appearance energies, taking also into account additional shifts due to energy distributions as is the case for thermionic emission (see e. g. Asundi and Kurepa 1963, Kieffer and Dunn 1966, Vriens *et al.* 1968, Field and Franklin 1970, Litzow and Spalding 1973, Beynon *et al.* 1975, Rosenstock 1976, Märk 1984 a). Moreover, the measured ion signal i_{ms} (see equation 5-18) is, for any nominal (or corrected) electron energy, proportional to the integral over the product of the electron energy distribution and the partial ionization cross section. It is clear that structures in the ionization cross section function comparable or less than the width of the electron energy distribution are obscured. Thus it is crucial, especially for the detection of structure close to threshold (see Chapter 3) to have as monoenergetic an electron beam as possible.

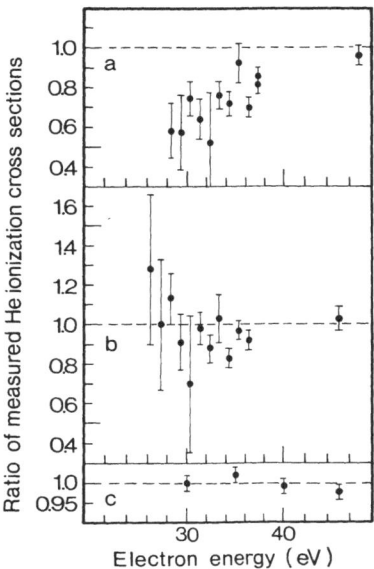

Fig. 5-28. Comparison of different experimental results of the electron impact ionization of He close to threshold. *a* ratio between $\sigma(\text{He}^+/\text{He})$ measured by Brooks *et al.* 1978 and $\sigma_t(\text{He})$ measured by Rapp and Englander-Golden 1965 (Note: in this energy regime $\sigma_i = \sigma(\text{He}^+/\text{He})$). *b* ratio between $\sigma(\text{He}^+/\text{He})$ measured by Brooks *et al.* 1978 with electron energy scale decreased by 2 eV (see Märk and de Heer 1979) and $\sigma_t(\text{He})$ measured by Rapp and Englander-Golden 1965. *c* ratio between $\sigma(\text{He}^+/\text{He})$ measured by Stephan *et al.* 1980 b and $\sigma_t(\text{He})$ measured by Rapp and Englander-Golden 1965

In addition, it should be mentioned, that a change of the electron beam size (or density distribution) with electron energy can introduce a possible error in measured partial ionization cross section functions due to a possible different extraction efficiency for ions originating at different positions in the ion source (see previous section). A related question is a possible variation of the electron path length with electron energy due to the guiding magnetic field (see also discussion in Chapter 3 and 7). Rapp and Englander-Golden (1965) have demonstrated in a careful error analysis based on the work of Massey *et al.* (1952, 1969) and Asundi (1963), that the energy dependence of total ionization cross sections (and this also applies to partial ionization cross sections) is affected by less than $\pm 1\%$ by this effect. In addition, a recent systematic study of the spiralling in a magnetically confined electron beam supports this view (Taylor *et al.* 1974).

b) Absolute Calibration

The absolute measurement of partial ionization cross sections not only necessitates the absolute determination of the quantities discussed so far (i.e., i_{ms}, $n(0)_e$, and L (see equation (5-18)), but also of the target gas density N_t. Usually, the perfect gas law is used to determine this quantity with help of a gas pressure measurement. In this case temperature differences between the collision regime and the pressure gauge have to be taken into account (Kieffer and Dunn 1966).

Gas pressures are typically measured with a McLeod gauge, although large errors can occur due to temperature and condensation effects (Ishii and Nakayama 1961, Meinke and Reich 1962, 1963, Rothe 1964, Rapp and Englander-Golden 1965). Recently, relative or absolute dynamic techniques have been used successfully (de Maria et al. 1963, Brunnee and Voshage 1964, Rapp and Englander-Golden 1965, Harrison et al. 1966, Beran and Kevan 1969, Kurepa et al. 1974, Märk 1975, 1982 b, c), using also capacitance manometers on the high pressure side.

However, because of the problems encountered with the collection efficiency in mass spectrometers, in most studies only relative partial ionization cross section functions and cross section ratios have been determined. If for a specific system all relative cross section curves and ratios have been measured it is possible with help of the measured ratios to calibrate the curves by normalization of the weighted sum of partial ionization cross sections (at a certain electron energy) versus the total ionization cross section ("summation method", see also Chapter 7). An alternate (seldomly used) method is the so called difference (Pavlov and Stotskii 1970) or fitting method (Stephan et al. 1978), which only needs the absolute total ionization cross section function and the respective relative partial ionization cross section functions. It should be checked in each case, however, whether the normalization factors obtained are independent of electron energy in order to exclude systematic errors.

Note: Closely related to this problem of absolute calibration (even for relative measurements) is the fact that discrimination may occur at the ion detector, falsifying i_{ms}. This is true especially if an ion multiplier is used as detector, since the multiplier response depends quite strongly on the mass to charge ratio (e.g. Higatsberger et al. 1954, Klein 1965, Dietz 1965, Schram et al. 1966 a, Collins 1969, Van Gorkom and Glick 1970, Van Gorkom et al. 1970, Pottie et al. 1973, Tuithof and Boerboom 1974, Märk 1977, Nagy et al. 1980, Beuhler 1983).

c) Other Methods

Some target species are not stable or at low pressure under normal experimental conditions (static gas target at room temperature); i.e. atomic and molecular radicals, alkalis, metals, excited and metastable species, ions, Van der Waals clusters. The method of choice for the study of such species is the (modulated) crossed beam technique, which is reviewed in detail by Kieffer and Dunn 1966 and in other chapters of this book (see also 5.2.3 e).

In addition, it is of great interest for certain applications to know the final states of the ions produced by electron impact ionization, i.e. to know the fraction of a partial ionization cross section corresponding to the production of the individual ion in a specific state. It is clear that the mass spectrometric analysis is not sufficient for this purpose. If the final state under study is radiative, it is possible to measure the emitted photons. Such a fluorescence technique involves the difficult task of absolute radiometry. This problem is especially difficult in the extreme ultraviolet region. For recent progress in this field see Beyer et al. 1979, Kühne et al. 1980, Mc Pherson et al.

1980, Einfeld 1980 and Bloemen *et al.* 1981. Closely related is the study of inner shell ionization, a subject which is covered in a separate chapter of this book (Chapter 6). Conversely, the production of ions excited to a metastable state necessitates a different experimental approach (e.g. Hagstrum 1956, Hofer *et al.* 1980, Varga *et al.* 1981), i.e. using the process of potential emission of electrons from a solid surface (e.g. Hagstrum 1956, 1960, Varga and Winter 1978). See also techniques used by Turner *et al.* 1968 and Hughes and Tiernan 1971.

5.2.3 Some Types of Experiments and Recent Techniques

As has been shown, there exist great difficulties and a number of possible pitfalls in the experimental determination of partial ionization cross sections. There exist, however, several recent experiments which have either employed new techniques or imaginative use of known techniques to overcome these difficulties. Some of these experiments and techniques will be discussed below:

a) Penetrating Field Extraction and Deflection Method

This technique, improving the operating conditions of a three-electrode, Nier type, ion source in combination with a sector field mass spectrometer (Stephan *et al.* 1980 a, b), has been discussed in detail in Section 5.2.1. It has been shown that the use of a penetrating field to extract ions out of the ion source ensures complete extraction (see Fig. 5-25), at least for thermal ions (for the extraction behavior of ions produced with excess kinetic energy see Kim 1981, Leiter 1983, Stephan *et al.* 1983 d, e, Leiter *et al.* 1984 a, b). Moreover, Stephan *et al.* (1980 a, b) showed that integration over

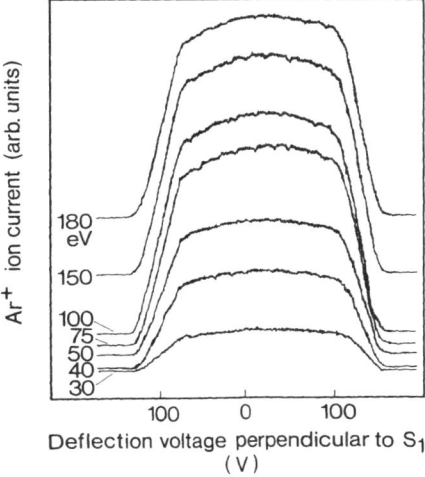

Fig. 5-29. Ion beam profiles (*y*-direction) of Ar$^+$ ions for different electron energies under "weak extracting and focussing" conditions (see text) after Stephan *et al.* 1980 a, b. The various curves are shifted vertically for the sake of clarity

extracted ion beam profiles (Figs. 5-26, 5-27) obtained with help of a deflection technique gives accurate ionization cross section ratios. The partial ionization cross section function, however, can be measured without this deflection and integration method based on the experimental observation that the ion-beam shape does not change with electron energy under certain experimental conditions ("weak extraction and focussing" (Stephan *et al.* 1980 a, b), e.g. see Fig. 5-29). In this case it is sufficient to center the ion beam approximately on the mass spectrometer entrance slit. The exact ion beam position was then found not to influence the accuracy of the partial ionization cross section function measured.

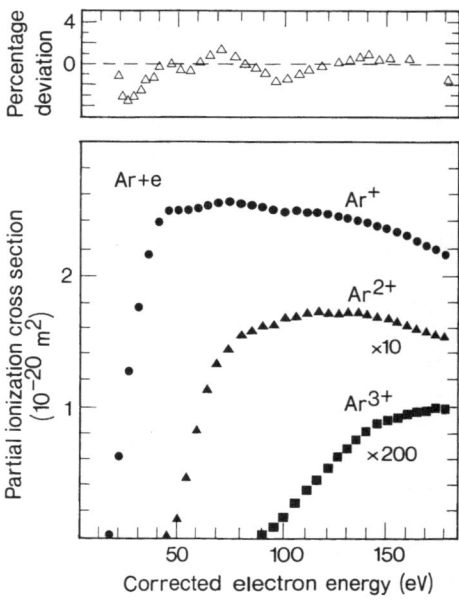

Fig. 5-30. *Lower part*: Absolute partial ionization cross section functions for the process $Ar + e \rightarrow Ar^+ + 2e$, $Ar + e \rightarrow Ar^{2+} + 3e$ and $Ar + e \rightarrow Ar^{3+} + 4e$ after Stephan *et al.* 1980 a, b. *Upper part*: Comparison (ratio) of the weighted sum of these partial ionization cross sections of Stephan *et al.* 1980 a, b with the total ionization cross section determined by Rapp and Englander-Golden 1965

Fig. 5-30 gives as an example partial ionization cross section functions obtained by this method by Stephan *et al.* (1980 a, b) for Ar. Also shown in this figure is a comparison with the total ionization cross section function showing the high accuracy of the mass spectrometer data obtained by this method. Märk (1982 b, c) has recently compared the experimental results for the rare gases of Stephan *et al.* (1980 a, b) with various theoretical results based on empirical, semiempirical, classical, semiclassical and quantal approximations. The agreement for single ionization in He, Ne and Ar between the experimental data and quantal calculations of Peach (1966, 1971) is noteworthy (see, e.g. Figs. 2-1, 2-2, 2-3, and 5-31). In case of double ionization large discrepancies exist (see, e.g. Figs. 2-4 and 2-5). Märk and co-workers have used this penetrating field extraction and deflection method in

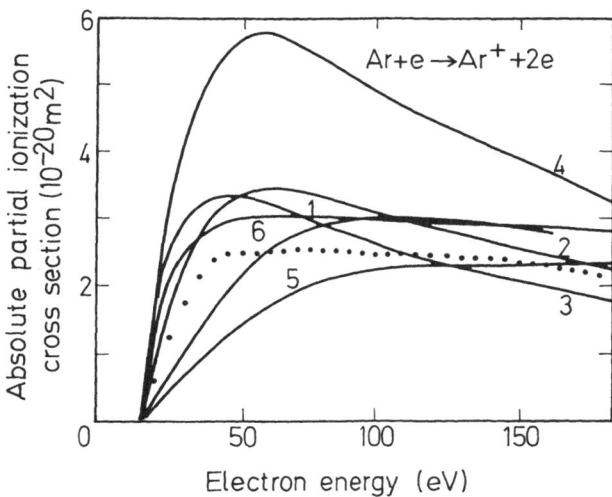

Fig. 5-31. Absolute partial ionization cross section functions for the process $Ar + e \rightarrow Ar^+ + 2e$ after Märk 1982 b. Full dots: experimental results of Stephan *et al.* (1980 a, b), *Curve 1*: calculated results by Pitchford *et al.* (1980) using the empirical formula of Elwert (1952), *Curve 2*: calculated results by Pitchford *et al.* (1980) using the formula given by Gryzinski (1965), *Curve 3*: calculated results by Pitchford *et al.* (1980) using the semiclassical formula given by Burgess (1963) and Vriens (1966), *Curve 4*: calculated results by Wallace *et al.* (1973) using the Born approximation, *Curve 5*: calculated results by Wallace *et al.* (1973) using the eikonal closure approximation, *Curve 6*: calculated results by Peach (1971) using the Ochkur approximation

addition to determine partial ionization cross section functions in He, Ne, Ar, Kr, Xe, H_2, N_2 (e.g. see Fig. 5-32), O_2, CO, NO, H_2O, D_2O, CO_2, N_2O (e.g. see Fig. 5-33), NO_2, NH_3 (e.g. see Fig. 5-8), ND_3, PH_3 (e.g. see Fig. 5-7), PD_3, CF_4 (e.g. see Fig. 5-34), CCl_4 and CF_2Cl_2 (Märk 1975, Märk and Egger 1976, 1977, Märk *et al.* 1977 a, Märk and Hille 1978, Hille and Märk 1978, Stephan *et al.* 1980 a, b, c, Märk *et al.* 1981 a, b Kim *et al.* 1981, Stephan and Märk 1981, 1984, Stephan *et al.* 1983 d, e, Leiter *et al.* 1984 a, b).

b) Large Acceptance Sector Field Mass Spectrometer

In order to avoid discrimination for ions with different mass to charge at the mass spectrometer entrance slit, Schmidt *et al.* (1976) and Nagy *et al.* (1980) used a conventional sector field mass spectrometer with a large acceptance (see also Hagstrum 1951, 1953, Giese 1959). For this purpose two electrostatic quadrupole lenses were installed in the field-free region (see Fig. 5-35) focussing the accelerated ion beam at the mass spectrometer entrance with variable astigmatism. In order to avoid discrimination in the z-direction (direction of the magnetic field) the electron beam traverses the ion source in y-direction. As only thermal ions at relatively high electron energy (> 500 eV) are studied, it is possible to extract all ions with help of a relatively large homogeneous extracting field without any discrimination. Using this

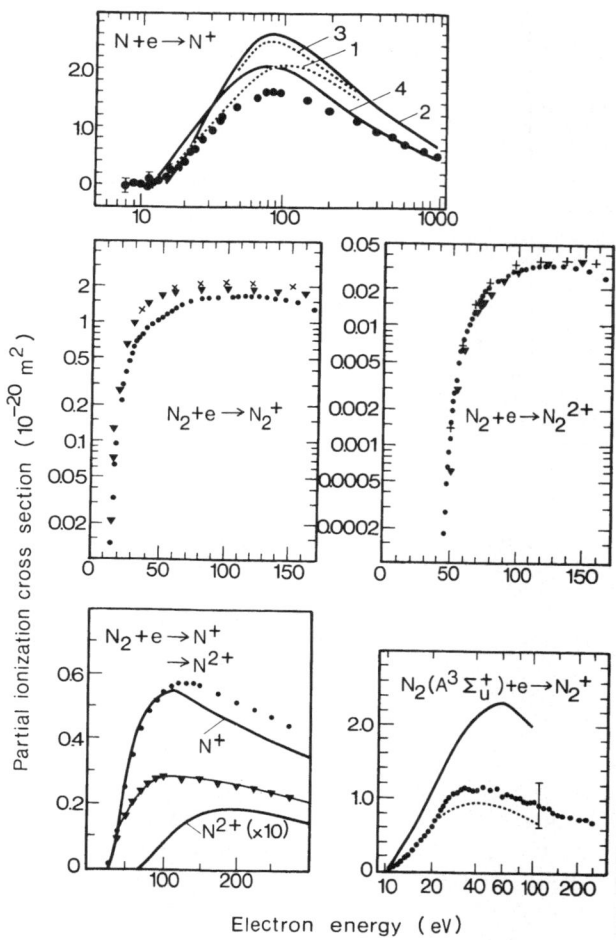

Fig. 5-32. Partial ionization cross sections for atomic and molecular nitrogen as a function of electron energy:

$N + e \rightarrow N^+$: Points: Brook *et al.* (1978), *Curve 1*: Seaton (1959), *Curve 2*: Peach (1970, 1971), *Curve 3*: Omidvar *et al.* (1972), *Curve 4*: Mc Guire (1972). Points are experimental, curve 1 to 4 are theoretical cross sections

$N_2 + e \rightarrow N_2^+$: ▼ ... Rapp *et al.* (1965), Rapp and Englander-Golden (1965) ($\sigma(N_2^+ + 2 N_2^{2+})$), × ... Halas and Adamczyk (1972), ● ... Märk (1975)

$N_2 + e \rightarrow N_2^{2+}$: + ... Daly and Powell (1966), ▼ ... Halas and Adamczyk (1972, ● ... Märk (1975)

$N_2 + e \rightarrow N^+$ and N^{2+}: ● ... Rapp *et al.* 1965, ▼ ... Halas and Adamczyk (1972), —— Crowe and Mc Conkey (1973 b)

$N_2 (A^3 \Sigma_u^+) + e \rightarrow N_2^+$: ● ... Armentrout *et al.* (1981). Theoretical cross section curves shown are: - - - semiclassical calculation by Tannen (1973) and —— binary-encounter approximation by Ton-That and Flannery (1977) (for the transition $N_2 (A^3 \Sigma_u^+) \rightarrow N_2^+ (A^2 \pi_u)$)

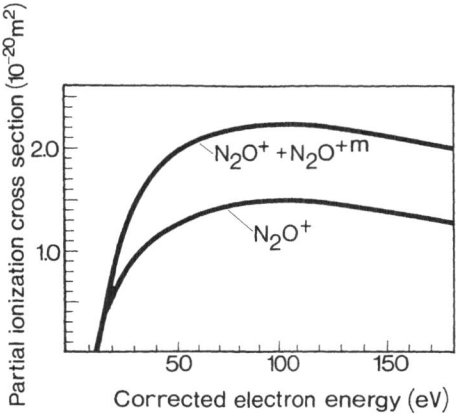

Fig. 5-33. Partial ionization cross section functions for N_2O after Märk *et al.* (1981a). Lower curve: production of stable N_2O^+, upper curve: production of stable and metastable N_2O^+

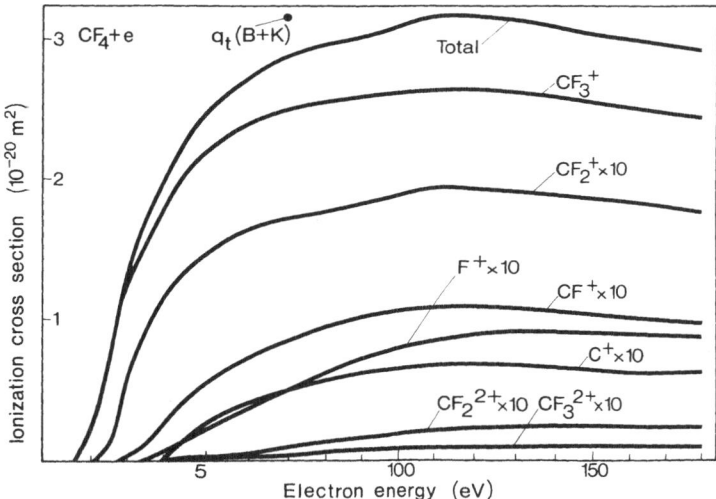

Fig. 5-34. Partial and total ionization cross section functions for fragment ions of CF_4 as measured by Stephan *et al.* (1983d, e). Full dot: total cross section at 70 eV determined by Beran and Kevan (1969) recalibrated to the Ar value of Rapp and Englander-Golden (1965)

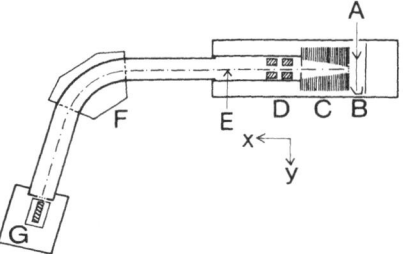

Fig. 5-35. Schematic view of the large acceptance sector field mass spectrometer after Nagy *et al.* (1980). *A* collimated electron beam (diameter ~ 1 mm), *B* electron trap, *C* accelerating section (~ 140 V/cm), *D* ion optics consist ng of two electrostatic quadrupole lenses of opposite sign, *E* focal point of the quadrupole ion optical system, *F* $60°$ magnetic sector field analyzer, *G* ion detecting system (channeltron)

experimental set-up Nagy *et al.* (1980) have made careful determinations of partial ionization cross sections for all rare gases between 500 eV and 5 keV. These authors give a thorough discussion of all possible errors and a comparison with other experimental and theoretical data. They showed that within the quoted error bars there is good agreement between their counting ionization cross sections and average experimental counting ionization cross sections as derived by de Heer and Jansen (1977) and de Heer *et al.* (1979) using a special weighting procedure of all previously existing experimental data. Moreover, the presented cross sections for He, Ne and Ar are close to theoretical calculations based on sum rules (e.g., Kim and Inokuti 1971, Kim *et al.* 1973, Eggarter 1975) or on direct calculations using good wavefunctions provided that exchange effects are taken into account (Knapp and Schulz 1974).

c] Field-Free Diffusive Extraction

Great care was recently taken by Crowe *et al.* 1972 and McConkey *et al.* 1972 to eliminate possible instrumental artefacts in measured partial ionization cross section functions, including those for dissociatively produced ions with excess kinetic energy. Their apparatus (see Fig. 5-36) consists of a modified Pierce-type (Pierce 1954) electron gun, capable of firing a beam of electrons with an extremely small beam divergence through a static gas target. The most important feature of this apparatus is the fact that the ions are produced in a field-free region. If the direction of an ion formed is such that it can drift through two apertures, defining an angular resolution of the system, it will then be accelerated and focussed by an ion lens system and mass analyzed with a quadrupole mass spectrometer. Because this ion lens has to be able to focus ions with a large range of excess kinetic energies to one point, a special design had to be used (see Fig. 3 in Crowe and McConkey 1973a). McConkey and co-workers used this apparatus to measure partial ionization cross sections (including the process of dissociative ionization) in Ar (e.g. see Fig. 5-3 and 5-24), H_2 (e.g. see Fig. 5-2), N_2 (e.g. see Fig. 5-32), CO_2 and NH_3 (Crowe *et al.* 1972, Crowe and McConkey 1973 a, b, 1974, 1977). In addition, this apparatus has been designed to investigate energy and angular distributions of dissociatively produced ions (see Chapter 4).

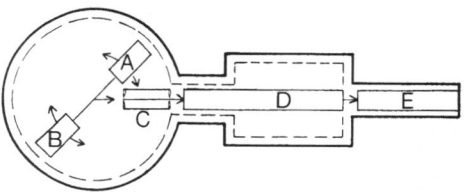

Fig. 5-36. Schematic view of field free diffusive extraction mass spectrometry after Crowe *et al.* 1972 and McConkey *et al.* 1972. *A* rotatable electron gun, *B* electron trap, *C* slit system with apertures defining the analyzer resolution, *D* ion lens with large ion energy acceptance, *E* quadrupole mass spectrometer and channeltron multiplier using single pulse counting

d) Open-Ion Source Cycloidal Mass Spectrometry

Whatever ion source is used for the determination of partial ionization cross sections, one is always faced with the fact of a velocity distribution and angular divergence of the ions entering the mass spectrometer. This leads to a mass-to-charge dependent discrimination unless a true double focussing analysator is used in combination with an appropriate ion source. To this end Schram *et al.* (1966 b), Schutten *et al.* (1966) and Adamczyk (1969) have put to use the principle of the cycloidal mass spectrometer combined with an open ion source without any slits (see Fig. 5-37). With this apparatus it is possible to achieve a complete collection of all (also energetic) ions produced in the source (Schutten *et al.* 1966, Adamczyk *et al.* 1966). A number of studies has been recently performed with this type of apparatus

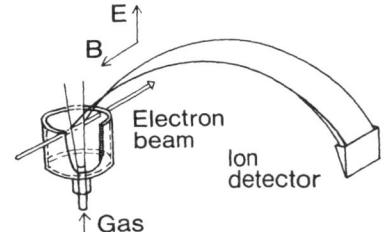

Fig. 5-37. Schematic view of cycloidal mass spectrometer with an open ion source after Schram *et al.* (1966 b)

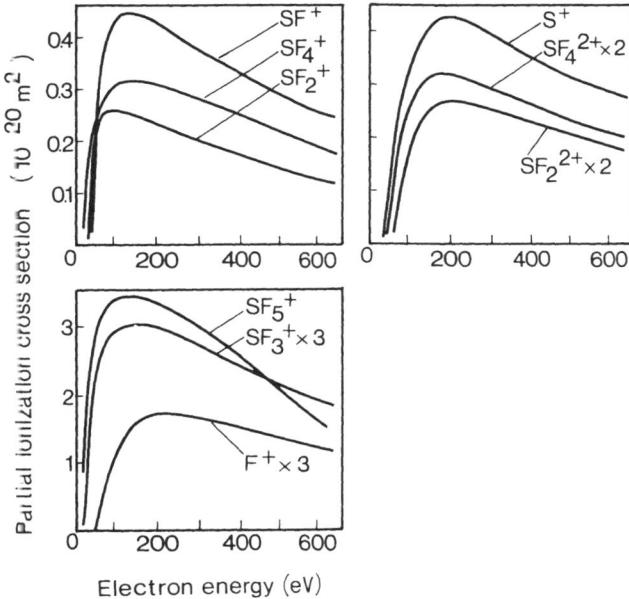

Fig. 5-38. Partial ionization cross section functions for fragment ions of SF_6 as measured by Stanski and Adamczyk 1982, 1983) with the open ion source cycloidal mass spectrometer (see 5.2.3 d)

:n order to determine partial ionization cross sections of parent (and fragment) ions of He, Ne, H_2, CH_4, H_2O, CO_2, NH_3 and SF_6 (see e.g. Fig. 5-38; Schutten *et al.* 1966, Adamzcyk *et al.* 1966, 1972; Halas and Adamczyk 1972, Bederski *et al.* 1980, Stanski and Adamczyk 1982, 1983).

ε) Advanced Crossed-Beam Experiments

The development and use of crossed-beam experiment with *thermal* beams (a technique introduced by Funk 1930 and pioneered by Boyd and Green 1958 and Fite and Brackmann 1958) was reviewed in detail by Fite (1962), Mc Daniel (1964), Kieffer and Dunn (1966), Mc Dowell (1969), Dolder (1980) and Märk and Castleman (1984) (see also details given in Chapter 7). An important advance was the introduction of *fast* neutral beams. This technique has been first utilized by Cook and Peterson (1962) and Peterson (1964), and more recently by Dixon and Harrison (1971), Dixon *et al.* (1973, 1975, 1976), Brook *et al.* (1978), and Armentrout *et al.* (1981). In this technique the fast beam target is prepared by charge transfer reactions of a fast ion beam in a static gas target cell filled with the charge transfer reactant (see e.g. Fig. 5-39). The major advantage of this method is the fact (1) that absolute partial ionization cross sections are obtained without recourse to normalization procedures (as is the case for thermal beam studies), because the neutral target beam is so energetic that the flux can be measured directly and (2) that it provides fully dissociated beams of atoms which cannot, in general, be produced by thermal or discharge dissociation. (Note: Atomic beams produced by heating an oven can contain unknown amounts of dimers, i.e. for Li at an oven temperature of 1000° K the ratio Li_2/Li may be as large as 5% (Mc Dowell 1969).) It should be mentioned in this conjunction, however, that in a series of recent studies by Karstensen and co-workers (Karstensen and Köster 1971, Schneider 1974, Karstensen and Schneider 1975, 1978, Dettmann and Karstensen 1982) on electron impact ionization of Ca, Mg (e.g. see Fig. 5-6) and Ba (e.g. see Fig. 5-4), some of the problems inherent to crossed beam studies with thermal beams have been solved (see also Orient and Srivastava 1983, 1984 and Mathur and Badrinathan 1984).

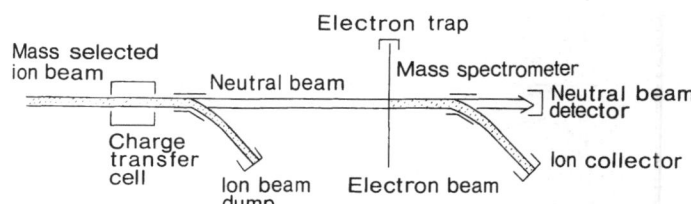

Fig. 5-39. Schematic view of the fast ion beam charge exchange technique employed for a crossed beam electron impact ionization study (see text)

The fast-ion beam charge technique has been successfully used after its introduction by Peterson and co-workers to measure partial ionization cross section functions for the following processes, i.e. $H(2S) + e \rightarrow H^+$ (Dixon and Harrison 1971, Dixon *et al.* 1975, e.g. see Fig. 5-40), $He(2^3S) + e \rightarrow He^+$ (Dixon *et al.* 1975, 1976, e.g. see Fig. 5-41), $Ne(^3P_2, {}^3P_0) + e \rightarrow Ne^+$ (Dixon *et al.* 1973), $Ar(^3P_2, {}^3P_0) + e \rightarrow Ar^+$ (Dixon *et al.*

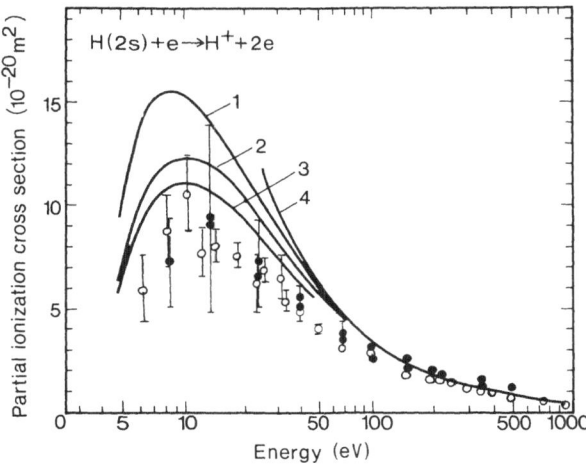

Fig. 5-40. Partial ionization cross section function for the process $H(2s)+e \rightarrow H^+ + 2e$. Full dots: experimental results by Dixon *et al.* (1976), open circles: experimental results by Defrance *et al.* (1981). Theoretical results (see also Kunc 1980): Curve *1*: Born-A results of Piraux and Joachain (1980), *Curve 2*: Born-B results of Prasad (1966), *Curve 3*: Born-exchange results of Prasad (1966), *Curve 4*: Born-Bethe approximation of Vriens and Bonsen (1968)

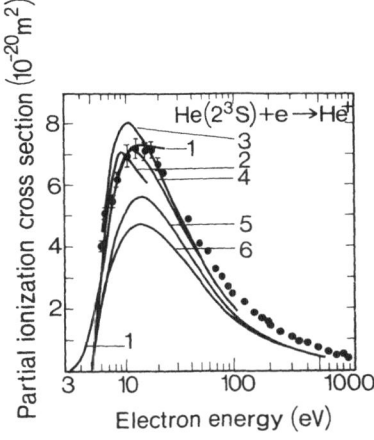

Fig. 5-41. Partial ionization cross section for metastable helium atoms, predominantly in the 2^3S state (in the experiments), as a function of electron energy

Points: Fast ion beam charge exchange technique by Dixon *et al.* (1976), Curve *1*: Thermal beam technique by Fite and Brackmann (1963) recalibrated by Dixon *et al.* (1976), Curve *2*: Electron beam excitation technique by Long and Geballe (1970), Curve *3*: Binary encounter approximation (Ton-That *et al.* 1976), Curve *4*: Born I approximation (Ton-That *et al.* 1976), Curve *5*: Born II approximation (Peach 1976), Curve *6*: Born-Ochkur approximation (Peach 1976)

1973), $N_2(4\Sigma_u^+)+e \rightarrow N_2^+$ (Armentrout *et al.* 1981, e.g. see Fig. 5-32), the single ionization of He, C, O and N (e.g. see Fig. 5-32) atoms (Brook *et al.* 1978) and ionization of highly reactive radicals (Baiocchi *et al.* 1984, Wetzel *et al.* 1984). One aspect of the experiments of Brook *et al.*, however, remains to be solved, e.g. the partial cross sections for single ionization of He as reported by Brook *et al.* 1978 are not zero below threshold. This was attributed to excited states present in the neutral beam and a correction was applied. With this correction the obtained results did not agree below 70 eV with other results. Märk and de Heer (1979) suggested that this can be improved if quite small changes are made to the below threshold correction and/or the energy scale of the fast beam measurements (see Fig. 5-28). Harrison *et al.* (1979) have replied to this proposal.

f. Trapped-Ion Mass Spectrometry

Another interesting new method for the study of electron impact ionization (as a function of time after formation of the ions) is based on the fact that positive ions may be trapped in the negative potential well produced by an electron space charge (Baker and Hasted 1966, 1968, Cuthbert *et al.* 1966, Redhead 1967, Bourne and Danby 1968, Herod and Harrison 1970). In this case a continuous electron beam from a thermionic filament (~ 5 eV, ~ 10 μA) is used to trap the ions produced when a short (~ 1 μs) voltage pulse (variable in height and negative with respect to the ion source chamber) is applied to this filament. At a known and variable time (up to 2 ms, Lifshitz and Gefen 1980) after this ionizing pulse, a positive voltage pulse is applied to the repeller (of this Nier type ion source) to extract the ions to the mass analyzer. Ion source concentrations have to be as low as possible in order to minimize contributions from ion molecule reactions. This technique has been recently used to measure time-dependent appearance potentials (to determine the kinetic shift) in the benzene system by measuring the partial ionization cross sections at various delay times (Lifshitz *et al.* 1974) and, similarly, to measure partial ionization cross sections for parent and fragment ions of 1.5 hexadiyne, cubane, aniline as a function of the time after the electron impact ionization event took place (Lifshitz and Gefen 1980, Lifshitz and Eaton 1983, Lifshitz *et al.* 1983); e.g. see Fig. 5-10 and discussion in Section 5.1.2 b.

It is interesting to note that this trapping technique has also been used extensively in recent years to study the energetics of consecutive ionization processes in a technique known as sequential mass spectrometry (Hasted 1967).

5.3 Results and Discussion

It is clearly outside the scope of this review to give a detailed account of all results available on partial ionization cross sections. Hence this section will be illustrative rather than exhaustive, but it will give the appropriate references to previous reviews, compilations and studies on partial ionization cross section functions to facilitate the access to the relevant data of interest to the reader.

The literature was for the first time (through 1968) searched systematically for ionization cross sections data by the JILA Information Centre, and comprehensive compilations have been issued by Kieffer and Dunn (1966), Kieffer (1968, 1969).

According to Kieffer and Dunn (1966) there were considerable differences in the partial ionization cross section data reported in the literature and in addition there were also large gaps in the data available. Moreover, it was concluded by Kieffer that almost all the measurements reported so far were defective, i.e. that apparently known sources of systematic errors (as described by Kieffer and Dunn, 1966) were not taken into account. According to Kieffer the range of uncertainties can be best illustrated by comparing different independent measurements, yielding differences varying from a minimum of about 10% up to factors of 2 and more. The situation has improved in the meanwhile (due to the advent of some new and sophisticated experimental studies, see Section 5.2.3), but the new data have not been collected in a new compilation (except for specific purposes, e.g. Drowart and Goldfinger 1967, Laborie *et al*. 1968, 1971, Kieffer 1973, Mc Daniel *et al*. 1977, Ciamda 1980, Barnett *et al*. 1980, de Heer 1981, Bell *et al*. 1983, Märk 1982 b, c, Märk 1984 a). These new data are referenced though in several bibliographies (e.g. Chamberlain and Kieffer 1970, Kieffer 1976, Gallagher *et al*. 1979, Gallagher and Beaty 1980) and bulletins (e.g., Atomic Data for Fusion (D. H. Crandall, C. F. Barnett and W. L. Wieser, eds.) and International Bulletin on Atomic and Molecular Data for Fusion (K. Katsonits, ed.)). In addition, a limited amount of data is summarized in previous monographies on various subjects in atomic and molecular collision physics (e.g., Massey 1956, Craggs and Massey 1959, Bates 1962, Mc Daniel 1964, von Engel 1965, Massey *et al*. 1969, Valyi 1975, Christophorou 1984).

The results available for absolute partial electron impact ionization cross sections are summarized in the following in the form of a data index, giving the respective cross sections measured, the electron energy range and the bibliographic citation. In addition, some actual data will be shown as illustrative examples. The main classification of this section is by the complexity of the neutral target, i.e. atoms, diatomics etc. In general the same criteria for selection of data as applied by the JILA Information Center have been used. For some gases only one study has been reported in the literature and is presented in the following despite large possible errors as in the case of fragmentation cross sections in some older studies. In addition, some references to measured ionization cross section ratios as a function of energy or clastograms will be included for the interested reader. In some cases partial ionization cross sections or (relative) ratios have been measured only at one particular electron energy or only in a limited range (e.g. Milne 1958, Colin 1961, Berkowitz *et al*. 1962, Brink 1964, Cooper *et al*. 1966, Kant 1966, Melton and Rudolph 1967, Drowart and Goldfinger 1967, Lin and Stafford 1967, 1968, Skudlarski *et al*. 1967, Fehlner and Callen 1968, Drewitz 1976, Egger and Märk 1978 b, Grimley *et al*. 1978, Raheja *et al*. 1983, Danchevskaya and Torbin 1984 and references given in Chapter 2, Section 2.2.4) and will not be included in the following index. In some of the studies great care was taken to eliminate all possible systematic errors (as summarized by Kieffer and Dunn 1966) and those studies are designated, in case of the rare gases, with an asterix.

5.3.1 Atoms

a) The Rare Gases

Since the rare gases are inert and act as ideal gases at the pressures used, they were the first and most frequent to be investigated quantitatively. Therefore, partial ionization cross section data of the rare gases appear to be the most reliable based on conclusions about likely systematic errors. Märk (1982b) has recently reviewed these measurements and given a set of recommended rare gas partial ionization cross sections (see Table 5-2) based on the data of Rapp and Englander-Golden (1965), Schram (1966), Schram et al. (1965, 1966 c, d), Stephan et al. 1980 a, b, Nagy et al. 1980 and Stephan and Märk (1981, 1984). Partial ionization cross section functions for Kr are shown as an example in Figs. 2-3, 2-5, for Ar in Figs. 5-3, 5-24, 5-30, 5-31, for Ne in Fig. 2-2 and for He in Figs. 5-28, 2-1, 2-4.

Table 5-2. *Set of recommended absolute partial ionization cross sections for the rare gases after Märk (1982b). Electron energy in keV and cross sections in 10^{-20} m²*

Electron energy	$\sigma(He^+/He)$	$\sigma(He^{2+}/He)$	$\sigma(Ne^+/Ne)$	$\sigma(Ne^{2+}/Ne)$	$\sigma(Ne^{3+}/Ne)$	$\sigma(Ar^+/Ar)$	$\sigma(Ar^{2+}/Ar)$
0.05	0.245	—	0.338	—	—	2.50	0.016
0.10	0.366	0.00014	0.655	0.0059	—	2.51	0.17
0.15	0.367	0.00073	0.730	0.021	0.00004	2.34	0.17
0.50	0.221	0.0013	0.534	0.025	0.0015	1.29	0.071
1.0	0.140	0.00067	0.355	0.014	0.00087	0.805	0.046
2.0	0.0790	0.00032	0.215	0.0075	0.00047	0.471	0.024
3.0	0.0568	0.00022	0.159	0.0051	0.00032	0.344	0.018
4.0	0.0452	0.00016	0.126	0.0039	0.00026	0.269	0.014
5.0	0.0372	0.00013	0.107	0.0033	0.00023	0.226	0.011
6.0	0.0324	0.00010	0.0932	0.0026	0.00015	0.200	0.011
8.0	0.0254	0.000071	0.0739	0.0020	0.00012	0.157	0.0085
10.0	0.0211	0.000060	0.0624	0.0016	0.000095	0.132	0.0072
12.0	0.0182	0.000050	0.0541	0.0013	0.000081	0.117	0.0065
14.0	0.0163	0.000045	0.0476	0.0012	0.000068	0.101	0.0055

Electron energy	$\sigma(Ar^{3+}/Ar)$	$\sigma(Kr^+/Kr)$	$\sigma(Kr^{2+}/Kr)$	$\sigma(Kr^{3+}/Kr)$	$\sigma(Xe^+/Xe)$	$\sigma(Xe^{2+}/Xe)$	$\sigma(Xe^{3+}/Xe)$
0.05	—	3.64	0.10	—	4.87	0.27	—
0.10	0.00075	3.56	0.31	0.0064	4.91	0.53	0.072
0.15	0.0044	3.20	0.28	0.021	4.34	0.49	0.18
0.50	0.0074	1.66	0.086	0.033	2.11	0.23	0.11
1.0	0.0064	1.10	0.069	0.034	1.25	0.16	0.073
2.0	0.0054	0.638	0.049	0.027	0.732	0.11	0.046
3.0	0.0040	0.458	0.036	0.021	0.533	0.084	0.036
4.0	0.0033	0.355	0.029	0.017	0.409	0.065	0.027
5.0	0.0030	0.299	0.025	0.015	0.336	0.054	0.023
6.0	0.0027	0.250	0.021	0.013	0.289	0.046	0.018
8.0	0.0022	0.197	0.017	0.011	0.227	0.037	0.015
10.0	0.0019	0.161	0.015	0.0094	0.186	0.032	0.012
12.0	0.0017	0.137	0.013	0.0085	0.159	0.027	0.011
14.0	0.0014	0.123	0.012	0.0072	0.140	0.024	0.0096

He: Bleakney and Smith (1936): $\sigma(He^{2+}/He)$ (threshold up to 500 eV)
Stanton and Monohan (1960): $\sigma(He^+/He) / \sigma(He^{2+}/He)$ (100 up to 2400 eV)
Schram et al. (1966 c): $\sigma(He^+/He)$, $\sigma(He^{2+}/He)$ (0.5 up to 16 keV)
Adamczyk et al. (1966)*: $\sigma(He^+/He)$, $\sigma(He^{2+}/He)$ (threshold up to 1000 eV)
Gaudin and Hagemann (1967): $\sigma(He^+/He)$, $\sigma(He^{2+}/He)$ (100 up to 2000 eV)

Brook et al. (1978)*: σ (He$^+$/He) (threshold up to 1000 eV, see also Märk and de Heer 1979)

Stephan et al. (1980 a, b)*: σ (He$^+$/He), σ (He^{2+}/He) (threshold up to 180 eV)

Nagy et al. (1980)*: σ (He$^+$/He), σ (He^{2+}/He) (0.5 up to 5 keV)

Ne: Bleakney (1930 a): σ (Ne$^+$/Ne), σ (Ne^{2+}/Ne), σ (Ne^{3+}/Ne) (threshold up to 500 eV)

Stevenson and Hipple (1942): σ (Ne$^+$/Ne) / σ (Ne^{2+}/Ne) (threshold up to 200 eV)

Ziese (1965): σ (Ne$^+$/Ne) / σ (Ne^{2+}/Ne) / σ (Ne^{3+}/Ne) / σ (Ne^{4+}/Ne) (250 to 2000 eV)

Schram et al. (1966 c): σ (Ne$^+$/Ne) through σ (Ne^{5+}/Ne) (0.5 up to 15 keV)

Adamczyk et al. (1966)*: σ (Ne$^+$/Ne), σ (Ne^{2+}/Ne), σ (Ne^{3+}/Ne) (threshold up to 2000 eV)

Gaudin and Hagemann (1967): σ (Ne$^+$/Ne) through σ (Ne^{4+}/Ne) (100 up to 2000 eV)

Stephan et al. (1980 a, b)*: σ (Ne$^+$/Ne), σ (Ne^{2+}/Ne), σ (Ne^{3+}/Ne) (threshold up to 180 eV)

Nagy et al. (1980)*: σ (Ne$^+$/Ne), σ (Ne^{2+}/Ne), σ (Ne^{3+}/Ne) (0.5 up to 5 keV)

Ar: Bleakney (1930 a): σ (Ar$^+$/Ar) through σ (Ar^{4+}/Ar) (threshold up to 500 eV)

Stevenson and Hipple (1942): σ (Ar$^+$/Ar) / σ (Ar^{2+}/Ar) (threshold up to 200 eV)

Fiquet-Fayard (1962), Fiquet-Fayard and Lahmani (1962), Fiquet-Fayard and Ziesel (1963 a): σ (Ar$^+$/Ar) / σ (Ar^{2+}/Ar) / σ (Ar^{3+}/Ar) / σ (Ar^{4+}/Ar) / σ (Ar^{5+}/Ar) (threshold up to 500 eV)

Schram (1966): σ (Ar$^+$/Ar) through σ (Ar^{7+}/Ar) (0.5 up to 18 keV)

Gaudin and Hagemann (1967): σ (Ar$^+$/Ar) through σ (Ar^{5+}/Ar) (100 to 2000 eV)

Okudaira et al. (1970): σ (Ar$^+$/Ar) / σ (Ar^{2+}/Ar) / σ (Ar^{3+}/Ar) / σ (Ar^{4+}/Ar) / σ (Ar^{5+}/Ar) (threshold up to 1000 eV)

Crowe et al. (1972)*: σ (Ar$^+$/Ar) / σ (Ar^{2+}/Ar) (threshold up to 300 eV)

Stephan et al. (1980 a, b)*: σ (Ar$^+$/Ar), σ (Ar^{2+}/Ar), σ (Ar^{3+}/Ar) (threshold up to 180 eV)

Nagy et al. (1980)*: σ (Ar$^+$/Ar), σ (Ar^{2+}/Ar), σ (Ar^{3+}/Ar) (0.5 up to 5 keV)

Kr: Tate and Smith (1934): σ (Kr$^+$/Kr) / σ (Kr^{2+}/Kr) / σ (Kr^{3+}/Kr) / σ (Kr^{4+}/Kr) (threshold up to 500 eV)

Fiquet-Fayard and Ziesel (1963 a), Ziesel (1967 a): σ (Kr$^+$/Kr) / σ (Kr^{2+}/Kr) / σ (Kr^{3+}/Kr) / σ (Kr^{4+}/Kr) / σ (Kr^{5+}/Kr) (threshold up to 500 eV)

Schram (1966): σ (Kr$^+$/Kr) through σ (Kr^{9+}/Kr) (0.5 up to 15 keV)

Stephan et al. (1980 a, b)*: σ (Kr$^+$/Kr) through σ (Kr^{4+}/Kr) (threshold up to 180 eV)

Nagy et al. (1980)*: σ (Kr$^+$/Kr), σ (Kr^{2+}/Kr), σ (Kr^{3+}/Kr) (0.5 up to 5 keV)

Xe: Tate and Smith (1934): $\sigma(Xe^+/Xe)$ / $\sigma(Xe^{2+}/Xe)$ / $\sigma(Xe^{3+}/Xe)$ / $\sigma(Xe^{4+}/Xe)$ / $\sigma(Xe^{5+}/Xe)$ / $\sigma(Xe^{6+}/Xe)$ (threshold up to 600 eV)

Schram (1966): $\sigma(Xe^+/Xe)$ through $\sigma(Xe^{13+}/Xe)$ (0.5 up to 15 keV)

Nagy et al. (1980)*: $\sigma(Xe^+/Xe)$ through $\sigma(Xe^{3+}/Xe)$ (0.5 up to 5 keV)

Stephan and Märk (1981, 1984)*: $\sigma(Xe^+/Xe)$ through $\sigma(Xe^{3+}/Xe)$ (threshold up to 180 eV)

b) Alkalis, Alkaline Earths and Metals

Mc Dowell (1969) (see also Mc Farland 1967) has recently reviewed electron impact ionization of the alkalis and tabulated the results for total and single ionization of Li through Cs, using the data of Tate and Smith (1934) and Mc Farland and Kinney (1965). See also Laborie et al. (1971).

Li: Mc Farland and Kinney (1965): $\sigma(Li^+/Li)$ (threshold up to 500 eV)

Jalin et al. (1973): $\sigma(Li^+/Li)$, $\sigma(Li^{2+}/Li)$ (100 up to 2000 eV, see Fig. 7-12)

Na: Tate and Smith (1934): $\sigma(Na^+/Na)$, $\sigma(Na^{2+}/Na)$ (threshold up to 700 eV)

Kaneko (1961): $\sigma(Na^+/Na)$ / $\sigma(Na^{2+}/Na)$ (threshold up to 100 eV, no multiplier correction)

Brink (1962): $\sigma(Na^+/Na)$ / $\sigma(Na^{2+}/Na)$ (threshold up to 500 eV no multiplier correction, see also Brink 1964)

Ziesel (1965): $\sigma(Na^+/Na)$ / $\sigma(Na^{2+}/Na)$ / $\sigma(Na^{3+}/Na)$ / $\sigma(Na^{4+}/Na)$ / $\sigma(Na^{5+}/Na)$ (250 up to 2000 eV)

K: Tate and Smith (1934): $\sigma(K^+/K)$ / $\sigma(K^{2+}/K)$ (threshold up to 500 eV)

Kaneko (1961): $\sigma(K^+/K)$ / $\sigma(K^{2+}/K)$ (threshold up to 100 eV, no multiplier correction)

Brink (1962): $\sigma(K^+/K)$ / $\sigma(K^{2+}/K)$ (threshold up to 500 eV, no multiplier correction, see also Brink 1964)

Fiquet-Fayard and Lahmani (1962), Fiquet-Fayard and Ziesel (1963 a): $\sigma(K^+/K)$ / $\sigma(K^{2+}/K)$ / $\sigma(K^{3+}/K)$ / $\sigma(K^{4+}/K)$ / $\sigma(K^{5+}/K)$ (threshold up to 500 eV)

Rb: Tate and Smith (1934): $\sigma(Rb^+/Rb)$ / $\sigma(Rb^{2+}/Rb)$ / $\sigma(Rb^{3+}/Rb)$ (threshold up to 700 eV)

Fiquet-Fayard and Ziesel (1963 a), Ziesel (1967 a): $\sigma(Rb^+/Rb)$ / $\sigma(Rb^{2+}/Rb)$ / $\sigma(Rb^{3+}/Rb)$ / $\sigma(Rb^{4+}/Rb)$ / $\sigma(Rb^{5+}/Rb)$ (threshold up to 500 eV)

Cs: Tate and Smith (1934): $\sigma(Cs^+/Cs)$ / $\sigma(Cs^{2+}/Cs)$ / $\sigma(Cs^{3+}/Cs)$ / $\sigma(Cs^{4+}/Cs)$ / $\sigma(Cs^{5+}/Cs)$ / $\sigma(Cs^{6+}/Cs)$ (threshold up to 100 eV)

Nygaard (1968): $\sigma(Cs^+/Cs)$ (threshold up to 100 eV) (see also Nygaard and Hahn 1973)

Mg: Kaneko (1961): $\sigma(Mg^+/Mg)$ / $\sigma(Mg^{2+}/Mg)$ (threshold up to 100 eV, no multiplier correction)

Ziesel (1965): $\sigma(Mg^+/Mg) / \sigma(Mg^{2+}/Mg) / \sigma(Mg^{3+}/Mg) / \sigma(Mg^{4+}/Mg)$ (250 up to 2000 eV)

Okudaira et al. (1970): $\sigma(Mg^+/Mg)$, $\sigma(Mg^{2+}/Mg)$, $\sigma(Mg^{3+}/Mg)$ (threshold up to 1000 eV)

Karstensen and Schneider (1978): $\sigma(Mg^+/Mg)$, $\sigma(Mg^{2+}/Mg)$ (threshold up to 280 eV, see as an example Fig. 5-6)

Ca: Fiquet-Fayard and Lahmani (1962), Fiquet-Fayard and Ziesel (1963 a): $\sigma(Ca^+/Ca) / \sigma(Ca^{2+}/Ca) / \sigma(Ca^{3+}/Ca) / \sigma(Ca^{4+}/Ca) / \sigma(Ca^{5+}/Ca)$ (threshold up to 500 eV)

Okudaira (1970): $\sigma(Ca^+/Ca)$, $\sigma(Ca^{2+}/Ca)$, $\sigma(Ca^{3+}/Ca)$ (threshold up to 1000 eV)

Sr: Fiquet-Fayard and Ziesel (1963 a), Ziesel (1967 a): $\sigma(Sr^+/Sr) / \sigma(Sr^{2+}/Sr) / \sigma(Sr^{3-}/Sr) / \sigma(Sr^{4+}/Sr) / \sigma(Sr^{5+}/Sr)$ (threshold up to 500 eV)

Okudaira (1970): $\sigma(Sr^+/Sr)$, $\sigma(Sr^{2+}/Sr)$, $\sigma(Sr^{3+}/Sr)$ (threshold up to 1000 eV)

Ba: Ziesel and Abouaf (1967): $\sigma(Ba^+/Ba) / \sigma(Ba^{2+}/Ba) / \sigma(Ba^{3+}/Ba) / \sigma(Ba^{4+}/Ba)$ (threshold up to 500 eV, no multiplier correction)

Okudaira (1970): $\sigma(Ba^+/Ba)$, $\sigma(Ba^{2+}/Ba)$, $\sigma(Ba^{3+}/Ba)$ (threshold up to 1000 eV)

Dettmann and Karstensen (1982): $\sigma(Ba^+/Ba)$ through $\sigma(Ba^{4+}/Ba)$ (threshold up to 600 eV, see Fig. 5-4)

Cu: Crawford (1967): $\sigma(Cu^+/Cu)$, $\sigma(Cu^{2+}/Cu)$ (threshold up to 800 eV)

Zn: Fiquet-Fayard and Ziesel (1963 a, b), Ziesel (1967 b): $\sigma(Zn^+/Zn) / \sigma(Zn^{2-}/Zn) / \sigma(Zn^{3+}/Zn) / \sigma(Zn^{4+}/Zn) / \sigma(Zn^{5+}/Zn) / \sigma(Zn^{6+}/Zn)$ (threshold up to 2000 eV)

Ag: Crawford and Wang (1967): $\sigma(Ag^+/Ag)$, $\sigma(Ag^{2+}/Ag)$ (threshold up to 900 eV)

Hg: Bleakney (1929, 1930 b): $\sigma(Hg^+/Hg)$ through $\sigma(Hg^{4+}/Hg)$ (threshold up to 400 eV)

Pb: Pavlov and Stotskii (1970): $\sigma(Pb^+/Pb)$ through $\sigma(Pb^{5+}/Pb)$ (threshold up to 400 eV, "difference method" (see also Stephan et al. 1978))

c) Other Atomic Species

H: The three studies on partial ionization cross section functions of the process $H + e \rightarrow H^+ + 2e$ (Fite and Brackmann 1958, Boyd and Boksenberg 1959, Rothe et al. 1962), for which the data are presented by Rothe et al. (1962) together with theoretical results in Fig. 3 and by Kieffer and Dunn (1966) in Fig. 3, are all relative because of the difficulty in determining the neutral beam density. The normalization procedures used and the associated problems are discussed by Fite (1962) and Kieffer and Dunn (1966). Experimental and theoretical results are summarized in Fig. 5-42.

Gowar and Clark (1968): $\sigma(H^+/H)$ (close to threshold)

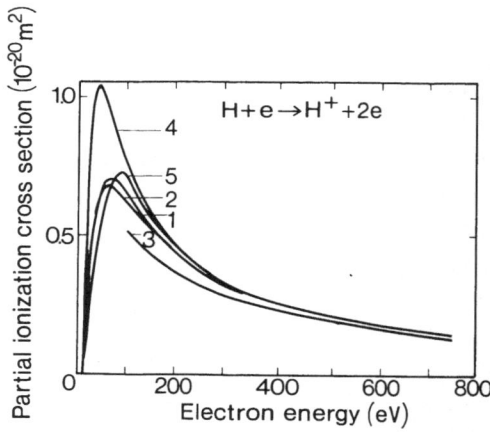

Fig. 5-42. Partial ionization cross sections for the process $H + e \rightarrow H^+ + 2e$ as a function of electron energy. *1* experimental results by Fite and Brackmann (1958); *2* experimental results by Boyd and Boksenberg (1959) normalized to the Born approximation at high electron energy; *3* experimental results by Rothe *et al.* (1962); *4* Born approximation (see Massey and Mohr (1933), Dalgarno (1953), Massey (1956), Peach (1965)); *5* Impulse approximation (Akerib and Borowitz 1961). See also Flannery (1970) and Kunc (1980)

C: Brook *et al.* (1978): $\sigma(C^+/C)$ (threshold up to 1000 eV)

N: Brook *et al.* (1978): $\sigma(N^+/N)$ (threshold up to 1000 eV, see Fig. 7-10)
       Results are shown as an example in Fig. 5-32

O: Fite and Brackmann (1959): $\sigma(O^+/O)$ (threshold up to 500 eV)
       Brook *et al.* (1978): $\sigma(O^+/O)$ (threshold up to 1000 eV, see Fig. 7-11)
       Ziegler *et al.* (1982): $\sigma(O^{2+}/O)$ (threshold up to 400 eV)

5.3.2 Diatomic Molecules

In case of diatomic molecules (and also for larger molecules), fragment ions may be produced by electron impact, and as mentioned above, these ions are difficult to collect quantitatively. Hence, only a few studies have been performed yielding reliable partial ionization cross section functions (see below). In addition Rapp *et al.* (1965) have measured without mass analysis the percentage of total ion current due to energetic (dissociative) ions with kinetic energies in excess of 0.25 eV, ranging from $\sim 7\%$ in H_2 to $\sim 35\%$ in N_2O, with other gases intermediate.

H_2:   Adamczyk *et al.* (1966): $\sigma(H^+/H_2)$, $\sigma(H_2^+/H_2)$ (threshold up to 1000 eV, see also Fig. 5-2)
       McGowan *et al.* (1968): $\sigma(H_2^+/H_2)$ (close to threshold)
       Crowe and McConkey (1973 a): $\sigma(H^+/H_2)/\sigma H_2^+/H_2)$ (threshold up to 25 eV, see also Fig. 5-2); $\sigma(H_2^+/H_2)$ and $\sigma(H^+/H_2)$ (close to threshold)

N_2:   Erhardt and Kresling (1967): $\sigma(N^+/N_2)$ (close to threshold)
       Halas and Adamczyk (1972): $\sigma(N_2^+/N_2)$, $\sigma(N^+/N_2)$, $\sigma(N_2^{2+}/N_2)$ (threshold up to 600 eV)

Crowe and Mc Conkey (1973 b): $\sigma(N_2^+/N_2)$, $\sigma(N^+/N_2)$, $\sigma(N^{2+}/N_2)$ (threshold up to 300 eV)

Märk (1975): $\sigma(N_2^+/N_2)$, $\sigma(N_2^{2+}/N_2)$ (threshold up to 170 eV)

Results for N_2 are shown as an example in Figs. 5-5 and 5-32

O_2: Erhardt and Kresling (1967): $\sigma(O^+/O_2)$ (close to threshold)

Märk (1975): $\sigma(O_2^+/O_2)$, $\sigma(O_2^{2+}/O_2)$ (threshold up to 170 eV)

CO: Hille and Märk (1978): $\sigma(CO^+/CO)$, $\sigma CO^{2+}/CO)$, $\sigma(CO^{2+m}/CO)$ (threshold up to 180 eV)

NO: Tate et al. (1935): $\sigma(NO^+/NO)$ / $\sigma(NO^{2+}/NO)$ (threshold up to 400 eV)

Kim et al. (1981): $\sigma(NO^+/NO)$, $\sigma(NO^{2+}/NO)$ (threshold up to 180 eV)

HCl: Nier and Hanson (1936): $\sigma(HCl^+/HCl)$ / $\sigma(H^+/HCl)$ / $\sigma(Cl^+/HCl)$ / $\sigma(HCl^{2+}/HCl)$ / $\sigma(Cl^{2+}/HCl)$ / $\sigma(Cl^{3+}/HCl)$ (threshold up to 500 eV)

P_2: Monnom et al. (1984): $\sigma(P_2^+/P_2)$, $\sigma(P^+/P_2)$ (threshold up to 200 eV)

As_2: Monnom et al. (1984): $\sigma(As_2^+/As_2)$, $\sigma(As^+/As_2)$ (threshold up to 200 eV)

5.3.3 Triatomic Molecules

H_2O: Schutten et al. (1966): $\sigma(H_2O^+/H_2O)$, $\sigma(OH^+/H_2O)$, $\sigma(O^+/H_2O)$, $\sigma(H_2^+/H_2O)$, $\sigma(H^+/H_2O)$, $\sigma(O^{2+}/H_2O)$ (threshold up to 2 keV)

Märk and Egger (1976): $\sigma(H_2O^+/H_2O)$, $\sigma(D_2O^+/D_2O)$ (threshold up to 170 eV)

CD_2: Baiocchi et al. (1984): $\sigma(CD_2^+/CD_2)$, $\sigma(CD^+/CD_2)$ (threshold up to 200 eV)

CO_2: Adamczyk et al. (1972): $\sigma(CO_2^+/CO_2)$, $\sigma(CO^+/CO_2)$, $\sigma(O^+/CO_2)$, $\sigma(C^+/CO_2)$, $\sigma(O_2^+/CO_2)$ (threshold up to 600 eV)

Crowe and Mc Conkey (1974): $\sigma(CO_2^+/CO_2)$, $\sigma(CO^+/CO_2)$, $\sigma(O^+/CO_2)$, $\sigma(C^+/CO_2)$ (threshold up to 300 eV)

Märk and Hille (1978): $\sigma(CO_2^+/CO_2)$, $\sigma(CO_2^{2+}/CO_2)$ threshold up to 170 eV)

N_2O: Märk et al. (1981 a): $\sigma(N_2O^+/N_2O)$, $\sigma(N_2O^{+m}/N_2O)$ (threshold up to 180 eV, see Fig. 5-33)

NO_2: Stephan et al. (1980 c): $\sigma(NO_2^+/NO_2)$, $\sigma(NO_2^{2+}/NO_2)$ (threshold up to 180 eV)

O_3: Siegel (1982): $\sigma(O_3^+/O_3)$, $\sigma(O_2^+/O_3)$, $\sigma(O^+/O_3)$ (threshold up to 1000 eV)

SO_2: Smith and Stevenson (1981): $\sigma(SO_2^+/SO_2)$, $\sigma(SO^+/SO_2)$, $\sigma(S^+$ plus $O_2^+/SO_2)$ (threshold up to 40 eV)

Orent and Srivastava (1984): $\sigma(SO_2^+/SO_2)$, $\sigma(SO^+/SO_2)$, $\sigma(S^+/SO_2)$, $\sigma(O^+/SO_2)$ (threshold up to 200 eV)

$HgBr_2$: Wiegand and Boedeker (1982): $\sigma(HgBr_2^+/HgBr_2)$, $\sigma(HgBr^+/HgBr_2)$, $\sigma(Br^+/HgBr_2)$ (threshold up to 70 eV)

5.3.4 Polyatomic Molecules

CD_3: Baiocchi *et al.* (1984): $\sigma(CD_3^+/CD_3)$, $\sigma(CD_2^+/CD_3)$ (threshold up to 200 eV)

C_2H_2: Tate *et al.* (1935): $\sigma(C_2H_2^+/C_2H_2)$ / $\sigma(C_2H^+/C_2H_2)$ / $\sigma(C_2^+/C_2H_2)$ / $\sigma(CH^+/C_2H_2)$ / $\sigma(C^+/C_2H_2)$ / $\sigma(H^+/C_2H_2)$ (threshold up to 400 eV)

C_2N_2: Tate *et al.* (1935): $\sigma(C_2N_2^+/C_2N_2)$ / $\sigma(C_2N^+/C_2N_2)$ / $\sigma(CN^+/C_2N_2)$ / $\sigma(C_2^+/C_2N_2)$ (threshold up to 400 eV)

Smith (1983): $\sigma(C_2N_2^+/C_2N_2)$, $\sigma(C_2N^+/C_2N_2)$, $\sigma(CN^+/C_2N_2)$, $\sigma(C_2^+/C_2N_2)$ (threshold up to 35 eV)

NH_3: Märk *et al.* (1977a): $\sigma(NH_3^+/NH_3)$, $\sigma(NH_2^+/NH_3)$, $\sigma(NH^+/NH_3)$, $\sigma(N^+/NH_3)$, $\sigma(H_2^+/NH_3)$, $\sigma(H^+/NH_3)$, $\sigma(NH_3^{2+}/NH_3)$ (threshold up to 180 eV, see Fig. 5-8)

Crowe and McConkey (1977): $\sigma(NH_3^+/NH_3)$, $\sigma(NH_2^+/NH_3)$, $\sigma(NH^+/NH_3)$ (threshold up to 300 eV)

Bederski *et al.* (1980): $\sigma(NH_3^+/NH_3)$, $\sigma(NH_2^+/NH_3)$, $\sigma(NH^+/NH_3)$, $\sigma(N^+/NH_3)$, $\sigma(H_2^+/NH_3)$, $\sigma(H^+/NH_3)$, $\sigma(NH_3^{2+}/NH_3)$ (threshold up to 1000 eV)

PH_3: Märk and Egger (1977): $\sigma(PH_3^+/PH_3)$, $\sigma(PH_2^+/PH_3)$, $\sigma(PH^+/PH_3)$, $\sigma(P^+/PH_3)$, $\sigma(H_2^+/PH_3)$, $\sigma(H^+/PH_3)$, $\sigma(PH_3^{2+}/PH_3)$, $\sigma(PH_2^{2+}/PH_3)$, $\sigma(PH^{2+}/PH_3)$, $\sigma(P^{2+}/PH_3)$ (threshold up to 180 eV, see as an example the clastogram in Fig. 5-7)

P_4: Monnom *et al.* (1984): $\sigma(P_4^+/P_4)$, $\sigma(P_3^+/P_4)$, $\sigma(P_2^+/P_4)$, $\sigma(P^+/P_4)$ (threshold up to 200 eV)

As_4: Monnom *et al.* (1984): $\sigma(As_4^+/As_4)$, $\sigma(As_3^+/As_4)$, $\sigma(As_2^+/As_4)$, $\sigma(As^+/As_4)$ (threshold up to 200 eV)

SO_3: Smith and Stevenson (1981): $\sigma(SO_3^+/SO_3)$, $\sigma(SO_2^+/SO_3)$, $\sigma(SO^+/SO_3)$, $\sigma((S^+ + O_2^+)/SO_3)$ (threshold up to 30 eV)

UO_3: Blackburn and Danielson (1972): $\sigma(UO_3^+/UO_3)$ / $\sigma(UO_2^+/UO_3)$ / $\sigma(UO^+/UO_3)$ (threshold up to 60 eV)

CH_4: Adamczyk *et al.* (1966): $\sigma(CH_4^+/CH_4)$, $\sigma(CH_3^+/CH_4)$, $\sigma(CH_2^+/CH_4)$, $\sigma(CH^+/CH_4)$, $\sigma(C^+/CH_4)$, $\sigma(H_2^+/CH_4)$, $\sigma(H^+/CH_4)$ (threshold up to 2 keV). See also results by Chatham *et al.* 1984

CF_4: Stephan *et al.* (1983d, e): $\sigma(CF_3^+/CF_4)$, $\sigma(CF_2^+/CF_4)$, $\sigma(CF^+/CF_4)$, $\sigma(C^+/CF_4)$, $\sigma(F^+/CF_4)$, $\sigma(CF_3^{2+}/CF_4)$, $\sigma(CF_2^{2+}/CF_4)$, $\sigma(CF^{2+}/CF_4)$ (threshold up to 180 eV, see Fig. 5-34)

CCl_4: Stephan *et al.* (1983e), Leiter *et al.* (1984a, b): $\sigma(CCl_3^+/CCl_4)$, $\sigma(CCl_2^+/CCl_4)$, $\sigma(CCl^+/CCl_4)$, $\sigma(C^+/CCl_4)$, $\sigma(Cl_2^+/CCl_4)$, $\sigma(Cl^+/CCl_4)$, $\sigma(CCl_3^{2+}/CCl_4)$, $\sigma(CCl_2^{2+}/CCl_4)$ (threshold up to 180 eV)

CF_2Cl_2: Leiter *et al.* (1984a): Partial ionization cross sections for singly and doubly charged ions from threshold up to 180 eV

SiH_4:   Turban *et al.* (1980): $\sigma(SiH_3^+/SiH_4)$, $\sigma(SiH_2^+/SiH_4)$, $\sigma(SiH^+/SiH_4)$, $\sigma(Si^+/SiH_4)$ (threshold up to 70 eV). See also results by Chatham *et al.* 1984

$TiCl_4$:   Kiser *et al.* (1968): Clastogram from threshold up to 70 eV

$TaCl_5$:   Kiser *et al.* (1968): Clastogram from threshold up to 70 eV

$POCl_5$:   Kiser *et al.* (1968): Clastogram from threshold up to 60 eV

SF_6:   Stanski and Adamczyk (1982, 1983): $\sigma(SF_5^+/SF_6)$, $\sigma(SF_4^+/SF_6)$, $\sigma(SF_3^+/SF_6)$, $\sigma(SF_2^+/SF_6)$, $\sigma(SF^+/SF_6)$, $\sigma(S^+/SF_6)$, $\sigma(F^+/SF_6)$, $\sigma(SF_4^{2+}/SF_6)$, $\sigma(SF_2^{2+}/SF_6)$ (threshold up to 600 eV, see Fig. 5-38)

$Cr(CO_6)$, $Mo(CO_6)$, $W(CO_6)$: Winters and Kiser (1965): Clastograms from threshold up to 70 eV

C_2H_6:   See results by Chatham *et al.* 1984

Si_2H_6: See results by Chatham *et al.* 1984

5.3.5 Clusters

In recent years, there has been a growing interest in a new category of molecules, i.e. the so called (Van der Waals) *clusters*. Neutral atomic and/or molecular clusters are produced in free jet nozzle expansion, and most experiments use electron impact ionization in combination with mass spectrometry for the detection of these weakly bound species. However, very little quantitative information is known yet in terms of partial ionization cross sections. This is mainly due to the fact that it is not possible to produce beams of neutral clusters of known density and defined cluster size (e.g. Andres 1969, Anderson 1974, Hagena 1974, Märk 1982 a, Märk and Castleman, jr. 1984).
Studies performed so far have obtained information on (see also Märk and Castleman, jr. 1984):

1. Fragmentation yields:
    Gramley *et al.* (1978) for $(LiF)_2$
    Lee and Fenn (1978) for Ar_2
    Gough and Miller (1982) for $(CO_2)_2$
    Geraedts *et al.* (1982) for $(SF_6)_2$, $(SF_6)_3$
    Buck and Meyer (1983) for Ar_2 and Ar_3

2. Relative partial ionization cross sections:
    Leckenby and Robbins (1966) for Ar_2, $(CO_2)_2$
    Milne *et al.* (1970) for $(H_2O)_x$
    Gspann and Körting (1973) for $(N_2)_{\overline{11838}}$ and $(H_2)_{\overline{65132}}$
    Helm *et al.* (1979) for Ar_2, Kr_2, Xe_2, ArKr, KrXe
    Castleman, jr. *et al.* (1981) for $(H_2O)_x$

3. Absolute partial cross sections:
In order to arrive at absolute cross sections, previous investigators have generally been forced to assume that encounters between an electron and a cluster, whether

monomor or polymer, have equal probability of producing the corresponding cluster ion. As e.g. a dimer is roughly twice the size of a monomer, the probability that a particular dimer will collide with an electron should be about twice the probability that a particular monomer will collide with an electron (additivity rule, see e.g. Otvos and Stevenson 1956, Lampe *et al.* 1957, Stevenson and Schissler 1961, Batabyal *et al.* 1965, Harrison *et al.* 1966, Pottie 1966, Beran and Kevan 1969, Grosse and Bothe 1970, Alberti *et al.* 1974. Bartmess and Georgiadis 1983, Märk 1984a and Chapter 2), hence yielding a cross section ratio of two. This working rule includes the assumption that there is no fragmentation of dimers (or polymers) during ionization, which has recently been shown to be incorrect. Märk (1982b) has recently used for rare gas dimers a modified additivity rule taking into account the dissociative ionization channel (see Table 5-3). Moreover, it has been shown that electron impact ionization of clusters can lead to a large fraction of metastable parent ions, the amount of metastable ions depending on the properties (e.g. temperature, see as an example Fig. 5-11) of the neutral precursor cluster (Stace and Shukla 1980, 1982 a, b, c; Stace and Moore 1982, 1982, Stephan and Märk 1982 a, b, Märk 1982a, Stephan and Märk 1983, Futrell *et al.* 1981, 1982, Stephan *et al.* 1983 a, b, c).

Table 5-3. *Estimated absolute partial ionization cross sections for the production of parent ions in* Ar_2, *ArKr, Kr_2, KrXe and Xe_2 after Märk 1982b. Electron energy in eV and cross sections in* $10^{-20} m^2$

Electron energy	$\sigma(Ar_2^+/Ar_2)$	$\sigma(ArKr^+/ArKr)$	$\sigma(Kr_2^+/Kr_2)$	$\sigma(KrXe^+/KrXe)$	$\sigma(Xe_2^+/Xe_2)$
12.5	–	–	–	0.21	0.25
15	–	0.24	0.51	1.53	2.91
20	1.49	1.43	2.21	4.03	6.14
25	2.66	2.54	3.59	5.74	7.49
30	3.17	3.44	4.73	6.17	7.69
35	3.40	3.82	4.96	6.36	7.40
40	3.46	3.98	5.11	6.15	7.27
45	3.43	4.10	5.32	6.09	7.43
50	3.38	4.03	5.29	5.92	7.18
60	3.41	4.08	5.58	6.05	7.43
70	3.43	4.15	5.57	6.13	7.58
80	3.38	4.08	5.23	6.24	7.27
90	3.33	3.99	4.93	5.71	6.84
100	3.28	3.97	4.67	5.61	6.54
110	3.30	3.89	4.96	5.64	6.56
120	3.32	4.07	4.90	5.71	6.63
130	3.32	4.09	4.69	5.52	6.51
140	3.29	4.05	4.49	5.28	6.09
150	3.25	3.99	4.31	5.01	6.06
160	3.17	3.98	4.10	4.77	5.48
170	3.10	3.90	3.71	4.41	5.11
180	3.02	3.71	3.62	4.06	4.68

4. Effective total and counting ionization cross sections:

Hagena and Henkes (1965) for $(CO_2)_x$
Falter *et al.* (1970) for Ar_x, $(CO_2)_x$
Tay *et al.* (1970) for $(H_2)_x$

Henkes and Mikosch (1974) for $(H_2)_x$, see as example Fig. 5-43
Bottiglioni *et al.* (1972) for $(H_2)_x$, $(N_2)_x$, $(CO_2)_x$ (theoretical)

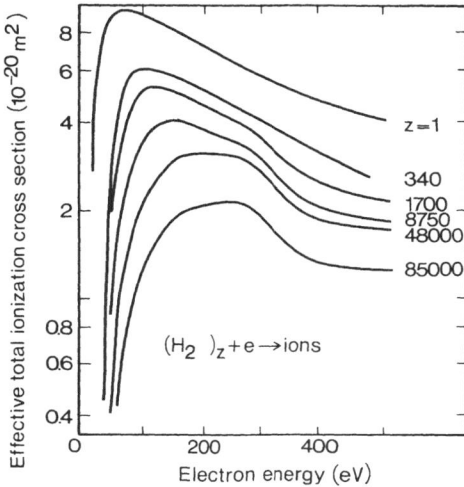

Fig. 5-43. Effective total ionization cross section (for a definition see Henkes and Mikosch 1974) as a function of electron energy for different average H_2 cluster sizes z as reported by Henkes and Mikosch (1974). In order to obtain the average total ionization cross section for a specific cluster size the effective cross section has to be multiplied by the average number of molecules per cluster. Henkes and Mikosch (1974) concluded from the observed decrease of the effective ionization cross section with average cluster size that the average escape depth of secondary electrons is 5.5 molecular diameters

5.3.6 Ionization of Excited Neutral Atoms and Molecules

Partial ionization cross section measurements discussed and presented so far describe electron impact ionization of neutral atoms and molecules in its electronic ground state (and in case of molecules in a vibrational and rotational distribution corresponding to the temperature of the target gas, usually room temperature or several $100°C$). Ionization in plasmas quite often proceeds via stepwise ionization mechanisms involving an intermediate excited atom or molecule. The lifetimes of excited states are usually so short that it is impossible to study directly (e.g. see the indirect determination of Korchevoi *et al.* (1977)) the electron impact ionization of excited (radiative) atoms or molecules; but the experimental difficulties become smaller if metastable targets are used.

There are several experimental results on partial ionization cross sections of metastable targets (most notably from the Culham group), but also one study of vibrationally excited oxygens (Evans *et al.* 1980). The results for ionization of $H(2s)$ are especially interesting, because according to Dolder (1980) classical scaling suggests that the partial cross section ratio $\sigma(H^+/H(2s)) / \sigma(H^+/H(1s))$ should be ~ 16 when the incident electron energy is expressed in units of the corresponding threshold energy. The measured cross section ratio is between 10 and 15 (see e.g. Fig. 5-40).

Available results include:

H: Dixon and Harrison (1971), Dixon *et al.* (1975): $\sigma(H^+/H\,(2\,s))$ (8.5 up to 500 eV)

       Defrance *et al.* (1981): $\sigma(H^+/H\,(2\,s))$ (6.3 up to 1000 eV)
       Results are shown in Fig. 5-40

He: Fite and Brackmann (1963): $\sigma(He^+/He\,(2^3S))$ (threshold up to 25 eV)

       Long and Geballe (1970): $\sigma(He^+/He\,(2^3S))$ (threshold up to 16 eV)

       Dixon *et al.* (1973), Dixon *et al.* (1976, see also reference to earlier work therein): $\sigma(He^+/He\,(2^3S))$ (threshold up to 1000 eV)
       Results are shown in Fig. 5-41. See also relative σ by Shearer-Izume and Botter (1974) and Vriens *et al.* (1968)

Ne: Dixon *et al.* (1973): $\sigma(Ne^+/Ne^m)$ (threshold up to 500 eV)

Ar: Dixon *et al.* (1973): $\sigma(Ar^+/Ar^m)$ (threshold up to 500 eV)

N_2: Armentrout *et al.* (1981): $\sigma\left(N_2^+/N_2\,(A^3\Sigma_u^+)\right)$ (threshold up to 240 eV, see Fig. 5-32)

It is interesting to mention in this section, that Hils *et al.* (1982) have recently studied electron impact ionization of polarized sodium atoms obtaining information on the spin dependent ionization asymmetry (see also Alguard *et al.* 1977, Hils and Kleinpoppen 1978, Baum *et al.* 1981).

5.3.7 Ionization of Atoms and Molecules into Specific (Excited) Ionic States

It is apparent that results mentioned in the preceding paragraphs have not included studies which determine the specific states (electronic, vibrational, fine structure) of ions produced by electron impact. There is only a small number of studies, all of which are difficult to evaluate, because of the difficulty of absolute detection of either radiation or metastable ions. Moreover, in case of radiative states the reported partial ionization cross sections are not cross sections for production of ions in a single specific state, but rather for the production of radiation of a specific wavelength emitted by this state (emission cross section) without taking into account cascading and branching ratios. It is outside the scope of this chapter to discuss in detail these measurements. Some results have been included in the review of Kieffer and Dunn (1966), some recent studies include, e.g. the work of Karstensen and Schneider (1970) on Ca, Walker and John (1972) on Ne, Mentall *et al.* (1973) on CO_2, Tan and McConkey on Ar (1974), Gerzanich *et al.* (1976) on N_2O, Mandelbaum and Feldman (1976) on N_2, Haasz and de Leeuw (1976) on N_2, O_2 and O, Van Sprang *et al.* (1978) on N_2O, Allison *et al.* (1979) on N_2, Dekoven *et al.* (1981) on N_2, Bloemen *et al.* (1981) on He, Hernandez *et al.* (1982) on N_2 and Perrin and Schmitt (1982) on SiH_4. Moreover, Eckhardt and Schartner (1983) have recently reported absolute single and double cross sections of neon 2 s- and 2 p-subshell ionization using photon spectroscopy (see also Chapter 6).

It is interesting to note that there exist several studies which have determined partial ionization cross sections functions for the production of singly charged metastable atomic ions in He, Ar, Kr, Xe and Cs (Hagstrum 1956, 1960, Novick and Commins 1958, Nygaard and Hahn 1973, Kadota and Kaneko 1975, Varga *et al.* 1981). According to Varga *et al.* (1981) the maximum cross sections are quite large ($\sim 10^{-21}$ m^2) (see e.g. Fig. 5-44), however, these cross sections decrease in agreement with predictions from Born-Bethe considerations at high electron energy faster than cross sections dominated by optically allowed transitions. Moreover, Adams *et al.* (1979) have determined the partial cross section ratios as a function of electron energy for the production of Xe^{2+} in the ^3P ground state and $^1D_2$ and $^1S_0$ excited state, respectively.

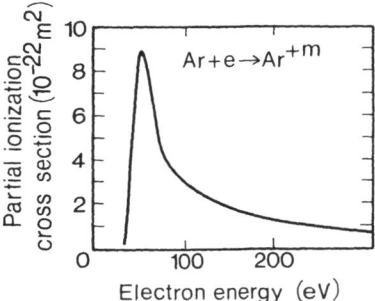

Fig. 5-44. Partial ionization cross section as a function of electron energy for the production of singly charged metastable Ar$^+$ ions via process, Ar $+e$, as measured by Varga *et al.* (1981)

In addition, some studies have estimated partial ionization cross sections for the production of long lived and/or metastable ions, however, without analysis of the states produced. These studies include the production of excited O$^+$ and O$_2^+$ (Turner *et al.* 1968, Hughes and Tiernan 1971), of N$_2$O^{+m} (Märk *et al.* 1981 a, see e.g. Fig. 5-33) and CO^{2+m} (Hille and Märk 1978). See also Chapter 9.4 for the production of metastable O$^+$ and N$^+$.

Acknowledgements

This work was supported by the Österreichischer Fonds zur Förderung der wissenschaftlichen Forschung under block grant S-18. It is a pleasure to thank M. Heigl, M. Placheta, and Dr. H. Märk-Zuegg for technical assistance.

References

Adamczyk, B. (1969): Ann. Univ. M. Curie-Sklodowska (Lublin) *24*, 141−156.
Adamczyk, B., Boerboom, A. J. H., Schram, B. L., Kistemaker, J. (1966): J. Chem. Phys. *44*, 4640−4648.
Adamczyk, B., Boerboom, A. J. H., Lukasiewicz, M. (1972): Int. J. Mass Spectrom. Ion Phys. *9*, 407−412

Adams, N. G., Smith, D., Grief, D. (1979): J. Phys. *B 12*, 791 – 800.
Akerib, R., Borowitz, S. (1961): Phys. Rev. *122*, 1177 – 1184.
Alberti, R., Genoni, M. M., Pascual, C., Vogt, J. (1974): Int. J. Mass Spectrom. Ion Phys. *14*, 89 – 98.
Alguard, M. J., Hughes, V. W., Lubell, M. S., Mainwright, P. F. (1977): Phys. Rev. Lett. *39*, 334 – 338.
Allison, J., Kondow, T., Zare, R. N. (1979): Chem. Phys. Lett. *64*, 202 – 204.
Anderson, J. B. (1974): In: Molecular Beams and Low Density Gas Dynamics (Wegener, P. P., ed.). 1 – 92. New York: Dekker.
Andres, R. P. (1969): In: Nucleation (Zettlemoyer, A. C., ed.), 69 – 108. New York: Dekker.
Appell, J., Durup, J. (1972): Int. J. Mass Spectrom. Ion Phys. *10*, 247 – 265.
Appell, J., Durup, J., Heitz, F. (1964): Adv. Mass Spectrom. *3*, 457 – 469.
Armentrout, P. B., Tarr, S. M., Dori, A., Freund, R. S. (1981): J. Chem. Phys. *75*, 2786 – 2794.
Ast, T. (1980): Adv. Mass Spectrom. *8A*, 555 – 576.
Asundi, R. K. (1963): Proc. Phys. Soc. *82*, 372 – 374.
Asundi, R. K., Kurepa, M. V. (1963): J. Sci. Instrum. *40*, 183 – 186.
Bainbridge, K. T. (1931): J. Frankl. Inst. *212*, 317 – 339.
Baiocchi, F. A., Wetzel, R. C., Freund, R. S. (1984): Phys. Rev. Lett. *53*, 771 – 774.
Baker, F. A., Hasted, J. B. (1966): Phil. Trans. Roy. Soc. *A 261*, 33 – 65.
Baker, F. A., Hasted, J. B. (1968): Adv. Mass Spectrom. *4*, 727 – 734.
Bauer, N., Beach, J. Y. (1947): J. Chem. Phys. *15*, 150 – 151.
Barnard, G. P. (1953): Modern Mass Spectrometry. London: The Institute of Physics.
Barnard, G. P. (1956): Mass Spectrometer Researches. London: HMSO.
Barnett, C. F., Ray, J. A., Ricci, E., Kirkpatrick, M. I., Mc Daniel, E. W., Thomas, E. W., Gilbody, H. B. (1980): ORNL-5207/R2.
Bartmess, J. E., Georgiadis, R. M. (1983): Vacuum TAIP *33*, 149 – 153.
Batabyal, A. K., Barna, A. K., Srivastava, B. N. (1965): Indian J. Phys. *39*, 219 – 226.
Bates, J. R. (ed.) (1962): Atomic and Molecular Processes. New York: Academic Press.
Baum, G., Kisker, E., Raith, W., Schroeder, W., Sillman, U., Zenses, D. (1981): J. Phys. *B 14*, 4377 – 4388.
Becker, S., Dietze, H. J. (1983): Int. J. Mass Spectrom. Ion Proc. *54*, 337 – 339.
Bederski, K., Wojcik, L., Adamczyk, B. (1980): Int. J. Mass Spectrom. Ion Phys. *35*, 171 – 178.
Bell, K. L., Gilbody, H. B., Hughes, J. G., Kingston, A. E., Smith, F. J. (1983): J. Phys. Chem. Ref. Data *12*, 891 – 916.
Beran, J. A., Kevan, L. (1969): J. Phys. Chem. *73*, 3866 – 3876.
Berkowitz, J., Tasman, H. A., Chupka, W. A. (1962): J. Chem. Phys. *36*, 2170 – 2179.
Berry, C. E. (1950): Phys. Rev. *78*, 597 – 605.
Bertein, F. (1950): C. R. Acad. Sci. (Paris) *231*, 1134 – 1135; 1448 – 1449.
Beuhler, R. J. (1983): J. Appl. Phys. *54*, 4118 – 4126.
Beyer, H. F., Hippler, R., Schartner, K. H. (1979): Z. Phys. *A 289*, 239 – 243.
Beynon, J. H. (1960): Mass Spectrometry and Its Applications to Organic Chemistry. Amsterdam: Elsevier.
Beynon, J. H., Cooks, R. G. (1976): Int. J. Mass Spectrom. Ion Phys. *19*, 107 – 137.
Beynon, J. H., Morgan, R. P. (1978): Int. J. Mass Spectrom. Ion Phys. *27*, 1 – 30.
Beynon, J. H., Caprioli, R. M., Richardson, J. W. (1971): J. Am. Chem. Soc. *93*, 1852 – 1857.
Beynon, J. H., Cooks, R. G., Jennings, K. R., Ferrer-Correia, A. J. (1975): Int. J. Mass Spectrom. Ion Phys. *18*, 87 – 99.
Beynon, J. H., Brenton, A. G., Harris, F. M. (1982): Int. J. Mass Spectrom. Ion Phys. *45*, 5 – 34.
Biemann, K. (1962): Mass Spectrometry, Organic Chemical Application. New York: McGraw-Hill.
Blackburn, P. E., Danielson, P. M. (1972): J. Chem. Phys. *56*, 6156 – 6164.
Blauth, E. W. (1965): Dynamische Massenspektrometer. Braunschweig: Vieweg.
Bleakney, W. (1929): Phys. Rev. *34*, 157 – 160.
Bleakney, W. (1930a): Phys. Rev. *36*, 1303 – 1308.
Bleakney, W. (1930b): Phys. Rev. *35*, 139 – 148.
Bleakney, W. (1936): Am. Phys. T. *4*, 12 – 23.
Bleakney, W., Smith, L. G. (1936): Phys. Rev. *49*, 402 – 402.
Bleakney, W., Condon, E. U., Smith, L. G. (1937): J. Phys. Chem. *41*, 197 – 208.
Bloemen, E. W. P., Winter, H., Märk, T. D., Dijkkamp, D., Barends, D., de Heer, F. J. (1981): J. Phys. *B 14*, 717 – 725.

Bottiglioni, F., Contant, J., Fois, M. (1972): Phys. Rev. *A 6*, 1830 – 1843.

Bourne, A. J., Danby, C. J. (1968): J. Sci. Instrum., Ser. 2, *1*, 155 – 155.

Boyd, R. L. F., Green, G. W. (1958): Proc. Phys. Soc. *71*, 351 – 356.

Bracher, J., Erhardt, H., Fuchs, R., Osberghaus, O., Taubert, R. (1963): Adv. Mass Spectrom. *2*, 285 – 295.

Brehm, B., De Frenes, G. (1978): Int. J. Mass Spectrom. Ion Phys. *26*, 251 – 266.

Brehm, B., De Frenes, G. (1980): Adv. Mass Spectrom. *8 A*, 138 – 143.

Brehm, B., Fröbe, U., Neitzke, H. P. (1984): Int. J. Mass Spectrom. Ion Proc. *57*, 91 – 102.

Brenton, A. G., Morgan, R. P., Beynon, J. H. (1979): Ann. Rev. Phys. Chem. *30*, 51 – 78.

Brink, G. O. (1962): Phys. Rev. *127*, 1204 – 1206.

Brink, G. O. (1964): Phys. Rev. *134*, A 345 – A 346.

Brion, C. E., Paddock, N. L. (1968): J. Chem. Soc. *A*, 388 – 392.

Brook, E., Harrison, M. F. A., Smith, A. C. H. (1978): J. Phys. *B 11*, 3115 – 3132.

Browning, R., Fryar, J. (1973): J. Phys. *B 6*, 364 – 371.

Brubaker, W. M. (1955): J. Appl. Phys. *26*, 1007 – 1012.

Brunnee, C., Voshage, H. (1964): Massenspektrometrie. München: Thiemig.

Buck, U., Meyer, H. (1983): Phys. Rev. Lett. *52*, 109 – 112.

Budzikiewicz, H. (1972): Massenspektrometrie. Weinheim: Verlag Chemie.

Budzikiewicz, H. (1981): Interpretation of Mass Spectra of Organic Compounds. Ann Arbor: University Microfilms International.

Burgess, A. (1963): Proc. 3rd ICPEAC, London, 237 – 242.

Cantone, B., Grasso, F., Pignataro, S. (1966): J. Chem. Phys. *44*, 3115 – 3120.

Carboneau, R., Marmet, P. (1972): Int. J. Mass Spectrom. Ion Phys. *10*, 143 – 155.

Careri, G., Nencini, G. (1950): J. Chem. Phys. *18*, 897 – 898.

Castleman, A. W., jr., Kay, B. D., Hermann, V., Holland, P. M., Märk, T. D. (1981): Surface Science *106*, 179 – 182.

Castro, M., Keller, J., Ventura, O. N. (1982): J. Chem. Phys. *77*, 6348 – 6350.

Chamberlain, G. E., Kieffer, L. J. (1970): JILA Information Center Report No. 10.

Chantreau, J. (1972): Proc. 2nd ICIS, Wien, 668 – 674.

Chantreau, J., Vauthier, R. (1971): In: Recent Developments in Mass Spectroscopy (Ogata, K., Hayakawa, T., eds.), 198 – 204. Baltimore: University Park Press.

Chatham, H., Hils, D., Robertson, R., Gallagher, A. (1984): J. Chem. Phys. *81*, 1770 – 1777.

Chatterjee, S. N., Kumar, A., Roy, B. N. (1982): J. Phys. *B 15*, 1415 – 1419.

Christophorou, L. G. (ed.) (1984): Electron-Molecule Interactions and Their Applications. New York: Academic Press.

CIAMDA (1980): An Index to the Literature on Atomic and Molecular Collision Data Relevant to Fusion Research. Wien: IAEA.

Coggeshall, N. D. (1942): J. Chem. Phys. *12*, 19 – 23.

Coggeshall, N. D. (1946): Phys. Rev. *70*, 270 – 280.

Coggeshall, N. D. (1962): J. Chem. Phys. *36*, 1640 – 1647.

Coggeshall, N. D., Jordan, E. B. (1943): Rev. Sci. Instr. *14*, 125 – 129.

Cohen, J. S., Bardsley, J. N. (1978): Phys. Rev. *18 A*, 1004 – 1011.

Colin, R. (1961): Ind. Chim. Belg. *26*, 51 – 52.

Collins, J. H., Winters, R. E., Engerholm, G. G. (1968): J. Chem. Phys. *49*, 2469 – 2472.

Collins, R. D. (1969): Vacuum *19*, 105 – 111.

Conrad, R. (1930): Physik. Zeitschrift *31*, 888 – 892.

Cook, C. J., Peterson, J. R. (1962): Phys. Rev. Lett. *9*, 164 – 166.

Cooks, R. G., Beynon, J. H., Caprioli, R. M., Lester, G. R. (1973): Metastable Ions. Amsterdam: Elsevier.

Cooper, J. L., jr., Pressley, G. A., Stafford, F. E. (1966): J. Chem. Phys. *44*, 3946 – 3949.

Craggs, J. D., Massey, H. S. W. (1959): In: Handbuch der Physik (Flügge, S., ed.), Vol. 37/1, 314 – 415. Berlin-Göttingen-Heidelberg: Springer.

Crawford, C. K. (1967) see Kieffer (1968), 27, 28, 92.

Crawford, C. K., Wang, K. I. (1967): J. Chem. Phys. *47*, 4667 – 4669.

Crowe, A., Mc Conkey, J. W. (1973 a): J. Phys. *B 6*, 2088 – 2107.

Crowe, A., Mc Conkey, J. W. (1973 b): J. Phys. *B 6*, 2108 – 2117.

Crowe, A., Mc Conkey, J. W. (1974): J. Phys. *B 7*, 349 – 361.

Crowe, A., Mc Conkey, J. W. (1977): Int. J. Mass Spectrom. Ion Phys. *24*, 181 − 189.
Crowe, A., Preston, J. A., Mc Conkey, J. W. (1972): J. Chem. Phys. *57*, 1620 − 1625.
Cuthbert, J., Farren, J., Prahallada Rao, B. S., Preece, E. R. (1966): Proc. Phys. Soc. *88*, 91 − 100.
Daly, R. N., Powell, R. E. (1966): Proc. Phys. Soc. *89*, 273 − 280.
Dalgarno, A. (1953, unpublished) quoted in Mc Carrol, R. (1957): Proc. Phys. Soc. *A 70*, 460 − 465.
Dawson, P. H. (1976): Quadrupole Mass Spectrometry and Its Applications. Amsterdam: Elsevier.
Defrance, P., Claeys, W., Cornet, A., Poulaert, G. (1981): J. Phys. *B 14*, 111 − 117.
Defrance, P., Claeys, W., Brouillard, F. (1982): J. Phys. *B 15*, 3509 − 3516.
de Heer, F. J. (1981): Physica Scripta *23*, 170 − 178.
de Heer, F. J., Jansen, R. H. J. (1977): J. Phys. *B 10*, 3741 − 3758.
de Heer, F. J., Jansen, R. H. J., van der Kaay, W. (1979): J. Phys. *B 12*, 979 − 1002.
De Koven, B. M., Levy, D. H., Harris, H. H., Zegarski, B. R., Miller, T. A. (1981): J. Chem. Phys. *74*, 5659 − 5668.
de Maria, G., Malaspina, L., Piacente, V. (1963): Ric. Sci. Rend. *A 3*, 681 − 688.
Dettmann, J. M., Karstensen, F. (1982): J. Phys. *B 15*, 287 − 300.
Deutsch, H., Leiter, K., Stephan, K., Märk, T. D. (1984): unpublished.
Dewar, M. J. S., Rona, P. (1965): J. Am. Chem. Soc. *87*, 5510 − 5510.
Dibeler, V. H., Wells, E. J., jr., Reese, R. M. (1950): Phys. Rev. *79*, 223 − 223.
Dibeler, V. H., Mohler, F. L., Reese, R. M. (1953): J. Chem. Phys. *21*, 180 − 181.
Dietz, L. A. (1965): Rev. Sci. Instrum. *36*, 1763 − 1770.
Dixon, A. J., Harrison, M. F. A. (1971): Proc. VIIth ICPEAC, Amsterdam, 892 − 894.
Dixon, A. J., Harrison, M. F. A., Smith, A. C. H. (1973): Proc. VIIIth ICPEAC, Belgrade, 405 − 406.
Dixon, A. J., von Engel, A., Harrison, M. F. A. (1975): Proc. Roy. Soc. *A 343*, 333 − 349.
Dixon, A. J., Harrison, M. F. A., Smith, A. C. H. (1976): J. Phys. *B 9*, 2617 − 2631.
Dolder, K. T. (1980): In: Atomic and Molecular Processes in Controlled Thermonuclear Fusion (Mc Dowell, M. R. C., Ferendec, A. M., eds.), 313 − 349. New York: Plenum.
Dole, M., Mack, L. L., Hines, R. L., Mobley, R. C., Ferguson, L. D., Alice, M. B. (1968): J. Chem. Phys. *49*, 2240 − 2249.
Dorman, F. H., Morrison, J. D. (1961): J. Chem. Phys. *35*, 575 − 581.
Drewitz, H. J. (1974): PTB Bericht, PTB-APh6, Braunschweig. 14.
Drewitz, H. J. (1976): Int. J. Mass Spectrom. Ion Phys. *19*, 313 − 325; *21*, 212 − 212.
Drewitz, H. J., Taubert, R. (1976): Int. J. Mass Spectrom. Ion Phys. *19*, 293 − 312.
Drowart, J., Goldfinger, P. (1967): Angew. Chemie *6*, 581 − 648.
Dunn, G. H. (1966): J. Chem. Phys. *44*, 2592 − 2594.
Durup, J., Heitz, F. (1964): J. Chim. Phys. Physiochim. Biol. *61*, 470 − 479.
Echt, O., Sattler, K., Recknagel, E. (1982): Phys. Lett. *90 A*, 185 − 189.
Eckhardt, M., Schartner, K. H. (1983): Zeitschr. f. Physik *A 312*, 321 − 328.
Eggarter, E. (1975): J. Chem. Phys. *62*, 833 − 847.
Egger, F., Märk, T. D. (1978a): Symp. Atomic Surface Phys., Maria Alm (Lindinger, W., *et al.*, eds.), Vol. 1, 51 − 57.
Egger, F., Märk, T. D. (1978b): Z. Naturforschg. *33 a*, 1111 − 1113.
Einfeld, D. (1980): Proc. 6th Int. Conf. Vacuum Ultraviolet Radiation Physics, Charlottesville, 65.
Elliott, R. M. (1963): In: Mass Spectrometry (Mc Dowell, C. A., ed.), 72. New York: McGraw-Hill.
Elwert, G. (1952): Z. Naturforschg. *7a*, 432 − 439.
Erhardt, H., Tekaat, T. (1964): Z. Naturforschg. *19 a*, 1382 − 1388.
Erhardt, H., Kresling, A. (1967): Z. Naturforschg. *22 a*, 2036 − 2043.
Evans, B., Hobson, R. M., Chang, J. S. (1980): Europhys. Conf. Abstracts *4 D*, 49 − 49.
Falter, H., Hagena, O. F., Henkes, W., Wedel, H. (1970): Int. J. Mass Spectrom. Ion Phys. *4*, 145 − 163.
Fehlner, T. P., Callen, R. B. (1968): Adv. Chem. Ser. *72*, 181 − 190.
Field, F. H., Franklin, J. L. (1970): Electron Impact Phenomena. New York: Academic Press.
Fiquet-Fayard, F. (1962): J. Chim. Phys. *59*, 439 − 441.
Fiquet-Fayard, F., Lahmani, M. (1962): J. Chim. Phys. *59*, 1050 − 1055.
Fiquet-Fayard, F., Ziesel, J. P. (1963a): Proc. VIth ICPEAC (Paris), 37 − 40.
Fiquet-Fayard, F., Ziesel, J. P. (1963b): C. R. Acad. Sci. *256*, 4885 − 4888.
Fiquet-Fayard, F., Chiari, J., Muller, F., Ziesel, J. P. (1968): J. Chem. Phys. *48*, 478 − 482.
Fite, W. L. (1962): In: Atomic and Molecular Processes (Bates, D. R., ed.), 421 − 492. New York: Academic Press.

Fite, W. L., Brackmann, R. T. (1958): Phys. Rev. *112*, 1141 – 1151.
Fite, W. L., Brackmann, R. T. (1959): Phys. Rev. *113*, 815 – 816.
Fite, W. L., Brackmann, R. T. (1963): Proc. 6th ICPIG (Paris).
Flannery, M. R. (1970): J. Phys. *B 10*, 1610 – 1619.
Fock, W. (1965): Int. J. Mass Spectrom. Ion Phys. *3*, 285 – 291.
Fox, R. E. (1960): J. Chem. Phys. *33*, 200 – 205.
Fuchs, R., Taubert, R. (1965): Z. Naturforschg. *20 a*, 823 – 826.
Funk, H. (1930): Ann. d. Physik *4*, 149 – 184.
Futrell, J. H., Stephan, K., Märk, T. D., Castleman, A. W., jr. (1981): Proc. 8th Int. Symp. Molecular Beams, Cannes, 262 – 266.
Futrell, J. H., Stephan, K., Märk, T. D. (1982): J. Chem. Phys. *76*, 5893 – 5901.
Gallagher, J. W., Rumble, J. R., jr., Beaty, E. C. (1979): NBS Special Publication 426, Suppl. *1*, 1 – 106.
Gallagher, J. W., Beaty, E. C. (1980): JILA Information Center Report No. 18, 1 – 142.
Gaudin, A., Hagemann, R. (1967): J. Chim. Phys. *64*, 1209 – 1221.
Gemmel, D. S., Kanter, E. P. (1984): Phys. Today, Jan. issue, S-27 – S-28.
Geraedts, J., Stolte, S., Reuss, J. (1982): Z. Phys. *A 304*, 167 – 175.
Gerzanich, G. J., Skubenich, V. V., Zapesochny, I. P. (1976): Opt. Spectry. *41*, 535 – 539.
Giese, C. F. (1959): Rev. Sci. Instr. *30*, 260 – 261.
Glasstone, S., Laidler, K. J., Eyring, H. (1941): Theory of Rate Processes. New York: McGraw-Hill.
Gough, T. E., Miller, R. E. (1982): Chem. Phys. Lett. *87*, 280 – 283.
Griffiths, I. W., Mukhtar, E. S., Harris, F. M., Beynon, J. H. (1982): Int. J. Mass Spectrom. Ion Phys. *43*, 283 – 292.
Grimley, R. T., Forsman, J. A., Grindstaff, Q. G. (1978): J. Phys. Chem. *82*, 632 – 639.
Grosse, H. J., Bothe, H. K. (1970): Z. Naturforschg. *25 a*, 1970 – 1976.
Gryzinski, M. (1965): Phys. Rev. *138*, A305 – A321, A322 – A335, A336 – A358.
Gspann, J., Körting, K. (1973): J. Chem. Phys. *59*, 4726 – 4734.
Haasz, A. A., de Leeuw, J. H. (1976): J. Geophys. Res. *81*, 4031 – 4034.
Hagena, O. F. (1974): In: Molecular Beams and Low Density Gas Dynamics (Wegener, P. P., ed.). New York: Dekker.
Hagena, O. F., Henkes, W. (1965): Z. Naturforschg. *20 a*, 1344 – 1348.
Hagstrum, H. D. (1951): Revs. Mod. Phys. *23*, 185 – 203.
Hagstrum, H. D. (1953): Rev. Sci. Instrum. *24*, 1122 – 1142.
Hagstrum, H. D. (1956): Phys. Rev. *104*, 309 – 318.
Hagstrum, H. D. (1960): J. Appl. Phys. *31*, 897 – 904.
Hagstrum, H. D., Tate, J. T. (1941): Phys. Rev. *59*, 354 – 370.
Halas, S., Adamczyk, B. (1972): Int. J. Mass Spectrom. Ion Phys. *10*, 157 – 160.
Harrison, A. G., Jones, E. G., Gupta, S. K., Nagy, G. P. (1966): Can. J. Chemistry *44*, 1967 – 1973.
Harrison, M. F. A., Smith, A. C. H., Brook, E. (1979): J. Phys. *B 12*, L433 – L435.
Hashizume, A., Wasada, N. (1980): J. Phys. *B 13*, 4865 – 4875.
Hasted, J. B. (1967): In: Some Newer Physical Methods in Structural Chemistry (Bonnett, J., Davis, J. G., eds.). London: United Trade Press.
Hellwinkel, D., Wünsche, C. (1969): Chem. Commun., 1412 – 1413.
Helm, H., Stephan, K., Märk, T. D. (1979): Phys. Rev. *A 19*, 2154 – 2160.
Helm, H., Stephan, K., Märk, T. D., Huestis, D. L. (1981): J. Chem. Phys. *74*, 3844 – 3851.
Henkes, W., Isenberg, G. (1970): Int. J. Mass Spectrom. Ion Phys. *5*, 249 – 254.
Henkes, W., Mikosch, F. (1974): Int. J. Mass Spectrom. Ion Phys. *13*, 151 – 161.
Hernandez, S. P., Dagdigian, P. J., Doering, J. P. (1982): J. Chem. Phys. *77*, 6021 – 6026.
Herod, A. A., Harrison, A. G. (1970): Int. J. Mass Spectrom. Ion Phys. *4*, 415 – 431.
Herzberg, G. (1967): Molecular Spectra and Molecular Structure III, 471 – 473. Princeton: V. Nostrand.
Higatsberger, M. J., Demorest, H. L., Nier, A. O. (1954): J. Appl. Phys. *25*, 883 – 886.
Hille, E., Märk, T. D. (1978): J. Chem. Phys. *69*, 4600 – 4605.
Hille, E., Märk, T. D. (1979): Unpublished.
Hille, E., Märk, T. D., Störi, H. (1978): Symp. Atomic Surface Phys., Maria Alm (Lindinger, W., et al., eds.), Vol. 1, 59 – 66.
Hils, D., Kleinpoppen, H. (1978): J. Phys. *B 11*, L283 – L287.
Hils, D., Jitschin, W., Kleinpoppen, H. (1982): J. Phys. *B 15*, 3347 – 3357.

Hipple, J. A. (1948): J. Phys. Colloid Chem. *52*, 456 – 462.
Hofer, W., Varga, P., Winter, H. (1980): Acta Physica Acad. Sci. Hung. *49*, 313 – 313.
Honig, R. E. (1948): J. Chem. Phys. *16*, 105 – 112.
Hubin-Franskin, M. J., Marmet, P., Huard, D. (1980): Int. J. Mass Spectrom. Ion Phys. *33*, 311 – 324.
Hughes, B. M., Tiernan, T. O. (1971): J. Chem. Phys. *55*, 3419 – 3426.
Illies, A. J., Jarrold, M. F., Bowers, M. T. (1982): J. Am. Chem. Soc. *104*, 3587 – 3593.
Ishii, H., Nakayama, K. (1961): Proc. 8th Natl. Vac. Symp., Vol. 1, 519 (1962).
Jackson, W. M., Brackmann, R. F., Fite, W. L. (1974): Int. J. Mass Spectrom. Ion Phys. *13*, 237 – 250.
Jalin, R., Hagemann, R., Botter, R. (1973): J. Chem. Phys. *59*, 952 – 959.
Jentsch, T., Drachsel, W., Block, J. H. (1982): Chem. Phys. Lett. *93*, 144 – 147.
Johnson, L. P., Morrison, J. D., Wahrhaftig, A. L. (1978): Int. J. Mass Spectrom. Ion Phys. *26*, 1 – 21.
Johnstone, R. A. W. (1980): Mass Spectrometry, Vol. 6. London: Royal Society of Chemistry.
Jordan, E. B., Coggeshall, N. D. (1942): J. Appl. Phys. *13*, 539 – 550.
Kadota, K., Kaneko, Y. (1975): J. Phys. Soc. (Japan) *38*, 524 – 531.
Kaneko, Y. (1961): J. Phys. Soc. (Japan) *16*, 2288 – 2293.
Kant, A. (1966): J. Chem. Phys. *44*, 2450 – 2456.
Karstensen, F., Schneider, M. (1970): Z. Physik *237*, 242 – 247.
Karstensen, F., Köster, H. (1971): Astron. Astrophys. *13*, 116 – 118.
Karstensen, F., Schneider, M. (1975): Z. Physik *A 273*, 321 – 323.
Karstensen, F., Schneider, M. (1978): J. Phys. *B 11*, 167 – 172.
Kellogg, G. L. (1981): Phys. Rev. *B 24*, 1848 – 1851.
Kellogg, G. L. (1982): Surf. Science *120*, 319 – 333.
Kerwin, L., Marmet, P., Carette, J. D. (1969): In: Case Studies in Atomic Collision Physics (Mc Daniel. E. W., Mc Dowell, M. R. C., eds.), Vol. 1, 527 – 581. Amsterdam: North-Holland.
Kieffer, L. J. (1968): JILA Information Center Report No. 6, 1 – 95.
Kieffer, L. J. (1969): Atomic Data *1*, 19 – 89.
Kieffer, L. J. (1973): JILA Information Center Report No. 13, 1 – 139.
Kieffer, L. J. (1976): NBS Special Publication 426, 1 – 212.
Kieffer, L. J., Dunn, G. H. (1966): Revs. Mod. Phys. *38*, 1 – 35.
Kienitz, H. (1968): Massenspektrometrie. Weinheim: Verlag Chemie.
King, R. B. (1969): Can. J. Chem. *47*, 559 – 568.
Kingston, E. E., Morgan, T. G., Harris, F. M., Beynon, J. H. (1982): Int. J. Mass Spectrom. Ion Phys. *43*. 261 – 272.
Kim, Y. B. (1981): Thesis, Universität Innsbruck.
Kim, Y. B., Stephan, K., Märk, E., Märk, T. D. (1981): J. Chem. Phys. *74*, 6771 – 6776.
Kim, Y. K., Inokuti, M. (1971): Phys. Rev. *3*, 665 – 678.
Kim, Y. K., Naon, M., Cornille, M. (1973): Argonne Natl. Lab. Report ANL 8060, 14 – 23.
Kiser, R. W., Dillard, J. G., Dugger, D. L. (1968): Advanc. Chem. Ser. *72*, 153 – 180.
Klein, H. J. (1965): Z. Phys. *188*, 78 – 92.
Knapp, E. W., Schulz, W. (1974): J. Phys. *B 7*, 1875 – 1890.
Korchevoi, Y. P., Lukashenko, V. I., Lukashenko, S. N., Khil'ko, I. N. (1977): Teplofizika Vysokikh Temp. *15*, 8 – 14.
Kovacs, I. (1969): Rotational Structure in the Spectra of Diatomic Molecules, 200 – 205. New York: American Elsevier.
Kühne, M., Stuck, D., Tegeler, E. (1980): Proc. 6th Int. Conf. Vacuum Ultraviolet Radiation Physics. Charlottesville, 15.
Kunc, J. A. (1980): J. Phys. *B 13*, 587 – 602.
Kurepa, M. V., Cadez, I. M., Pejcev, V. M. (1974): Fizika *6*, 185 – 209.
Laborie, P., Rocard, J. M., Rees, J. A., Delcroix, J. L., Craggs, J. D. (1968): Electronic Cross Sections and Macroscopic Coefficients. 1. Hydrogen and Rare Gases. Paris: Dunod.
Laborie, P., Rocard, J. M., Rees, J. A. (1971): Electronic Cross Sections and Macroscopic Coefficients. 2. Metallic Vapours and Molecular Gases. Paris: Dunod.
Lampe, F. W., Franklin, J. L., Field, F. H. (1957): J. Am. Chem. Soc. *79*, 6129 – 6132.
Lee, N., Fenn, J. B. (1978): Rev. Sci. Instrum. *49*, 1269 – 1272.
Lefaivre, D., Marmet, P. (1978): Can. J. Phys. *56*, 1549 – 1558.
Leckenby, R. E., Robbins, E. J. (1966): Proc. Roy. Soc. (London) *A 291*, 389 – 412.
Leiter, K. (1983): Thesis. Universität Innsbruck.

Leiter, K., Stephan, K., Deutsch, H., Märk, T. D. (1984a): Proc. 4th Symposium Atomic Surface Physics, Maria Alm (Howorka, F., et al., eds.), 39 – 44.

Leiter, K., Stephan, K., Märk, E., Märk, T. D. (1984b): Plasma Chem. Plasma Proc. 4, 235 – 249.

Levsen, K. (1978): Fundamental Aspects of Organic Mass Spectrometry. Weinheim: Verlag Chemie.

Lifshitz, C. (1971): J. Chem. Phys. 55, 4155 – 4156.

Lifshitz, C. (1978): Adv. Mass Spectrom. 7A, 3 – 18.

Lifshitz, C., Eaton, P. E. (1983): Int. J. Mass Spectrom. Ion Phys. 49, 337 – 345.

Lifshitz, C., Gefen, S. (1980): Int. J. Mass Spectrom. Ion Phys. 35, 31 – 37.

Lifshitz, C., Long, F. A. (1964): J. Chem. Phys. 41, 2468 – 2471.

Lifshitz, C., Mac Kenzie Peers, A., Weiss, M., Weiss, M. J. (1974): Adv. Mass Spectrom. 6, 871 – 875.

Lifshitz, C., Gotchiguian, P., Roller, R. (1983): Chem. Phys. Lett. 95, 106 – 108.

Lin, S. S., Stafford, F. E. (1967): J. Chem. Phys. 47, 4664 – 4666.

Lin, S. S., Stafford, F. E. (1968): J. Chem. Phys. 48, 3885 – 3890.

Litzow, M. F., Spalding, T. R. (1973): Mass Spectrometry of Inorganic and Organometallic Compounds. Amsterdam: Elsevier.

Locht, R., Momigny, J. (1969): Int. J. Mass Spectrom. Ion Phys. 2, 425 – 440.

Locht, R., Olivier, J. L., Momigny, J. (1979): Chem. Phys. 43, 425 – 432.

Long, D. R., Geballe, R. (1970): Phys. Rev. A 1, 260 – 265.

Lorquet, J. C. (1980): Adv. Mass Spectrom. 8A, 3 – 16.

Lorquet, J. C. (1981): Org. Mass Spectrom. 16, 469 – 482.

Lotz, W. (1970): Z. Phys. 232, 101 – 107.

Maccoll, A. (1975): Mass Spectrometry. London: Butterworths.

Maccoll, A. (1982): Org. Mass Spectrom. 17, 1 – 9.

Maeda, K., Semeluk, G. P., Lossing, F. P. (1968): Int. J. Mass Spectrom. Ion Phys. 1, 395 – 407.

Märk, E., Märk, T. D., Kim, Y. B., Stephan, K. (1981a): J. Chem. Phys. 75, 4446 – 4453.

Märk, T. D. (1975): J. Chem. Phys. 63, 3731 – 3736.

Märk, T. D. (1977): Z. Naturforschg. 32a, 1559 – 1560.

Märk, T. D. (1982a): Properties and Reactions of Cluster Ions. Book of Invited Papers, 4th Symp. Elementary Processes and Chemical Reactions in LTP, Stara Lesna (Martisovits, V., Lukac, P., eds.), 55 – 73.

Märk, T. D. (1982b): Beitr. Plasmaphysik 22, 257 – 294.

Märk, T. D. (1982c): Int. J. Mass Spectrom. Ion Phys. 45, 125 – 145.

Märk, T. D. (1984a): In: Electron-Molecule Interactions and Their Applications (Christophorou, L. G., ed.), Chapter 3. New York: Academic Press.

Märk, T. D. (1984b): Int. J. Mass Spectrom. Ion Proc. 55, 325 – 327.

Märk, T. D., Egger, F. (1976): Int. J. Mass Spectrom. Ion Phys. 20, 89 – 99.

Märk, T. D., Egger, F. (1977): J. Chem. Phys. 67, 2629 – 2635.

Märk, T. D., Hille, E. (1978): J. Chem. Phys. 69, 2492 – 2496.

Märk, T. D., de Heer, F. J. (1979): J. Phys. B 14, L429 – L432.

Märk, T. D., Castleman, A. W., jr. (1980): J. Phys. E 13, 1121 – 1124.

Märk, T. D., Castleman, A. W., jr. (1984): Adv. Atomic Molecular Physics 20, 65 – 172.

Märk, T. D., Egger, F., Cheret, M. (1977a): J. Chem. Phys. 67, 3795 – 3802.

Märk, T. D., Egger, F., Hille, E., Cheret, M., Störi, H., Stephan, K. (1977b): Proc. Xth ICPEAC, Paris, 1070 – 1071.

Märk, T. D., Märk, E., Stephan, K. (1981b): J. Chem. Phys. 74, 3633 – 3634.

Mandelbaum, D, Feldman, P. D. (1976): J. Chem. Phys. 65, 672 – 677.

Marchand, P., Pacquet, C., Marmet, P. (1969): Phys. Rev. 180, 123 – 132.

Marcus, R. A., Rice, O. K. (1951): J. Phys. and Colloid. Chem. 55, 894 – 908.

Marmet, P., Bolduc, E., Quemener, J. J. (1972): J. Chem. Phys. 56, 3463 – 3468.

Massey, H. S. W, Mohr, C. B. O. (1933): Proc. Roy. Soc. A 140, 613 – 636.

Massey, H. S. W. (1956): In: Handbuch der Physik (Flügge, S., ed.), Vol. 36, 307 – 408. Berlin-Göttingen-Heidelberg: Springer.

Massey, H. S. W., Burhop, E. H. S., Gilbody, H. B. (1952, 1969): Electronic and Ionic Impact Phenomena. Oxford: Clarendon Press.

Mathur, B. P., Abbey, L. E., Burgess, E. M., Moran, T. F. (1980): Org. Mass Spectrom. 15, 312 – 316.

Mathur, D. (1981a): Chem. Phys. Lett. 81, 115 – 118.

Mathur, D. (1981b): Int. J. Mass Spectrom. Ion Phys. 40, 235 – 239.

Mathur, D., Badrinathan, C. (1984): Int. J. Mass Spectrom. Ion Proc. 57, 167 – 178.
Mathur, D., Frost, D. C. (1981): J. Chem. Phys. 75, 5381 – 5384.
Mc Conkey, J. W., Crowe, A., Hender, M. A. (1972): Phys. Rev. Lett. 29, 1 – 4.
Mc Daniel, E. W. (1964): Collision Phenomena in Ionized Gases. New York: Wiley.
Mc Daniel, E. W., Flannery, M. R., Ellis, H. W., Eisele, F. L., Pope, W., Roberts, T. G. (1977): Technical
    Report H-78-1 (US Army Missile Research, Redstone Arsenal, AL 35809) II, 526 – 587.
Mc Dowell, M. R. C. (1969): In: Case Studies in Atomic Collision Physics (Mc Daniel, E. W.,
    Mc Dowell, M. R. C., eds.), Vol. 1, 47 – 97. Amsterdam: North-Holland.
Mc Farland, R. H. (1967): Phys. Rev. 159, 20 – 26.
Mc Farland, R. A., Kinney, J. D. (1965): Phys. Rev. 137, A 1058 – A 1061.
Mc Gowan, W., Clarke, E. M. (1968): Phys. Rev. 167, 43 – 51.
Mc Gowan, W., Fineman, M. A., Clarke, E. M., Hanson, H. P. (1968): Phys. Rev. 167, 52 – 60.
Mc Guire, E. J. (1972): Phys. Rev. A 3, 267 – 279.
Mc Guire, E. J. (1979): Phys. Rev. A 20, 445 – 456.
Mc Lafferty, F. W. (1963): Mass Spectrometry of Organic Ions. New York: Academic Press.
Mc Lafferty, F. W. (1966): Interpretation of Mass Spectra. New York: Benjamin.
Mc Pherson, A., Westerveld, W., Risley, J. S. (1980): Proc. 6th Int. Conf. Vacuum Ultraviolet Radiation
    Physics, Charlottesville, 62.
Meinke, C., Reich, G. (1962): Vakuum Technik 11, 86 – 88.
Meinke, C., Reich, G. (1963): Vakuum Technik 12, 79 – 82.
Meisels, G. G. (1982): Radiat. Phys. Chem. 20, 1 – 6.
Melton, C. E. (1970): Principles of Mass Spectrometry and Negative Ions. New York: Dekker.
Melton, C. E., Rudolph, P. S. (1967): J. Chem. Phys. 47, 1771 – 1774.
Mentall, J. E., Coplan, M. A., Kushlis, R. J. (1973): J. Chem. Phys. 59, 3867 – 3868.
Meyerson, S., Van der Haar, R. W. (1962): J. Chem. Phys. 37, 2458 – 2462.
Miletic, M., Eres, D., Veljkovic, M., Zmbov, K. F. (1980): Int. J. Mass Spectrom. Ion Phys. 35,
    231 – 242.
Milne, T. A. (1958): J. Chem. Phys. 28, 717 – 718.
Milne, T. A., Beachey, J. E., Greene, F. T. (1970): Air Force Cambridge Research Laboratories, Rep.
    AFCR L-70-0341.
Mohler, F. L., Bloom, E. G., Wells, E. J., Lengel, J. H., Wise, C. E. (1949): Phys. Rev. 74, 1332 – 1332.
Momigny, J., Wankenne, H., Krier, C. (1980): Int. J. Mass Spectrom. Ion Phys. 35, 151 – 170.
Monnom, G., Gaucherel, P., Paparodilis, C. (1984): J. Phys. 45, 77 – 84.
Morrison, J. D. (1964): J. Chem. Phys. 40, 2488 – 2492.
Morrison, J. D., Traeger, J. C. (1970): J. Chem. Phys. 53, 4053 – 4058.
Morrison, J. D., Traeger, J. C. (1973): Int. J. Mass Spectrom. Ion Phys. 11, 277 – 288.
Nagy, P., Skutlartz, A., Schmidt, V. (1980): J. Phys. B 13, 1249 – 1267.
Naidu, P. S., Westphal, K. O. (1966): Brit. J. Appl. Phys. 17, 645 – 656.
Nesbet, R. K. (1964): J. Chem. Phys. 40, 3619 – 3633.
Newton, A. S. (1964): J. Chem. Phys. 40, 607 – 608.
Nier, A. O. (1940): Rev. Sci. Instrum. 11, 212 – 216.
Nier, A. O. (1947): Rev. Sci. Instrum. 18, 398 – 411.
Nier, A. O. (1950): Phys. Rev. 77, 789 – 793; 79, 450 – 454.
Nier, A. O., Hanson, E. E. (1936): Phys. Rev. 50, 722 – 726.
Nikulin, V. K., Samoylov, A. V. (1982): Phys. Lett. 89 A, 225 – 228.
Nygaard, K. J. (1968): J. Chem. Phys. 49, 1995 – 2002.
Nygaard, K. J. (1975): Phys. Rev. A 11, 1475 – 1478.
Nygaard, K. J., Hahn, Y. B. (1973): Phys. Rev. A 8, 151 – 156.
Novick, R., Commins, E. D. (1958): Phys. Rev. 111, 822 – 840.
Okudaira, S. (1970): J. Phys. Soc. (Japan) 29, 409 – 415.
Okudaira, S., Kaneko, Y., Kanomata, I. (1970): J. Phys. Soc. (Japan) 28, 1536 – 1541.
Omidvar, K., Kyle, H. L., Sullivan, E. C. (1972): Phys. Rev. A 5, 1174 – 1187.
Orient, O. J., Srivastava, S. K. (1983): J. Chem. Phys. 78, 2949 – 2952.
Orient, O. J., Srivastava, S. K. (1984): J. Chem. Phys. 80, 140 – 143.
Osberghaus, O., Taubert, R. (1951): Angew. Chem. 63, 287 – 287.
Otvos, J. W., Stevenson, D. P. (1956): J. Am. Chem. Soc. 78, 546 – 551.
Paul, W. (1948): Z. Physik 124, 244 – 257.

Pavlov, S. I., Stotskii, G. I. (1970): Soviet Physics JETP *31*, 61 – 64.
Peach, G. (1965): Proc. Phys. Soc. *85*, 709 – 718.
Peach, G. (1966): Proc. Phys. Soc. *87*, 381 – 391.
Peach, G. (1970): J. Phys. *B 3*, 328 – 349.
Peach, G. (1971): J. Phys. *B 4*, 1670 – 1677.
Peach, G. (1976): Private Communication, see Dixon *et al.* (1976).
Peart, B., Walton, D. S., Dolder, K. T. (1971): J. Phys. *B 4*, 88 – 93.
Perrin, J., Schmitt, J. P. M. (1982): Chem. Phys. *67*, 167 – 176.
Peterson, J. R. (1963): Proc. 3rd ICPEAC, London, 465 – 473.
Pichou, F., Hall, R. I., Landau, M., Schermann, C. (1983): J. Phys. *B 16*, 2445 – 2456.
Pierce, J. R. (1954): Theory and Design of Electron Beams. Princeton: Van Nostrand.
Piraux, B., Joachain, C. (1980): Private Communication, see Defrance *et al.* (1981).
Pitchford, L. C. Märk, T. D., Castleman, A. W., jr. (1980): Proc. Xth SPIG, Dubrovnik, 20 – 21.
Pople, J. A. (1975): Int. J. Mass Spectrom. Ion Phys. *19*, 89 – 106.
Pottie, R. F. (1966): J. Chem. Phys. *44*, 916 – 922.
Pottie, R. F., Cocke, D. L., Gingerich, K. A. (1973): Int. J. Mass Spectrom. Ion Phys. *11*, 41 – 48.
Prasad, S. S. (1966): Proc. Phys. Soc. *87*, 393 – 398.
Quinn, E. I., Mohler, F. L. (1959): J. Research Natl. Bur. Standards *62*, 39 – 42.
Rabrenovic, M., Brenton, A. G., Beynon, J. H. (1983): Int. J. Mass Spectrom. Ion Phys. *52*, 175 – 182.
Raheja, U. T., Badrinathan, C., Mathur, D. (1983): Indian J. Phys. *57 B*, 27 – 31.
Rapp, D. (1971): J. Chem. Phys. *55*, 4154 – 4155.
Rapp, D., Englander-Golden, P. (1965): J. Chem. Phys. *43*, 1464 – 1479.
Rapp, D., Englander-Golden, P., Briglia, D. D. (1965): J. Chem. Phys. *42*, 4081 – 4085.
Redhead, P. A. (1967): Can. J. Phys. *45*, 1791 – 1812.
Reese, R. M., Eipple, J. A. (1949): Phys. Rev. *75*, 1332 – 1332.
Remberg, G., Remberg, E., Spiteller-Friedman, M., Spiteller, G. (1968): Org. Mass Spectrom. *1*, 87 – 113.
Rose, M. E., Johnstone, R. A. W. (1982): Mass Spectrometry for Chemists and Biochemists. Cambridge Cambridge University Press.
Rosenstock, H. M. (1968): Adv. Mass Spectrom. *4*, 523 – 546.
Rosenstock, H. M. (1976): Int. J. Mass Spectrom. Ion Phys. *20*, 139 – 190.
Rosenstock, H. M., Krauss, M. (1963): Adv. Mass Spectrom. *2*, 251 – 284.
Rosenstock, H. M., Wallenstein, M. B., Wahrhaftig, A. L., Eyring, H. (1952): Proc. Natl. Acad. Sci. U.S.A. *38*, 667 – 678.
Rothe, E. W. (1964): J. Vac. Sci. Techn. *1*, 66 – 68.
Rothe, E. W., Marino, L. L., Neynaber, R. H., Trujillo, S. M. (1962): Phys. Rev. *125*, 582 – 583.
Rowland, C. G. (1971): Int. J. Mass Spectrom. Ion Phys. *7*, 70 – 87.
Rowland, C. G., Eland, J. H. D., Danby, C. J. (1969): Int. J. Mass Spectrom. Ion Phys. *2*, 457 – 469.
Rüdenauer, F. G (1970): Rev. Sci. Instr. *41*, 1487 – 1488.
Rüdenauer, F. G (1971): J. Vac. Sci. Technol. *9*, 215 – 215.
Sattler, K., Mühlbach, J., Echt, O., Pfau, P., Recknagel, E. (1981): Phys. Rev. Lett. *47*, 160 – 163.
Schaeffer, O. A. (1950): J. Chem. Phys. *18*, 1681 – 1682.
Schaeffer, O. A. (1954): Rev. Sci. Instrum. *25*, 660 – 662.
Schaeffer, O. A. (1955): J. Chem. Phys. *23*, 1309 – 1313.
Schaeffer, O. A., Hastings, J. M. (1950): J. Chem. Phys. *18*, 1048 – 1050.
Schmidt, V., Sandner, N., Kuntzemüller, H. (1976): Phys. Rev. *A 13*, 1743 – 1747.
Schneider, M. (1974): J. Phys. *D 7*, L83 – L86.
Schram, B. L. (1966): Physica *32*, 197 – 208.
Schram, B. L., de Heer, F. J., van der Wiel, M. J., Kistemaker, J. (1965): Physica *31*, 94 – 112.
Schram, B. L., Boerboom, A. J. H., Kleine, W., Kistemaker, J. (1966a): Physica *32*, 749 – 761.
Schram, B. L., Adamczyk, B., Boerboom, A. J. H. (1966b): J. Sci. Instrum. *43*, 638 – 640.
Schram, B. L., Boerboom, A. J. H., Kistemaker, J. (1966c): Physica *32*, 185 – 196.
Schram, B. L., Moustafa, H. R., Schutten, J., de Heer, F. J. (1966d): Physica *32*, 734 – 740.
Schulz, W., Drost, H., Klotz, H. D. (1968): Exper. Technik d. Physik *16*, 16 – 22.
Schutten, J., de Heer, F. J., Moustafa, H. R., Boerboom, A. J. H., Kistemaker, J. (1966): J. Chem. Phys. *44*, 3924 – 3928.
Seaton, M. J. (1959): Phys. Rev. *113*, 814 – 814.

Selim, E. T. M. (1980): Ind. J. Pure Appl. Phys. *18*, 31 – 35.
Sen Sharma, D. K., Franklin, J. L. (1974): Int. J. Mass Spectrom. Ion Phys. *13*, 139 – 150.
Shearer-Izumi, W., Botter, R. (1974): J. Phys. *B 7*, L 125 – L 128.
Siegel, M. W. (1982): Int. J. Mass Spectrom. Ion Phys. *44*, 19 – 36.
Skudlarski, K., Drowart, J., Exsteen, G., Van Der Anwera-Mahieu, A. (1967): Trans. Faraday Soc. *63*, 1146 – 1151.
Smith, O. I. (1963): Int. J. Mass Spectrom. Ion Phys. *54*, 55 – 59.
Smith, O. I., Stevenson, J. S. (1981): J. Chem. Phys. *74*, 6777 – 6783.
Solomon, M., Mandelbaum, A. (1969): Chem. Commun., 890 – 890.
Spiteller, G. (1968): Massenspektrometrische Strukturanalyse organischer Verbindungen. Weinheim: Verlag Chemie.
Stace, A. J., Moore, C. (1982): J. Phys. Chem. *86*, 3681 – 3683.
Stace, A. J., Moore, C. (1983): Chem. Phys. Lett. *96*, 80 – 84.
Stace, A. J., Shukla, A. K. (1980): Int. J. Mass Spectrom. Ion Phys. *36*, 119 – 122.
Stace, A. J., Shukla, A. K. (1982a): Chem. Phys. Lett. *85*, 157 – 160.
Stace, A. J., Shukla, A. K. (1982b): J. Phys. Chem. *86*, 865 – 867.
Stace, A. J., Shukla, A. K. (1982c): J. Am. Chem. Soc. *104*, 5314 – 5318.
Stanski, T., Adamczyk, B. (1982): 9th Int. Mass Spectrom. Conference, Wien.
Stanski, T., Adamczyk, B. (1983): Int. J. Mass Spectrom. Ion Phys. *46*, 31 – 34.
Stanton, H. E., Monahan, J. E. (1960): Phys. Rev. *119*, 711 – 715.
Stanton, H. E., Monahan, J. E. (1964): J. Chem. Phys. *41*, 3694 – 3702.
Stephan, K. (1979): Thesis, Universität Innsbruck.
Stephan, K., Märk, T. D. (1981): Proc. XIIth ICPEAC, Gatlinburg, 265 – 266.
Stephan, K., Märk, T. D. (1982a): Chem. Phys. Lett. *90*, 51 – 54.
Stephan, K., Märk, T. D. (1982b): Chem. Phys. Lett. *87*, 226 – 228.
Stephan, K., Märk, T. D. (1983): Int. J. Mass Spectrom. Ion Phys. *47*, 195 – 198.
Stephan, K., Märk, T. D. (1984): J. Chem. Phys. *81*, 3116 – 3117.
Stephan, K., Märk, T. D., Helm, H. (1978): Symp. Atomic Surface Phys., Maria Alm (Lindinger, W., et al., eds.), 77 – 90.
Stephan, K., Helm, H., Märk, T. D. (1980a): Adv. Mass Spectrom. *8A*, 122 – 132.
Stephan, K., Helm, H., Märk, T. D. (1980b): J. Chem. Phys. *73*, 3763 – 3778.
Stephan, K., Helm, H., Kim, Y. B., Sejkora, G., Ramler, J., Grössl, M., Märk, E., Märk, T. D. (1980c): J. Chem. Phys. *73*, 303 – 308.
Stephan, K., Futrell, J. H., Peterson, K. I., Castleman, A. W., jr., Wagner, H. E., Djuric, N., Märk, T. D. (1981): Proc. 8th Intern. Symp. Molec. Beams, Cannes, 211 – 215.
Stephan, K., Stamatovic, A., Märk, T. D. (1982a): ÖGV Tagung, Balzers. See also: Stephan, K., Märk, T. D., Futrell, J. H., Castleman, A. W., jr. (1983): Vacuum TAIP *33*, 77 – 85.
Stephan, K., Helm, H., Märk, T. D. (1982b): Phys. Rev. *A 26*, 2981 – 2982.
Stephan, K., Futrell, J. H., Peterson, K. I., Castleman, A. W., jr., Märk, T. D. (1982c): J. Chem. Phys. *77*, 2408 – 2415.
Stephan, K., Futrell, J. H., Peterson, K. I., Castleman, A. W., jr., Wagner, H. E., Djuric, N., Märk, T. D. (1982d): Int. J. Mass Spectrom. Ion Phys. *44*, 167 – 181.
Stephan, K., Märk, T. D., Castleman, A. W., jr. (1983a): J. Chem. Phys. *78*, 2953 – 2956.
Stephan, K., Märk, T. D., Märk, E., Stamatovic, A., Djuric, N., Castleman, A. W., jr. (1983b): Beitr. Plasmaphysik *23*, 175 – 179.
Stephan, K., Stamatovic, A., Märk, T. D. (1983c): Phys. Rev. *A 28*, 3105 – 3108.
Stephan, K., Deutsch, H., Märk, T. D. (1983d): Proc. Annual Conference on Mass Spectrometry and Allied Topics, Boston, 734.
Stephan, K., Leiter, K., Deutsch, H., Märk, T. D. (1983e): Proc. XIIIth ICPEAC, Berlin, 291.
Stephan, K., Märk, T. D., Futrell, J. H., Castleman, A. W., jr. (1984): J. Chem. Phys. *80*, 3185 – 3188.
Stevenson, D. P. (1947): J. Chem. Phys. *15*, 409 – 411.
Stevenson, D. P. (1949): J. Chem. Phys. *17*, 101 – 102.
Stevenson, D. P. (1960): J. Am. Chem. Soc. *82*, 5961 – 5965.
Stevenson, D. P., Hipple, J. A. (1942): Phys. Rev. *62*, 237 – 240.
Stevenson, D. P., Schissler, D. O. (1961): In: The Chemical and Biological Action of Radiation (Harssinsky, M., ed.), Vol. 5, 181 – 192. London: Academic Bks.
Tan, K. H., Mc Conkey, J. W. (1974): Phys. Rev. *A 10*, 1212 – 1222.

Tannen, P. D. (1973): Thesis (Air Force Institute of Technology), Univ. Microfilms, Ann Arbor Michigan, Order No. 74-14940.

Tate, J. T., Smith, P. T., Vaughan, A. L. (1935): Phys. Rev. *48*, 525 − 531.

Taubert, R. (1959): Adv. Mass Spectrom. *1*, 489 − 503.

Taubert, R. (1964): Z. Naturforschg. *19a*, 484 − 493, 494 − 506, 911 − 925.

Tay, E. S., Dawson, P. G., Mosson, G. A. G. (1970): Proc. 6th Europ. Symp. Fusion Techn., Aachen, 51 − 58.

Taylor, P. O., Dolder, K. T., Kauppila, W. E., Dunn, G. H. (1974): Rev. Sci. Instrum. *45*, 538 − 544.

Teleshefsky, L. A., Jones, B. E., Abbey, L. E., Bostwick, D. E., Burgess, E. M., Moran, T. F. (1982): Org. Mass Spectrom. *17*, 481 − 492.

Thomson, J. J. (1912): Phil. Mag. *24*, 668 − 672.

Thomson, J. J. (1921): see Reference in Vaughan (1931).

Ton-That, D. Manson, S. T., Flannery, R. M. R. (1976): Private Communication, see Dixon *et al.* (1976).

Ton-That, D.. Flannery, M. R. (1977): Phys. Rev. *A 15*, 517 − 526.

Tuithof, H. F., Boerboom, A. J. H. (1974): Int. J. Mass Spectrom. Ion Phys. *15*, 105 − 109.

Turban, G., Catherine, Y., Grolleau, B. (1980): Thin Solid Films *67*, 309 − 320.

Turner, B. R., Rutherford, J. A., Compton, D. M. J. (1968): J. Chem. Phys. *48*, 1602 − 1608.

Valyi, L. (1975): Atom and Ion Sources. London: Wiley.

Van Brunt, R J., Wacks, M. E. (1964): J. Chem. Phys. *41*, 3195 − 3199.

Van Gorkom, M., Glick, R. E. (1970): Int. J. Mass Spectrom. Ion Phys. *4*, 203 − 218.

Van Gorkom, M., Beggs, D. P., Glick, R. E. (1970): Int. J. Mass Spectrom. Ion Phys. *4*, 441 − 450.

Van Sprang, H. A., Möhlmann, G. R., de Heer, F. J. (1978): Chem. Phys. *33*, 65 − 72.

Varga, P., Winter, H. (1978): Phys. Rev. *A 18*, 2453 − 2458.

Varga, P., Hofer, W., Winter, H. (1981): J. Phys. *B 14*, 1341 − 1351.

Vaughan, A. L. (1931): Phys. Rev. *38*, 1687 − 1695.

Vauthier, R. (1950): C. R. Acad. Sci. (Paris) *231*, 764 − 765, 1218 − 1220.

Vauthier, R. (1955): C. R. Acad. Sci. (Paris) *241*, 1033 − 1036; Ann. Phys. (Paris) *10*, 968 − 1025.

Vestal, M. L. (1968): In: Fundamental Processes in Radiation Chemistry (Ausloos, P., ed.), 59 − 118. New York: Interscience.

von Engel, A (1965): Ionized Gases. Oxford: Clarendon Press.

Vriens, L. (1966): Phys. Rev. *141*, 88 − 92.

Vriens, L., Bonsen, T. F. M. (1968): J. Phys. *B 1*, 1123 − 1130.

Vriens, L., Bonsen, T. F. M., Smit, J. A. (1968): Physica *40*, 229 − 252.

Wahrhaftig, A. L. (1972): In: Mass Spectrometry (Maccoll, E. A., ed.), 1 − 24. London: Butterworths.

Waight, E. S. (1969): Chem. Commun., 1258 − 1258.

Waldron, J. D., Wood, K. (1952): Mass Spectrometry, Institute of Petroleum, London, 16.

Walker, K. C., John, R. M. S. (1972): Phys. Rev. *A 6*, 240 − 250.

Wallace, S. J., Berg, R. A., Green, A. E. S. (1973): Phys. Rev. *A 7*, 1616 − 1629.

Wallington, M. J. (1970): J. Phys. *E 3*, 599 − 604.

Wallington, M. J. (1971): J. Phys. *E 4*, 1 − 8.

Washburn, H. W., Berry, C. E. (1946): Phys. Rev. *70*, 559 − 559.

Werner, H. W. (1974): J. Phys. *E 7*, 115 − 121; Int. J. Mass Spectrom. Ion Phys. *14*, 189 − 204.

Werner, H. W., Linssen, A. J. (1974): J. Vac. Sci. Technol. *11*, 843 − 847.

Werner, H. W., Venema, A., Linssen, A. J. (1972): J. Vac. Sci. Technol. *9*, 216 − 219.

Wetzel, R. C., Baiocchi, F. A., Freund, R. S. (1984): 37th GEC, Boulder.

White, F. A. (1968): Mass Spectrometry in Science and Technology. New York: Wiley.

Wiegand, W. J., Boedeker, L. R. (1982): Appl. Phys. Lett. *40*, 225 − 227.

Winters, R. E., Kiser, R. W. (1965): Inorg. Chemistry *4*, 157 − 161.

Winters, R. E., Collins, J. H., Courchenne, W. L. (1966): J. Chem. Phys. *45*, 1931 − 1937.

Ziegler, D. L. Newman, J. H., Smith, K. A., Stebbings, R. F. (1982): Planet. Space Sci. *30*, 451 − 456.

Ziesel, J. P. (1965): J. Chim. Phys. *62*, 328 − 335.

Ziesel, J. P., Abouaf, R. (1967): J. Chim. Phys. *64*, 702 − 705.

Ziesel, J. P. (1967a): J. Chim. Phys. *64*, 695 − 701.

Ziesel, J. P. (1967b): J. Chim. Phys. *64*, 706 − 707.

6

Innershell Ionization Cross Sections*

C. J. Powell

Surface Science Division, National Bureau of Standards, Gaithersburg, Md., U.S.A.

6.1 Introduction

Previous chapters of this volume have described in some detail calculations and measurements of ionization cross sections. The discussion has been almost exclusively devoted to the ionization of atoms, molecules, or ions in the gas phase by electron impact and to processes involving the removal of one or more *valence* electrons.

The present chapter is concerned with the cross sections for removal of *inner-shell* or core electrons by electron impact. Measurements of cross sections for removal of inner-shell levels have been made both with gas-phase atoms and with solids. In addition to the fundamental interest in determining physical mechanisms for inner-shell excitation and de-excitation and in obtaining basic cross-section data for comparisons with theory, cross-section data are needed in three types of materials characterization: electron-probe microanalysis (EPMA), Auger-electron spectroscopy (AES), and electron energy-loss spectroscopy (EELS). Cross sections for inner-shell ionization are also required in the modelling of the interactions of ionizing radiation with matter. Calculations and some measurements of inner-shell ionization cross sections have been made for free atoms and are expected (for reasons to be discussed later) to be useful guides for inner-shell ionization in solids.

In the use of EPMA, AES, and EELS for materials characterization, data are needed for the production of inner-shell *vacancies*. The term ionization cross section will refer here to the cross section for the production of a vacancy in a specified inner-shell or subshell even though, for a solid, an ionized state rather than a free ion is produced. Cross sections for inner-shell ionization are typically much less than those for the excitation (in a solid) or ionization (in a gas) of valence electrons. For an atom or molecule, the inner-shell ionization cross section as just defined will be greater than the cross section for the production of a free ion by the cross section for

---

excitation of an inner-shell electron to unoccupied discrete states that do not decay by autoionization; this latter cross section is expected to be small (Codling, 1973).

The present author has published two reviews that summarize cross-section data and related information for inner-shell ionization in atoms and solids (Powell, 1976a; Powell, 1976b). The present chapter will summarize, update, and extend these previous reviews. Emphasis will be placed on cross sections for inner-shell ionization with incident electron energies less than 50 keV since information in this range is useful in EPMA, AES, and EELS as well as for other purposes. For such incident energies, relativistic corrections are usually small compared to other uncertainties. (Note that $mc^2 = 510$ keV.) EELS measurements, however, are often made at incident energies of 100 keV and above and in these cases relativistic corrections should be considered. Investigations of inner-shell ionization by relativistic electrons have been published by Quarles (1976), Hahn (1976), Scofield (1978), Tawara (1978), Hoffmann et al. (1980), Palinkas and Schlenk (1980), Eschwey and Manakos (1982), Genz (1982), and Müller et al. (1983). We note here that there are two types of relativistic corrections. First, there are the kinematic modifications to the cross-section formulas (Inokuti, 1971). Second, there are modifications to the atomic charge distributions and core-electron binding energies for which corrections are difficult.

It will be convenient here to review and compare cross-section formulas and data in terms of Bethe's (1930) formula for inner-shell ionization. The Bethe theory has been successfully employed for many applications [see Chapters 2 and 7 and Inokuti and Manson (1934)] as it provides a convenient and consistent quantum-mechanical basis for electronic excitation and ionization in atoms and molecules (Inokuti, 1971; Inokuti et al., 1978). The theory also provides a connection to electronic excitation and inner-shell ionization in condensed matter. The Bethe theory is only expected to be valid if the incident electron energy is sufficiently high. We will discuss later the expected consequences of the incident energy not being sufficiently high and the use of the Bethe formula, albeit empirical, at low incident energies and of certain proposed modifications and alternatives to the Bethe formula. Relativistic corrections can be made readily to the Bethe formula (Inokuti, 1971).

The dielectric theory of inelastic electron scattering in solids is reviewed in Section 6.2 and related to the Bethe (1930) theory for atomic excitation and ionization. Reference is made to recent calculations of inner-shell ionization cross sections. Section 6.3 contains a summary of semi-empirical and empirical formulas for ionization cross sections. The several techniques for measuring ionization cross sections are reviewed in Section 6.4 and examples of cross-section data given. Measurements made since the previous review (Powell, 1976a) are surveyed. A comparison of calculations and measurements of inner-shell ionization cross sections is made in Section 6.5 with emphasis given to phenomena that limit the range of validity of particular formulas. Finally, examples are given in Section 6.6 of applications of cross-section data to bulk analysis of materials by EPMA, surface analysis by AES, and thin-film analysis by EELS.

6.2 Theory

6.2.1 Dielectric and Bethe Theories of Inelastic Electron Scattering

We begin by discussing the theory of inelastic scattering of electrons in solids (Schnatterly, 1979; Raether, 1980; Powell, 1984). The differential inelastic scattering cross section, per atom or molecule, for energy loss E and momentum transfer q in an infinite medium is

$$\frac{d^2\sigma}{dE\,dq} = \frac{2\,e^2}{\pi\,N\,v^2}\,\text{Im}\left[\frac{-1}{\varepsilon(\omega,q)}\right]\frac{1}{q} \tag{6-1}$$

where N is the density of atoms, e is the electronic charge, and v is the velocity of the incident electrons. For small scattering angles, $q \approx P(\theta^2 + \theta_E^2)^{1/2}$, P is the momentum of the incident electrons, θ is the scattering angle, $\theta_E = E/2\,E_0$, and E_0 is the incident energy. The term $\text{Im}\,[-1/\varepsilon(\omega, q)]$ in Eq. (6-1) is the so-called energy-loss function which is defined in terms of the complex dielectric constant $\varepsilon(\omega, q)$, where $E = \hbar\,\omega$. For $q = 0$, the dielectric constant is related to the conventional optical constants, the refractive index n and the extinction coefficient k, by

$$\varepsilon(\omega, 0) = (n + i\,k)^2 = \varepsilon_1 + i\,\varepsilon_2 \tag{6-2a}$$

where

$$\varepsilon_1 = n^2 - k^2, \tag{6-2b}$$

$$\varepsilon_2 = 2\,n\,k = \rho\,c\,\mu_m/\omega \tag{6-2c}$$

and

$$\text{Im}(-1/\varepsilon) = \varepsilon_2/(\varepsilon_1^2 + \varepsilon_2^2). \tag{6-2d}$$

In Eq. (6-2 c), ρ is the density of the solid, c is the velocity of light, and μ_m is the optical (often x-ray) mass absorption coefficient.

For Eq. (6-1) to be valid, E_0 should be "much greater" than E. The extent to which E_0 should be greater than E will be discussed later. Equation (6-1) can be extended to materials with crystalline anisotropies and to include the terms associated with specific surface excitations but these complications are not relevant here (Raether, 1980). Equation (6-1) can also be utilized for inelastic electron scattering in liquids.

Equations (6-1) and (6-2) indicate the similarities and differences that occur for inelastic electron scattering in condensed matter and in free atoms or molecules. The denominator in Eq. (6-2 d) is essentially unity for free atoms and molecules ($\varepsilon_1 \approx 1$, $\varepsilon_2 \ll 1$), and the energy loss function is then proportional to ε_2 and to the optical absorption coefficient. For condensed matter, the denominator in Eq. (6-2 d) may depart significantly from unity; this situation generally occurs for inelastic scattering processes involving valence-electron excitations. For inner-shell ionization in solids, however, and for energy losses $\gtrsim 100\,\text{eV}$, $\varepsilon_1 \approx 1$ and $\varepsilon_2 \ll 1$ so that $\text{Im}(-1/\varepsilon) \approx \varepsilon_2$. Certain specific "solid-state" effects (extended x-ray absorption fine-structure or EXAFS oscillations, near-edge density-of-states effects, and two-electron or many-electron excitation satellites) can lead to small modulations of the intensity of inelastically scattered electrons associated with inner-shell excitations or to specific local modulations of the scattered intensity for excitation energies close to the threshold for core-level ionization (Powell, 1984). These effects arise when relatively slow electrons are ejected and they are subject to delicate forces in the solid state. It has recently been shown that "solid-state" effects can also modify the form of photoionization cross-section data for some transition metals (Abbati *et al.*, 1983). With the exception of these specific effects, the energy-loss function for inner-shell excitation in solids will be generally similar to the corresponding data (ε_2 or μ_m) for free atoms and molecules. Thus, cross-section measurements that represent integrals of Eq. (6-1) over appropriate limits of E and q for a particular element can be expected to yield similar results whether or not measurements are made in the gas phase or the solid phase (Codling, 1973; Kunz, 1973; Ing and Pendry, 1975). Similarly, cross-section measurements for an element in different chemical states should give similar results.

It is convenient to define the differential oscillator strength

$$\frac{df(E,q)}{dE} = \frac{2\,E\,\mathrm{Im}\,[-1/\varepsilon(\omega,q)]}{\pi\,\Omega_p^{\,2}} \tag{6-3}$$

where $\Omega_p = (4\pi N e^2/m)^{1/2}$ and m is the electron rest mass. The differential oscillator strength satisfies the Thomas-Reiche-Kuhn sum rule for each value of q:

$$\int\limits_0^\infty [df(E,q)/dE]\,dE = Z \tag{6-4}$$

where Z is the atomic number.

Equations (6-1) and (6-3) can be combined to yield

$$\frac{d^2\sigma}{dE\,dq} = \frac{4\pi e^4}{m v^2} \cdot \frac{df(E,q)}{dE} \cdot \frac{1}{q}. \tag{6-5}$$

Equation (6-5) is in the form of equations derived by Bethe (1930) to describe electronic excitation and ionization of atoms and molecules by electron impact. For excitation to a discrete state n of an atom, the function $f_n(q)$ is termed the generalized oscillator strength (Bethe, 1930).

The determination of either a partial or total integrated cross section (see Chapter 1) for inner-shell ionization from Eq. (6-5) requires knowledge of the function $df(E,q)/dE$. With the increasing use of synchrotron radiation to investigate atomic, molecular and solid-state properties, there is a growing amount of optical absorption data for photon energies that can excite inner-shell electrons (e.g., Weaver *et al.*, 1981). Data of this type give information on $df(E,0)/dE$. Very similar data are obtained in measurements of electron energy-loss spectra in transmission through thin specimen films for $\theta = 0$; although the minimum momentum transfer q_{min} for a particular excitation is not zero, $\mathrm{Im}\,[-1/\varepsilon(\omega,q_{min})] \approx \mathrm{Im}\,[-1/\varepsilon(\omega,0)]$. Information on $df(E,q)/dE$ for small q can be obtained from EELS experiments in electron microscopes over a range of scattering angles. Over the last several years, calculations of $df(E,q)/dE$ have been reported for a moderate number of atoms (McGuire, 1977; Egerton, 1979 and 1981; Leapman *et al.*, 1980; Inokuti and Manson, 1983; Rez, 1984). These calculations are useful guides to the q-dependence of $df(E,q)/dE$ for the excitation of electrons from specific shells of representative elements.

We will assume for the moment that $df(E, q)/dE$ is a slowly varying function of q for small q where the differential cross section is large. That is,

$$\frac{df(E, q)}{dE} \approx \frac{df(E, q_{min})}{dE} \approx \frac{df(E, 0)}{dE} \tag{6-6}$$

where $df(E, 0)\, dE$ is simply related to the optical absorption coefficient. Equation (6-5) can now be integrated from $q_{min} = E/v$ to an "effective" upper limit

$$q_{max} = (m\, c\, (E)\, E/2)^{1/2} \tag{6-7}$$

where $c(E)$ is a term that depends on the specific q-dependence of $df(E, q)/dE$. From calculations based on the hydrogenic approximation, Bethe (1930) estimated $c(E)$ for discrete excitations to be approximately 4 although it is clear from recent calculations (Leapman et al., 1980; Inokuti and Manson, 1983; Rez, 1984) that $c(E)$ can depart appreciably from 4 for inner-shell ionization of certain shells of some atoms. In such cases, there is substantial oscillator strength for momentum transfers greater than 1 Å^{-1}. Equations (6-5) − (6-7) can be combined to give the differential cross section for energy loss E:

$$\frac{d\sigma}{dE} \approx \frac{2\pi e^4}{m v^2} \cdot \frac{1}{E} \cdot \frac{df(E, 0)}{dE} \ln\left[\frac{c(E)\, m v^2}{2E}\right]. \tag{6-8}$$

It should be noted that $c(E)$ is, in general, strongly dependent on E although $d\sigma/dE$ is usually only mildly dependent on E (Inokuti, private communication).

The total integrated cross section σ_{nl} per atom or molecule for ionization of the nl shell can now be found by integration of Eq. (6-8) from a lower limit E_{nl}, the binding energy of electrons in the nl shell, to some upper limit E_{max}. If the differential oscillator strength $df(E, q)/dE$ is known from calculations, the integrations of Eq. (6-5) over q and E can be performed numerically and the limit E_{max} can be chosen to be infinity. If, however, $df(E, 0)/dE$ is obtained from experimental (e.g., optical absorption) data, the upper limit E_{max} must be chosen carefully since the experimental data for any excitation energy may contain significant contributions from the ionization of two or more shells. While the customary assumption that E_{max} can be chosen to the binding energy of the next most tightly bound shell may be a satisfactory approximation to infinity in some cases, in other cases there may be appreciable error (Powell, 1984). In the latter situations, one cannot determine σ_{nl} without developing a satisfactory algorithm for determining the component of $df(E, 0)/dE$ associated with nl-shell ionization over a sufficiently large range of excitation energies.

Bethe (1930) has expressed σ_{nl} in the form

$$\sigma_{nl} = \frac{2\pi e^4}{m v^2} \frac{Z_{nl}\, b_{nl}}{E_{nl}} \ln\left[\frac{c_{nl}\, E_0}{E_{nl}}\right] \tag{6-9}$$

where Z_{nl} is the number of electrons in the nl shell, and b_{nl} and c_{nl} can be regarded as parameters for a specific shell and element. Equation (6-9) is in the form described and discussed in Section 7.2.2 of Chapter 7; see also Inokuti (1971). Comparison of Eqs. (6-8) and (6-9) indicates that

$$b_{nl} = \frac{E_{nl}}{Z_{nl} \ln(c_{nl} E_0/E_{nl})} \int\limits_{E_{nl}}^{E_{\max}} \frac{1}{E} \frac{df(E,0)}{dE} \ln\left[\frac{c(E)E_0}{E}\right] dE$$

$$\approx \frac{E_{nl}}{Z_{nl}} \int\limits_{E_{nl}}^{E_{\max}} \frac{1}{E} \frac{df(E,0)}{dE} dE. \tag{6-10}$$

Note that the determination of b_{nl} from Eq. (6-10) (e.g., with use of photoabsorption data) is subject to uncertainty concerning the extent to which the integral is dominated by contributions to the differential oscillator strength from only the nl-subshell, as just discussed.

Equation (6-9) can be rewritten:

$$\sigma_{nl} E_{nl}^2 = \pi e^4 Z_{nl} b_{nl} \ln(c_{nl} U_{nl})/U_{nl} \tag{6-11 a}$$

or, with values of constants inserted,

$$\sigma_{nl} E_{nl}^2 = 6.51 \times 10^{-14} Z_{nl} b_{nl} \ln(c_{nl} U_{nl})/U_{nl} \quad cm^2\, eV^2 \tag{6-11 b}$$

where U_{nl}, often referred to as the overvoltage, is equal to E_0/E_{nl}. Equation (6-11) indicates that plots of $\sigma_{nl} E_{nl}^2$ versus U_{nl} for the ionization of a given shell are useful in determining whether the parameters b_{nl} and c_{nl} are a function of Z; if not, values of σ_{nl} for the ionization of a particular shell in one element can readily be scaled to derive values of σ_{nl} for ionization of the same shell in other elements. In addition, a plot of $\sigma_{nl} E_n^2 U_{nl}/\pi e^4 Z_{nl}$ versus $\ln U_{nl}$ for a given element shows directly the range of U_{nl} for which Eq. (6-11) is valid and allows a convenient means of determining the parameters b_{nl} and c_{nl}. Such a plot is referred to as a Fano plot (see Chapter 7). Equation (6-11), like Eq. (6-1), has been derived with the use of the first Born approximation (Inokuti, 1977); that is, it has been assumed that U_{nl} is sufficiently large. This requirement will be discussed below.

6.2.2 Atomic Calculations of Inner-Shell Ionization Cross Sections

We will now discuss several recent quantum-mechanical calculations of cross sections for inner-shell ionization in atoms (see also Chapters 1 and 2).

McGuire (1977 and 1979) has reported extensive calculations of cross sections for inner-shell ionization of various atomic subshells using the plane-wave Born approximation (PWBA). His results have been presented in scaled form, that is, values of $\sigma_{nl}(E_{nl})^\alpha$ as a function of U_{nl}. For a particular subshell and a selected value of U_{nl}, it would be expected from Eq. (6-11) that $\alpha = 2$ if b_{nl} or c_{nl} was not a function of atomic number (with appropriate consideration of the Z_{nl} term). McGuire found that the parameter α often departed substantially from the expected value of 2 for subshells with fairly low binding energy (i.e., for E_{nl} typically less than 500 to 1000 eV, depending on the subshell). For example, McGuire found $\alpha = 1.70$ and $\alpha = 1.655$ for ionization of the $2s$ and $2p$ subshells, respectively, of the elements Na through Ar; $\alpha = 2$ for ionization of both subshells for elements with $Z > 18$. Similarly, the cross-section calculations indicated values of α of 1.67, 1.69, and 1.48 for $3s$, $3p$, and $3d$ subshell ionization for the elements Mn-I, Cr-I, and Ga-In, respectively. Some measurements of inner-shell ionization cross sections support the trends found in McGuire's calculations although calculations of $b_{L_{23}}$ from photoabsorption data with the use of Eq. (6-10) do not show the strong Z-dependence found by McGuire (Powell, 1976a).

Hippler and Jitschin (1982) have extended the PWBA for K-shell ionization of light atoms to relatively low values of U_K. They took account of electron exchange with the Ochkur approximation and found good agreement with experimental data $(Z=6$ to $18)$ for $U_K > 2$.

Rudge and Schwartz (1966) derived an analytic formula based on a calculation of σ_K and σ_{L_1} for a fictitious hydrogenic ion with $Z=128$ using the Born-exchange approximation. Their result is:

$$\sigma_{nl} E_{nl}^2 = \frac{1.626 \times 10^{-14} Z_{nl} S_{nl} \ln U_{nl}}{U_{nl}} \quad \text{cm}^2 \text{ eV}^2 \qquad (6\text{-}12\,a)$$

where

$$S_K(U_K) = 2.799 - \frac{0.218}{U_K} + \frac{0.047}{U_K^2} \qquad (6\text{-}12\,b)$$

and

$$S_{L_1}(U_{L_1}) = 2.168 + \frac{1.147}{U_{L_1}} - \frac{0.212}{U_{L_1}^2} \qquad (6\text{-}12\,c)$$

and has been obtained for $1.25 \leq U_{nl} \leq 4$. The approach of Rudge and Schwartz has been extended by Golden, Sampson and co-workers in a series of papers which are summarized by Moores *et al.* (1980). They have fitted total cross sections for the ionization of electrons from the K, L_1, L_{23}, M_1, M_{23}, and M_{45} levels of a heavy hydrogenic ion for $1.125 \leq U_{nl} \leq 6$ to the formula:

$$\sigma_{nl}^R (U_{nl}) = [A \ln U + B(1 - U^{-1})^2 + (CU^{-1} + DU^{-2})(1 - U^{-1})]/U \qquad (6\text{-}13\,a)$$

where σ_{nl}^R is a reduced cross section defined by

$$\sigma_{nl}^R = \frac{Z^4 \sigma_{nl}}{n^4 \pi a_0^2 Z_{nl}}. \qquad (6.13\,b)$$

In Eq. (6-13 b), a_0 is the Bohr radius and Z is the nuclear charge. If $(Z/n)^2$ is assumed to be equal to E_{nl}/E_R where E_R is the Rydberg energy (i.e., no correction is made for screening), Eqs. (6-13 a) and (6-13 b) can be rewritten as

$$\sigma_{nl} E_{nl}^2 = 1.628 \times 10^{-14} Z_{nl} \sigma_{nl}^R \quad \text{cm}^2 \text{ eV}^2. \qquad (6\text{-}13\,c)$$

Values of the parameters for Eq. (6-13 a) are given in Table 6-1.

Egerton (1979) has calculated generalized oscillator strengths for K-shell ionization of low-Z elements. He has utilized a simple hydrogenic model and has obtained partial and total K-shell ionization cross sections for $E_0 > 30 \text{ keV}$ (conditions appropriate for materials microanalysis by EELS). The calculated cross sections agree quite well with experimental measurements. He has used the same approach to compute L-shell ionization cross sections (Egerton, 1981).

Table 6.1. *Values of the parameters in Eq. (6.13) for ionization of the indicated levels [from Moores et al.* (*1980*)]

Shell	A	B	C	D
K	1.13	4.41	-2.00	3.80
L_1	0.823	3.69	0.62	1.79
L_{23}	0.530	5.07	1.20	2.50
M_1	0.652	3.83	0.64	2.10
M_{23}	0.551	4.38	1.83	1.90
M_{45}	0.280	5.70	2.21	2.65

Leapman, Rez, and Mayers (1980) have performed calculations of generalized oscillator strengths and differential scattering cross sections as a function of excitation energy for K-, L-, and M-shell ionization of selected low- and medium-Z elements. These authors employed a Hartree-Slater potential for determining the generalized oscillator strengths. The general features of the calculated differential scattering cross sections agreed with experimental data for solids although, not surprisingly, there were discrepancies near the thresholds for core-electron excitation. Further calculations of the same type, including L- and M-shell ionization of

Table 6-2. *Values of the Bethe parameters b_{nl} and c_{nl} determined by Rez (1984)*

Element	b_{nl}	c_{nl}
(a) K-shell ionization		
Al	0.709	1.113
P	0.913	0.933
K	1.195	0.706
Ti	0.972	0.873
Cr	0.980	0.864
Fe	1.102	0.846
Cu	1.037	0.831
Ge	1.052	0.822
Y	1.081	0.804
(b) L_3-subshell ionization		
Zr	0.958	0.862
Ag	0.972	0.846
Sb	0.972	0.854
Ba	1.007	0.828
Gd	1.045	0.801
W	1.094	0.744
(c) M_4-subshell ionization		
W	1.211	0.700
Au	1.251	0.688
Pb	1.209	0.704
U	1.223	0.717

These values were obtained from Fano plots [Eq. (6-11)] and calculated cross sections for $U_{nl} > 2$.

a few heavy elements, have been reported by Rez (1984). The differential and total cross sections for K-shell ionization calculated by Leapman et al. (1980) and Rez (1984) agreed well (typically better than $5-10\%$) with the cross sections calculated using the hydrogenic approximation. Rez has also fitted his calculated cross sections to the Bethe equation via Fano plots [cf. Eq. (6-11)] and has determined values of the Bethe parameters for $U_{nl} > 2$ (for which the Fano plots are "reasonably" linear). Values of the Bethe parameters determined by Rez are given in Table 6-2.

Similar calculations of generalized oscillator strengths for K- and L-shell ionization of atoms with $Z \leq 30$ have been reported by Inokuti and Manson (1984). The generalized oscillator strengths have been computed both with the use of the Hartree-Slater potential and with the hydrogenic approximation for a wide range of excitation energies. For the L_1 and L_{23} subshells of elements such as Na, Al, and Cl, the Hartree-Slater and hydrogenic results differ by up to a factor of five for excitation energies close to the relevant subshell binding energy. In an earlier calculation, Omidvar (1977) has found that total cross sections for ionization of the krypton $3d$ subshell differ by $5-13\%$ when computed with the Hartree-Slater and hydrogenic approaches.

Inokuti and Manson (1984) discuss mechanisms by which generalized oscillator strengths for atoms may be modified in molecules and solids. Shape resonances occur widely in the photoionization spectra of molecules and so-called EXAFS (extended x-ray absorption fine structure) modulations of x-ray absorption and EELS data occur in molecules and solids.

A specific calculation for M_{45}-shell ionization of Kr has been reported by Omidvar (1977). With the use of Hartree-Slater wave functions, he has calculated cross sections for $1.2 < U_{M_{45}} < 36.3$. The maximum cross section occurs for $U_{M_{45}} \approx 8$ rather than $3-4$ found for K- and L-shell ionization.

Finally, Gryzinski (1965) has developed a classical description of excitation and ionization that has been widely applied largely because of its simplicity and general utility. His expression for inner-shell ionization is

$$\sigma_{nl}^2 E_{nl}^2 = 6.51 \times 10^{-14} Z_{nl} g(U_{nl}) \quad \text{cm}^2 \text{eV}^2 \tag{6-14a}$$

where

$$g(U_{nl}) = \frac{1}{U} \left[\frac{U-1}{U+1} \right]^{3/2} \left[1 + \frac{2}{3} \left(1 - \frac{1}{2U} \right) \ln[2.7 + (U-1)^{1/2}] \right]. \tag{6-14b}$$

A review of classical and binary-encounter collision theories has been given by Vriens (1969). Tung (1980) has pointed out recently that the constraints of several oscillator-strength sum rules can yield more accurate generalized oscillator strengths for small momentum transfers than is possible with the classical binary-collision model. His calculated cross sections for K-shell ionization in C and O agree quite well with experimental measurements.

6.3 Semi-Empirical and Empirical Formulas

A large number of formulas have been proposed to represent calculated and measured inner-shell ionization cross sections. Many of these formulas are modifications of the Bethe formula [Eqs. (6-9) and (6-11)] either to gain analytic simplicity or to extend the applicability of the Bethe formula to low incident electron energies. We will summarize here briefly the older formulas that have been utilized in data analysis (e.g., in the quantification of EPMA, AES, and EELS data) and the newer formulas that are generally more accurate or more useful (e.g., over a wide range of Z and U_{nl}); comparisons with experimental data will be made in Section 6.5.

de la Ripelle (1949) has modified the Bethe formula to be useful near the threshold for ionization:

$$\sigma_{nl} E_{nl}^2 = \frac{6.51 \times 10^{-14} Z_{nl} \ln(U_{nl})}{k_{nl}(U_{nl} + \chi_{nl})} \quad \text{cm}^2 \text{eV}^2 \tag{6-15}$$

where k_{nl} and χ_{nl} are parameters; k_{nl} is equivalent to $1/b_{nl}$ in Eq. (6-9). de la Ripelle (private communication) has fitted experimental K-shell ionization cross sections to Eq. (6-15) and finds that $k_K = 1.18$ and $\chi_K = 1.32$.

Worthington and Tomlin (1956) have also modified the Bethe formula in order to give a convenient (but nevertheless arbitrary) analytic description of the cross section near threshold. Their formula is

$$\sigma_{nl} E_{nl}^2 = \frac{6.51 \times 10^{-14} Z_{nl} b_{nl}}{U_{nl}} \ln\left[\frac{4 U_{nl}}{1.65 + 2.35 \exp(1 - U_{nl})}\right] \quad \text{cm}^2 \text{eV}^2. \tag{6-16}$$

Green and Cosslett (1961) assumed c_K in Eq. (6-11) to be unity and adjusted b_K to agree with experimental measurements of σ_K for Ni and Ag in the vicinity of $U_K = 3$. Their result,

$$\sigma_K E_K^2 = 7.92 \times 10^{-14} \ln(U_K)/U_K \quad \text{cm}^2 \text{eV}^2 \tag{6-17}$$

corresponds to $b_K = 0.61$.

Kolbenstvedt (1967) has obtained the following approximate formula for K-shell ionization:

$$\sigma_K E_K^2 = \frac{3.590 \times 10^{-14} \ln(2.38 U_K)}{U_K} +$$
$$+ \frac{1.293 \times 10^{-13}}{U_K}\left[1 - \frac{1 + \ln U_K}{U_K}\right] \quad \text{cm}^2 \text{eV}^2. \tag{6-18}$$

· Drawin (1961) has reviewed earlier semi-empirical cross-section formulas and has proposed

$$\sigma_{nl} E_{nl}^2 = 4.32 \times 10^{-14} Z_{nl} f_1 (U_{nl} - 1) \ln(1.25 f_2 U_{nl})/U_{nl}^2 \quad \text{cm}^2 \text{eV}^2 \tag{6-19}$$

where f_1 and f_2 are parameters estimated to have values in the ranges 0.7 to 1.3 and 0.8 to 3.0, respectively, but which are often assumed to be unity. For $U_{nl} \gg 1$, Eq. (6-19) reduces to Eq. (6-11) with $b_{nl} = 0.66$ and $c_{nl} = 1.25$.

Lotz (1970) has suggested the formula

$$\sigma_{nl} E_{nl}^2 = a_{nl} Z_{nl} \ln U_{nl} \{1 - b_{nl} \exp[-c_{nl}(U_{nl} - 1)]\}/U_{nl} \tag{6-20}$$

where a_{nl}, b_{nl} and c_{nl} are parameters (not to be confused with the Bethe parameters

b_{nl} and c_{nl} in Eq. (6-9)). Values of the parameters in Eq. (6-20) were selected by Lotz on the basis of cross sections for the removal of valence electrons of atoms to produce ions. Equation (6-20), however, has been found useful in describing inner-shell ionization cross sections (Pessa and Newell, 1971; Szajman and Leckey, 1981). A slightly modified version of the Lotz formula has been proposed by Campeanu and Koch (1981) to fit McGuire's (1977 and 1979) calculated ionization cross sections. Casnati et al. (1982) have analyzed a large number of measured K-shell ionization cross sections and have proposed a rather complex expression:

$$\sigma_K E_K^2 = (Z_K a_0^2 R E_R^2 \psi \phi \ln U)/U \tag{6-21 a}$$

where

$$\psi = (E_K/E_R)^d \tag{6-21 b}$$

$$d = -0.0318 + 0.3160/U - 0.1135/U^2 \tag{6-21 c}$$

$$\phi = 10.57 \exp[(-1.736/U) + (0.317/U^2)] \tag{6-21 d}$$

and a_0 is the Bohr radius $(5.29 \times 10^{-11} \text{ m})$, E_R is the Rydberg energy (13.606 eV), and R is a complex relativistic correction factor (taken to be unity here). Equation (6-21) was found to fit cross section data typically better than $\pm 10\%$ over the range $1 \le U_K \le 20$ and $6 \le Z \le 79$.

Many of the formulas listed here have been incorporated into data-reduction algorithms (e.g., the determination of elemental concentrations from EPMA measurements). The formulas have served a useful function although they have often been derived from a much more limited base of measured and calculated cross sections than is available now. The apparent success of a particular formula in one application, often over a limited range of U_{nl} and with a variety of additional sources of uncertainty in the data analysis, should not necessarily suggest that the same formula is useful in other applications under different conditions. A particular hazard with empirical and semi-empirical formulas is their possible lack of validity beyond the range of conditions for which they were initially developed.

6.4 Experimental Measurements

6.4.1 Techniques for Inner-Shell Cross-Section Measurements

Three techniques have been used to measure inner-shell ionization cross sections for both gas and solid targets. With one technique, measurements are made of the number of electrons that have excited a particular core level while for the other two techniques, measurements are made of the decay products, either characteristic x-rays or Auger electrons, from a specific core level. The latter approach is of particular value since data on the cross sections for the yields of x-rays or Auger electrons, relevant to EPMA and AES, respectively, are acquired directly. These cross sections, for other than K-shell excitation, are generally not the same as ionization cross sections since an ionization in some initial shell or subshell may be rapidly transferred to another by one or more Auger or Coster-Kronig processes (Bambynek et al., 1972; Krause, 1979). In general, Coster-Kronig processes are rapid so that, for example, an ionization in the L_1-subshell of Al is rapidly transferred to the L_2- or L_3-subshells and the resulting emission of Auger electrons from each subshell (by $L_1 VV$, $L_2 VV$, or $L_3 VV$ transitions) does not reflect the relative number of initial ionizations in the L_1, L_2, or L_3 subshells.

Data for fluorescence yields, Auger, and Coster-Kronig probabilities are available for K-shell excitation and to a lesser extent for L- and M-shell excitation (Bambynek et al., 1972 Krause, 1979). The increasing complexities of the multiple decay channels following M- and higher-shell ionization is largely responsible for the limited data on innershell ionization cross sections as well as on the associated fluorescence yields, Auger, and Coster-Kronig probabilities. A difficult, though potentially valuable method of identifying specific decay channels is the use of coincidence techniques. By this approach, Haak et al. (1978) have separated the two components of the $L_3 VV$ Auger emission from copper due to initial ionizations of the L_3 and L_2 subshells. The initial ionizations in this experiment were created with x-rays but the method could be extended to electron excitation in a manner similar to that developed for other purposes (Celotta and Huebner, 1979; Bonham, 1979).

The first measurements of inner-shell ionization cross sections were made by measuring the absolute yield of characteristic x-rays from a target under electron bombardment (Webster et al., 1933). The principal corrections and calibrations required for valid cross-section measurements include: slowing down of the incident beam in the target; x-ray absorption in the target; the fluorescence yield; efficiency of the detector; solid angle; and bremsstrahlung background (Bambynek et al., 1972; Jesserberger and Hink, 1975).

For core electrons with low binding energy ($\lesssim 1\,\text{keV}$), the fluorescent yield is small ($\lesssim 0.01$) and often not well known (Bambynek et al., 1972; Krause, 1979), and thus inner-shell ionization cross sections determined from x-ray yields can have large uncertainties. For this situation, of particular interest to AES, it is better to measure instead the yield of characteristic Auger electrons that result from the decay of a particular level (Glupe and Mehlhorn, 1967 and 1971). Corrections and calibrations for such measurements include: slowing down of the incident beam; attenuation of the Auger-electrons in the target; analyzer transmission; detector efficiency; solid angle; secondary-electron background; and additional ionizations due to bremsstrahlung and characteristic x-rays emitted by the target (Glupe and Mehlhorn, 1971; Powell et al., 1977; Cazaux and Mouton, 1984).

A more direct approach for the determination of inner-shell ionization cross sections is to measure the electron energy-loss spectra associated with the excitation of electrons from a particular shell (Swanson and Powell, 1968). This type of measurement, now often made in a transmission electron microscope equipped with an electon energy analyzer (Joy and Maher, 1980), can give differential cross sections integrated experimentally over limited ranges of energy loss and scattering angle or momentum transfer (cf. Eqs. (6-5)–(6-8) and the discussion in Section 6.2) of particular relevance to EELS measurements.

A total inner-shell ionization cross section requires both the determination of a varying background due to other inelastic scattering processes (e.g., multiples of the much more probable valence excitations) and the separation, in general, of the overlapping contributions to an energy-loss spectrum from the ionization of two or more shells. Methods for making such a separation have not been developed although it is possible that an algorithm could be developed with the use of calculations of $df(E, q)/dE$ (cf. Section 6.2). The signal of interest also has to be corrected for plural inelastic scattering and for different angular collection efficiencies associated with single and plural scattering (Swanson and Powell, 1966 and 1968; Stephens, 1980). Other aspects of data analysis have been discussed by Joy and Maher (1981).

The three techniques for the measurement of inner-shell ionization cross sections described above can be applied to both solid and gaseous targets. Nevertheless, essentially all of the cross sections determined by the x-ray yield and electron energy-loss methods have been obtained using solid targets while the cross sections determined by the Auger yield method have been almost entirely obtained with gaseous targets. It would be valuable to apply the x-ray and electron energy-loss methods to gases in which case the experimental requirements discussed by Kieffer and Dunn (1966), also summarized in Chapters 5 and 7, Kuyatt (1968), Langenberg et al. (1975), Celotta and Huebner (1979) and Bonham (1979) should be considered. Similarly, in applications of the Auger-yield method to solids, the experimental details discussed by Powell (1968), Powell et al. (1977), and Zaporozhchenko et al. (1979) should be examined.

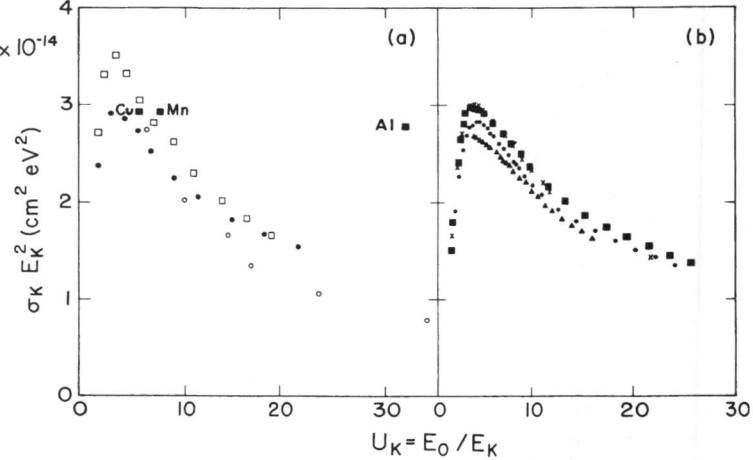

Fig. 6-1. Experimental values of $\sigma_K E_K^2$ as a function of U_K. a Data for C (open circles), Al (open squares). and Ni (solid circles); the solid squares represent data for Al, Mn, and Cu from Fischer and Hoffmann (1967). b data for C (triangles), Ne (crosses), N (squares), and O (circles) (Powell, 1976 a)

Figs. 6-1 and 6-2 show cross-section data for K- and L-shell ionization (Powell, 1976 a). These plots are in the form of $\sigma_{nl} E_{nl}^2$ versus U_{nl}, as suggested by Eq. (6-11), for a number of elements. The data in Fig. 6-1 a were obtained with the x-ray yield method while those in Fig. 6.1 b were obtained using the Auger yield method; it is clear that good consistency has been obtained with these two methods. Most of the data in Fig. 6-1 appears to lie close to a common curve and there is thus not a substantial variation of the Bethe parameters b_K and c_K with atomic number Z. In contrast, Fig. 6-2 shows a larger variation of cross sections for L_{23}-shell ionization amongst the several elements (all measured with the Auger yield method) and it is found that the parameters $b_{L_{23}}$ and $c_{L_{23}}$ increase with Z (Powell, 1976 a).

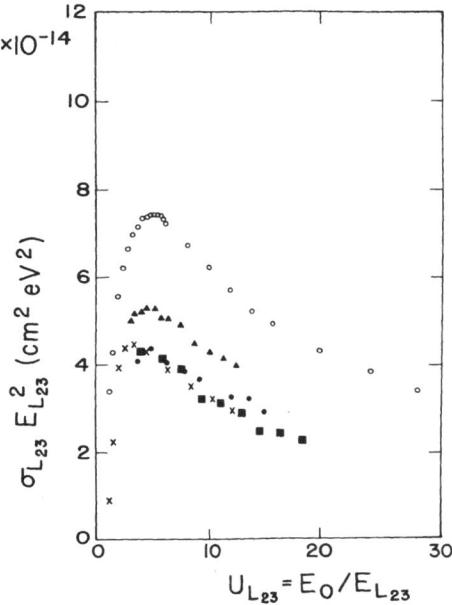

Fig. 6-2. Experimental values of $\sigma_{L_{23}} E_{L_{23}}{}^2$ as a function of $U_{L_{23}}$. Data shown are for P (squares), S (circles), Cl (triangles), and Ar (crosses and open circles (Powell, 1976 a)

Cross-section measurements published since the previous review (Powell, 1976 a) was prepared will now be briefly described. This summary will discuss in turn measurements made with the x-ray yield, Auger-electron yield, and electron energy-loss methods.

6.4.2 X-Ray Yield Measurements

Jessenberger and Hink (1975) measured K-shell ionization cross sections of Ti and Ni for U_K in the ranges 1 to 10 and 1 to 6, respectively, using the x-ray yield technique. Their cross sections were larger, when scaled [Eq. (6-11)], than those shown in Fig. 6-1 and about 20% higher than expected from the results of Rudge and Schwartz [Eq. (6-12)] and Gryzinski [Eq. (6-14)]. The data could be fitted well to Drawin's formula [Eq. (6-19)] with $f_1 = 1.2$ (Ti), $f_1 = 1.3$ (Ni) and $f_2 = 1.2$ (Ni and Ti).

Subsequent measurements have been made with incident energies close to threshold ($U_{nl} < 5$) for which the Bethe theory described in Section 6.2 would not be expected to be applicable. Shima (1980) and Shima et al. (1981) have measured cross sections for K-shell ionization of Mn, Cu, Ge, and Ag for $1 < U_K < 2.5$ and found close agreement with the Green and Cosslett (1961) empirical formula [Eq. (6-17)]. In this range of U_K, the Green-Cosslett formula predicts cross sections very similar to those of the Rudge-Schwartz [Eq. (6-12)] and Worthington-Tomlin [Eq. (6-16)] formulas but appreciably larger (by $25 - 100\%$ or more, depending on U_K) than cross sections predicted by the Gryzinski formula [Eq. (6-14)]. For $U_K < 1.2$, Shima et al. (1981) find better agreement with the Green-Cosslett formula [Eq. (6.17)] with the constant reduced from 7.92×10^{-14} to 7.2×10^{-14}.

Quarles and Semaan (1982) have reported K-shell ionization cross sections of argon for $1.3 < U_K < 3.2$ which agreed well with McGuire's (1977) calculations. These authors found that their data could be fitted to the Bethe (1930) equation [Eq. (6-9)] with parameters $b_K = 0.63$ and $c_K = 0.82$; the value of b_K is about 30% less and the value of c_K about 30% more than values found previously (Powell, 1976 a) for $U_K > 4$. K-shell ionization cross sections of argon have also been measured by Hippler et al. (1982) for $U_K < 5$; the data of these authors agree well with the data of Quarles and Semaan (1982).

L-subshell ionization cross sections have been obtained for W for $1 < U_L < 4$ by Chang (1979) with the use of the x-ray yield method. Similar measurements on gas samples have been made for Xe by Hippler et al. (1981) for $1 < U_L < 3$ and total L-shell cross sections have been measured by Quarles and Semaan (1982) for Kr ($2.1 < U_L < 4$) and Xe ($1.2 < U_L < 2$). Both Chang and Hippler et al. find their measured cross sections are comparable to the calculated results of Gryzinski (1965) and McGuire (1977) for $U_L > 2$ but the measurements decrease less rapidly with U_L for $U_L < 2$ than indicated by the calculations; the calculations would not, however, be expected to be valid at such low values of U_L. The Kr data of Quarles and Semaan (1982) is found to be about 40% less than the values calculated by McGuire (1977) while the Xe data of Hippler et al. (1981) are about $20 - 30\%$ larger than McGuire's calculations; this apparent discrepancy is due to the different methods used to select the fluorescent yields in the two experiments (Hippler, private communication).

Hippler et al. (1983) have measured ionization cross sections with incident energies from about 10 eV to 1 keV above threshold. They have examined the cross-section dependence for K-shell ionization in Ar and L_3-shell ionization in Xe on excess electron energy above the threshold for ionization, namely

$$\sigma_{nl} \propto (E_0 - E_{nl})^n. \tag{6.22}$$

The exponent n was found to be 1.10 ± 0.04 for the Ar K-shell cross section, a value consistent with that predicted (1.127) by Wannier (1953) from a consideration of the correlated motion of the two slow electrons escaping the ion. For L_3 ionization in Xe, $n = 0.96 \pm 0.04$ and this value was interpreted by Hippler et al. (1983) to correspond to uncorrelated electron escape for excess energies above 10 eV. Under these conditions, a linear dependence of the cross section on excess energy is expected.

6.4.3 Auger-Electron Yield Measurements

We turn now to the use of the Auger-yield method for the measurement of inner-shell ionization cross sections. There are only two known recent measurements. Hink et al. (1981) have reported new measurements of K-shell ionization cross sections for Ne in the range $1 < U_K < 5.2$; these measurements agree closely with the earlier data of Glupe and Mehlhorn (1971). Hink et al. (1981) have also determined the exponent in relation (6.22) to be 1.13 ± 0.02. Yagishita (1981) has determined cross sections for the ionization of the M_2, M_3, and M_{45} subshells of Kr for $1 < U_M < 34$. These measurements are typically about one-half of the values calculated for each subshell by McGuire (1977) and also show a less rapid decrease with decreasing U_M for $U_M \lesssim 3$.

6.4.4 Electron Energy-Loss Measurements

Finally, we mention use of the EELS method to determine K-shell ionization cross sections of 30 keV electrons in graphite and aluminium nitride by Rossouw and Whelan (1979). They measured differential K-shell excitation cross sections for C and N at small scattering angles which were then integrated and fitted to the Bethe (1930) equation [Eq. (6-9)]. Their values of the Bethe parameters were $b_K \sim 1$ (C), $b_K = 0.92$ (N), and $c_K \sim 2.9$ (C and N).

6.5 Comparison of Theory and Experiment

In this section we will discuss measured inner-shell ionization cross sections and compare these data both with calculated cross sections and with semi-empirical and empirical formulas that have been presented in Sections 6.2 and 6.3. We will discuss separately data for K-and L-shell ionization and then indicate how the data' can be interpreted in terms of the Bethe equation [Eq. (6-11)] and the likely (empirical) dependence of the parameters in this equation on incident electron energy.

6.5.1 K-Shell Data

Fig. 6-1 shows representative measurements of K-shell ionization cross sections for a number of elements. With the exception of one data point in Fig. 6-1 a, most of the data appears to lie close to a common curve within the expected accuracy of individual measurements. That is, the plots of $\sigma_K E_K^2$ as a function of U_K do not appear to vary with Z for $Z \leq 29$.

Fig. 6-3 a shows again the data of Fig. 6-1 b. These data for C, N, O, and Ne have been selected for further analysis since they represent the most comprehensive set of measurements from a single laboratory [Glupe and Mehlhorn (1971)] and there should therefore be minimal relative error. In addition, the cross-section measurements for C have been subsequently confirmed by Hink et al. (1981).

The smooth curve in Fig. 6-3 a has been drawn through the experimental points and has been also replotted in the other panels as a solid line. The dashed line in Fig. 6-3 a is the Bethe (1930) equation [Eq. (6-11)] with $b_K = 0.9$ and $c_K = 0.65$; the dashed line coincides with the solid line for $U_K \geq 4$.

Figs. 6.3 b, c and d show comparisons of cross sections predicted by a number of formulas with experiment (as represented by the data in Fig. 6-3 a). Plots of the formulas of Worthington and Tomlin (1965) [Eq. (6-16)] and of Green and Cosslett (1961) [Eq. (6-17)] in Fig. 6-3 b do not agree at all well with the experimental data. The formulas of Gryzinski (1965) [Eq. (6-14)], Drawin (1961) [Eq. (6-19)], and Lotz

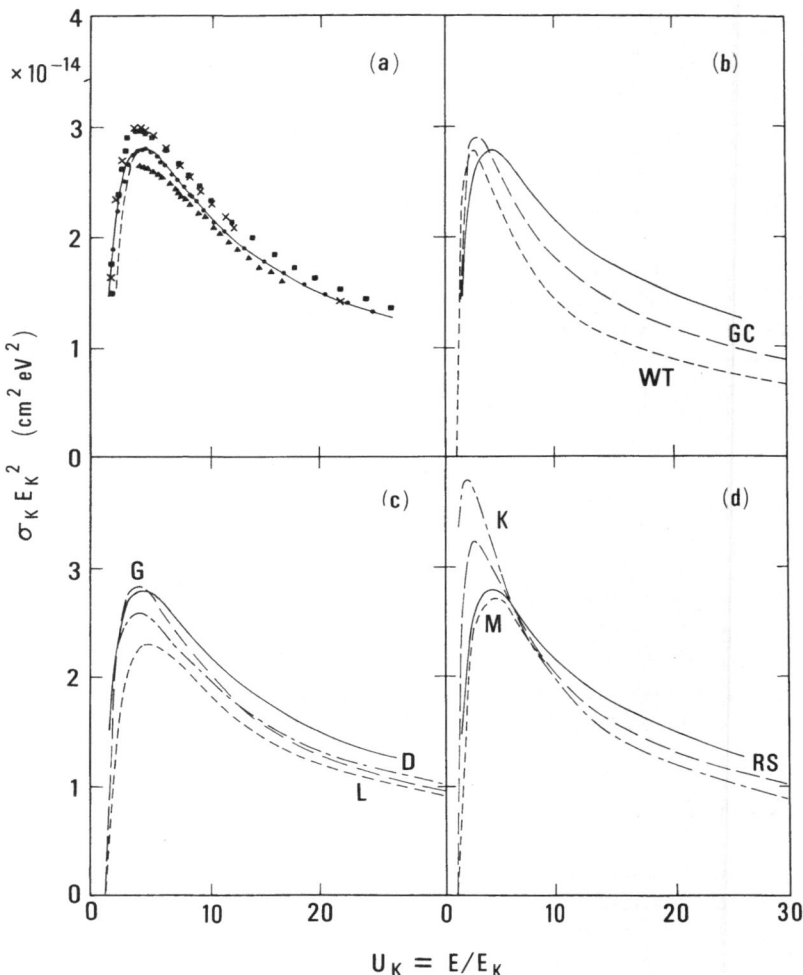

Fig. 6-3. Plot of $\sigma_K E_K{}^2$ versus U_K (Powell, 1976 b). *a* Experimental values for C (triangles), Ne (crosses), N (squares), and O (circles) (also shown in Fig. 6-1 b). The solid line is a smooth curve through the experimental points (and is replotted in the other panels) and the dashed curve is the Bethe (1930) equation [Eq. (6-11)] with $b_K = 0.9$ and $c_K = 0.65$. *b* The short-dashed curve (*WT*) is the Worthington and Tomlin (1956) equation [Eq. (6-16)] and the long-dashed line (*GC*) is the Green and Cosslett equation [Eq. (6-17)]. *c* The short-dashed line (*L*) is the Lotz (1970) equation [Eq. (6-20)], the long-dashed curve (*G*) is the result of Gryzinski (1965) [Eq. (6-14)], and the dot-dashed curve (*D*) is the result of Drawin (1961) [Eq. (6-19)]. *d* The short-dashed curve (*M*) represents calculations of McGuire (1971) for Be, C, and O, the long-dashed line (*RS*) is the result of Rudge and Schwartz (1966) [Eq. (6-12)], and the dot-dashed curve (*K*) is the result of Kolbenstvedt (1967) [Eq. (6-18)]

(1979) [Eq. (6-20)] are shown in Fig. 6-3 c. The Gryzinski formula agrees well with experiment for $U_K < 6$. The Drawin formula would agree well with experiment if the parameter a_K was increased by about 25%.

Fig. 6.3 d shows a curve based on McGuire's (1971) calculations for Be, C, and O, the result of Rudge and Schwartz (1966) [Eq. (6-12)], and Kolbenstvedt's (1967) formula

[Eq. (6-18)] McGuire's result agrees reasonably with experiment, particularly near threshold. The Rudge and Schwartz equation is close to experiment for $U_K \gtrsim 5$ but closer to threshold (the region for which the calculations were made), the agreement is less satisfactory. Kolbenstvedt's formula does not agree particularly well with experiment.

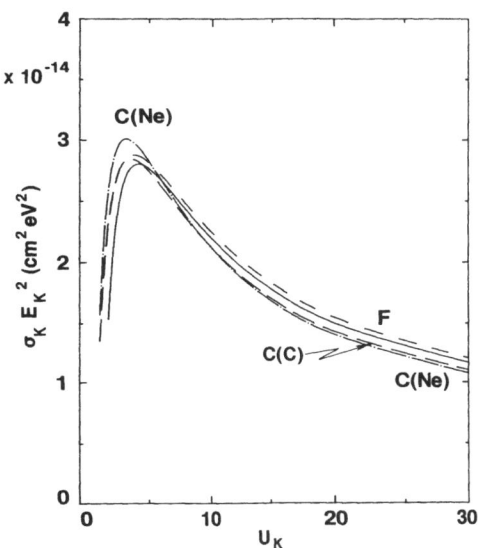

Fig. 6-4. Plot of $\sigma_K E_K^2$ versus U_K. The solid curve is the Bethe (1930) equation [Eq. (6-11)] with $b_K = 0.9$ and $c_K = 0.65$ (cf. Fig. 6-3a), the short-dashed curve (F) is the result of Fabre de la Ripelle (1949) [Eq. (6-15)], and the long-dashed curve [C(C)] and the dot-dashed curve [C(Ne)] are the results of Casnati et al. (1982) [Eq. (6-21)] for C and Ne, respectively

Fig. 6-4 is a comparison of the Bethe equation [Eq. (6-11)] with $b_K = 0.9$ and $c_K = 0.65$, as shown also in Fig. 6-3a, with the predictions of two other formulas, those of de la Ripelle (1949) [Eq. (6-15)] and of Casnati et al. (1982) [Eq. (6-21)]; Eq. (6-21) has been evaluated for carbon and neon. With consideration of the difference between the Bethe equation and the experimental data in Fig. 6-3a for $U_K < 4$, Fig. 6-4 indicates that both the de la Ripelle and Casnati et al. results represent the experimental data well.

Fig. 6-5 is a comparison of the calculations of Rez (1984) with the Bethe equation in which $b_K = 0.9$ and $c_K = 0.65$, as in Fig. 6-4. Rez fitted his calculated cross sections for a number of elements to the Bethe equation and the derived parameters, for $U_K \geq 2$, are listed in Table 6-2. The cross sections of Rez in Fig. 6-5 vary significantly with Z although the experimental data in Fig. 6-1 do not show any Z dependence.

Fig. 6-6 shows measured and calculated cross-section data for the threshold region ($U_K < 4$). The experimental results show a high degree of consistency, particularly for $U_K < 1.3$, even though the measurements were made with different techniques (x-ray yield and Auger-electron yield) and with both gaseous and solid samples. The solid line in Fig. 6-6 is the result of Moores et al. (1980), Eq. (6-13), the dashed line is

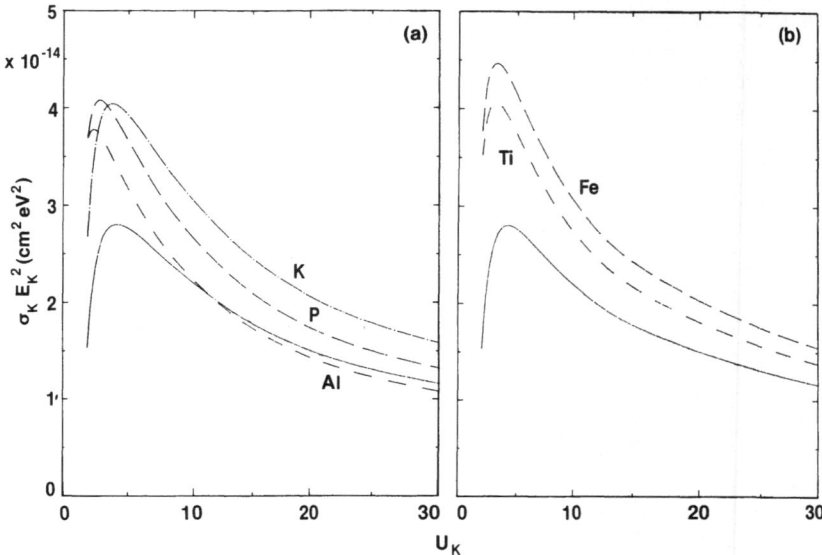

Fig. 6-5. Plot of $\sigma_K E_K^2$ versus U_K. The solid curve in both panels is the Bethe (1930) equation [Eq. (6-11)] with $b_K = 0.9$ and $c_K = 0.65$ (cf. Figs. 6-3a and 6-4). The other curves are the results of Rez (1984) for the indicated elements. Rez fitted calculated cross sections to the Bethe equation and his parameters are listed in Table 6-2

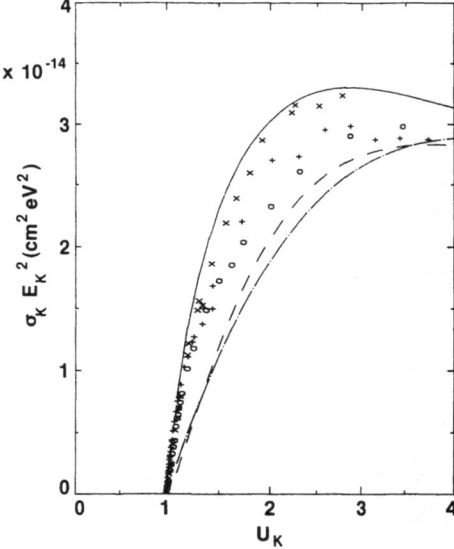

Fig. 6-6. Plot of $\sigma_K E_K^2$ versus U_K. The point show the measurements of: Hink et al. (1981) for neon ($\bigcirc$): Hippler et al. (1982 and 1983) for argon ($+$); and of Shima et al. (1981) for Cu ($\times$). The lines are plots of the results of: Moores et al. (1980), Eq. (6-13), solid line; Gryzinski (1965), Eq. (6.14), dashed line; and Lotz (1970), Eq. (6-20), dot-dashed line. The Lotz plot has been computed with the parameter a_K in Eq. (6-20) equal to 5.0×10^{-14} cm^2 eV^2, an increase of 25% over the value recommended by Lotz and used in Fig. 6-3c, as suggested in Section 6.5.1 of the text

Gryzinski's (1965) formula, Eq. (6-14), and the dot-dashed line is Lotz's semi-empirical result, Eq. (6-20), with his a_K parameter increased by 25% as indicated by Fig. 6-3 c. The Gryzinski and Lotz formulas in Fig. 6-6 approach the experimental data in the vicinity of $U_K = 4$ but both are appreciably less than the measurements for $U_K > 2$. The calculations of Moores et al. agree well with the measurements for $U_K < 1.5$ but exceed the data for $U_K > 1.5$. A correction for screening (Hippler et al., 1983) improves the agreement in the latter region.

6.5.2 L-Shell Data

Fig. 6-2 shows measured values of $\sigma_{L_{23}} E_{L_{23}}^2$ for low-Z elements as a function of $U_{L_{23}}$ (Powell, 1976 a). These data indicate that the Bethe parameter $b_{L_{23}}$ is increasing with Z, a result expected from the calculations of McGuire (1977). The observed variation with Z, however, is much less than that expected from the calculations but is greater than that calculated from photoabsorption data by use of Eq. (6-10), (Powell, 1976 a). In view of this inconsistency and the extremely limited number of cross section measurements, the discussion of L-shell data will be brief.

Rez (1984) has recently calculated cross sections for L_3-shell ionization of medium- and high-Z elements (Table 6-2). In this range of Z, the Bethe parameters $b_{L_{23}}$ and $c_{L_{23}}$ do not vary appreciably, a result consistent with McGuire's (1977) calculations, and are comparable to Rez's values of b_K and c_K for Ti. A comparison of Rez's Ti cross section curve in Fig. 6-5 with the experimental data for low-Z elements in Fig. 6-2 together with consideration of the Z_{nl} term in Eq. (6-11) indicates that the calculated L_3-shell cross sections are approximately double those measured.

The shape and magnitude of the L_3-shell cross sections calculated from the Worthington and Tomlin (1956) [Eq. (6-16)] and Lotz (1970) [Eq. (6-20)] expressions do not agree at all well with the data in Fig. 6.2 (Powell, 1976 a). The Gryzinski (1965) [Eq. (6-14)] and Drawin (1961) [Eq. (6-19)] formulas, however, agree reasonably well with the Ar data in Fig. 6-2.

6.5.3 Analysis of Experimental Cross-Section Data with the Use of the Bethe Equation

The Bethe equation, Eq. (6-11), is convenient and appropriate for the analysis of experimental cross-section data. Within its range of expected validity, to be discussed shortly, it is expected to describe inner-shell ionization in atoms, molecules, and solids, as discussed in Section 6.1 and 6.2.

A simple and effective means of analyzing cross-section data is the Fano plot based on Eq. (6-11) (see also Chapter 7). A Fano plot is made from experimental (or calculated) cross-section data by plotting $\sigma_{nl} E_{nl}^2 U_{nl}/\pi e^4 Z_{nl}$ versus $\ln U_{nl}$. Linearity of the plot defines the range of U_{nl} for which Eq. (6-11) describes the data and enables values of the Bethe parameters b_{nl} and c_{nl} to be determined by a linear least-squares fit to the plotted data.

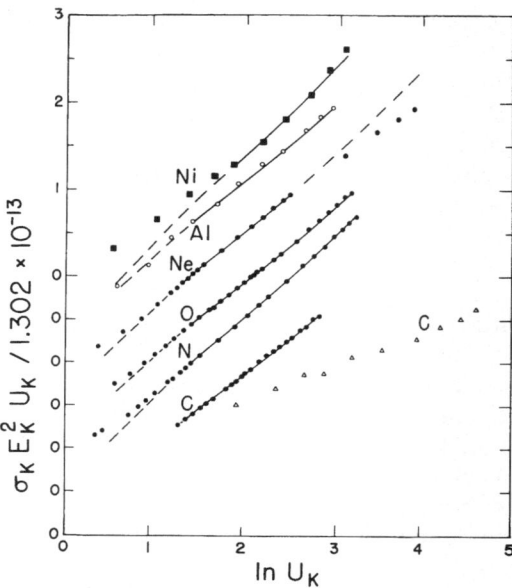

Fig. 6-7. Plot of experimental values of $\sigma_K E_K^2 U_K/1.302 \times 10^{-13}$ versus $\ln U_K$ [a Fano plot based on Eq. (6-11)]. Successive plots have been displaced vertically for clarity. The solid lines represent the range of U_K for which linear least-squares fits were made; the dashed lines are extrapolations. The derived Bethe parameters are shown in Table 6-3 (Powell, 1976 a)

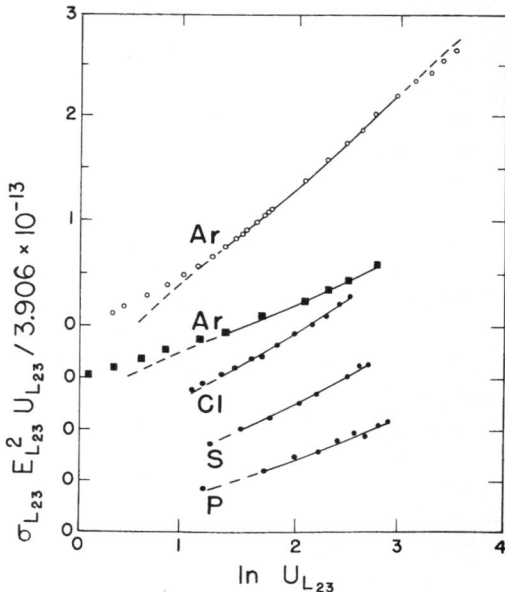

Fig. 6-8. Plot of experimental values of $\sigma_{L_{23}} E_{L_{23}}^2 U_{L_{23}}/3.906 \times 10^{-13}$ versus $\ln U_{L_{23}}$ [a Fano plot based on Eq. (6-11)]. See also caption to Fig. 6-7 (Powell, 1976 a)

Table 6-3. *Effective values of the Bethe parameters found from the Fano plots shown in Figs. 6-7 and 6-8 (Powell, 1976a)*

Shell	Range of U_{nl}	Bethe parameters
K-shell	$4 \lesssim U_K \lesssim 25$	$b_K \approx 0.9$
		$c_K \approx 0.65$
L_{23}-shell	$4 \lesssim U_{L_{23}} \lesssim 20$	$b_{L_{23}} \approx 0.5 - 0.9$
		$c_{L_{23}} \approx 0.6$

Examples of Fano plots based on the cross-section data of Figs. 6-1 and 6-2 are shown in Figs. 6-7 and 6-8 for K- and L_{23}-shell ionization, respectively. Linear regions are typically found for $4 \lesssim U_{nl} \lesssim 25$. Values of the Bethe parameters derived from the linear least-squares fits to the plots in Figs. 6-7 and 6-8 are summarized in Table 6-3 (Powell, 1976a). The range of values for the parameter $b_{L_{23}}$ is associated with the Z-variation of the cross section data in Fig. 6-2.

The fact that the Fano plots of Figs. 6-7 and 6-8 are linear only over limited ranges of U_{nl} is not surprising. The derivation of the Bethe formula is based on the validity of the first Born approximation and other kinematic approximations justifiable at high U_{nl}. At low values of U_{nl}, however, the first Born approximation is not valid and exchange effects also have to be considered (Chapter 5). It is clear from the discussion in Section 6.2 that a total inner-shell ionization cross section at a given incident energy is derived from an appropriate integration of the generalized oscillator strength [Eq. (6-5)] or the energy-loss function [Eq. (6-1)]. Such an integration has to be made over the kinematically allowed range of momentum transfer and energy transfer. For some elements, the oscillator strength for inner-shell ionization may be concentrated in a relatively small range of excitation energies (say from E_{nl} to 3 E_{nl}) while for other elements the oscillator strength may be distributed over a much larger range of excitation energies (say, from E_{nl} to 10 E_{nl} or more) (Powell, 1984). As the cross section for a *specific* excitation energy or energy loss is similar in shape to those shown in Fig. 6-1, it is clear that the incident electron may have to be quite high such that U_{nl} is greater than, say, 20 or 30 before it can be expected that the available oscillator strength for the ionization of a particular shell has saturated at close to its maximum value and thus the parameter b_{nl} has reached its maximum value [Eq. (6-10)]. Similarly, the term $c(E)$ in Eq. (6-8) depends on the distribution of oscillator strength for a specific excitation energy as a function of momentum transfer.

The Fano plot is an extremely useful means of assessing experimental cross-section data (Chapter 7). If U_{nl} is sufficiently high, the Fano plot should be linear; if necessary, the relativistic (kinematic) modification to the Bethe formula should be made (Inokuti, 1971). For U_{nl} smaller than the values for the linear region, the parameters b_{nl} and c_{nl} will, in an empirical sense, be a function of U_{nl}. That is, these parameters would not reach their "saturation" values until U_{nl} is sufficiently large. The Fano plot can also be used to determine the reliability of cross-section data. For example, two sets of cross-section data for carbon are shown in Fig. 6-7; the data shown as solid circles have a "reasonable" dependence on $\ln U_K$ whereas the data shown as triangles are more erratic.

A surprising feature of the Fano plots of Figs. 6-7 and 6-8 is that they are linear over a substantial range of U_{nl} extending down to $U_{nl}=4$. This result is certainly useful since the Bethe equation with parameters derived from the Fano plots can be conveniently used over a substantial range of U_{nl}, that is, the range of U_{nl} for which the Fano plot is linear. It was found in previous analysis (Powell, 1976 a), however, that values of b_K from Fig. 6-7 were larger and that values of c_K were smaller than those expected (e. g., by calculation of b_K from photoabsorption data with the use of Eq. (6-10)). This result could be understood from a consideration of the expected distributions of oscillator strength, as just discussed. If c_K was arbitrarily assumed to be a larger value and constant (in this case 2.42, a value proposed earlier), the "effective" value of b_K could be calculated as a function of U_K from the cross-section data. As expected, b_K increased and reached a saturation value for $U_K \gtrsim 25$ (Powell, 1976 a). A similar result was found for the variation of $b_{L_{23}}$ with $U_{L_{23}}$. The saturation values of b_{nl} were then in closer agreement with values derived from photo-absorption data. Although the magnitudes and variations of the derived b_{nl} values with U_{nl} appear to be reasonable, they are of limited significance since an arbitrary value of c_{nl} was assumed. It is possible that c_{nl} could also be a function of excitation energy and thus also of U_{nl}. Unfortunately, knowledge of the momentum-dependence of the generalized oscillator strength is still limited (Leapman et al., 1980; Inokuti and Manson, 1984). Measurements of the generalized oscillator strength have not been made over a sufficiently large range of energy and momentum transfers for either atoms or solids to validate the calculations and to permit a more detailed analysis and correlation of the photoabsorption and ionization cross-section data. The cross-section calculations of McGuire (1977) and Rez (1984), for example, appear to agree only qualitatively with experiment. There are also inconsistencies between experimental L_3-shell cross-section data and photoabsorption data (Powell, 1976 a).

6.6 Applications

As noted in the introduction, inner-shell ionization cross-section data are required in materials characterization by electron probe microanalysis (EPMA), Auger-electron spectroscopy (AES), and electron energy-loss spectroscopy (EELS). These techniques are utilized for bulk analysis, surface analysis, and thin-film analysis, respectively, and are schematically illustrated in Fig. 6-9; EPMA is also used for thinfilm analysis in the transmission and scanning transmission electron micro-scopes. An electron beam bombards the specimen of interest in each case, and although the experimental arrangements are different in detail for each technique there are considerable conceptual similarities.

We will describe briefly the use of each of the three techniques. Some remarks will then be made concerning how all three techniques (and others) may be combined in what has become known as analytical electron microscopy. Finally, we will comment on the common problem of electron-beam damage during these analytical measurements.

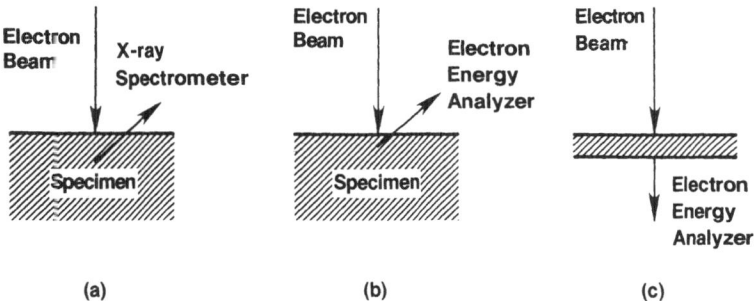

Fig. 6-9. Schematic experimental arrangements for (a) electron probe microanalysis, (b) Auger-electron spectroscopy, and (c) electron energy loss spectroscopy

6.6.1 Bulk Analysis (Electron-Probe Microanalysis)

An electron beam of energy usually between 10 and 50 keV bombards what is often a macroscopic specimen (Fig. 6-9 a). The incident electrons are scattered elastically and inelastically many times by atoms of the specimen. As long as the electron energy is greater than the binding energy of one or more atomic core levels, inner-shell ionization of specimen atoms may occur. Some of these inner-shell ionizations decay by characteristic x-ray emission so that measurement of the x-ray spectrum can give information on the elements present in the specimen. Further details of the principles of instrument operation and use are given by Heinrich (1981).

The average penetration depth of the incident electron beam in the specimen depends considerably on the incident energy and the atomic number of the specimen constituents but is typically about 1 μm. Inner-shell ionizations may be produced not only directly by the incident electron beam but also by photoionization from x-rays produced elsewhere in the specimen (i.e., fluorescence). The intensity of x-rays reaching the x-ray spectrometer will depend on the amount of absorption in the specimen at the particular x-ray energies of interest and the geometry. These factors together with the spreading of the incident beam (and of course the beam geometry) determine the specimen volume that produces the measured x-ray spectrum. This volume will typically have dimensions of about 1 μm and gives rise to the "bulk" analysis by EPMA. By rastering the electron beam across the specimen surface, compositional information can be obtained as a function of position.

Quantitative analysis by EPMA has been described in some detail by Heinrich (1981). An extensive procedure has been developed, the so called ZAF method, in which separate corrections are made to the measured x-ray intensities for atomic-number (Z) effects, absorption (A), and fluorescence (F). The atomic number correction includes factors describing the inner-shell ionization cross section as a function of electron energy, electron backscattering, and the stopping power of the specimen. If, as is often the case, measurements from an "unknown" specimen are compared with measurements from a standard of similar composition, the atomic-number correction is small so that accurate information concerning inner-shell ionization cross sections is not required. On the other hand, if the specimen is not flat, homogeneous over the volume probed by the incident beam, and electron-opaque, the ZAF method cannot be applied. For specimens such as thin films, inclusions, fine particles, or sections of biological tissue, the modelling of the electron transport in the specimen is more complex and data for inner-shell ionization cross section is required. For this non-ideal type of specimen, modelling using Monte Carlo methods is useful (Heinrich et al., 1976; Heinrich, 1981).

Bulk analysis by EPMA is based on a reasonable physical model of the microscopic processes that give rise to the measurement of an x-ray spectrum for a given specimen. Although certain simplifying approximations have been made in the development of the ZAF method and knowledge of some of the parameters is still limited, this approach has been used widely. The use of a standard of similar composition to the unknown specimen serves to minimize uncertainties in the ZAF correction procedure as well as, in effect, to calibrate the instrument for the particular measurement conditions. Alternatively, it is possible to make use of empirical sensitivity factors for the elements if the instrument is operated under carefully controlled conditions and if only semi-quantitative analytical information (i.e., without corrections for possible matrix or geometrical effects) is required (Heinrich, 1981).

6.6.2 Surface Analysis (Auger-Electron Spectroscopy)

The specimen is typically bombarded by electrons of energy between 3 and 25 keV (Fig. 6-9 b). Inner-shell ionizations are created as the incident electrons are scattered elastically and inelastically, as in EPMA. Some of the atomic ionizations decay by the emission of Auger electrons, particularly for shells with binding energies less than 5 keV (Bambynek et al., 1972). These Auger electrons can be detected with an electron energy analyzer and elements in the specimen can be identified from the characteristic energies of the Auger electrons. Details on the practice of AES are given by Joshi et al. (1975).

The surface sensitivity of AES arises from the high inelastic scattering cross sections for low-energy (e.g., $50-2000$ eV) electrons in solids. For Auger electrons with energy in this range, the inelastic mean free paths are generally between 3 and 30 Å (Seah and Dench, 1979). The measured Auger-electron signal originates mostly from a depth of several multiples of the inelastic mean free path and an area determined by the incident beam and the subsequent multiple scattering, as in EPMA.

Quantitative surface analysis by AES has been described by Joshi et al. (1975), Powell (1978), and Holloway (1980). The procedures have not been as well developed as for EPMA and there are a number of additional complications (Powell, 1980). It is not always easy to determine whether the specimen is homogeneous over the volume being probed (or to correct for inhomogeneities), it is difficult to measure intensities reliably, and there are significant variations in measurements made by different instruments (Powell et al., 1982). While the imprecision or repeatability of a surface analysis by AES may be satisfactory (e.g., a few per cent) and comparable to bulk analysis by EPMA, the accuracy of the surface analysis may be much worse.

There are three approaches utilized in surface analysis by AES. First, measurements from an "unknown" specimen can be compared with a standard of similar composition. Although simple in concept and execution, this method suffers from the difficulty of ensuring that the standard can be regenerated reliably with the same composition and topography. Second, measured Auger intensities can be normalized by published atomic sensitivity factors determined from measurements with pure elements or compounds under the same conditions. With this approach, it is assumed that the matrix corrections associated with possible variations of inelastic mean free path and electron backscattering are insignificant and that the local instrument has a sufficiently similar performance to those on which the sensitivity factors were determined. Third, measured Auger intensities can be corrected with the use of a physical model, as for the ZAF method in EPMA. This approach requires data for a number of parameters including inner-shell ionization cross sections, much of which are not well known for a wide range of elements and conditions.

Auger-electron spectroscopy has been developed much more recently than EPMA so it is not surprising that the analytical methodology is at a more primitive stage. Surface analyses are most often made with the use of sensitivity factors because of their simplicity and convenience. Hall et al. (1977) have shown that improved accuracy in the analysis of binary alloys can be obtained if matrix corrections are made to the sensitivity factors as indicated by the physical model. A test of the physical model has shown that it appears to be sound (Powell, 1977).

6.6.3 Thin-Film Analysis (Electron Energy-Loss Spectroscopy)

Electron energy-loss spectroscopy is a technique developed recently for thin-film analysis in the electron microscope. A thin specimen, of typically $50 - 100$ nm thickness, is irradiated by an electron beam of energy often between 50 and 250 keV.

Electrons transmitted through a selected region of the specimen film are energy analyzed, as indicated in Fig. 6-9 c. Structure in the energy-loss spectrum at characteristic values of energy loss can be associated with core-electron excitation in particular elements. Joy and Maher (1980, 1981) have summarized how EELS experiments are performed.

One of the principal advantages of EELS measurements in the electron microscope is the capability for microanalysis; this capability exists in both the conventional and the scanning transmission electron microscopes. A region of interest can be selected in an electron microscope image and composition data can be obtained from a specimen area as small as $10^{-16} - 10^{-17}$ m^2. Information on the chemical state of an element can in favorable cases be obtained from small "chemical" shifts in the threshold energy loss and from the shape of the energy loss spectrum. It is also possible to perform EXAFS (extended x-ray absorption fine structure) analyses of the energy-loss data to obtain local structural information for selected elements. EELS is particularly valuable for biological samples since light elements can be readily detected, in contrast to EPMA.

Quantitative analysis by EELS has been described by Joy and Maher (1981) and by Egerton (1982). The energy-loss signal of interest is usually measured on a background due to another core excitation or to a core excitation plus a valence-electron excitation. Although the intensity associated with inner-shell excitation will generally extend over a large range of energy loss (Powell (1984) and Section 6.5.3 above), for example 1000 eV, the signal can only be separated meaningfully from the background over a limited energy-loss range, for example 100 eV. The measured energy-loss spectrum is also associated with a limited range of scattering angles. An analysis therefore requires correction of the intensities of different elements for the particular limits of energy loss and scattering angle for the measurement. This correction depends on adequate knowledge of the generalized oscillator strength [Eq. (6-5)]. Use has been made of the calculations of Egerton (1979, 1981) for K- and L-shell ionization based on a hydrogenic model and of Leapman et al. (1980) based on a Hartree-Slater model. These calculations have been discussed in Section 6.2. For K-shell ionization, the simpler calculations of Egerton are satisfactory but for L-shell ionization appreciable errors are to be expected (Sklad et al., 1981). For incident electron energies greater than about $10-20$ keV, relativistic corrections should be applied to the cross-section formulas (Inokuti, 1971).

A comparison of EELS measurements in five laboratories has been reported by Joy and Newbury (1981). Although the data base is limited, the variation in the results indicates that better control of the instrumental operation and the measurements is required.

6.6.4 Materials Analysis (Analytical Electron Microscopy)

The three methods of materials analysis just described, EPMA, AES, and EELS, are frequently applied separately in individual instruments. Recently, these and other techniques have become available in a single instrument, generically termed the analytical electron microscope (Hren et al., 1979; Geiss, 1981). For characterization of thin-film specimens using the scanning transmission electron microscope (STEM) and of macroscopic specimens using the scanning electron microscope (SEM), it is possible to detect emitted characteristic x-rays (EPMA) as well as to generate images and diffraction patterns. With the STEM, EELS is possible while in the SEM, various electron signals can be measured: backscattered electrons (Z contrast) secondary electrons (topography and Z contrast), and Auger electrons (surface composition). With multiple techniques, different microscopic properties of a selected region of a specimen can be probed. Furthermore, a consistent physical model can be used to describe the interaction and transport of the incident beam in the specimen, the generation of the signal(s) of interest, and the subsequent scattering or absorption of the signal radiation. The Monte Carlo method shows promise for this purpose as it can be readily applied to complex specimen geometries or morphologies (for example, Heinrich et al., 1976). Appropriate inner-shell ionization cross sections are needed in such calculations in order to determine elemental concentrations from observed x-ray intensities (EPMA). The same procedure could be used for surface analysis by AES in the SEM although data for electron backscattering effects at low electron energies has only recently become available (Ichimura *et al.*, 1983). Additional data on ionization cross sections and Coster-Kronig transition rates is still needed for medium- and high-Z elements so that useful surface analyses can be made. Cazaux (1983) has recently published expressions for the minimum detectable mass that may be achievable with EPMA, AES, and EELS.

6.6.5 Radiation Damage

The interaction of electron beams with solids can lead to various kinds of damage or change to the specimen during an analysis. Damage processes in the electron microscope have been described by Isaacson (1977) and by Hren *et al.* (1979) while Pantano and Madey (1981) have reviewed damage processes important in AES. Different types of damage (dissociation, desorption, reduction, polymerization, oxidation, carbonization, diffusion, etc.) have been reported in different specimens but the damage rates are not sufficiently documented to make quantitative predictions for particular materials.

One type of electron-beam damage is the desorption of ions from surfaces. One mechanism for the desorption of ions is the ionization of an inner-shell electron followed by an Auger transition after which two ionic fragments dissociate due to Coulomb repulsion (Tolk *et al.*, 1983; Madey *et al.*, 1983). For specimens in which this type of damage predominates, the damage rate and a possible signal (an Auger electron or a characteristic x-ray) intensity may both be proportional to the cross section for inner-shell ionization under the particular excitation conditions (i.e., for incident electrons of a particular energy and for a particular energy distribution of backscattered electrons). For other types of damage, it may be possible to optimize the experimental conditions to minimize specimen damage (e.g., by cooling the specimen, lowering the partial pressure of hydrocarbons in the specimen chamber). In general, however, the analyst has limited time to acquire information from a particular specimen volume on account of damage and tradeoffs must be made between spatial resolution, accuracy and precision of analysis, sensitivity for the detection of particular elements and chemical state, and beam damage.

A final form of damage not associated with materials analysis concerns the interaction of ionizing radiation with living tissue for diagnosis or therapy. If the radiation is used for therapy, the nature of the radiation exposure has to be selected so as to maximize the desired effects and to minimize unwanted side effects. This optimization can be performed with Monte Carlo calculations of the type described in Section 6.6.1 for which pertinent inner-shell ionization and other cross-section data are required. Similar analyses are made to predict damage and to protect semiconductor electronic devices exposed to ionizing radiation.

6.7 Summary

Measurements and calculations of inner-shell ionization cross sections by electron impact have been reviewed with emphasis on developments since the preparation of two previous reviews by the author (Powell, 1976 a and 1976 b). Although there have been a number of new calculations of generalized oscillator strengths since 1976, the agreement between calculated and measured cross sections is still only semi-quantitative. Most measurements of inner-shell ionization have been made for K- and L-shell ionization and there is virtually no data for the ionization of other shells.

It is convenient to examine whether measured cross sections can be described adequately by the Bethe (1930) expression [Eq. (6-9)] since this formula is expected to be valid for sufficiently high incident electron energies. This formula can also be readily applied to all elements and is analytically simple for many applications. The extent to which the Bethe formula describes measured cross-section data can be established easily with the use of a Fano plot and experimental cross-section data. Although reasons were given for expecting the Bethe formula not to be valid for overvoltages U_{nl} less than about 20 to 30 with parameters consistent with other data, it was found empirically that the Bethe formula described cross-section data well for overvoltages in the range $4 \leq U_{nl} \leq 25$. The values of the parameter b_{nl} found in these empirical results were different from those expected from photoabsorption data but this apparent inconsistency could be resolved by a consideration of the likely variations of generalized oscillator strengths as a function of excitation energy. An important conclusion is that the values of the parameters in the Bethe equation for different elements should be regarded only as convenient, empirical data that should not be applied outside the range of U_{nl} for which the Bethe equation with these parameters has been shown to be a valid description. In addition, the Bethe formula with parameters determined from theory or with b_{nl} determined from photoabsorption data should not be used unless U_{nl} is at least 30; it has not been possible here to establish a lower limit for U_{nl} since the asymptotic regions of the Fano plots have not been reached with the available cross-section data.

We now present recommendations and comments concerning cross-section data:

(1) Cross Sections for K-Shell Ionization. The Bethe formula describes measured K-shell cross sections for $6 < Z < 28$ in the overvoltage range $4 \lesssim U_K \lesssim 25$ with parameters $b_K \approx 0.9$ and $c_K \approx 0.65$ (Powell, 1976 a). The cross section data is also fitted well by the empirical formulas of de la Ripelle [Eq. (6-15)] and of Casnati et al. [Eq. (6-21)]. Other formulas have been frequently utilized in EPMA and other applications, particularly those of Gryzinski [Eq. (6-14)], Worthington and Tomlin [Eq. (6-16)], Green and Cosslett [Eq. (6-17)], Drawin [Eq. 6-19)], and Lotz [Eq. (6-20)]. Of these formulas, that due to Lotz agrees best with experiment over the range $4 < U_K < 25$ if the parameter a_K in the formula is increased by 25%. For near-threshold excitation ($U_K < 4$), the Bethe formula is definitely not expected to be valid and should not be used. For such values of U_K, the formula of Moores et al., [Eq. (6-13)] agrees reasonably well with experiment and better than the formulas of Gryzinski [Eq. (6-14)] and Lotz [Eq. (6-20)]. The recent calculations of K-shell ionization cross sections by Rez (1984) show appreciable variations in the effective Bethe parameters with Z that are not observed in the experimental data.

(2) Cross Sections for L-Shell Ionization. The data for L-shell ionization are much more limited than for K-shell ionization. Scaled values of L_{23}-shell ionization cross sections show a variation with Z that is qualitatively similar to that expected from several calculations; there are, however, inconsistencies between the limited

experimental cross-section data and experimental photoabsorption data. The Bethe formula can be used nevertheless to fit cross-section data for $15 < Z < 18$ in the range $4 < U_{L_{23}} < 20$ (Powell, 1976 a). These results together with photoabsorption data suggest that reasonable values of the Bethe parameters are $b_{L_{23}} \approx 0.5 - 0.9$ and $c_{L_{23}} \approx 0.6$; it should be emphasized that these values are estimates due to limitations and uncertainties of the cross-section data. The shape of the cross-section curve calculated from the L_3-shell formula of Lotz [Eq. (6-20)] differs significantly from the experimental results (Powell, 1976 a).

(3) **Cross Sections for Other Shells.** There are very few measurements and calculations of cross sections for ionization of M and N shells. For these shells, it would be expected that "delayed onsets" of oscillator strength in photoabsorption data would lead to qualitatively different curves of ionization cross section versus incident electron energy than found for K-shell ionization. Experimental evidence for this expectation was found by Smith et al. (1974) from Auger-yield measurements associated with N_{67}-shell ionization in Au, Pb, and Bi. The Bethe formula should therefore not be used for $U_{nl} \lesssim 10$; likewise, other formulas for the ionization cross section should not be used near threshold until further data and better guidance is available.

Finally, we make some remarks concerning the adequacy of the information now available on inner-shell ionization cross sections. Data for K-shell ionization cross sections exist for a wide range of elements and for a wide range of incident electron energies. Unfortunately, there are relatively few measurements or calculations for other shells. For situations in which there may not be adequate data available, it may still be necessary to make reasonable estimates. Nevertheless, it should not be assumed that an empirical formula found to be useful under some conditions is necessarily valid under other conditions. Due consideration should be given to the obvious benefits associated with analytical convenience of a particular formula versus the possible inaccuracies.

There is a strong need for both measurements and calculations of inner-shell ionization cross sections. The measurements of cross sections for other than K-shells become more difficult due to the redistribution of vacancies by Coster-Kronig transitions; coincidence techniques could be helpful in determining the various atomic rearrangement rates. Knowledge of generalized oscillator strengths is required both for the calculation of total ionization cross sections as well as for intensity-correction algorithms in EELS measurements. It would be valuable to have measurements of generalized oscillator strengths and total ionization cross sections for gas-phase elements as a function of incident energy so that the energy range over which the Bethe theory is valid can be determined (as has been demonstrated for valence-electron excitations by Vriens et al. (1969)).

Information on inner-shell ionization cross sections is required in bulk analysis, surface analysis, and thin-film analysis of materials by EPMA, AES, and EELS, respectively. A qualitatively similar physical model for each technique is used to describe electron transport in the specimen, inner-shell ionization, and the eventual detection of x-rays, Auger electrons, or scattered electrons. The complexity of atomic rearrangements following the ionization of $L, M, \ldots$ shells is such that it may be difficult to make effective use of cross-section data even if it were available. That is, the prediction of the yields of characteristic x-rays or **Auger** electrons (or the correction of measured intensities) depends on knowledge of relevant Coster-Kronig and Auger transition rates as well as on ionization cross sections. In EELS, measurements of loss intensities over limited ranges of energy loss and scattering angle may be perturbed by "solid-state" effects (Powell, 1984). In such cases it may be necessary to determine "effective" yields of the quantity of interest for a range of elements rather than to rely on ionization cross-section data. Comparative measurements with a limited number of reference materials may be sufficient for determining needed instrumental parameters. Otherwise, measurements of the specimen may need to be compared with measurements made using a standard of similar composition.

Acknowledgements

The author is indebted to Drs. R. J. Celotta, R. Hippler, M. Inokuti, Y. K. Kim, R. D. Leapman. D. E. Newbury, and P. Rez who have provided information and many useful comments and suggestions.

References

Abbati, I., Braicovich, L., Rossi, G., Lindau, I., del Pennino, U., Nannarone, S. (1983): Phys. Rev. Lett. *50*, 1799 – 1802.
Bambynek, W., Crasemann, B., Fink, R. W., Freund, H.-U., Mark, H., Swift, C. D., Price, R. E., Rao. P. V. (1972): Rev. Mod. Phys. *44*, 716 – 813.
Bethe, H. (1930): Ann. Phys. (Leipzig) *5*, 325 – 400.
Bonham, R. A. (1979): High-Energy Electron Impact Spectroscopy. In: Electron Spectroscopy: Theory, Techniques and Applications (Brundle, C. R., Baker, A. D., eds.), Vol. 3, 127–187. New York: Academic Press.
Campeanu, R. I., Koch, S. (1981): Z. Physik *A 299*, 95 – 96.
Casnati, E., Tartari, A., Baraldi, C. (1982): J. Phys. *B 15*, 155 – 167.
Cazaux, J. (1983): Ultramicroscopy *12*, 83 – 86.
Cazaux, J., Moutou, S. (1984): Surface and Interface Anal. *6*, 62 – 67.
Celotta, R. J., Huebner, R. H. (1979): Electron Impact Spectroscopy: An Overview of the Low-Energy Aspects. In: Electron Spectroscopy: Theory, Techniques and Applications (Brundle, C. R., Baker. A. D., eds.), Vol. 3, 41 – 125. New York: Academic Press.
Chang, C.-N. (1979): Phys. Rev. *A 19*, 1930 – 1935.
Codling, K. (1973): Reports on Progress in Physics *36*, 541 – 624.
de la Ripelle, F. M. (1949): J. Phys. (Paris) *10*, 319 – 329.
Drawin, H.-W. (1961): Z. Physik *164*, 513 – 521.
Egerton, R. F. (1979): Ultramicroscopy *4*, 169 – 179.
Egerton, R. F. (1981): Proceedings of the 39th Annual Meeting of the Electron Microscopy Society of America (Bailey, G. W., ed.), 198 – 199. Baton Rouge: Claitors Press.
Egerton, R. F. (1982): Principles and Practice of Quantitative Electron Energy-Loss Spectroscopy. In: Microbeam Analysis – 1982 (Heinrich, K. F. J., ed.), 43 – 53. San Francisco: San Francisco Press.
Eschwey, P., Manakos, P. (1982): Z. Physik *A 308*, 199 – 207.
Fischer, B., Hoffmann, K.-W. (1967): Z. Physik *204*, 122 – 128.
Geiss, R. H. (ed.) (1981): Analytical Electron Microscopy 1981. San Francisco: San Francisco Press.
Genz, H. (1982): AIP Conference Proceedings 94: X-ray and Atomic Inner-Shell Physics 1982 (Crasemann, B., ed.), 85 – 99. New York: American Institute of Physics.
Glupe, G., Mehlhorn, W. (1967): Phys. Letters *25 A*, 274 – 275.
Glupe, G., Mehlhorn, W. (1971): J. Phys. (Paris) *C 4*, 40 – 43.

Green, M., Cosslett, V. E. (1961): Proc. Phys. Soc. (London) *78*, 1206 – 1214.

Gryzinski, M. (1965): Phys. Rev. *138*, A336 – A358.

Haak, H. W., Sawatzky, G. A., Thomas, T. D. (1978): Phys. Rev. Letters *41*, 1825 – 1827.

Hahn, Y. (1975): Phys. Rev. *A 13*, 1326 – 1333.

Hall, P. M., Morabito, J. M., Conley, D. K. (1977): Surf. Science *62*, 1 – 20.

Heinrich, K. F. J., Newbury, D. E., Yakowitz, H. (eds.) (1976): Use of Monte Carlo Calculations in Electron Probe Microanalysis and Scanning Electron Microscopy. Washington, D.C., U.S.A.: U.S. National Bureau of Standards Special Publication 460.

Heinrich, K. F. J. (1981): Electron Beam X-Ray Microanalysis. New York: Van Nostrand-Reinhold Co.

Hink, W., Kees, L., Schmitt, H.-P., Wolf, A. (1981): Near K-Ionization Threshold Auger Electron Measurements for Neon under Electron Impact. In: Inner-Shell and X-Ray Physics of Atoms and Solids (Fabian, D. J., Kleinpoppen, H., Watson, L. M., eds.), 327 – 330. New York: Plenum Press.

Hippler, R., McGregor, I., Aydinol, M., Kleinpoppen, H. (1981): Phys. Rev. *A 23*, 1730 – 1736.

Hippler, R., Jitschin, W. (1982): Z. Physik *A 307*, 287 – 292.

Hippler, R., Saeed, K., McGregor, I., Kleinpoppen, H. (1982): Z. Physik *A 307*, 83 – 87.

Hippler, R., Klar, H., Saeed, K., McGregor, I., Duncan, A. J., Kleinpoppen, H. (1983): J. Phys. *B 16*, L617 – L621.

Hoffmann, D. H. H., Brendel, C., Genz, H., Löw, W., Müller, S., Richter, A. (1979): Z. Physik *A 293*, 187 – 201.

Holloway, P. H. (1980): Fundamentals and Applications of Auger Electron Spectroscopy. In: Advances in Electronics and Electron Physics (Marton, L., Marton, C., eds.), Vol. 54, 241 – 298. New York: Academic Press.

Hren, J. J., Goldstein, J. I., Joy, D. C. (eds.) (1979): Introduction to Analytical Electron Microscopy. New York: Plenum Press.

Ichimura, S., Shimizu, R., Langeron, J. P. (1983): Surf. Science *124*, L49 – L54.

Ing, B. S., Pendry, J. B. (1975): J. Phys. *C 8*, 1087 – 1098.

Inokuti, M. (1971): Rev. Mod. Phys. *43*, 297 – 347.

Inokuti, M., Itikawa, Y., Turner, J. E. (1978): Rev. Mod. Phys. *50*, 23 – 35.

Inokuti, M., Manson, S. T. (1984): Cross Sections for Inelastic Scattering of Electrons by Atoms – Selected Topics Related to Electron Microscopy. In: Electron Beam Interactions with Solids for Microscopy, Microanalysis, and Microlithography (Kyser, D. F., Newbury, D. E., Niedrig, H., Shimizu, R., eds.), 1 – 17. Chicago: Scanning Electron Microscopy.

Isaacson, M. (1977): Specimen Damage in the Electron Microscope. In: Principles and Techniques of Electron Microscopy (Hayat, M. A., ed.), Vol. 7, 1 – 78. New York: Van Nostrand-Reinhold Co.

Ito, S., Shimizu, S., Kawaratani, T., Kubota, K. (1980): Phys. Rev. *A 22*, 407 – 412.

Jessenberger, J., Hink, W. (1975): Z. Physik *A 275*, 331 – 337.

Joshi, A., Davis, L. E., Palmberg, P. W. (1975): Auger Electron Spectroscopy. In: Methods of Surface Analysis (Czanderna, A. W., ed.). New York: American Elsevier.

Joy, D. C., Maher, D. M. (1980): J. Phys. *E 13*, 260 – 270.

Joy, D. C., Maher, D. M. (1981): J. Microscopy *124*, 37 – 48.

Joy, D. C., Newbury, D. E. (1981): A "Round Robin" Test on ELS Quantitation. In: Analytical Electron Microscopy 1981 (Geiss, R. H., ed.), 178 – 180. San Francisco: San Francisco Press.

Kieffer, L. J., Dunn, G. H. (1966): Rev. Mod. Phys. *38*, 1 – 35.

Kolberstvedt, H. (1967): J. Appl. Phys. *38*, 4785 – 4787.

Krause, M. O. (1979): J. Phys. Chem. Ref. Data *8*, 307 – 327.

Kunz, C. (1973): Comments on Solid State Physics *5*, 31 – 40.

Kuyatt, C. E. (1968): Measurement of Electron Scattering from a Static Gas Target. In: Atomic and Electron Physics: Atomic Interactions, Part A (Bederson, B., Fite, W. L., eds.), 1 – 43 (Methods of Experimental Physics, Vol. 7) (Marton, L., ed.). New York: Academic Press.

Langenberg, A., de Heer, F. J., van Eck, J. (1975): J. Phys. *B 8*, 2079 – 2108.

Leapman, R. D., Rez, P., Mayers, D. F. (1980): J. Chem. Phys. *72*, 1232 – 1243.

Lotz, W. (1970): Z. Physik *232*, 101 – 107.

Madey, T. E., Doering, D. L., Bertel, E., Stockbauer, R. (1983): Ultramicroscopy *11*, 187 – 198.

Maher, D. M. (1979): Elemental Analysis Using Inner-Shell Excitations: A Microanalytical Technique for Materials Characterization. In: Introduction to Analytical Electron Microscopy (Hren, J. J., Goldstein, J. I., Joy, D. C., eds.), 259 – 294. New York: Plenum Press.

McGuire, E. J. (1971): J. Physique (Paris) *C 4*, 37 – 39.

McGuire, E. J. (1977): Phys. Rev. *A 16*, 73 – 79.

McGuire, E. J. (1979): Phys. Rev. *A 20*, 445 – 456.

Moores, D. L., Golden, L. B., Sampson, D. H. (1980): J. Phys. *B 13*, 385 – 395.

Müller, A., Groh., W., Kneissl, U., Heil, R., Ströher, H., Salzborn, E. (1983): J. Phys. *B 16*, 2039 – 2052.

Omidvar, K. (1977): J. Phys. *B 10*, L55 – L61.

Palinkas, J., Schlenk, B. (1980): Z. Physik *A 297*, 29 – 33.

Pantano, C. G., Madey, T. E. (1981): Appl. of Surf. Science *7*, 115 – 141.

Pessa, V. M., Newell, W. R. (1971): Physica Scripta *3*, 165 – 168.

Powell, C. J. (1968): Interaction of Electrons with Solids. In: Atomic and Electron Physics: Atomic Interactions, Part B (Bederson, B., Fite, W. L., eds.), 275 – 305 (Methods of Experimental Physics. Vol. 7) (Marton, L., ed.). New York: Academic Press.

Powell, C. J. (1976a): Rev. Mod. Phys. *48*, 33 – 47.

Powell, C. J. (1976b): Evaluation of Formulas for Inner-Shell Ionization Cross Sections. In: Proceedings of a Workshop on the Use of Monte Carlo Calculations in Electron Probe Microanalysis and Scanning Electron Microscopy (Heinrich, K. F. J., Newbury, D. E., Yakowitz. H., eds.), 97 – 104. Washington, D.C.: U.S. National Bureau of Standards Special Publication 460.

Powell, C. J. (1977): Proceedings of the Seventh Int. Vacuum Congress and the Third Int. Conference on Solid Surfaces (Dobrozemsky, R., ed.), Vol. 3, 2319 – 2322. Vienna: R. Dobrozemsky.

Powell, C. J., Stein, R. J., Needham, P. B., Driscoll, T. J. (1977): Phys. Rev. *B 16*, 1370 – 1379.

Powell, C. J. (1978): The Physical Basis for Quantitative Surface Analysis by Auger Electron Spectroscopy and X-Ray Photoelectron Spectroscopy. In: Quantitative Surface Analysis of Materials (McIntyre, N. S., ed.), 5 – 30. Philadelphia, Pa.: American Society for Testing and Materials Special Technical Publication 643.

Powell, C. J. (1980): Appl. of Surf. Science *4*, 492 – 509.

Powell, C. J., Erickson, N. E., Madey, T. E. (1982): J. Electron Spect. and Rel. Phen. *25*, 87 – 118.

Powell, C. J. (1984): Inelastic Scattering of Electrons in Solids. In: Electron Beam Interactions with Solids for Microscopy, Microanalysis, and Microlithography (Kyser, D. F., Newbury, D. E., Niedrig, H., Shimizu, R., eds.), 19 – 32. Chicago: Scanning Electron Microscopy.

Quarles, C. A. (1976): Phys. Rev. *A 13*, 1278 – 1280.

Quarles, C., Semaan, M. (1982): Phys. Rev. *A 26*, 3147 – 3151.

Raether, H. (1980): Springer Tracts in Modern Physics, Vol. 88, 1 – 196. Berlin-Heidelberg-New York: Springer.

Rez, P. (1984): X-Ray Spectrometry *13*, 55 – 59.

Roussouw, C. J., Whelan, M. J. (1979): J. Phys. *D 12*, 797 – 807.

Rudge, M. R. H., Schwartz, S. B. (1966): Proc. Phys. Soc. (London) *88*, 563 – 578.

Schnatterly, S. E. (1979): Inelastic Electron Scattering Spectroscopy. In: Solid State Physics (Ehrenreich, H., Seitz, F., Turnbull, D., eds.), Vol. 34, 275 – 358. New York: Academic Press.

Scofield, J. H. (1978): Phys. Rev. *A 18*, 963 – 970.

Seah, M. P., Dench, W. A. (1979): Surf. and Interface Anal. *1*, 2 – 11.

Shima, K. (1980): Phys. Letters *77A*, 237 – 239.

Shima, K., Nakagawa, T., Umetani, K., Mikumo, T. (1981): Phys. Rev. *A 24*, 72 – 78.

Sklad, P. S., Bentley, J., Lehman, G. L. (1981): Quantification of Energy Loss Measurements with the Use of K and L Absorption Edges. In: Analytical Electron Microscopy 1981 (Geiss, R. H., ed.). 173 – 175. San Francisco: San Francisco Press.

Smith, D. M., Gallon, T. E., Matthew, J. A. D. (1974): J. Phys. *B 7*, 1255 – 1261.

Stephens, A. P. (1980): Ultramicroscopy *5*, 343 – 349.

Swanson, N., Powell, C. J. (1966): Phys. Rev. *145*, 195 – 208.

Swanson, N., Powell, C. J. (1968): Phys. Rev. *167*, 592 – 600.

Szajman, J., Leckey, R. C. G. (1981): J. Electron Spect. and Rel. Phen. *23*, 83 – 96.

Tawara, H. (1978): Innershell Ionization by Relativistic Electron, Positron, and Proton Impact. In: Electronic and Atomic Collisions (Watel, G., ed.), 311 – 329. Amsterdam-New York: North-Holland.

Tolk, N. H., Traum, M. M., Tully, J. C., Madey, T. E., eds. (1983): Desorption Induced by Electronic Transitions DIET I. New York: Springer.

Tung, C. J. (1980): Phys. Rev. *A 22*, 2550 – 2555.

Vriens, L., Simpson, J. A., Mielczarek, S. R. (1968): Phys. Rev. *165*, 7 – 15.

Vriens, L. (1969): Binary-Encounter and Classical Collision Theories. In: Case Studies in Atomic Collision Physics (McDaniel, E. W., McDowell, M. R. C., eds.), 335 – 398. New York: Elsevier.

Wannier, G. H. (1953): Phys. Rev. *90, 817* – 825.

Weaver, J. H., Krafka, C., Lynch, D. W., Koch, E. E. (1981): Optical Properties of Metals. Part I: The Transition Metals ($0.1 < h\nu < 500$ eV). Physik Daten (Physics Data), Karlsruhe: Fachinformationszentrum, Number 18-1, 1 – 302.

Webster, D. L., Hansen, W. W., Duveneck, F. B. (1933): Phys. Rev. *43, 839* – 858.

Worthington, C. R., Tomlin, S. G. (1956): Proc. Phys. Soc. (London) *A 69*, 401 – 412.

Yagishita, A. (1981): Phys. Letters *87 A*, 30 – 32.

Zaporozhchenko, V. I., Kantsel, V. V., Kashin, G. N., Lyubimova, T. A. (1979): Sov. Phys. Tech. Phys. *24*, 821 – 823.

7

Total Ionization Cross Sections*

F. J. de Heer and *M. Inokuti*

FOM Institute for Atomic and Molecular Physics,
Amsterdam, The Netherlands
Argonne National Laboratory, Argonne, Illinois, U.S.A.

7.1 Experimental Techniques

7.1.1 General

The measurement of the total ionization cross section, σ_t, for a gaseous sample requires the determination of four quantities. They are the current $i_e(T)$ of a beam of electrons having incident energy T, the total ion current $i_i(T)$ produced by the electron beam, the number density ρ of atoms or molecules in the gas, and the collision pathlength L over which the produced ions are collected. Then, σ_t may be obtained from the relation

$$i_i(T)/i_e(T) = \rho L \sigma_t(T). \tag{7-1}$$

The *total* ionization cross section (or electron-production cross section) may be expressed as

$$\sigma_t = \Sigma_z z \sigma_z, \tag{7-2}$$

where σ_z represents the *partial* ionization cross section for the z-fold ionization, i.e., for the production of z electrons plus ion (or ions) having total charge $+ze$. In the measurement of σ_t, no information is obtained about individual σ_z in general, except, of course, at energies below the threshold for multiple ionization. [Another

* This work is performed under the auspices of the U.S. Department of Energy, and is also part of the research program of the Stichting voor Fundamenteel Onderzoek der Materie (Foundation for Fundamental Research on Matter) and was made possible by financial support from the Nederlandse Organisatie voor Zuiver-Wetenschappelijk Onderzoek (Netherlands Organization for the Advancement of Pure Research).

exception may occur at electron energies T slightly above a multiple-ionization threshold, for instance, the double-ionization threshold. Then, one may be justified to assume either a threshold behavior for the double ionization cross section $\sigma_2(T)$ or a smooth continuation of the single-ionization cross section $\sigma_1(T)$; to the extent that such an assumption is valid, one may infer from measured $\sigma_t(T)$, $\sigma_1(T)$ and $\sigma_2(T)$ separately.]

The total ionization cross section $\sigma_t(T)$, as discussed above, is sometimes called the *gross* ionization cross section. There are also a few experiments that determine another kind of cross sections, namely, the *counting* ionization cross sections (or ion-production cross sections) given as

$$\sigma_c = \Sigma_z \, \sigma_z. \tag{7-3}$$

For comparison with certain theories, it is sometimes more convenient to treat σ_c. Under certain circumstances, the single ionization is dominant, and then one may write

$$\sigma_t = \sigma_c = \sigma_1. \tag{7-4}$$

This applies rigorously at electron energies below the double ionization threshold. It also applies approximately for single-shell systems such as He and H_2.

Kieffer and Dunn (1966) have extensively discussed most of the problems encountered in obtaining accurate experimental σ_t values. Indeed, Table 1 of their review article is a valuable checklist for judging the reliability of given σ_t values from a purely experimental point of view. As we shall see in Section 7.2.2, it is also possible, and is indeed useful, to apply some criteria based on theory for judging the data reliability; this is expecially true at high incident-electron energies.

In what follows we shall discuss various methods for determining *absolute* values of σ_t, as opposed to *relative* values (which represent the T-dependence of σ_t up to an overall normalization factor). We restrict our discussion to the most salient points, and will not repeat all the details given in earlier excellent treatments by Kieffer and Dunn (1966), by Massey and Burhop (1969), by Massey (1969), by Kerwin *et al.* (1969), by Field and Franklin (1970), and by Märk (1984).

7.1.2 The Condenser-Plate Method

This method has been used most often for measuring σ_t by use of a static-gas target. One of the earliest measurements with this method was carried out by Compton and Van Voorhis (1925), who successfully dealt with many problems. In their experiment, however, it was difficult to prevent those electrons scattered out of the primary beam from reaching the ion-collecting electrode. Therefore, a large negative potential had to be applied, leading to a bad definition of the energies of the electrons producing ionization.

Tate and co-workers [as seen in Smith (1930, 1931) and in Tate and Smith (1932) for example] avoided the difficulty by introducing a longitudinal magnetic field of a few hundred Gauss to keep the scattered electrons close to the primary-beam direction. The design of an apparatus by Tate and co-workers, shown in Fig. 7-1, has been used in many later experiments, and is still used now with only slight modifications. In

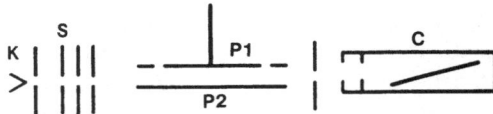

Fig. 7-1. Schematic diagram of the apparatus used by Tate and Smith. K cathode, S collimeter, P_1 and P_2 condenser plates, C Faraday cage

view of the excellence of that design, it is not surprising to find that the results of these workers compare very well with other more recent results for some gases.
In the apparatus shown in Fig. 7-1, electrons emerge from cathode K, pass through the collimator S, receive acceleration, and finally enter the collision chamber filled with a gas. An electric field of 5 V/cm is applied across the condenser plates P_2 and P_1 (together with the latter's guard plates), so that the positive ions may be collected at P_1 and its current $i_i(T)$ determined. An axial magnetic field prevents the primary electron beam from deflecting, and suppresses secondary-electron emission from P_1. The primary electron beam is eventually trapped in cage C, and its current $i_e(T)$ is measured. The pathlength L, to be used in Eq. (7-1), follows from the length of plate P_1 in the beam direction. The two guard plates serve to establish a homogeneous electric field in the region of ion collection. The number density ρ (cm^{-3}) is evaluated from the gas pressure p_c (Torr) and temperature T_c (K) of the collision chamber from the relation

$$\rho = 3.535 \times 10^{16} p_c (273.2/T_c).\tag{7-5}$$

So far we have presented principles involved in the measurements of the four quantities necessary for the evaluation of $\sigma_t(T)$. Now we shall discuss circumstances that influence the accuracy of $\sigma_t(T)$. The following discussion is limited, because fuller treatments have been given by Kieffer and Dunn (1966), by Massey and Burhop (1969), and by Massey (1969).

1. Measurement of the ion current $i_i(T)$

It is essential that the electric field between plates P_1 and P_2 is strong enough to ensure complete collection of all the positive ions formed between the plates. In a monatomic gas, electric fields of about 5 V/cm appear to be sufficient. In a molecular gas, some of the ions produced through dissociative ionization may have appreciable kinetic energies (as seen in Chapter 4) and therefore stronger electric fields (of about 30 V/cm) are often necessary to realize the saturation of the ion current, taken as an indication of complete ion collection.
For measurements at higher incident energies (at which the total ionization cross section becomes small), one must avoid effects of additional ionization by energetic secondary electrons, which may arise from an ionizing collision of an electron with a gas molecule or from electron bombardment of a slit. These effects can be reduced by a system of electrode potentials chosen such that the ionization region between the condenser plates has the largest negative potential, as described for instance by Schram et al. (1965). The longitudinal axial magnetic field also helps to reduce the same effects. The stronger the magnetic field is, the more efficient is the suppression of the effects.

2. Measurements of the Electron Current $i_e(E)$

One must make certain that every electron in the primary beam enters the collector and no reflected electron escapes. Furthermore, there should be a negligible number of secondary electrons and ions that enter or leave the collector. In other words, the goal is to build the apparatus so that no charged particles, apart from the primary electrons, either enter or leave the collector. To meet the requirements for good $i_e(T)$ measurement, one usually applies appropriate electric fields in the collector and uses low enough gas pressures to allow only a small number of secondary electrons and ions in the collision region.

3. The Pathlength L

To a first approximation, we may set L equal to the length l of plate P_1 in the beam direction. However, the presence of the axial magnetic field necessitates some corrections. An electron in the collimating magnetic field has a helical path. Moreover, in the crossed electric and magnetic field a trochoidal drift is possible, but its effect on L is usually negligible, as seen in Schram et al. (1965). For electrons of velocity v with transverse component v_1, the pathlength is given by

$$L = l\,[1 - (v_\perp/v^2)]^{-1/2}. \tag{7-6}$$

It was noted early that a maximum pathlength could be associated with $v_\perp$ as limited by the electron-gun collimating apertures. However, Asundi (1963) argued that more realistic limits on the transverse velocity were set by the transverse momentum received in the electron gun in the lens system; thus, he found that $(v_\perp/v)^2 = 0.001$ for conditions used by Tate and Smith. Sources of transverse velocities in a magnetically confined electron gun have been studied in detail by Taylor et al. (1974). Craggs et al. (1975) pointed out that the scattering of electrons by the gas molecules also contributes to the transverse component of the electron velocity. According to Kieffer and Dunn (1966), it might well be that the systematic differences in the ionization cross sections of rare gases between Smith (1930) and Rapp and Englander-Golden (1965) are connected to the gas scattering effect. In other words, Smith used higher pressures, but neglected a correction to l for the gas-scattering effect, which may have been appreciable. Kurepa et al. (1974) paid much attention to various effects that may affect L; however, these effects appear to be negligibly small in their experiment.

4. Measurement of the Number Density ρ

A variety of methods have been used to measure the gas pressure and thence to determine ρ. It is important to consider the difference between the temperature T_G in the gauge and the temperature T_C in the collision chamber. This difference leads to the difference between the pressure p_G in the gauge and the pressure p_C in the collision chamber. For an ideal gas, there is the relation

$$p_C/p_G = T_C/T_G. \tag{7-7}$$

In the past a McLeod gauge was most often used to measure pressure. The measurement with a McLeod gauge in general is no more accurate than a few percent, because of the capillary depression effect. To inhibit contamination by mercury, it is necessary to introduce a cold trap between the McLeod gauge and the collision chamber. However, as Ishii and Nakayama (1961) first showed, this arrangement gives rise to a steady mercury stream from the mercury reservoir to the cold trap, thus simulating a diffusion pump and leading to incorrect results in the pressure measurement. This effect (now called the Ishii effect) had been predicted by Gaede (1915), and was considered in detail by the Vries and Rol (1965). Schram et al. (1965) followed suggestions by Ishii and Nakayama (1961), and cooled the walls of the McLeod gauge so that the temperature there was only slightly above the reservoir temperature; then, the pressure readings changed to higher values, ranging from 1% for H_2 and 14% for Xe.

Rapp and Englander-Golden (1965) used a McLeod gauge for H_2 only. For each sample gas, an effusive flow was generated through a leak between the high-pressure reservoir and the collision chamber, and also out of the collision chamber through another aperture leading to pumps. Rapp and Englander-Golden showed that the collision-chamber pressure should be proportional to the reservoir pressure; therefore, using the knowledge of the pressure of the collision chamber when it is filled with H_2, it was possible to evaluate the pressure of other gases in the collision chamber once the reservoir pressure is measured.

As the foregoing discussion illustrates, the measurement of the gas pressure may give rise to substantial systematic errors in the total ionization cross section. Thus, measurements done before 1961, especially on heavier gases, are likely to suffer from errors due to the Ishii effect. The seriousness of this effect depends on the dimension of a McLeod gauge used, and its magnitude can be evaluated absolutely for any target as soon as the magnitude for one heavier gas, say Xe, may be considered as known. Formulas for the evaluation of the Ishii-effect are given by de Vries and Rol (1965). A key number here is the coefficient D_{12} of diffusion of the gas in mercury in the McLeod gauge. By use of the method of de Vries and Rol, de Heer (1981) evaluated the Ishii-effect correction for the measurement on CH_4 by Tozer (1958) and for those on C_2H_2 by Tate and Smith (1932), considering the ionization cross-section values reported by the same workers for He, Ne, Ar, Kr, and Xe.

In newer experiments, gas pressures are mostly measured with a membrane manometer calibrated by a continuous-flow method, as described for instance by Bannenberg and Tip (1968) and by Bannenberg et al. (1969). In this way, accuracies as good as $1-2\%$ have been accomplished at pressures above 10^{-3} Torr. At these high pressures, one may calibrate an ionization manometer against a membrane manometer; then, one may use the ionization manometer at lower pressures. For instance, Cowling and Fletcher (1973) used an AEI VH ion gauge below 10^{-3} Torr and an NRC Alphatron gauge above 10^{-4} Torr. Both of these gauges were calibrated against an MKS Baratron membrane manometer as a secondary standard; thus, the calibration with the use of a successive-expansion technique, led to errors of only 0.5% in the Alphatron gauge and of about 2% in the ion gauge at lower pressures. The Baratron calibration was checked against an oil manometer and was found to be accurate to within 0.2%. For discussion of typical problems with the use of a membrane manometer, see for instance, Blaauw et al. (1980).

Kurepa et al. (1974) measured gas pressures with great care, using calibration with a constant-gas-flow method, in their determination of the total ionization cross section of Ar by electrons.

5. The Electron Energy T

Upon consideration of the definition of T, one must first note that ions are collected at P_1 by means of an electric field that is due to a positive voltage at P_2 and the earth voltage at F_1. As a result, there is a voltage drop across the electron-beam width, leading to a spread in the electron energies across to the beam. Of course, one can apply corrections to T for the nonzero voltage at the beam axis. The field in the ionizing region is avoided in certain apparatuses, e.g., in the Lozier tube to be described in the next subsection. To avoid the field in the condenser-plate method, Cowling and Fletcher (1973) used a pulsed electron beam and a pulsed ion collector voltage. The duration of the pulsed beam was $0.5\,\mu s$. For slow-ion collection, a $6\text{-}\mu s$ pusher voltage was applied across the condenser plates $0.5\,\mu s$ after each electron-beam pulse.

Another critical element in the definition of T, especially important near the ionization threshold, is the potential drop across the filament. Still another element is the contact-potential differences between electrodes, which can make the actual electron energy T quite different from the value given by the voltage applied between the cathode and the collision region; the difference may amount to one eV or greater. This difference is unimportant at high energies, where the total ionization cross section varies slowly with T in general. But the same difference is significant near the threshold, where the total ionization cross section either rises steeply with T (as a result of direct ionization) or has complicated structures (as a result of auto-ionization). Measurements near the threshold are treated by Read in Chapter 3, and therefore we shall present here a few remarks only. One method of absolute calibration of the electron energy is to introduce a gas for which the (first) ionization threshold energy is well known, say, from some spectroscopic studies. One may observe the appearance of the single charge ions of this gas in a region slightly above the threshold, and may extrapolate the ionization signal as a function of the electron energy. So long as direct ionization dominates over autoionization, the signal should depend on the excess energy (i.e., the electron energy minus the threshold energy) almost linearly. Then, if one extrapolates the signal to the abscissa, one may equate the intercept with the known threshold energy.

7.1.3 The Lozier Tube

Since an apparatus was first described by Lozier (1934), there have been several measurements made by use of similar apparatuses [e.g., Asundi *et al.* (1963) and Craggs *et al.* (1975)], and of a variant, i.e., a cylindrical tube that has grids instead of vanes in the Lozier tube [e.g., Schulz (1962)]. In this kind of measurement, only relative ionization cross sections (i.e., only the energy dependences of the total ionization cross section) are determined; the data thus obtained are subsequently normalized by use of the results such as those by Smith (1930) and by Tate and Smith (1932). Even then, the method is subject to substantial errors in measuring total cross sections for molecules because of the anisotropy in the angular distribution of dissociation products (Dunn 1962).

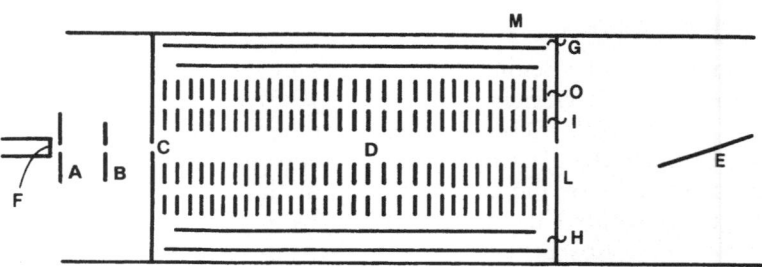

Fig. 7-2. Schematic diagram of the Lozer tube. *F* tungsten filament, *A*, *B*, and *C* apertures, *D* collision chamber, *E* trap, *I* inner vanes, *O* outer vanes, *L* rear plate, *H* ion collector, *G* guard electrode

Fig. 7-2 shows the apparatus used by Tozer (1958). Electrons from a hot tungsten filament *F* first pass through apertures *A*, *B*, and *C*, and then enter a collision chamber *D*. In this chamber there is no electric field, but an axial magnetic field is applied to confine the electron beam, as it is done in the condenser-plate method. The electrons are eventually collected on trap *E*, which is set at 50 V positive. A set of vanes *I* are placed so that they are concentric with the axis of the chamber and are perpendicular to it. The vanes, as well as the front plate *C* and the rear plate *L*, are set at the earth potential. A number of outer vanes *O* are set at a negative voltage when one wants to draw out positive ions formed by electron collisions; they are set at positive voltage when one wants to draw out negative ions. The ions are collected on cylinder *H*, which is surrounded by guard electrode *G*. The whole apparatus is shielded by cylinder *M*, which is set at the earth potential. By applying an appropriate voltage between *H* and *O*, one can prevent *H* from collecting those ions which are initially formed with less than a fixed amount of kinetic energy. In this way, one can measure the energy distribution of ions resulting from dissociative ionization, although it may be distorted because of the antisotropy effects (Dunn 1962).

By varying the primary energy of the electrons emerging from the gun, one can determine the relative total ionization cross section. To determine the absolute ionization cross section, one must resort to a normalization procedure, because the efficiency of collection of the ions is smaller than unity and is unknown in general. In his measurements on methane, Tozer (1958) normalized the signal against that due to Ar, for which he used the total ionization cross section given by Tate and Smith (1932). As pointed out by Tozer (1958), the method of normalization may be questionable for polyatomic gases, because dissociative ionization processes may lead to ions of differing masses and energies, and therefore of differing collection efficiencies. In addition to the ion-collection problem, one also faces the same problems as discussed in the case of the condenser-plate method.

The Lozier tube has been used in the determination of appearance energies and of the kinetic energies of different ionic species resulting from molecules, as seen in Massey (1969) and in Field and Franklin (1970).

7.1.4 The Summation Method

This method was introduced in the present form (i.e., in the determination of total cross sections) by Märk and Egger (1977), and the appellation is due to Märk (private communication). Like the Lozier-tube method, this method gives only the relative values of the total ionization cross section. But the experimental approach is quite different. Following Bleakney (1929) and Tate and Smith (1934), one uses a mass-spectrometric technique to separate ions of different ratio q/M of the charge q and the mass M of ions. This enables one to determine the partial ionization cross section σ_z, as used in Eqs. (7-2) and (7-3). For an atom, σ_z is characterized by the charge $q = ze$, where $-e$ is the charge on the electron, M being practically the same for all ionization produced (apart from an isotope effect). For Ar as an example, we obtain Ar^{+z} ions, where m takes a sequence of integer values starting with unity. For a molecule, the mass M of the ionic products also varies. For the PH_3 molecule treated by Märk and Egger, one may obtain the parent ion PH_3^+, fragments such as PH_2^+, PH^+, P^+, H_2^+, and H^+, as well as a substantial number of doubly charged ions of the parent and fragment species.

In several experiments, care has been taken to make certain that the ion-collection efficiency is independent of q/M, and also of T. Then, every partial ionization cross section for a given gas is obtained on the same scale, viz., it differs from the absolute value by a common constant factor. One may evaluate the factor by the use of Eq. (2); one may sum up the relative partial ionization cross sections as on the right-hand side of Eq. (2) and set the result equal to the total ionization cross section determined by the condenser-plate method. (See also Schutten *et al.*, 1966, and Adamczyk *et al.*, 1972.)

When Märk and Egger (1977) treated PH_3, no value of the total ionization cross section was available. They measured relative partial ionization cross sections for PH_3 and Ar summed the values for each gas, and normalized the sum for PH_3 against the sum for Ar at 90 eV by use of the total ionization cross section of Ar as measured by Rapp and Englander-Golden (1965). As long as one may assume the same normalization factor for PH_3, this summation method should lead to reliable values of the total ionization cross section for this species.

Märk and co-workers carefully considered the influence of the ion-collection efficiency on the measurement of partial ionization cross sections. This topic is extensively treated in Chapter 5 by Märk. Stephan *et al.* (1980 b) have demonstrated the reliability of the summation method for noble gases He, Ne, Ar, and Kr at electron impact energies ranging from the threshold to 180 eV. The results obtained by the summation agree well in the impact-energy dependence with the total ionization cross sections measured by Rapp and Englander-Golden (1965), apart from a constant factor for each gas.

7.1.5 Gas-Filled Counters

Another kind of measurement was introduced by Graf (1939), and was conducted by McClure (1953). It has been carried out most extensively by Rieke and Prepejchal (1972) on forty gaseous species and for electrons of kinetic energies between 0.1 and 2.7 MeV.

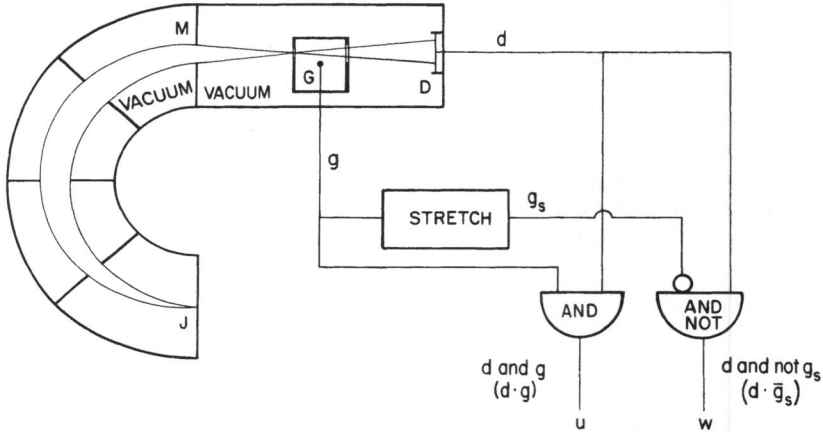

Fig. 7-3. Schematic diagram of the apparatus used by Rieke and Prepejchal (1972). J electron source (β-ray emitter, M double focusing magnetic analyzer, G gas-filled chamber, D silicon detector, d channel that registers a primary electron, g channel that registers an ionization event in G, g_s channel that follows g with stretch, u counts in channel $d \cdot g$ (i.e., d and g), w counts in channel $d \cdot \bar{g}_s$ (i.e., d and not g_s)

An electron passing through a chamber of thickness L filled with a gas of molecular number density ρ produces the average number of α of ions as given by

$$\alpha = \rho L \sigma_c \cdot \tag{7-8}$$

The probability that the transit of an electron produces no ionization in the chamber is $\theta = e^{-\alpha}$, according to the Poisson statistics. Thus, measuring θ and knowing ρ and L, one can determine σ_c. Fig. 7-3 shows the apparatus of Rieke and Prepejchal. Electrons from beta emitter J are selected for energy by the magnetic analyzer M, pass through the gas-filled chamber G, and eventually reach the silicon barrier-layer detector D. A pulse in channel $d \cdot g$ indicates that a primary electron passed through G and caused an ionization event. However, one obtains no information of about the kind of that event (viz., whether it was a single ionization or a multiple ionization), because the signal in g is greatly amplified to give rise to a pulse. A pulse in channel $d \cdot \bar{g}_s$ indicates that a primary electron passed without causing any ionization. The "stretch" is introduced to exclude dead-time effects. If there are u counts in channel $d \cdot g$ and w counts in channel $d \cdot \bar{g}_s$, the probability ϕ that the transit of an electron through G gives rise to no count in G is given by

$$\phi = w/(u + w). \tag{7-9}$$

If there were no wall effects, one could equate ϕ with $\theta = e^{-\alpha}$. Wall effects may occur because some of the electrons due to the gas ionization generated near either window of G are lost by diffusion to the window and because there is some secondary-electron emission from the window into the gas. Thus, the effective number of ionizations per primary electron is not quite α of Eq. (7-8), but is given by $(1 - \lambda/\rho) \alpha$, where λ is a positive constant. Suppose that the secondary emission gives rise to z_s electrons that reach the sensitive volume of G. Then, one may write

$$\phi = w/(u + w) = \exp[-\rho L \sigma_c (1 - \lambda/\rho) - z_s]. \tag{7-10}$$

By measuring ϕ at a series of pressures (i.e., ρ values), one can derive σ_c at each electron energy from a straight-line fit of $\ln \phi$ against ρ, provided z_s may be taken as independent of ρ.

This experiment, conducted at high impact energies, is particularly suited to study the cross-section expression given by the Bethe (1932, 1933) theory:

$$\sigma_c = \beta^{-2}(Ax + B),\tag{7-11}$$

where A and B are constants,

$$x = \ln[\beta^2/(1-\beta^2)] - \beta^2,\tag{7-12}$$

and β is the speed of the primary electron measured in the light speed c in vacuum. The formula may be written in the form

$$\sigma_c = 4\pi(\hbar/mc)^2 \beta^{-2}(M_{ion}^2 x + C_{ion}),\tag{7-13}$$

where M_{ion}^2 and C_{ion}^2 are the quantities whose meanings are extensively discussed by Inokuti (1971). In particular, M_{ion}^2 is the total dipole matrix element squared measured in units of the Bohr radius squared, $a_0^2 = (\hbar^2/me^2)^2$.

In order to make connection between experimental results and Eq. (7-13), values of $\beta^2 \sigma_c$ obtained for a series of primary electron energies may be plotted against x, as first suggested by Fano (1954). The points should fall on a straight line with slope M_{ion}^2 and intercept C_{ion} with the horizontal axis, apart from the vertical scale corresponding to the universal constant $4\pi(\hbar/mc)^2$.

Rieke and Prepejchal (1972) discuss several critical aspects of the experiment. The counter G has been used in the Geiger-Müller mode in most of the cases, and in a proportional-counter mode in a few instances, depending on the occurrence of electron avalanches. Pressures in the counter were much higher than those used in other methods, and were measured by use of a Wallace and Tierman gauge that was calibrated against a Texas-Instrument fused-quartz Bourdon gauge. Gas pressures were of the order of tens of Torr and the measurements were accurate to 0.05 Torr. The electron pathlength L was determined by the counter design to an accuracy of half a percent. The results of Rieke and Prepejchal, pertaining to forty molecular species, are highly important for the establishment of the asymptotic behavior of the ionization cross section at high primary electron energies.

7.1.6 Crossed-Beam Methods

This method was introduced for measuring the ionization cross sections of atomic or molecular species that are not chemically stable at ordinary temperature and pressure; they include, for instance, atomic hydrogen, atomic oxygen or nitrogen, alkali metals, atoms or molecules in metastable states, and ionic species. Funk (1930) first used the method to study the ionization of sodium vapor by electrons. The use of an atomic or molecular beam as target necessarily means a low number density ρ, and thence weak signals of ionization, which must be distinguished from the signals due to the ionization of the background gas. In order to obtain a significantly strong signal, one may modulate the target beam and use a highly sensitive detection method, as discussed by Boyd and Green (1958) and by Fite and Brackmann (1958).

In a series of studies on the ionization of initially ionic species by electrons, Dolder *et al.* (1961) introduced the use of the modulation of both a target beam and an electron beam. Furthermore, it is worth mentioning the time-of-flight coincidence method, which was used by Shah and Gilbody (1981) for measuring the ionization cross section for ion-atom collisions more accurately than had been done with the use of a modulating-beam technique; certainly, the same approach will be effective in electron-atom collisions. In this approach, one prepares a pulsed primary beam and analyzes the time-of-flight of the ions produced. Thus, one distinguishes the atomic-beam ions from the background-gas ions by means of their different times of flight resulting from their different masses. Shah and Gilbody (1981) applied the method to study the ionization of atomic hydrogen.

The basics of the crossed-beam technique are explained in an excellent discussion by Kieffer and Dunn (1966). Fig. 7-4 is a reproduction of their block diagram for explanation. For discussion of the crossed-beam method, one must modify Eq. (7-1), which pertains to the static-gas target. Suppose that a beam of electrons traveling at speed v_e crosses a beam of target particles traveling perpendicularly at speed v_t. Let S

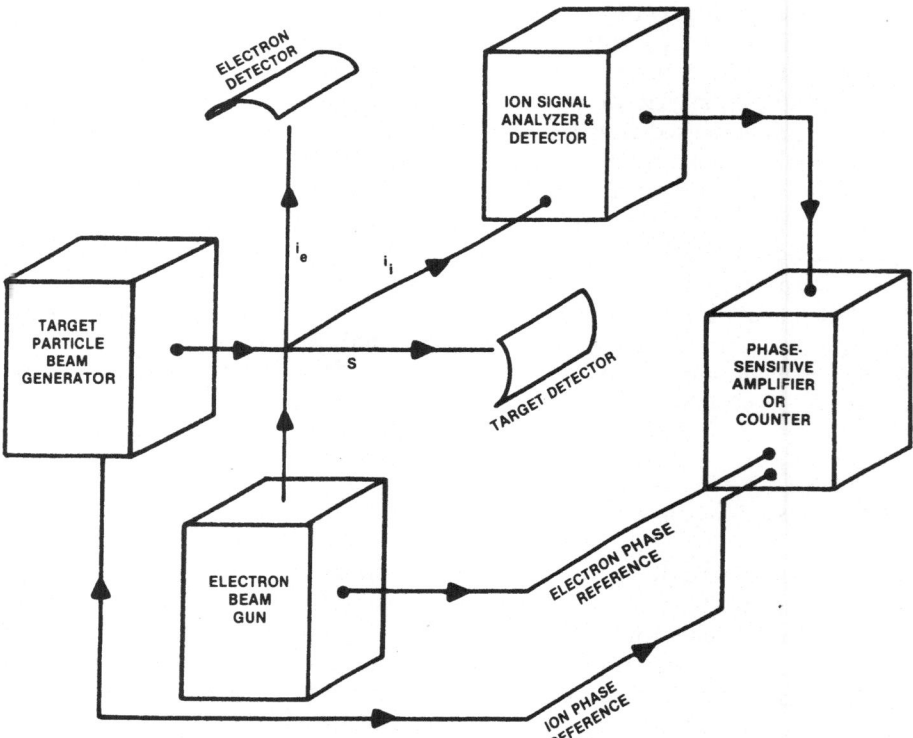

Fig. 7-4. Schematic diagram of a typical crossed-beam apparatus, taken from Kieffer and Dunn (1966). I_e the electron-beam current, I_i the ion current generated by electron impact, R the number of atomic particles per second arriving at the target detector

be the number of the target particles arriving at the detector in each second. The relative collision energy T is of the order of $T_e + (m/M_t)\, T_t$, where T_e is the electron kinetic energy, T_t is the target kinetic energy, and M_t is the target mass. The electron beam current i_e may be measured in the usual way be means of a Faraday cage. Let i_i be the ion current. Then, the cross section $\sigma_t(T)$ at the relative collision energy T may be determined from the relation

$$i_i/i_e = \sigma_t(T)\, S\, [(v_e{}^2 + v_t{}^2)^{1/2}/v_e v_t]\, F, \tag{7-14}$$

where F is the beam overlap factor to be discussed later.

When i_i is the *total* ion current, one obtains from Eq. (7-14) the total ionization cross section. When the produced ions are selected for charge or mass, one obtains a partial ionization cross section.

To determine i_i, one may have to apply an electric field to extract the ions from the collision region. In several experiments, a field is applied across the condenser plate, as in the condenser-plate method. Sometimes, a hole is made in one of the plates to allow mass analysis of the ions, as was done, for instance, by Fite and Brackmann (1958, 1959). In their work on atomic hydrogen, the mass analysis was necessary because the target beam consisted mainly of H, but also contained a small amount of H_2, and the electron impact led to both H^+ and H_2^+.

The factor F in Eq. (7-14) represents the overlap of the beams in the collision region. Suppose that $j_e(z)$ is the spatial distribution of the electron beam at distance z perpendicular to the two beam directions, and that $j_t(z)$ is the spatial distribution of the target beam at the same distance. The distributions can be determined by scanning each beam with a movable slit. Then, F may be given as

$$F = \int j_e(z) j_t(z)\, dz \left/ \left[\int j_e(z)\, dz \int j_t(z)\, dz \right] \right. . \tag{7-15}$$

For neutral targets, there are a number of difficulties involved in obtaining accurate values of cross sections from crossed-beam experiments. For instance, it is particularly difficult to determine S for neutral particles in the ground state. Therefore, a normalization procedure had been often used to put experimental data on an absolute scale. For instance, Fite and Brackmann (1958, 1959) normalized the signal from an atom (e.g. H or O) against that from the corresponding parent molecule (H_2 or O_2) which had been studied under the same crossed-beam condition. As a reference on absolute values, results of measurements on H_2 by the condenser-plate method could be used. The atomic beams were obtained by heating the gas in a tungsten oven [Fite and Brackmann (1958)] or by radio-frequency discharges [Boyd and Green (1958)]. As pointed out earlier, mass analysis of the ions produced was needed to determine the degree of dissociation in the target beam. For further details of the procedure, the reader is referred to the original article or to the discussion by Massey and Burhop (1969). An alternative is to use some theoretical results for data normalization. For instance, Fite and Brackmann (1958) normalized a measured relative value of the total ionization cross section of atomic hydrogen to the result of the first Born approximation at 500 eV. The two sets of results obtained by Fite and Brackmann (1958) using the two methods of normalization agree extremely well with each other.

As pointed out by Kieffer and Dunn (1966) and by McDowell (1969), there are certain questions concerning S in the measurements on alkali metals by Brink (1964) and by McFarland and Kinney (1965). In these measurements, a hot wire was used as a surface ionization detector for the determination of the neutral-beam intensity. In the determination of the efficiency of ionization at the hot wire, it was assumed that the reflection coefficient of the incident beam was zero. It was also assumed that the ratio $(1 - r_a)/(1 - r_i)$ was unity, where r_a is the reflection coefficient of the atom and r_i is that of the ion, both under thermal equilibrium on the hot wire. These assumptions are questionable (Datz and Taylor, 1956), and the uncertainties in the S determination may have caused uncertainties as large as 30% in the ionization cross sections reported by Brink (1964) and by McFarland and Kinney (1965).

There are other methods for the S determination. Zapesochnyĭ and Aleksakhin (1969) determined the total mass of alkali atoms condensed on the surface of a piezoelectric quartz crystal (cooled by liquid nitrogen), from the change of the natural oscillation frequency.

In the work on Li, Jalin et al. (1973) collected a neutral beam on a tantalum plate cooled by liquid nitrogen, and measured after each experiment the amount of material deposited on the plate by use of an isotope dilution technique or by atomic absorption spectroscopy. They also undertook independent experiments to demonstrate negligible reflection of the atomic beam at the surface kept at low temperatures.

Nygaard (1968) studied the surface ionization by comparing the measured number density of Cs atoms with the value given by the formula of Taylor and Langmuir (1937). Then, he measured the ionization cross section of Cs in both the vapor phase and an atomic beam; results from the two cases agreed within 3%. In the vapor phase, the number density obtained by the surface ionization was consistent with the formula of Taylor and Langmuir (1937), which should apply to a Knudsen flow. Nygaard and Hahn (1973) analyzed data on the surface ionization of K and Rb and concluded that S values for these species should be (98 ± 2)% of the value for Cs. This probably means that the same S value applies to all alkalis except for Li. [For fuller discussion, see Nygaard (1974).]

In studies on $He^{2+} - Li$ collisions, Kadota et al. (1982) used an optical method for the density determination. They observed the resonance line of Li I at 670.8 nm produced by impact of 500-eV electrons, and used the emission cross section reported by Leep and Gallagher (1974) for the density determination. They obtained the density profile by scanning the position of the atomic beam by means of a turnable mirror and of optical lenses so that a one-to-one image of the intersecting region was generated at the entrance slit of the monochrometer used for the detection of the radiation. Such scanning can be accomplished also with a movable slit.

An optical method for normalization has also been used recently by Dettmann and Karstensen (1982) in a crossed-beam measurement on the ionization of Ba by electrons. They used a calibrated filtered photomultiplier to measure the yield of photons emitted from the collision region, and used the results of Chen and Gallagher (1976) for data normalization. Chen and Gallagher studied the resonance line of Ba I at 553.5 nm, and normalized the excitation cross section to the Bethe-theory prediction at high electron energies; the uncertainty in the absolute scale is estimated at about 5%. Dettmann and Karstensen (1982) used the relation

$$\sigma_z(E)/\sigma_{ex}(E) = r_z/r_{ex}, \tag{7-16}$$

where r_z is the rate of z-fold ionization and r_{ex} is the rate of excitation, both to be measured. In the experiment, the electron beam was pulsed periodically and the ions were separated for different charges by means of a time-of-flight analyzer. Measurements were made on the fraction of the ions that passed into the analyzer. A circular opening on a metal sheet in the path of the ions defined the aperture of the analyzer. First, the opening was shut, and the total charge on the first collector was measured; yet, different degrees of ionization were not distinguished. Second, in a separate experiment the ions passed through the opening, and were detected in a Faraday cup after the time-of-flight analysis. The sum of the intensities of all major ionic species thus obtained agrees within a few percent with the total charge determined in the first experiment, indicating that there is no appreciable loss or gain of charge in the analyzer. The experiment, like that of Cowling and Fletcher (1973), uses a pulse technique, so that the collision region is field-free at the moment of ionization because the electric field for ion extraction is applied when the electron beam is off.

Experiments with modulated crossed-beams have been carried out by Halle *et al.* (1981) for electron-impact ionization of U atoms. An absolute scale for the cross section was set by comparison of the electron impact with associative ionization on thermal collisions

$$U + O_2 \rightarrow UO_2^+ + e, \tag{7-17}$$

the cross section for which had been determined earlier [Halle *et al.* (1980)].

So far we have discussed various problems in crossed-beam experiments in which the target particles have thermal speeds and are in the ground electronic state. Now we take up experiments in which target particles have thermal speeds and are in a metastable (or otherwise long-lived) state. A typical example is the electron-impact ionization of the $2\,^3S$ and $2\,^1S$ states of He, studied by Vriens *et al.* (1968), Koller (1969), Long and Geballe (1970), Fite and Brackmann (1964), and Shearer-Izumi and Botter (1974). Problems in these experiments are difficult to overcome, as may be readily recognized from the large discrepancies in the results given by different workers [see Shearer-Izumi and Botter (1974)].

Vriens *et al.* (1968) used a crossed-beam apparatus, in which the atomic beam was crossed perpendicularly by two electron beams. A schematic drawing of the apparatus is seen in Fig. 7-5. The first electron beam produces the metastable states, and the second electron beam ionizes some of them. The ions thus produced are extracted by an electric field, accelerated, analyzed for mass, and finally detected with a Daly (1960) detector. Relative values of the ionization cross section were obtained at various electron impact energies. The metastable states were produced by electrons of $20.5 - 23$ eV, which may be compared with the threshold 19.82 eV for the $2\,^3S$ state and the threshold 20.3 eV for the $2\,^1S$ state. However, no difference was found in the relative ionization cross section with varying electron energies for the metastable production.

Fite and Brackmann (1964) studied metastable states of He, Ne, Ar, and O_2. A neutral beam emerging from a 30-MHz radio-frequency gas discharge contained the

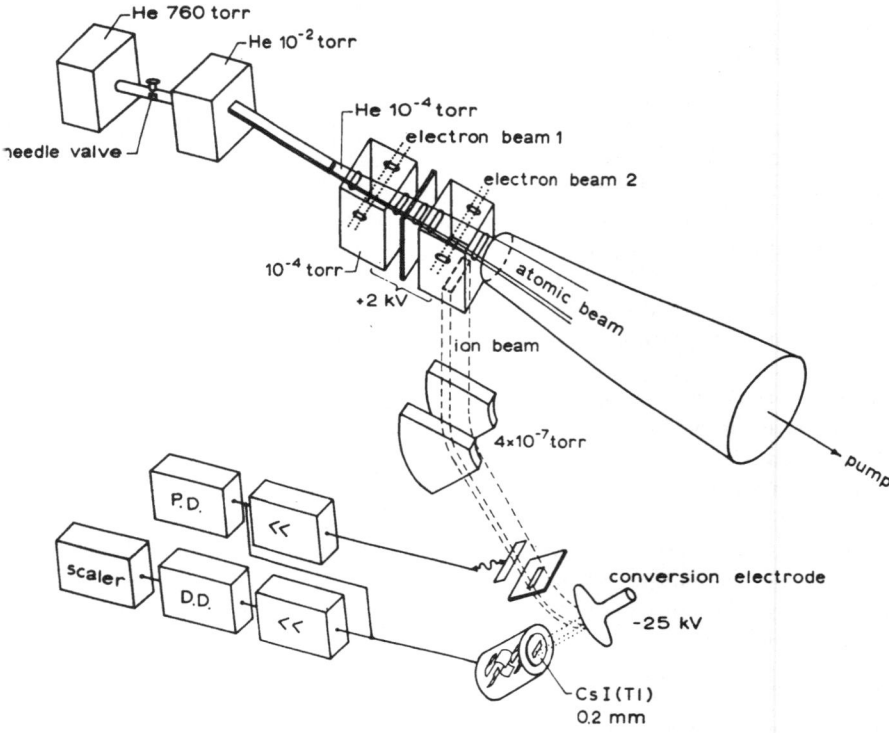

Fig. 7-5. Schematic diagram of the crossed-beam apparatus used by Vriens *et al.* (1968) for measurements of the ionization cross sections of metastable atoms

metastable states as well as the ground state. Thus, the electron-impact ionization of the metastable states could be studied unambiguously only at electron energies below the ionization threshold of the ground state. Only for He was it possible to use a surface electron-ejection detector to determine the number density of the metastable atoms [Stebbings (1957)], and in this way an approximate estimate of the absolute ionization cross section was obtained. Long and Geballe (1970) undertook a more difficult task of determining the ionization cross sections of the $2\,^3S$ state and the $2\,^1S$ state separately. The first electron beam crossed perpendicularly the neutral He beam, and the yield of metastable states was measured as a function of the electron impact energy, the metastable states being recorded on an electrode by means of secondary-electron emission. The result was compared with that of finer energy-resolution measurements by Schulz and Fox (1957) and by Fleming and Higginson (1964); eventually, Long and Geballe set the energy of the exciting electron beam at 20.4 eV to produce the $2\,^3S$ only. As in other experiments, a second electron beam was used to study the ionization of the $2\,^3S$ state. The He$^+$ ions produced were detected and analyzed by use of electron multiplier techniques as discussed by Shearer-Izumi and Botter (1974). Thus, the relative ionization cross section was determined as a function of the electron impact energy. The result was normalized on an absolute scale at 12.0 eV by use of data from earlier work [Fleming and Higginson (1964)]. Long and Geballe (1970) also discuss problems that may have given rise to the differences of their results from those of Fite and Brackmann (1958) and of Vriens *et al.* (1968).

In the latest work of this kind by Shearer-Izumi and Botter (1974), a thermal beam of He atoms is produced by a multichannel jet, and encounters two beams of electrons in succession. Electrons from an exciting electron gun (EE) pass collinearly through the neutral He beam, and excite some of the atoms to the 2^3S and 2^1S states. Electrons from an ionizing electron gun (IE) perpendicularly cross the atomic beam downstream and ionize some of the atoms. Electrons and ions generated by the EE gun are removed from the beam by an electrostatic field before it enters the region of collision with electrons from the IE gun. Ions generated in the second collision region are accelerated, magnetically analyzed, and then detected with a secondary-electron multiplier. The number of neutral excited atoms is monitored by another secondary-electron multiplier placed further downstream on the neutral-beam axis. This arrangement is suitable for measuring the ions produced by electrons with kinetic energies ranging from the threshold to about 22.6 eV (i.e., about 2 eV below the ionization energy of He in its ground state). The cross-section result thus obtained is relative (i.e., represents the relative dependence on the electron kinetic energy), and pertains to a mixture of the 2^3S and 2^1S states in an unknown proportion. During a study of background noises produced by the EE gun, it was noted that the signal began to increase abruptly as the EE energy exceeded 24 eV. This increase is attributable to the ionization of highly excited Rydberg states of He, which are long-lived; after this recognition, relative cross sections for the ionization of these species were measured.

Next, we discuss another kind of crossed-beam experiment in which the target particles are energetic (i.e., in the keV region), because they are produced by charge exchange if ions coming out of a source and passing through a cell. The first experiment of this kind was conducted by Petersen (1964), who studied the ionization of N, N_2, Ar, and Ne by electron impact. The intensity of the neutral beam was determined by stopping the particles in a thin nickel film and by measuring the temperature rise with a thermocouple. The detector was calibrated with the use of a corresponding ion beam; however, according to Kieffer and Dunn (1966), the calibration is subject to the assumption that the accommodation coefficient of the neutral on the surface be the same as that of fast ions, for which no experimental evidence is available. As pointed out by Dolder (1980), the method is superior to the thermal-beam method because full dissociation is accomplished to provide a pure beam of atoms and also because the flux of energetic neutral particles can be measured so that one dispenses with a normalization procedure as needed in the experiments of Fite and Brackmann (1958) and of Boyd and Green (1958), for example.

The method was improved by Harrison and co-workers [see, for instance, Brook *et al.* (1978)], who studied electron-impact ionization of He, C, O, and N for electron energies between 7 eV and 1 keV. The apparatus of Brook *et al.* (1978) is seen in Fig. 7-6. Ions from the source are selected for energy by the magnet M, and then pass through a charge-exchange cell filled with a gas, which is chosen to neutralize the ions with the minimum energy defect so that the production of metastable states is avoided and thus the neutral beam contains predominantly atoms in the ground state. For instance, He$^+$ may be neutralized by He (in which case the energy defect is zero), and C$^+$ by Kr (in which case the energy defect is 0.38 eV). Lower excited states of atoms (having the principal quantum number $n \leq 8$), apart from metastable states, decay before reaching the electron-collision region, while higher excited states (with $n \geq 13$) are field-ionized and are deflected from the beam by an electric field (see DP in Fig. 7-6).

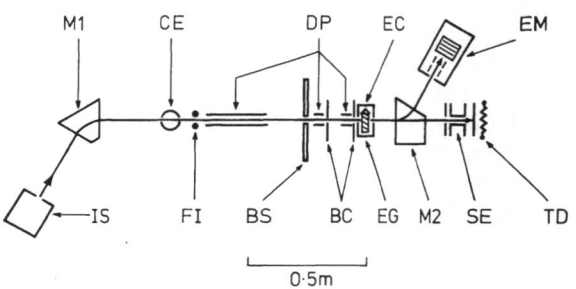

Fig. 7-6. Schematic diagram of the crossed-beam apparatus used by Brook *et al.* (1978). *IS* ion source. *M1* selector magnet, *CE* charge-exchange gas cell. *FI* field-ionizer wires. *DP* deflector plates, *BS* beam stop. *BC* beam collimators, *EG* electron gun, *EC* electron collector, *M2* analyzer magnet, *SE* secondary-electron detector, *TD* thermophile detector, *EM* electron multiplier

Ions formed in the collision region are analyzed by the magnet *M* 2, and are counted by a calibrated electron multiplier EM [Dance *et al.* (1967)], and the neutral-beam flux was measured by means of a thermopile TD. A beam-modulation technique was used to discriminate against background signals [Harrison (1968)]. The thermopile was subject to small thermal drift during the cross-section measurement. For providing an index of the drift, measurement was made of the current of secondary electrons produced by the neutral particles on a thin aluminum film that was in good thermal contact with the thermopile but was electrically isolated from it. Equal accomodation coefficients for ions and atoms were assumed for the purpose of calibration. The method proved to be reliable in the sense that it reproduced the result of Rapp and Englander-Golden (1965) for He at electron energies above 100 eV. This is discussed further in Chapter 5.

There are certain problems in the definition of the neutral beam. When one produces N from N$^+$, there may be contamination due to N$_2$ produced from N$_2{}^{2+}$ through double electron capture. This effect was probably negligible because very little N$_2^+$ was found in the electron-multiplier detector EM. Brook *et al.* (1978) showed, however, the presence of a small amount of longlived excited states in the neutral beam. With the He beam, for instance, they detected ionization at electron energies below the ionization threshold of the ground state. As suggested by Brook *et al.* (1978), this effect is attributable to long-lived Rydberg states of He ($8 \leq n \leq 12$), and not to metastable states. Thus, substantial corrections were applied to data near the threshold. For electron energies above the threshold, ionization due to the long-lived states was estimated under the assumption that the cross section for the process was proportional to T^{-1} over the entire range of electron energy T.

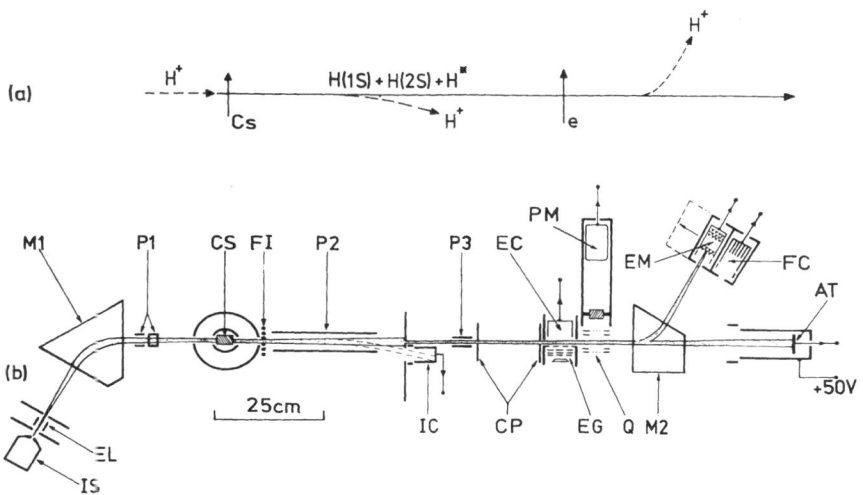

Fig. 7-7. The principle (a) and the schematic diagram (b) of the apparatus used by Dixon *et al.* (1975) for measurements of the ionization cross section of H (2 ^{2}S) by electron impact, *IS* duoplasmatron ion source, *EL* einzel lens, *M1* 60° electromagnet to select 2 keV protons, *P1* electrodes for beam-profile adjustment, *FI* field-ionizer wires, *P2* proton deflector plates (which generates an electrostatic field of 10 V/cm), *IC* proton collector, *P3* electrodes for a pulsed electrostatic quench field of 100 V/cm, *CP* beam-collimeter apertures, *EG* electron gun, *EC* electron collector, *Q* tubular electrodes for a quench field of a Lyman-alpha photomultiplier, *M2* 60° beam-analyzer electromagnet, *EM* proton detector, which can be moved so that the Faraday cup collector (*FC*) intercepts the product-proton beam, *AT* secondary-electron detector for monitoring the H-atom beam

Crossed-beam techniques have been used to study the ionization of fast beams of metastable atoms. Indeed, Harrison and co-workers at Culham started this kind of experiment before the work on ionization of ground-state neutrals [Brook *et al.* (1978)]. The first experiment dealt with ionization of the 2 ^{2}S *state of* H *by electrons.* Dixon *et al.* (1975) let protons pass through a cell filled with cesium to generate a beam of hydrogen atoms having a large fraction of the metastable state. Cesium was chosen because the charge-transfer reaction

$$H^+ + Cs \rightarrow H(2\,^2S) + Cs^+ \tag{7-18}$$

has a minute energy defect of 0.51 eV. The setup of Dixon *et al.* (1975) is seen in Fig. 7-7. Naturally, it is similar in part to the setup of Brook *et al.* (1978) seen in Fig. 7-6. However, additional corrections for excited species in the beam were necessary. Highly excited atoms produced after charge exchange were field-ionized (as indicated by "FI" in Fig. 7-7), and ions were deflected from the beam by a weak electric field across plates (as indicated by "P2" in Fig. 7-7). Electrons from the gun (EG) ionized some of the H(2 ^{2}S) atoms, and the resulting protons were detected at the Faraday-cup collector (FC). The H(2 ^{2}S) component of the beam was monitored with quench plates (Q) and with a photomultiplier (PM) that detected Lyman-alpha photons emitted. To separate the ionization of H(2 ^{2}S) from the ionization of the ground state, the 2 ^{2}S component in the beam was modulated by use of a pulsed quench field, and the protons resulting from ionization by electron impact were detected with a phase-sensitive technique. The total intensity of the beam including both the $2\,{}^2$S state and the ground state was determined by means of secondary-electron emission from an aluminum surface. This determination relies on the ratio of the secondary-electron yield for hydrogen atoms to that for protons, as measured by Chambers (1966). It is implicitly assumed here that the $2\,{}^1$S state has the same secondary-electron emission yield as the ground state, which is highly questionable. Later, Dixon et al. (1975) measured the secondary-electron yield for fast atoms by use of a sensitive thermopile. There remains the question as to what is the fraction, f, of the metastable state in the beam. Dixon et al. (1975) used the value $f = 0.25 \pm 0.01$ as determined earlier by Donnally and O'Dell (1971), by Spiess et al. (1972), and by Schlachter et al. (1973).

Dixon et al. (1973, 1976) extended the work to metastable He (mainly the $2\,{}^3$S state), Ne $({}^3P_2, {}^3P_0)$, and Ar $({}^3P_2, {}^3P_0)$. Again, the fraction of metastables in the neutral beam had to be estimated. For the preparation of a He beam, the collision of He$^+$ with Cs was used. It was assumed on the basis of some evidence [Dixon et al. (1976)] that the fraction of metastables (all in the $2\,{}^3$S state) was 0.8 and the fraction of the ground state was 0.2. The signal for electron-impact ionization was composed of contributions from the metastable state and those from the ground state. For deconvolution of the signal into the two contributions, use was made of the ionization cross section of the ground state, determined with He in the charge-exchange cell; the results were apparently consistent within 2% with those of Rapp and Englander-Golden (1965).

7.2 Data: Survey and Critique

7.2.1 General

Total and partial ionization cross sections have been measured for a great number of atomic and molecular species by many workers, but the results for the same species do not always agree. How do we decide which of those results is right, or is more likely to be right?

One might be tempted to answer this question by comparison with theory. However, the comparison is not at all straightforward because different theoretical methods lead to widely different values. This is true for most of the cross sections for atomic and molecular collisions in general; but it is especially so for ionization cross sections for any atomic or molecular species by electron impact. In other words, we have no universal scheme for theoretically predicting the ionization cross section at all electron energies, as may be seen from discussions in Chapters 1 and 2 of the present volume.

Whenever one examines cross-section data concerning ionization, one should always distinguish many different kinds of quantities. In the very beginning of the present chapter, we have discussed the meanings of terms such as absolute, relative, total, partial, gross, counting, single, double, and multiple (each of which may become and adjective modifying the term "ionization"). Clear perception of all these terms is important in the use of any of the data in the literature; misunderstanding of the meaning of the data has caused serious confusion in the minds of data users, and also unproductive controversies or useless polemics.

Another point one should keep in mind is the increasing volume of ionization cross-section data and the resulting need for data management. Owing in part to the strong demand for the data in many applications including astrophysics, atmospheric research, plasma physics, and radiation physics (as seen in Chapter 9), many data centers have been established. The data centers collect the cross-section values from the literature, and make them available for use by the public in various forms. Some of the data centers not only compile the data, but also conduct critical assessment of the data; thus, they issue tabulations and graphs showing the cross-section values that are deemed most reliable or worth recommendation for use in applications. Different data centers have different goals and emphasis, and treat different classes of cross-section data. However, the total cross section for ionization by electron impact is of interest in many applications, and thus is being treated by many data centers. Table 7-1 is a listing of some of the better known data centers.

Table 7-1. *Some of the data centers that treat ionization cross sections*

Name and address	Newsletter or publications	Remarks
Atomic Collision Information Center Joint Institute for Laboratory Astrophysics University of Colorado Boulder, Colorado 80309, U.S.A.	JILA Information Center Report	Atomic collisions in general, with emphasis on atoms and small molecules and on astrophysical applications
Controlled Fusion Atomic Data Center Oak Ridge National Laboratory Oak Ridge, Tennessee 37830, U.S.A.		Applications to fusion research
Atomic and Molecular Data Unit Nuclear Data Section International Atomic Energy Agency A-1400 Vienna, Austria	International Bulletin on Atomic and Molecular Data for Fusion (4 times a year)	Applications to fusion research
GAPHYOR, Laboratoire de Physique des Gaz et des Plasmas Université Paris-Sud, F-91405 France, Orsay	GAPHYOR (Bulletin Signaletique 166) (4 times a year)	Gas and plasma physics in general
Research Information Center Institut of Plasma Physics Nagoya University Nagoya 464, Japan	IPPJ-AM Reports	Applications to fusion research
Belfast Database on Atomic and Molecular Physics Department of Computer Science Queen's University Belfast, BT7 1NN Northern Ireland		Atomic and molecular data in general

The foregoing discussion should have conveyed an idea of the many-fold difficulties in ever recommending a particular set of cross-section data as "best". By the term "best" we mean here the most reliable among published results, as judged from the current status of knowledge including theory and experiment. There are many approaches to arrive at the best values. One way is to follow Kieffer and Dunn (1966), who based their judgement largely on experimental grounds. (While they pointed out strengths and weaknesses of experiments, they never actually ended up with "best" values.) Another way is to use general theoretical criteria, as will be explained in the next section.

7.2.2 *The Use of the Bethe Theory and Other Theoretical Constraints*

One of the general and simple ways to examine the ionization cross section is to use some results of the Bethe theory [Bethe (1930, 1932, 1933), Inokuti (1971)]. This approach was first advanced by Platzman [as discussed by Fano (1975)], and is discussed in detail by Inokuti (1971, 1980). In this approach, one chooses a target gas and collects all available experimental and theoretical results concerning inelastic scattering of electrons. According to the Bethe theory, cross sections for various inelastic scattering processes are related among themselves, and are also connected with other properties of the target species. This makes it possible to test consistency of the cross-section data among themselves or with other independent information, either theoretical or experimental.

One of the results of the Bethe theory concerns the high-energy asymptotic behavior of the cross section σ_n for the excitation to a given state n, or of the total ionization cross section σ_t, as a function of the electron impact energy T. The excitation cross section may be written as

$$\sigma_n = A_n T^{-1} \ln T + B_n T^{-1} + C_n T^{-2}, \tag{7-19}$$

where A_n, B_n, and C_n are properties of the target species. In particular A_n is equal to f_n/E_n apart from a universal constant, where f_n is the optical (dipole) oscillator strength for the transition to state n and E_n is the excitation energy for that transition. The universal constant is $4 \pi a_0^2 R^2$, where a_0 is the Bohr radius and R is the Rydberg energy. The first term on the right-hand side of Eq. (7-19) is dominant at high impact energies, and represents the contributions from glancing collisions, which are closely related to photoabsorption. The coefficients B_n and C_n depend on close collisions, and thus are not reducible to f_n alone.

As Fano (1954) first suggested, it is good to plot $T\sigma_n$ against $\ln T$ using a given set of data. For sufficiently high T, one should then obtain a straight line; its slope gives A_n, and its intercept with either axis determines B_n. Fig. 7-8 includes an example of the Fano plot, for the ionization of Ar by electrons. It also shows two other ways of plotting the same data. For electron kinetic energies not negligible compared to the electron rest energy $m_e c^2 = 511 \text{ keV}$, one should use a relativistic expression given by Eqs. (7-11)−(7-13), instead of the non-relativistic expression, Eq. (7-19).

The ionization cross section σ_t is also given in the same analytic form as Eq. (7-19), or its relativistic version, Eqs. (7-11)−(7-14). That is to say,

$$\sigma_t = A_{\text{ion}} T^{-1} \ln T + B_{\text{ion}} T^{-1} + C_{\text{ion}} T^{-2}, \tag{7-20}$$

where the coefficient of the first term is given by

$$A_{\text{ion}} = 4 \pi a_0^2 R M_{\text{ion}}^2, \tag{7-21}$$

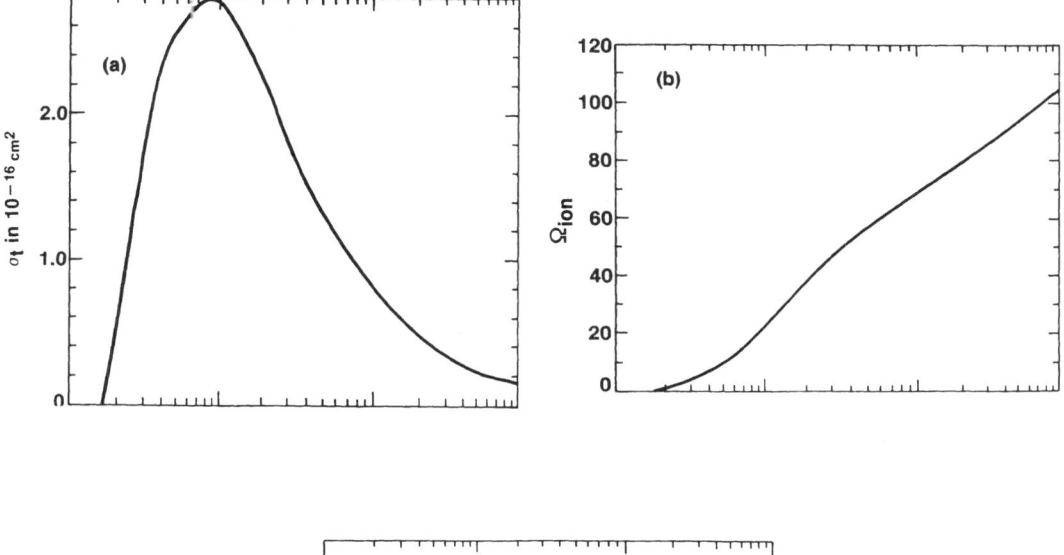

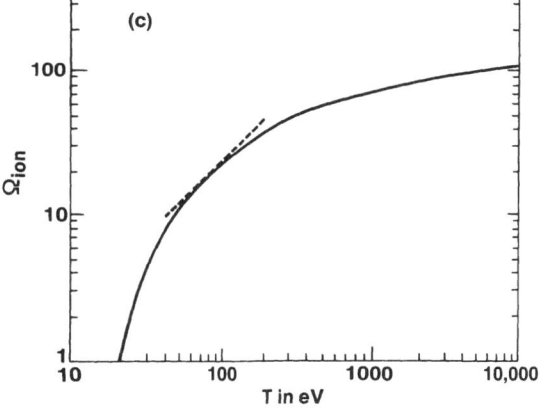

Fig. 7-8. Three plots of the total (gross) ionization cross section σ_t of Ar for electron collision. The data are taken from de Heer and Jansen (1975), who critically examined various experimental results and arrived at recommended values of σ_t and of other cross sections. The horizontal axis is common to the three plots, and shows the incident kinetic energy T in eV on a logarithmic scale. In contrast, the vertical axis differs for each plot. In plot (a), σ_t itself in untis of 10^{-16} cm^2 is shown on a linear scale. One sees the familiar bell-shaped curve with a maximum at $T \cong 90$ eV. In plot (b), the vertical axis represents the collision strength for total ionization $\Omega_{ion} = (T/R)(\sigma_t/\pi a_0{}^2)$. The curve of Ω_{ion} versus log T gradually rises from zero at threshold, and approaches the straight-line asymptotic behavior in accordance with Eq. (24). The asymptotic slope corresponds to a theoretically expected value $M_{ion}{}^2 = 3.50$ [Kim et al. (1973)] within several percent. Finally, plot (c) shows Ω_{ion} now on a logarithmic vertical axis whose scale is the same as the horizontal axis. The curve rapidly rises near threshold, and becomes less and less steep at higher and higher T, but never flattens out. The continuous curve must take a tangent making 45° with either axis at a point, at which σ_t attains the maximum (at $T \cong 90$ eV in this example). The tangent is indicated by the broken line. This figure is taken from Fano and Inokuti (1976)

where M_{ion}^2 is the dipole matrix element squared for total ionization (measured in a_0^2). To be more precise, M_{ion}^2 is given in terms of the oscillator-strength density df/dE per unit range of excitation energy E, which in turn is related to the absorption cross section $\sigma_{ph}(E)$ of the same target species for a photon of energy E as

$$df/dE = (m_e c/2\pi e^2 \hbar)\sigma_{ph}(E). \tag{7-22}$$

There is another element that enters into the precise definition of M_{ion}^2, namely, the ionization efficiency, $\eta_{ion}(E)$, which means the total number of electrons released upon transfer of energy E to the target species. As Platzman (1962, 1963) first pointed out, $\eta_{ion}(E)$ is often less than unity, especially at E not greatly exceeding the (first) ionization threshold I. In terms of df/dE and $\eta_{ion}(E)$ thus defined, one may express M_{ion}^2 as

$$M_{ion}^2 = \int_I^\infty (R/E)(df/dE)\eta_{ion}(E)\,dE. \tag{7-23}$$

Inokuti (1971) writes the ionization cross-section in the form

$$\sigma_t = (4\pi a_0^2 R/T)M_{ion}^2 \ln(4c_{ion}T/R), \tag{7-24}$$

where the use of the notation c_{ion} follows the tradition since Bethe (1930); it is related to A_{ion} and B_{ion} of Eq. (7-20) as

$$\ln(4c_{ion}/R) = B_{ion}/A_{ion}. \tag{7-25}$$

It is also useful to consider the total inelastic-scattering cross section $\sigma_{tot.\,inel}$ defined as the sum of σ_c and all the excitation cross sections σ_n [Inokuti et al. (1967)]. One may write

$$\sigma_{tot.\,inel} = (4\pi a_0^2 R/T)M_{tot}^2 \ln(4c_{tot}T/R), \tag{7-26}$$

where

$$M_{tot}^2 = M_{ion}^2 + \Sigma_n M_n^2, \tag{7-27}$$

$$M_n^2 = (R/E_n)f_n, \tag{7-28}$$

and c_{tot} is another constant representing a target property. If the ground state-wave function is sufficiently well known, one can calculate M_{tot}^2 through a sum-rule expression

$$M_{tot}^2 = \Sigma_j \Sigma_k \langle x_j x_k \rangle/a_0^2 - (\Sigma_j \langle x_j \rangle)^2/a_0^2, \tag{7-29}$$

where x_j is a Cartesian component of the j-th electron in an atom or molecule, and the bracket $\langle\ \rangle$ represents the ground-state expectation value. So long as the atom or molecule is randomly oriented, one implies in Eq. (7-29) an average over the orientation. Then, for an atom or a nonpolar molecule, the second term vanishes; however, for a polar molecule, it is nonvanishing. The other constant c_{tot} in Eq. (7-26) is also calculable to some accuracy, as seen in Liu (1973), Saxon (1973), Inokuti et al. (1975), and Inokuti et al. (1981).

The total inelastic-scattering cross section is sometimes amenable to direct measurement. Examples may be seen in the measurements on He by Rieke and

Prepejchal (1972) and in the measurements on CF_4 and other fluoroalkanes by Winters and Inokuti (1982).

An alternative method of evaluation of M_{tot}^2 or M_{ion}^2 is to use the oscillator-strength distribution (i.e., df/dE as a function of E and a set of E_n and f_n values) derived from measurements. They include photoabsorption and photoionization cross sections, now available for many atoms and molecules, as seen in Berkowitz (1979), Starace (1982), and Samson (1982), for example. Another source of data on the oscillator-strength distribution is the energy-loss spectra for the forward scattering of fast electrons. The oscillator-strength distribution is also subject to many other sum rules. In other words, some values of the moments

$$S(\mu) = \Sigma_n (E_n/R)^\mu f_n + \int_I^\infty (E/R)^\mu (df/dE) \, dE \qquad (7\text{-}30)$$

are derivable from theory or from independent measurements. The best known example is the Thomas-Kuhn-Reiche sum rule, i.e., $S(0) = N$, N being the total number of electrons in an atom or molecule. Another example is $S(-1)$, which is identical with M_{tot}^2 by definition, and may be expressed as Eq. (7-29). Further, $S(-2) = \alpha_d/(4 \, a_0^3)$, where α_d is the electrostatic dipole polarizability, which is usually known for common atoms and molecules. Berkowitz (1979) has discussed the values of $S(\mu)$ as evaluated from the oscillator-strength spectra for many atoms and molecules.

An earlier study by Berkowitz *et al.* (1973) concerned M_{ion}^2 values of molecules. One problem in the evaluation of Eq. (7-23) is that sometimes $\eta_{ion}(E)$ is known only poorly or in a limited range of E. Under such circumstances, one may evaluate

$$M_I^2 = \int_I^\infty (R/E) \, (df/dE) \, dE, \qquad (7\text{-}31)$$

which serves as an upper limit on M_{ion}^2. For atoms $\eta_{ion}(E)$ may be set equal to unity below the double ionization threshold in general. Although Platzman (1962, 1963) was right in first pointing out that $\eta_{ion}(E) < 1$, recent data tend to indicate that $\eta_{ion}(E)$ is much smaller than unity only in a small range of E immediately above the first ionization threshold, as seen for instance in Tan *et al.* (1978).

In discussing $\eta_{ion}(E)$ and M_{ion}^2, one must make the distinction between the gross ionization and the counting ionization, which we referred to in Section 7.1.1. If one follows precisely the definition of $\eta_{ion}(E)$ (in terms of the total number of electrons) we made below Eq. (7-22), then the resulting M_{ion}^2 of Eq. (7-23) will pertain to the gross ionization cross section. If we define $\eta_{ion}(E)$ as the probability that ionization of any multiplicity takes place upon energy transfer E, then the resulting M_{ion}^2 will pertain to the counting ionization cross section.

Examples of the analysis with the use of sum rules may be seen in the work of Gerhart (1975) and of Douthat (1979) on H_2 and in the work of Eggarter (1975, 1976) on Ar.

Finally, we should note that the appropriate use of the Bethe theory demands a great deal of mature judgement. In the first place, the evaluation of M_{ion}^2 and other quantities that characterize the ionization cross section from first principles requires good knowledge of atomic or molecular eigenfunctions in the initial state and in the final state. However, this knowledge is available only under certain special circumstances, as pointed out by Inokuti (1971) with reference to many examples. In addition, the range of validity of the first Born approximation has not been completely elucidated. In other words, although we are certain that the first Born approximation (with the use of good enough atomic or molecular eigenfunctions) should give right values of cross sections at sufficiently high impact electron energies, we do not know very well how departures from that approximation occur as the impact electron energy goes down. Inokuti (1971) discussed this issue in some detail in his Section 5.2. Bell and Kingston (1974) also show extensive data concerning the same issue, and present many useful remarks.

In the use of the Fano plot, some caution is necessary. Quite often, cross-section data result in a straight-line behavior in the Fano plot, at electron impact energies below the true asymptotic region. In other words, the Fano plot should show a straight-line behavior at high enough electron-impact energies, but the plot may show a straight-line behavior at much lower electron impact energies as well with a slope much greater than that represented by the correct value of the M_{ion}^2 value. An example may be seen in plot (b) of Fig. 7-8. The curve behaves like a straight line for electron energies between 100 eV and 400 eV. However, this is not the true asymptotic behavior, which occurs at electron energies above a few keV. For more detailed discussion, see Section III C of Winters and Inokuti (1982).

Another remark may be useful about departures from the true asymptotic behavior of the Fano plot. There are in general many reasons for the departures as one goes down from high energies to low energies. For the total ionization cross section (and for the total inelastic-scattering cross section) in particular, the initial departures seem to be negative; that is to say, the data points on the Fano plot begin to lie on the lower side of the asymptotic straight line. Moreover, the resulting curve is convex upwards. This trend is true in every Fano plot for the total ionization cross section, to the best of our knowledge. Most likely, the reason for this trend is that the dominant cause for the initial departure is the electron-exchange effect in close collisions. This effect arises when an incident electron transfers a large amount of momentum to an atomic or molecular electron. The contribution of this effect to the total ionization cross section may be estimated by use of the Mott formula (which applies to collisions of two free electrons) with minor modifications to account for some binding effects. As Kim and Inokuti (1971) show in several examples, the exchange effects thus estimated gives about the right magnitude and the trend of the initial departure. Notice that the foregoing discussion applies strictly to incident electrons. In other words, the departures from the asymptotic behavior should be different for positrons, protons, and other charged particles; this expectation is supported by much evidence. Finally, the role of the electron-exchange effect in close collisions is more important for stopping power, as first shown by Bethe (1932, 1933).

7.2.3 Atoms

The Belfast Database on Atomic and Molecular Physics has started providing recommended data in a paper by Bell *et al.* (1983), who treated the single ionization cross sections for the isolelectronic sequences of hydrogen through oxygen atoms. The authors begin with high impact energies at which the Born approximation (for neutral atoms) or the Coulomb-Born approximation (for ion targets) should give reliable results provided accurate wave functions are used for both initial and final states of the target. At lower energies, neither approximation is valid and experimental data are the main basis for recommended values. For establishing a connection between low-energy data and high-energy data, the elements of the Bethe theory as outlined in the previous section are often useful. Indeed, when different experiments yield conflicting cross sections, Bell *et al.* (1983) adopt the data set that agrees best with theory at high energies. For cases in which no experimental result was available, they used theoretical results.

For each target species the recommended cross section has been fitted to the expression

$$\sigma_t = (I\,T)^{-1}\left[a\ln\left(T/I\right) + \sum_{i=1}^{N} b_i(1-I/T)^i\right],\tag{7-32}$$

where I is the first ionization threshold energy, b_i are coefficients determined through a least-squares fitting, a is a quantity closely related to A_{ion} of Eq. (7-20), and N is an integer not exceeding 6. The expression gives cross sections behaving reasonably both at high energies and at low energies. For many instances, two or three coefficients appear to be sufficient. Near the threshold energy, the expression is approximately proportional to the excess energy $T-I$. This agrees with the theoretical threshold law apart from electron correlation effects that are important only in the close vicinity of the threshold and for neutral atoms, as discussed by Read in Chapter 3. Furthermore, Eq. (7-32) conforms to the scaling rule that $I^2\sigma$ be a function of the reduced variable T/I, which follows certain general considerations [Tawara *et al.* (1973)].

The review of data by Bell *et al.* (1983) is extremely useful for applications such as plasma modeling in controlled thermonuclear fusion research, as discussed by Katsonis and Lorenz (1982). For neutral atoms, the recommended data are claimed to be accurate to 5%, except for Li (10%), Be (20%), and B (20%).

De Heer and co-workers (de Heer *et al.* 1977, de Heer and Jansen 1977, and de Heer *et al.* 1979) analyzed experimental data for noble-gas atoms and the hydrogen atom through a different procedure. They were motivated to the work by studies on dispersion relations for electron-atom scattering cross sections. In particular, they needed to determine the total cross section σ_{tot} for scattering (including both elastic and inelastic scattering) of an electron with the atoms at all impact energies. The total cross section for each atom has been evaluated for $0 < T \le 3\,\mathrm{keV}$ through an extensive analysis of experimental and theoretical data, which include the elastic scattering cross section σ_{el}, the total ionization cross section σ_t, the total excitation cross section σ_{exc}, as well as differential cross sections for elastic and inelastic scattering.

Table 7-2. *Total cross sections for ionization, σ_t, of He by electrons in units of a_0^2. The numbers in parentheses are the total errors in the last significant digits*

E(eV)	Experimental average de Heer and Jansen (1977)	Recommended Bell *et al.* (1982)	Theory Kim and Inokuti (1971)
30	0.2321 (86)	0.2353	
40	0.6030 (225)	0.5880	
50	0.8445 (315)	0.8189	
60	0.9984 (315)	0.9793	
70	1.097 (46)	1.095	
80	1.175 (44)	1.178	
90	1.211 (56)	1.235	
100	1.229 (58)	1.273	1.417
150	1.191 (92)	1.297	1.414
200	1.126 (51)	1.217	1.286
300	0.9437 (468)	1.031	1.049
400	0.8084 (416)	0.8815	0.8796
500	0.7023 (399)	0.7676	0.7580
600	0.6210 (362)	0.6798	0.6671
700	0.5592 (330)	0.6105	0.5966
800	0.5102 (308)	0.5544	0.5403
900	0.4681 (513)	0.5082	0.4943
1000	0.4300 (273)	0.4694	0.4559
2000	0.2403 (192)	0.2712	0.2625
3000	0.1710 (124)	0.1938	0.1877
4000	0.1346 (85)	0.1520	0.1473

For He for instance, σ_{tot} was determined to an accuracy of perhaps 5% by adding experimental values of σ_{el}, σ_t, and σ_{exc}, chosen after careful assessment of the data reliability. Indeed, when σ_{tot} was later measured as seen in the work by Blaauw et al. (1980) and by Wagenaar and de Heer (1980), the measured result was found to be consistent with the earlier semi-empirical determination by de Heer and Jansen (1977) and de Heer et al. (1979). From this finding, it is probably safe to conclude that the experimental value of σ_t used in the analysis should have been correct. This is so because σ_t for those atoms (and indeed for many other atoms and molecules) is a dominant fraction of σ_{tot}; thus, any appreciable errors in σ_t must have led to appreciable errors in σ_{tot}. The total ionization cross section σ_t for He was determined by de Heer and Jansen (1977) and de Heer et al. (1979) by using experimental results of six groups of workers together with stated errors (5–10%). Table 7-2 shows the average of the experimental values derived from a procedure described by Langenberg and van Eck (1976); for the calculation of the average, use was made of weights depending on the stated errors. The calculation led to estimates of external errors and internal errors. Table 7-2 includes the larger of the two estimates at each T which ranges from about 4% to 10%. Table 7-2 shows the gross ionization cross section as given by Eq. (7-2), whereas the counting ionization cross section as given by Eq. (7-3) or the single ionization cross section may be needed for comparison with theory or for some application. However, for He, the gross ionization cross section differs from the counting ionization cross section by less than one percent, as seen in de Heer and Jansen (1977). The theoretical value represents the single ionization cross section evaluated by Kim and Inokuti (1971) within the Born approximation but with allowance for electron-exchange effects. Above 500 eV, where the theory should begin to be valid, experimental data differ from theory by 5.8% – 10.9%, i.e., by an amount comparable with estimates of errors in the experimental data. Table 7-2 also shows the cross-section values recommended by Bell *et al.* (1983), which are due to Montague *et al.* (1983) and represent the result of a smooth fit to a large number of measurements. At 750 eV, the values merge smoothly with the Born-approximation results by Bell and Kingston (1969).

Let us now discuss briefly the analysis of de Heer *et al.* (1979) for other noble gases. The averaging procedure according to Langenberg and van Eck (1976) allows one to include relative cross sections in the analysis. De Heer *et al.* (1979) had to take this approach because no estimate of the correction was made in this work for the so-called Ishii effect in the gas pressure measurement with the McLeod gauge, as we discussed in Section 7.1.2. [See Ishii and Nakayama (1961) and de Vries and Rol (1965).] For noble gases other than He, the gross ionization cross section appreciably differs from the counting ionization cross section. For comparison with theory, the single ionization cross section was needed; it was derived from the gross ionization cross section and the ratios of the partial ionization cross sections at each electron energy. Table 7-3 illustrates the difference between the gross ionization cross section and the counting ionization cross section; it also shows for comparison the Bethe cross section given by Kim *et al.* (1973).

Table 7-3. *Total cross sections for ionization (in units of a_0^2) for electrons incident on Ar. The numbers in parentheses represent the total errors in the last significant digits [de Heer et al. (1979)]*

E (eV)	$\sigma_{\text{gross ion}}$ experimental average	$\dfrac{\sigma_{\text{count ion}}}{\sigma_{\text{gross ion}}}$	$\sigma_{\text{count ion}}$ experimental	$\sigma_{\text{count ion}}$ theory Kim *et al.* (1973)
20	2.299 (63)	1	2.299 (63)	
30	6.565 (174)	1	6.565 (174)	
40	8.797 (252)	1	8.797 (252)	
50	9.471 (304)	0.995	9.424 (304)	
60	9.756 (278)	0.986	9.619 (290)	
70	10.32 (30)	0.977	10.08 (31)	
80	10.31 (32)	0.968	9.980 (326)	
90	10.50 (30)	0.959	10.07 (31)	
100	10.21 (26)	0.9498	9.697 (265)	18.72
150	9.468 (283)	0.9396	8.896 (280)	14.09
200	8.566 (256)	0.9353	8.012 (252)	11.43
300	6.999 (255)	0.9342	6.538 (247)	8.427
400	5.942 (232)	0.9305	5.529 (223)	6.751
500	5.126 (222)	0.9297	4.766 (212)	5.668
600	4.380 (164)	0.9280	4.065 (159)	4.906
700	3.926 (147)	0.9278	3.643 (141)	4.337
800	3.573 (138)	0.9275	3.314 (132)	3.894
900	3.297 (134)	0.9277	3.059 (128)	3.539
1000	3.036 (130)	0.9269	2.814 (124)	3.248
2000	1.741 (77)	0.9225	1.606 (73)	1.831
3000	1.267 (63)	0.9216	1.168 (59)	1.301
4000	0.9854 (488)	0.9197	0.9063 (458)	1.019

Another approach for the presentation of inelastic-collision cross sections including the total ionization cross section has been taken by Green and co-workers [see, for instance, Green and Dutta (1967), Jusick *et al.* (1967), and Miles *et al.* (1972)]. They used analytical expressions for the energy spectrum of the secondary electrons resulting from an ionizing collision. The energy-spectrum expression was determined in part to conform with the Bethe-theory result. The integral of the energy spectrum gives the total ionization cross section, which was also expressed in an analytic form. At lower energies, modifications for departures from the first Born approximation was introduced, again in terms of an analytic expression. The procedure is comparable to some extent with that of Bell *et al.* (1983).

Kieffer and Dunn (1966) gave a critical review of experimental data on the total ionization cross section, published up to the time of their work. Their Table 1 summarizes experimental techniques and the necessary conditions under which they lead to accurate ionization cross-section values. Table 2 of Kieffer and Dunn (1966) shows how well these conditions were met in experiments. No theoretical criterion was used to assess the data reliability. Kieffer and Dunn (1966) presented many graphs showing results reported by different groups, and pointed out some of the data sets as unreliable. Discussion there concerns not only the total ionization cross section but also the ratios of partial ionization cross sections. The data up to that time are extensively compiled and numerically presented in Kieffer (1965, 1966). More recent data are seen in the reports of the Information Center of the Joint Institute for Laboratory Astrophysics (Kieffer, 1976; Gallagher *et al.*, 1979; Gallagher and Beaty, 1980; Gallagher and Beaty, 1981).

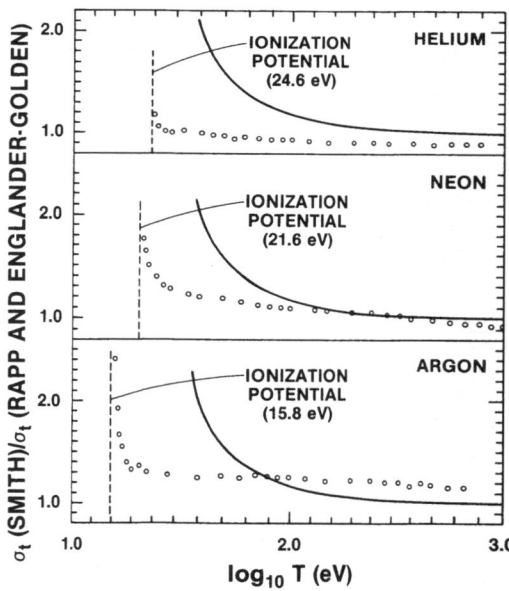

Fig. 7-9. Comparison of the data of Smith (1930) with those of Rapp and Englander-Golden (1965), for the total ionization cross sections of He, Ne, and Ar for electron impact. The solid curve shows the maximum ratio expected according to Eq. (7-6), if the only systematic error in either experiment was a pathlength error in Smith's. This figure was taken from Kieffer and Dunn (1966)

For He, Ne. and Ar, Kieffer and Dunn (1966) compared the two major sets of data, viz., one by Smith (1930) and the other by Rapp and Englander-Golden (1965), as seen in Fig. 7-9. According to Kieffer and Dunn (1966), an increase in the pathlength in Smith's apparatus due to the transverse velocity component might have led to systematic errors increasing with lower impact energies. For Ar, however, the difference between the two data sets is larger than the maximum possible as a consequence of the pathlength effect. It is probable that much of the difference is due to the Ishii effect in the pressure measurement in Smith's experiment, as we discussed earlier. The presence of the Ishii effect is likely also in the data by other groups of workers, e.g., Tozer and Craggs (1960) and Asundi et al. (1963); indeed, their data show a constant difference from the data of Rapp and Englander-Golden (1965) over a wide range of impact energy. The data by Schram et al. (1965) for $0.6 - 20$ keV are generally lower than the data by other groups. Although there is no clear indication of systematic errors, analysis suggests that the cross sections reported by Schram et al. (1965) are probably too low, especially for lighter species such as He, H_2, and Ne. Kieffer and Dunn (1966) then conclude that the cross sections for the noble gases are no more accurate than $20 - 30\%$. However, the analysis by de Heer and co-workers (1977, 1979), which we discussed earlier in the present section, tends to indicate that the cross sections for the noble gases should be accurate to $5 - 10\%$.

Kieffer and Dunn (1966) also discuss the data for atomic hydrogen by Fite and Brackmann (1958), by Boksenberg (1961), and by Rothe et al. (1962). Here the main problem concerns the procedure by which the data were put on an absolute scale. Fite and Brackmann (1958) used two methods for data normalization. The first is to normalize the data to the result of the first Born approximation at 500 eV. The other is to compare signals due to H^+ and H_2^+ in a mass spectrometer resulting from the target beam containing H and H_2 at presumably known proportions and to attribute the H_2^+ signal to the ionization of H_2, the cross section for which was taken as known from the measurement by Tate and Smith (1932). Results of the two methods agreed very well. Boksenberg (1961) normalized his results to the first Born approximation at 300 eV. Rothe et al. (1962) paid special attention to the complete collection of ions including energetic protons resulting from dissociative ionization, and used for normalization the data of Tate and Smith (1932). This procedure assumes that Tate and Smith (1932) also accomplished complete ion collection; however, Kieffer and Dunn (1966) consider this assumption questionable. Notwithstanding, most of the data on atomic hydrogen are close among themselves. Bell et al. (1983) recommend the data by Fite and Brackmann (1958) on some theoretical grounds.

Kieffer and Dunn (1966) also discuss atomic nitrogen and oxygen. Smith et al. (1962) used a pulsed dc discharge, and measured the total ionization cross section of atomic nitrogen. They compared signals due to N and N_2, and normalized the N signal by use of the cross section for N_2 measured by Tate and Smith (1932). Peterson (1964) used a crossed beam of fast N atoms, and determined the ionization cross section for N absolutely. He selected the ions by use of a mass spectrometer and thereby determined the single ionization cross section. Therefore, the cross section reported by Peterson (1964) should be smaller than that by Smith et al. (1962). Actually, Peterson's results are larger by a factor of about two; therefore, they are most likely to be incorrect. Indeed, the cross section $\sigma_t(N)$ for atomic nitrogen by Smith et al.

(1962) and the cross section $\sigma_t(N_2)$ by Tate and Smith (1932) are related to each other by $2\sigma_t(N) \cong \sigma_t(N_2)$ at impact energies around the cross-section maximum. This is an example of the common observation referred to as the additivity relation between atomic and molecular cross sections, which is discussed further in Section 7.2.5, and in Chapter 2. Kieffer and Dunn (1966) present also other evidence indicating that Peterson's data are too high. At the same time, questions may be raised about the accuracy of the results of Smith et al. (1962); the data normalization with the use of the data of Tate and Smith (1932) should have introduced some influence of the Ishii effect. More recently Brook et al. (1978) repeated measurements of the ionization cross section for N using Peterson's method but avoiding some of the instrumental problems in Peterson's apparatus. Fig. 7-10 shows various experimental data and also theoretical results obtained from the first Born approximation or its variants. The single ionization cross section obtained by Brook et al. (1978) differs from the result of Smith et al. (1962) only by 7% at the peak, and may be considered to be the most reliable [see Bell et al. (1983)].

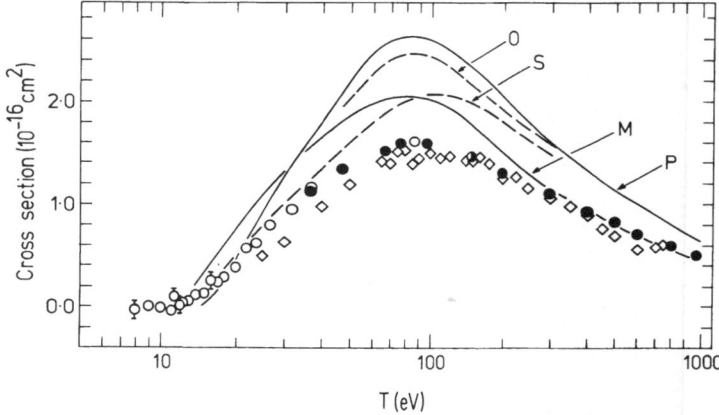

Fig. 7-10. The ionization cross section of the nitrogen atom in the ground state. The results obtained by Brook et al. (1978) using 2-keV atomic beams are shown by open circles (O), and those obtained by the same workers using 4-keV atomic beams are shown by filled circles (●). The error bars represent 90% confident limit and are attached to data points only when the error limits exceed the symbol size. The results of Smith et al. (1962) are shown by open diamonds (◇). The curves show theoretical results. M McGuire (1971), O Omidvar et al. (1972), P Peach (1970, 1971), S Seaton (1959). The figure is taken from Brook et al. (1978)

As for O, Kieffer and Dunn (1966) point out the agreement between the results of Fite and Brackmann (1959) and those of Rothe et al. (1962), although there is some room for doubt about the complete collection of ions. There are some indications that the cross sections given by Boksenberg (1961) are too large. All these measurements were relative, and the results were normalized to the cross-section values for O_2 reported by Tate and Smith (1932), which must include some influence of the Ishii effect. Again, Brook et al. (1978) used Peterson's method and measured the ionization cross section of O. Fig. 7-11 shows experimental and theoretical results. Bell et al. (1983) recommend the results of Brook et al. (1978) as best.

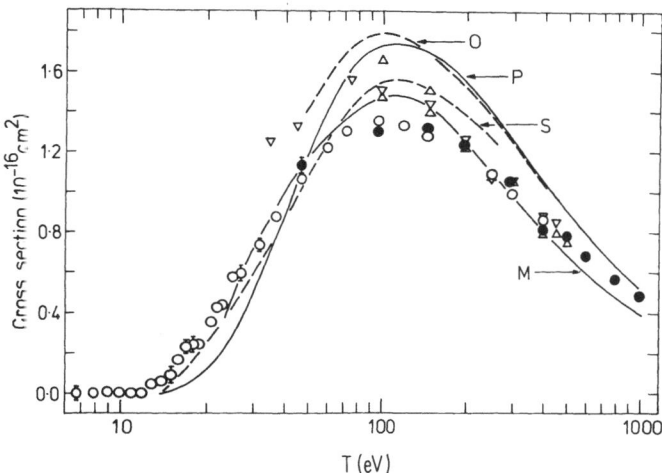

Fig. 7-11. The ionization cross section of the oxygen atom in the ground state. The results of Fite and Brackmann (1959) are shown by inverted triangles (∇), and those of Rothe et al. (1962) by triangles ($\triangle$). All the other symbols and notations are the same as those in Fig. 10. The figure is taken from Brook et al. (1978)

For alkali-metal atoms Li, Na, K, Rb, and Cs, the first measurement of the total ionization cross section was carried out by McFarland and Kinney (1965). Their results are shown numerically in Table 3 of Kieffer and Dunn (1966). McFarland and Kinney used crossed beams, and determined the neutral atomic-beam intensity by means of a surface ionization detector. The efficiency of this detector had not been established specifically for the alkali-metal atoms, and probably made the largest contributions to systematic errors in the result. Zapesochnyĭ and Aleksakhin (1969) also measured the total ionization cross sections for the alkali-metal atoms from threshold to 70 eV, again with the use of crossed beams. To obtain the neutral beam intensity, they let the atoms condense on the surface of a piezoelectric quartz crystal (cooled with liquid nitrogen) and determined the mass of the condensed material from the change in the natural oscillation frequency of the crystal. As Zapesochnyĭ and Aleksakhin (1969) state, errors in their absolute values of the total ionization cross sections are 15%, while the root mean squared deviation of the data is as large as 12%. Electron beams of energy spread of about 0.1 eV were used in the experiment

Table 7-4. *Total ionization cross sections (in 10^{-16} cm^2) of alkali-metal atoms at the maximum [taken from Zapesochnyĭ and Aleksakhin (1960)]*

Atomic species	Zapesochnyĭ and Aleksakhin (1069)	Brink (1962, 1964)	McFarland and Kinney (1965)
Li	4.2		4.9
Na	6.8	8.6	7.6
K	7.9	9.6	8.2
Rb		9.6	8.2
Cs	10.2	11.2	9.4

for studying the threshold behavior and any structure in the cross section as a function of impact energy. Table 7-4 is a summary of the results compared with those of McFarland and Kinney (1965) and of Brink (1962, 1964); specifically, it shows the maximum values of the total ionization cross sections.

Jalin *et al.* (1973) measured absolute values of the ionization cross section for Li at electron energies ranging from 100 eV to 2 keV, again by the use of crossed beams. The neutral-beam atoms condensed on a cold trap cooled with liquid nitrogen. The amount of metal deposited was measured by use of an isotopic dilution technique and of atomic absorption; thence, the intensity of the neutral beam was evaluated. The total ionization cross section for Li at 500 eV turned out to be 0.356×10^{-16} cm², which is about half of the value given by McFarland and Kinney (1965). Jalin *et al.* (1973) pointed out that their results at high energies agreed with calculations within the Born-Bethe approximation and argued that their measurements should be right.

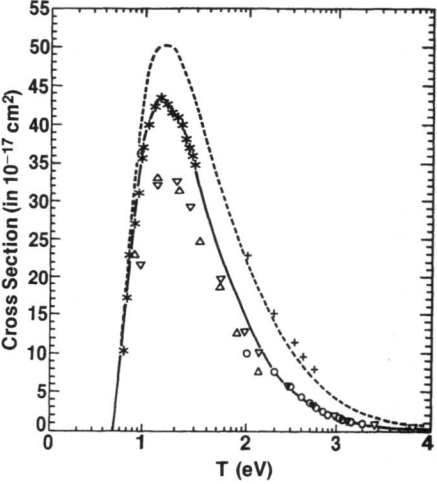

Fig. 7-12. The single ionization cross section of Li for electron impact. The experimental results of McFarland and Kinney (1965) are shown by crosses (+), those of Zapesochnyi and Aleksakhin (1969) by asterisks (*), and those of Jalin *et al.* (1973) by open circles (O). The Born-approximation results by McDowell *et al.* (1965) are shown by triangles (△), and those by McGuire (1971) by inverted triangles (▽). The broken curve represents values from the expression of Lotz (1968). The solid curve represents the values recommended by Bell *et al.* (1983). The figure is taken from Bell *et al.* (1983)

McDowell (1969) extensively discussed theoretical and experimental results for alkali metals, with particular emphasis on Li and Na. It is seen that the experimental results of McFarland and Kinney (1965) are always about twice as large as calculated values. Fig. 7-12 is a summary of various results for single ionization of Li, taken from Bell *et al.* (1983). The curve recommended by Bell *et al.* (1983) follows the data points of Zapesochnyï and Aleksakhin (1969) up to about 30 eV, and then those of Jalin *et al.* (1973) at high energies, which are consistent with the calculation by McGuire (1971, 1977) within the first Born approximation. This curve may be questioned at low energies because the results of Zapesochnyǐ and Aleksakhin (1969) around the cross-section maximum are rather high in the sense they come close to the results of McFarland and Kinney (1969), as seen in Table 7-4. It is not surprising that the empirical formula of Lotz (1968) gives large cross section values; the choice of the parameters in the formula was clearly dictated by the only experimental data then available, i.e., the results of McFarland and Kinney (1965).

The difference between the single ionization cross section and the total ionization cross section is small for Li [Jalin *et al.* (1973)], but it is appreciable for the other alkali-metal atoms. [See Tate and Smith (1934) and also Tables 2-3-1 and 2-3-2 of McDowell (1969).] For those alkali-metal atoms again, the cross sections given by McFarland and Kinney (1965) appear much too large compared to the Born-approximation results of McGuire (1971, 1977). A thorough analysis of all the data is necessary for recommending reasonably reliable cross sections for those atoms. Some of the more recent measurements include those on Rb by Nygaard and Hahn (1973) and by Korchevoi and Przonski (1966), as well as those on Cs by Heil and Scott (1966), by Nygaard (1968), and by Korchevoi and Przonski (1967).

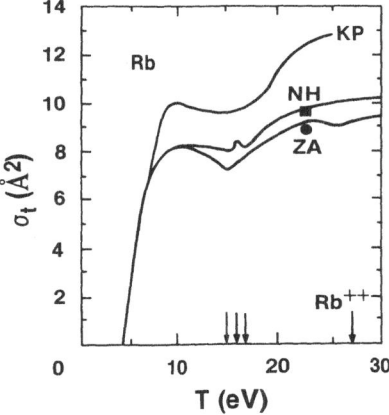

Fig. 7-13. The total ionization cross section of Rb at low electron energies. The top curve marked *KP* shows the results of Korchevoi and Przonski (1966), and the bottom curve marked *ZA* shows the results of Zapesochnyǐ and Aleksakhin (1968), which were normalized to the values by McFarland and Kinney (1965), indicated here by the filled circle. The filled square shows the measurements by Brink (1964). The middle curve marked *NH* shows the results of Nygaard and Hahn (1973). The three vertical arrows on the horizontal axis indicate the energies of autoionizing levels. The threshold for the Rb^{++} production is also indicated. The figure is taken from Nygaard and Hahn (1973)

Let us review the data on Rb. Korchevoi and Przonski (1967) covered the impact energies from threshold to 25 eV, and Zapesochnyǐ and Aleksakhin (1969) from threshold to 30 eV. Brink (1964) studied the region between 50 eV and 500 eV. Tate and Smith (1934) measured relative and partial ionization cross sections for 4 eV to 700 eV. The data of Nygaard and Hahn (1973) cover the region between 4.18 eV and 250 eV, which overlaps with the other measurements mentioned earlier. Nygaard and Hahn (1973) used a crossed-beam apparatus; their surface ionization detector (made of a tungsten wire) had an efficiency of $98 \pm 2\%$. They claim an uncertainty of

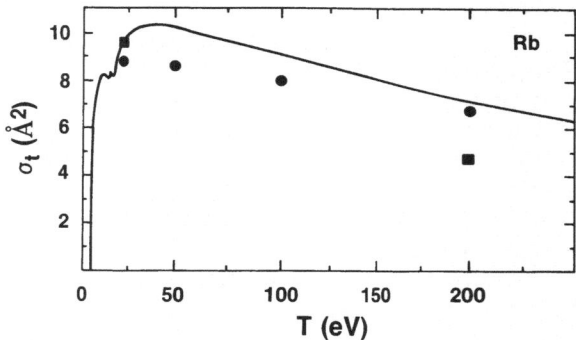

Fig. 7-14. The total ionization cross section of Rb for electron impact. The filled circles show the results of McFarland and Kinney (1965), and the filled squares those of Brink (1964). The curve shows the results of Nygaard and Hahn (1973). The figure is taken from Nygaard and Hahn (1973)

12%. Figs. 7-13 and 7-14 show various results on Rb. The structure at low energies (seen in Fig. 7-13) is due to autoionization. Korchevoi and Przonski (1967) used an ion-trap method, as described by Korchevoi and Przonski (1966) and by Schulz (1958); the pressure of alkali-metal vapor in a collision cell was determined from the thermoionic saturation current on the surface of an incandescent tungsten filament.

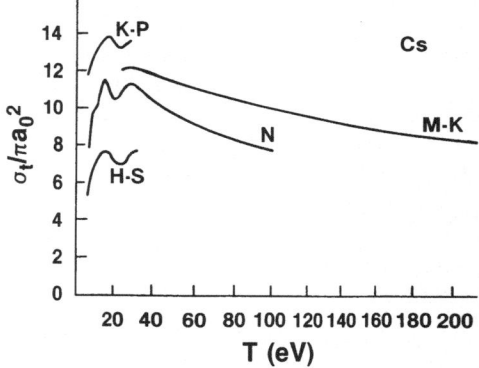

Fig. 7-15. The total ionization cross section of Cs for electron impact. The curves show experimental results of the four groups of workers. N Nygaard (1968), M-K McFarland and Kinney (1965), H-S Heil and Scott (1966), K-P Korchevoi and Przonski (1966). The figure is taken from McDowell (1969)

Fig. 7-15 shows results of various measurements on Cs. The cross-section curve due to Nygaard (1968) is close in shape to that of Tate and Smith (1934). The structure due to autoionization is seen in all low-energy data. However, there are large discrepancies in the absolute value. Clearly, recommendation of a reliable data set would await further analysis.

Table 7-5. *Some of the recent measurements of the total ionizations of neutral atoms*

Authors (year)	Species	Electron energies	Remarks
Crawford and Wang (1967)	Ag	≤ 150 eV	single, double*
Korchevoi and Przonskii (1967)	Cs, Rb, K	near threshold	
McFarland (1967)	Ca, Sr, Ba, Tl	≤ 500 eV	
Pavlov et al. (1967)	Ag, Cu, Pb	≤ 150 eV	
Nygaard (1968)	Cs	≤ 100 eV	
Zapesochnyi and Aleksakhin (1969)	Li, Na, K, Rb, Cs	≤ 30 eV	
Okuno et al. (1970)	Mg	≤ 1130 eV	
Pavlov and Stoiskii (1970)	Pb	≤ 400 eV	single, multiple
Okuno (1971)	Ca, Sr, Ba	≤ 1 keV	
Vainshtein et al. (1971)	Mg, Ca, Sr, Ba	≤ 200 eV	
Rieke and Prepejchal (1972)	Ne, Ar, Kr, Xe, Hg	$0.1 - 2.7$ MeV	
Fletcher and Cowling (1973)	Ne, Ar	≤ 500 eV	
Jalin et al. (1973)	Li	100 eV $- 2$ keV	
Nygaard and Hahn (1973)	Rb	< 250 eV	
Schroeer et al. (1973)	Cu, Au	$40 - 250$ eV	
Kurepa et al. (1974)	Ar		
Dixon et al. (1975)	H ($2\,^2S$)		
Karstensen and Schneider (1975)	Mg	$7 - 60$ eV	
Shimon et al. (1975)	Al, Ga, In, Tl		
Brook et al. (1978)	He, C, O, N	≤ 997 eV	
Karstensen and Scheider (1978)	Mg	≤ 280 eV	single, double
Nagy et al. (1980)	He, Ne, Ar, Kr, Xe	500 eV $- 5$ keV	
Stephan et al. (1980 b)	He, Ne, Ar, Kr	≤ 180 eV	
Defrance et al. (1981)	H ($2\,^2S$)	$6.3 - 998.3$ eV	
Halle et al. (1981)	U	$7.5 - 500$ eV	
Dettmann and Karstensen (1982)	Ba	≤ 600 eV	single, multiple

* This means that the cross sections for single ionization and for double ionization are also presented.

Finally, in the literature there are ionization cross-section data concerning atoms other than those we have discussed so far; indeed, Kieffer and Dunn (1966) already discussed data on Hg. Table 7-5 shows some of the more recent measurements on those atoms. The increased variety of the atoms treated represents major progress. Space limitation prevents us from presenting here a detailed commentary.

7.2.4 Molecules

Kieffer and Dunn (1966) also discuss data on molecular gases H_2, D_2, N_2, O_2, CO, and NO. They point out that the σ_t values of Tate and Smith (1932) for molecules decrease more rapidly with increasing impact energies above 100 eV than those of

Rapp and Englander-Golden (1965). This trend of the data is similar to what one sees in data for He, Ne, and Ar. The cross sections for molecules as reported by Schram *et al.* (1965, 1966) are again lower than results of other groups of workers. For diatomic molecules, dissociative ionization is most likely to make complete ion collection difficult, because there may be an appreciable number of ions produced with sizable kinetic energies. Rapp *et al.* (1965) estimated the fraction of the energetic ions resulting from dissociative ionization and concluded that it ranged from 36% for O_2 to 7% for H_2 at maximum. According to Kieffer and Dunn (1966), Tate and Smith (1932) may not have used high enough electric fields to ensure complete collection of the energetic ions.

For H_2, various cross-section data have been compiled and examined by Miles *et al.* (1972), by Gerhart (1975), by Douthat (1979), and by van Wingerden *et al.* (1980). In particular, Gerhart (1975) and Douthat (1979) made full use of the Bethe-theory constraints on the cross-section data (as outlined in Section 7.2.2). The work of van Wingerden *et al.* (1980) belongs to the same series of studies by de Heer and co-workers [de Heer *et al.* (1977), de Heer and Jansen (1977), and de Heer *et al.* (1979)]. and consists of a semi-empirical determination of σ_{tot} and a measurement of the same cross section. As for σ_t, van Wingerden *et al.* (1980), as well as the other authors cited above, adopt the results of Rapp and Englander-Golden (1965); one reason for this is the close agreement of these results with the old results by Tate and Smith (1932) and also with the newest measurements by Cowling and Fletcher (1973). Table 7-6 shows the values of σ_t thus recommended by van Wingerden *et al.* (1980), together with the elastic scattering cross section σ_{el} and the total discrete-excitation cross section σ_{exc}. Notice the near consistency of the semi-empirically determined σ_{tot} with the experimental results σ_{tot}^{exp}. Also notice that σ_t is the dominant fraction of σ_{tot}, and is almost always the case with any atom or molecule. For further discussion on σ_t of H_2 and D_2, the reader is referred to the recent review by de Heer (1981).

Table 7-6. *The total ionization cross section σ_t, the elastic scattering cross section σ_{el}, the total discrete-excitation cross section σ_{exc}, and the total scattering cross section σ_{tot} for electron impact on H_2. The values except those in the rightmost column) are based on the semi-empirical analysis of various data by van Wingerden et al. (1980). The rightmost column shows experimental results by the same authors. All cross-section values are in units of a_0^2*

E (eV)	σ_t	+	σ_{el}	+	σ_{exc}	=	σ_{tot}		σ_{tot}^{exp}
100	3.30		3.17		2.78		9.25	(4)	9.01
150	2.90		1.85		2.35		7.10	(4)	7.13
200	2.55		1.28		2.06		5.89	(4)	5.76
300	2.05		0.809		1.60		4.48	(4)	4.36
400	1.70		0.543		1.30		3.56	(4)	3.45
500	1.45		0.420		1.11		2.98	(4)	2.97
600	1.27		0.340		0.965		2.57	(4)	2.55
700	1.13		0.260		0.859		2.24	(4)	2.28
750	1.07		0.248		0.814		2.13	(4)	2.15
800	1.01		0.235		0.775		2.00	(4)	
900	0.930		0.211		0.707		1.85	(4)	
1000	0.858		0.186		0.651		1.70	(4)	
2000	0.475		0.097		0.375		0.947	(4)	

As for N_2, O_2, CO, and NO, there have been numerous papers published since the article by Kieffer and Dunn (1966), as one can readily see in the JILA tabulations [Kieffer (1976); Gallagher et al. (1979); Gallagher and Beaty (1980); Gallagher and Beaty (1981)]. Despite numerous papers, the situation with respect to the total ionization cross section σ_t appears to be basically unchanged. That is to say, the results by Rapp and Englander-Golden (1965) are most likely to be right within stated limits of errors. More recent work apparently has not substantially reduced the error limit.

The main progress in the study of the total ionization cross section in the past decade seems to be the great increase in the variety of molecular species treated. To illustrate, Rapp and Englander-Golden (1965) studied CO_2, N_2O, CH_4, SF_6, and C_2H_4, in addition to rare gases and commonly available diatomic molecules such as H_2 and N_2. Other molecules recently studied include diatomics such F_2 and Cl_2, as well as polyatomics such as H_2O, BF_3, NH_3, UF_6, O_3, and many hydrocarbons.

Table 7-7. Some of the recent measurements of the total ionization cross sections of molecules

Authors (year)	Species	Electron energies	Remarks
Schutten et al. (1966)	H_2O	100 eV – 20 keV	partial also
Schram et al. (1966 b)	CH_4, CD_4, and 16 other hydrocarbons	600 eV – 12 keV	
Gaudin and Hagemann (1976)	C_2H_2	100 eV – 12 keV	
Adamczyk et al. (1972)	CO_2	25 – 600 eV	partial also
Center and Mandl (1972)	F_2, Cl_2	15 – 100 eV	
Halas et al. (1972)	N_2	16 – 600 eV	partial also
Rieke and Prepejchal (1972)	H_2 and 33 molecular species	0.1 – 2.7 MeV	
Cowling and Fletcher (1973)	H_2, D_2	16 – 500 eV	
Gomet (1975)	H_2O, NH_3	20 – 250 eV	partial also
Winters (1975)	CH_4	$\leq$ 500 eV	total dissociation also
Kurepa et al. (1976)	BF_3	16 – 250 eV	
Compton (1977)	UF_6	$\leq$ 1 keV	
Märk and Egger (1977)	PH_3	$\leq$ 180 eV	partial also
Märk et al. (1977)	NH_3	$\leq$ 180 eV	partial also
Gomet (1978)	H_2	20 – 350 eV	partial also
Hille and Märk (1978)	CO	$\leq$ 180 eV	partial also
Kurepa and Belić (1978)	Cl_2	10 – 100 eV	
Märk and Hille (1978)	CO_2	$\leq$ 180 eV	partial also
Pejcev et al. (1979)	CCl_2F_2	13 – 250 eV	
Stephan et al. (1980 a)	NO_2	$\leq$ 180 eV	partial also
Märk et al. (1981)	N_2O	$\leq$ 180 eV	partial also
Stevie and Vasile (1981)	F_2, Cl_2	$\leq$ 102 eV	
Monnom et al. (1982)	AS_4, P_4	10 – 200 eV	
Siegel (1982)	O_3	$\leq$ 100 eV	partial also
Wiegand and Boedeker (1982)	$HgBr_2$	10.7 – 70 eV	

Table 7-7 is a selected list of the measurements reported in recent years. The list includes only those papers that give absolute values of the total ionization cross sections, but is by no means exhaustive even within this limitation.

7.2.5 Systematics of the Total Ionization Cross Section

How does the total ionization cross section at a fixed electron energy depend upon atomic or molecular species? This question has been treated many times, but has been answered only partially.

For electron energies greatly exceeding the (first) ionization energy, one may consider the question from the point of view of the Bethe theory. The value of σ_t [Eq. (7-24)] is practically governed by M_{ion}^2, the second constant c_{ion} playing a much lesser role. The dominant role of M_{ion}^2 applies also to the expression for relativistic electrons, i.e., Eq. (7-13), in which the constant C_{ion} should behave in nearly the same way as M_{ion}^2 when one goes from one species to another, on theoretical grounds. Thus, the systematics of σ_t is in effect the systematics of M_{ion}^2.

Sometimes it is sensible to use an approximation that the ratio M_{ion}^2/M_{tot}^2 is nearly constant for different species. This is usually acceptable at least at modest accuracy because the preponderance of the oscillator strength lies in the ionization continua rather than in the discrete spectra for all neutral species (and for negative ions), and then M_{ion}^2 is the dominant part of M_{tot}^2, as given by Eq. (7-27). The systematics of M_{tot}^2 have been extensively discussed by Inokuti et al. (1981) for atoms. The approximation is better justified if we compare a series of species with similar electronic structure (and hence, with similar oscillator-strength distributions). Examples of such a series may be rare gases (except for He), ten-electron systems Ne, HF, H_2O, NH_3, CH_4, and BH_5, and saturated hydrocarbons. Once we thus relate σ_t with M_{tot}^2 approximately, then we see in Eq. (7-29) a theoretical expression that strongly appeals to intuition. That equation states roughly that M_{tot}^2 (and thence σ_t in our approximation) is given by the mean squared radius of the total electron density in an atom or a molecule. In other words, σ_t is roughly governed by the geometrical area so long as glancing collisions dominate (as is indeed the case for sufficiently high electron impact energies).

Experimental values of M_{ion}^2 have been most extensively presented by Rieke and Prepejchal (1972). Some of these values have been compared with values derived from the oscillator-strength spectra by Berkowitz et al. (1973) and by Berkowitz (1979). Schram et al. (1966 b) have treated the systematics of σ_t for hydrocarbons from a similar point of view; more recently, de Heer (1981) has given a concise review of the topic.

Let us recall a well-known fact that the total electron density of a molecule may be approximately expressed as the sum of the electron densities associated with constituent atoms. Obviously, the formation of molecular bonds implies the presence of corrections to the sum; however, these corrections stem mainly from valence electrons only and are not overwhelming. Consequently, once we relate σ_t of a molecule with the total electron density, one readily recognizes a notable consequence, i.e., the approximate additivity of σ_t. In other words, σ_t of a molecule

XYZ may be expressed as

$$\sigma_t = \sigma_t(X) + \sigma_t(Y) + \sigma_t(Z). \tag{7-33}$$

Here, X, Y, and Z stand for constituent atoms.

An example of studies on the additivity and the systematics is due to Grosse and Bothe (1968), who used β rays from ^{85}Kr (i.e., electrons having a wide range of kinetic energies around 135 keV). They present relative values of σ_t for nearly fifty organic molecules. As for the additivity, they state that, with suitably chosen atomic cross sections, the molecular σ_t can be expressed in the form of Eq. (7-33) to a precision of 7% or better. In addition, if they choose for the constituents X, Y, and Z bonds (or groups of bound atoms) such as $C-H$ and $C-C$ rather than atoms, then Eq. (7-33) holds to a precision of 2% (which is smaller than estimated limits of the accuracy of the measurements).

All the above experimental data seem to confirm the theoretical expectations, certainly within modest precision; in other words, no gross inconsistency with the theoretical expectations has been found so far.

The question of the systematics has most often been raised in the context of the electron-impact mass spectrometry for analytical chemistry, a subject fully treated by Field and Franklin (1970) (see also Chapter 9.6). Because σ_t is largest at electron energies around 75 eV $-$ 100 eV, commercially available machines always used impact electron having kinetic energies in that range. A pioneering work in this field of research was done by Otvos and Stevenson (1956), who were stimulated by the famous paper by Bethe (1930). Mann (1967) gave theoretical results pertinent to atoms. However, they applied the conclusions of the Bethe theory (as we discussed earlier in the present section) to their cross-section data they obtained with 75-eV electrons. The energy of 75 eV is only a few times the (first) ionization threshold, and is too low for us to expect that the Bethe theory might even serve as a guide. Nevertheless, the systematics of σ_t at such low electron energies roughly agrees with the Bethe-theory predictions, although a full explanation of this finding must await for further theoretical studies, and perhaps experimental studies as well.

The systematics and the additivity of σ_t are qualitatively understandable also from the point of view of the binary-encounter (or impulse) approximation (discussed in detail in Chapter 2 by Younger and Märk). In this approximation one begins with the picture in which the incident electron interacts with each of atomic or molecular electrons individually. The cross section for the energy transfer from the incident electron to an *unbound* electron is completely known (in the form of the Mott formula). A major effect of the electron binding comes from the instantaneous velocity of a struck electron, and may be taken into account by use of the velocity distribution derivable from the atomic or molecular eigenfunction in the initial state. Thus, σ_t is evaluated as a sum of contributions from individual atomic or molecular electrons, weighted with their velocity distributions, which in turn are closely related to electron densities. From this point of view, the additivity and the scaling of σ_t are natural consequences. It is empirically known that σ_t calculated with the binary-encounter approximation is often quite reasonable (compared to experiment) at electron energies around 100 eV. In this way, one sees some justification for the systematics and the additivity of σ_t at those lower electron energies.

Finally, Center and Mandl (1972) point out (in their Fig. 2) the maximum values of σ_t (which usually occurs at an electron impact energy around 100 eV) were proportional to the square root of the dipole polarizability α_d. Hotta et al. (1975 a, b) and Arai and Hotta (1975) used electrons in the MeV range and confirmed the systematics that had been found by Rieke and Prepejchal (1972). In addition, Arai and Hotta (1975) show in their Fig. 2 that σ_t at high electron energies was proportional to α_d. The correlation of σ_t with α_d is not surprising because the two quantities are roughly governed by the geometrical size of atomic or molecular species. However, it is difficult to say either α_d or $\alpha_d^{1/2}$ should be more nearly proportional to σ_t on any theoretical ground. This illustrates how the question of the systematics remains only incompletely answered.

Acknowledgements

We thank Jean W. Gallagher for computer search of the literature needed in the preparation of Tables 7-5 and 7-7.

References

Adamczyk, B., Boerboom, A. J. H., Lukasiewicz, M. (1972): Int. J. Mass Spectrom. Ion Phys. 9, 407–412.

Arai, H., Hotta, H. (1975): Radiat. Res. 64, 407–415.

Asundi, R. K. (1963): Proc. Phys. Soc. (London) 82, 372–374.

Asundi, R. K., Craggs, J. D., Kurepa, M. V. (1963): Proc. Phys. Soc. (London) 82, 967–978.

Bannenberg, J. G., Saris, F. W., Tip, A. (1969): Ned. Tijdschr. Vacuumtech. 7, 81–85.

Bannenberg, J. G., Tip, A. (1968): Proceedings of the Fourth International Vacuum Congress, Part II. 609–612 (Conference Series No. 5). London: Institute of Physics and Physical Society.

Bell, K. L., Gilbody, H. B., Hughes, J. G., Kingston, A. E., Smith, F. J. (1983): J. Phys. Chem. Ref. Data 12, 891–916.

Bell, K. L., Kingston, A. E. (1969): J. Phys. B: Atom. Molec. Phys. 2, 1125–1130.

Bell, K. J., Kingston, A. E. (1974): In: Advances in Atomic and Molecular Physics (Bates, D. R., Bederson, B., eds.), Vol. 10, 53–130. New York: Academic Press.

Berkowitz, J. (1979): Photoabsorption, Photoionization and Photoelectron Spectroscopy. New York-San Francisco-London: Academic Press.

Berkowitz, J., Inokuti, M., Person, J. C. (1973): In: Abstracts of Papers, XIIIth International Conference on the Physics of Electronic and Atomic Collisions (Cobić, B. C., Kurepa, M. V., eds.). Vol. 2, 561–562. Beograd: Institute of Physics.

Bethe, H. (1930): Ann. Physik (Leipzig) 5, 325–400.

Bethe, H. (1932): Z. Phys. 76, 293–299.

Bethe, H. (1933): In: Handbuch der Physik (Geiger, H., Scheel, K., eds.), Vol. 24/1, 491–523. Berlin: Springer.

Blaauw, H. J., Wagenaar, R. W., Barends, D. H., de Heer, F. J. (1980): J. Phys. B: Atom. Molec. Phys. 13, 359–376.

Bleakney, W. (1929): Phys. Rev. 34, 157–160.

Boksenberg, A. (1961): Thesis, University of London.

Boyd, R. L. F., Green, G. W. (1958): Proc. Phys. Soc. (London) 71, 351–356.

Brink, G. O. (1962): Phys. Rev. 127, 1204–1206.

Brink, G. O. (1964): Phys. Rev. 134, A345–A346.

Brook, E., Harrison, M. F. A., Smith, A. C. H. (1978): J. Phys. B: Atom. Molec. Phys. 11, 3115–3132.

Center, R. E., Mandl, A. (1972): J. Chem. Phys. 57, 4104 – 4106.

Chambers, E. S. (1964): Phys. Rev. 133, A1202 – A1207.

Chen, S. T., Gallagher, A. (1976): Phys. Rev. A 14, 593 – 601.

Compton, K. T., Van Voorhis, C. C. (1925): Phys. Rev. 26, 436 – 453.

Compton, R. N. (1977): J. Chem. Phys. 66, 4478 – 4485.

Cowling, I. R., Fletcher, J. (1973): J. Phys. B: Atom. Molec. Phys. 6, 665 – 674.

Craggs, J. D., Thorburn, R., Tozer, B. A. (1957): Proc. Roy. Soc. (London) A 240, 473 – 483.

Crawford, C. K., Wang, K. L. (1967): J. Chem. Phys. 47, 4667 – 4669.

Daly, N. R. (1960): Rev. Sci. Instr. 31, 264 – 267.

Dance, D. F., Harrison, M. F. A., Rundel, R. D. (1967): Proc. Roy. Soc. (London) A 299, 525 – 537.

Datz, S., Taylor, E. H. (1956): J. Chem. Phys. 25, 389 – 394.

De Heer, F. J. (1981): Physica Scripta 23, 170 – 178.

De Heer, F. J., McDowell, M. R. C., Wagenaar, R. W. (1977): J. Phys. B: Atom. Molec. Phys. 10, 1945 – 1953.

De Heer, F. J., Jansen, R. H. J. (1977): J. Phys. B: Atom. Molec. Phys. 10, 3741 – 3758.

De Heer, F. J., Jansen, R. H. J., van der Kaay, W. (1979): J. Phys. B: Atom. Molec. Phys. 12, 979 – 1002.

De Vries, A. E., Rol, P. K. (1965): Vacuum 15, 135 – 139.

Defrance, P., Claeys, W., Cornet, A., Poulaert, G. (1981): J. Phys. B: Atom. Molec. Phys. 14, 111 – 117.

Dettmann, J. M., Karstensen, F. (1982): J. Phys. B: Atom. Molec. Phys. 15, 287 – 300.

Dixon, A. J., Harrison, M. F. A., Smith, A. C. H. (1973): In: Abstracts of Papers, VIIIth International Conference on the Physics of Electronic and Atomic Collisions (Cobić, B. C., Kurepa, M. V., eds.), Vol. 1, 405 – 406. Beograd: Institute of Physics.

Dixon, A. J., Harrison, M. F. A., Smith, A. C. H. (1976): J. Phys. B: Atom. Molec. Phys. 9, 2617 – 2631.

Dixon, A. J., von Engel, A., Harrison, M. F. A. (1975): Proc. Roy. Soc. (London) A 343, 333 – 349.

Dolder, K. T. (1980): In: Atomic and Molecular Processes in Controlled Thermonuclear Fusion (McDowell M. R. C., Ferendeci, A. M., eds.), 313 – 349. New York-London: Plenum Press.

Dolder, K. T., Harrison, M. F. A., Thonemann, P. C. (1961): Proc. Roy. Soc. (London) A 264, 367 – 378.

Donnally, B. L., O'Dell, J. E. (1971): In: Abstracts of Papers, VIIth International Conference on the Physics of Electronic and Atomic Collisions (Govers, T. R., de Heer, F. J., eds.), Vol. 2, 821 – 822. Amsterdam: North-Holland.

Douthat, D. A. (1979): J. Phys. B: Atom. and Molec. Phys. 12, 663 – 678.

Dunn, G. H. (1962): Phys. Rev. Lett. 8, 62 – 64.

Eggarter, E. (1975): J. Chem. Phys. 62, 833 – 847.

Eggarter, E. (1976): J. Chem. Phys. 65, 2044.

Fano, U. (1954): Phys. Rev. 95, 1198 – 1200.

Fano, U. (1975): Radiat. Res. 64, 217 – 232.

Field, F. H., Franklin, J. L. (1970): Electron Impact Phenomena and the Properties of Gaseous Ions, revised edition. New York-London: Academic Press.

Fite, W. L., Brackmann, R. T. (1958): Phys. Rev. 112, 1141 – 1151.

Fite, W. L., Brackmann, R. T. (1959): Phys. Rev. 113, 815 – 816.

Fite, W. L., Brackmann, R. T. (1964): In: Proceedings of the VIth International Conference on Ionization Phenomena in Gases, Paris, 1962 (Hubert, P., ed.), Vol. 1, 21 – 25. Paris: SERMA.

Fleming, R. J., Higginson, G. S. (1964): Proc. Phys. Soc. (London) 84, 531 – 538.

Fletcher, J., Cowing, I. R. (1973): J. Phys. B: Atom. Molec. Phys. 6, L258 – L261.

Funk, H. (1930): Ann Physik (Leipzig) 4, 149 – 184.

Gaede, W. (1915): Ann. Physik (Leipzig) 46, 357 – 392.

Gallagher, J. W., Beaty, E. C. (1980): Bibliography of Low Energy Electron and Photon Cross Section Data (1978). JILA Information Center Report No. 18. Joint Institute for Laboratory Astrophysics, University of Colorado, Boulder, Colorado.

Gallagher, J. W., Beaty, E. C. (1981): Bibliography of Low Energy Electron and Photon Cross Section Data (1979). JILA Information Center Report No. 21. Joint Institute for Laboratory Astrophysics, University of Colorado, Boulder, Colorado.

Gallagher, J. W., Rumble, J. R., jr., Beaty, E. C. (1979): Bibliography of Low Energy Electron and Photon Cross Section Data (January 1975 through December 1977). NBS Special Publication 426, Suppl. 1. Washington, D.C.: U.S. Government Printing Office.

Gaudin, A., Hagemann, R. (1967): J. Chim. Phys. 64, 1209 – 1221.

Gerhart, D. E. (1975): J. Chem. Phys. 62, 821 – 832.

Gomet, J.-C. (1975): C. R. Acad. Sci. Ser. B 281, 627−630.
Gomet, J.-C. (1978): C. R. Acad. Sci. Ser. B 287, 77−79.
Graf, T. (1939): J. Phys. Radium 10, 513−518.
Green, A. E. S., Dutta, S. K. (1967): J. Geophys. Res. 72, 3933−3941.
Grosse, H.-J., Bothe, H. K.: Z. Naturforschg. 23a, 1583−1590.
Halas, St., Adamczyk, B. (1972): Intern. J. Mass Spectrom. Ion Phys. 10, 157−160.
Halle, J. C., Lo, H. H., Fite, W. L. (1980): J. Chem. Phys. 73, 5681−5683.
Halle, J. C., Lo, H. H., Fite, W. L. (1981): Phys. Rev. A 23, 1708−1716.
Harrison, M. F. A. (1968): In: Methods of Experimental Physics, Vol. 7, Part A: Atomic and Electron Physics. Atomic Interactions (Bederson, B., Fite, W. L., eds.), 95−115. New York-London: Academic Press.
Heil, H., Scott, B. (1966): Phys. Rev. 145, 279−284.
Hille, E., Märk, T. D. (1978): J. Chem. Phys. 69, 4600−4605.
Hotta, H., Tanaka, R., Sunaga, H., Arai, H. (1975a): Radiat. Res. 63, 24−31.
Hotta, H., Tanaka, R., Arai, H. (1975b): Radiat. Res. 63, 32−41.
Inokuti, M. (1971): Rev. Mod. Phys. 43, 297−347.
Inokuti, M. (1980): In: Invited Papers and Progress Reports, XIth International Conference on the Physics of Electronic and Atomic Collisions, Kyoto, 1979 (Oda, N., Takayanagi, K., eds.), 31−45. Amsterdam-New York-Oxford: North-Holland.
Inokuti, M., Kim, Y.-K., Platzman, R. L. (1967): Phys. Rev. 164, 55−61.
Inokuti, M., Saxon, R. P., Dehmer, J. L. (1975): Int. J. Radiat. Phys. Chem. 7, 109−120.
Inokuti, M., Dehmer, J. L., Hanson, J. D. (1981): Phys. Rev. A 23, 95−109.
Ishii, H., Nakayama, K.: In: Transactions of the VIIIth National Vacuum Symposium, combined with the IInd International Congress on Vacuum Science and Technology (Preuss, L. E., ed.) 519−524. Oxford-London-New York-Paris: Pergamon Press.
Jalin, R., Hagemann, R., Botter, R. (1973): J. Chem. Phys. 59, 952−959.
Jusick, A. T., Watson, C. E., Peterson, L. R., Green, A. E. S. (1967): J. Geophys. Res. 72, 3943−3951.
Kadota, K., Dijkkamp, D., van der Woude, R. L., de Boer, A., Yan, P. G., de Heer, F. J. (1982): J. Phys. B: Atom. Molec. Phys. 15, 3275−3296.
Karstensen, F., Schneider, M. (1975): Z. Phys. 273, 321−323.
Karstensen, F., Schneider, M. (1978): J. Phys. B: Atom. Molec. Phys. 11, 167−172.
Katsonis, K., Lorenz, A. (1982): First Coordinated Research Meeting on Atomic Collision Data for Diagnostics of Magnetic Fusion Plasmas, Report No. INCD(NDS)-136/GA. Wien: IAEA Nuclear Data Section.
Kerwin, L., Marmet, P., Carette, J. D. (1969): In: Case Studies in Atomic Physics I (McDaniel, E. W., McDowell, M. R. C., eds.), 525−581. Amsterdam-London: North-Holland.
Kieffer, L. J. (1965): A compilation of critically evaluated electron impact ionization cross section data for atoms and diatomic molecules, JILA Report No. 30, Joint Institute for Laboratory Astrophysics, University of Colorado, Boulder.
Kieffer, L. J. (1966): Addendum to JILA Report No. 30, Joint Institute for Laboratory Astrophysics, University of Colorado, Boulder.
Kieffer, L. J. (1976): Bibliography of Low Energy Electron and Photon Cross Section Data (through 1974), NBS Special Publication 426. Washington, D.C.: U.S. Government Printing Office.
Kieffer, L. J., Dunn, G. H. (1966): Rev. Mod. Phys. 38, 1−35.
Kim, Y.-K., Inokuti, M. (1971): Phys. Rev. A 3, 665−678.
Kim, Y.-K., Naon, M., Cornille, M. (1973): Argonne National Laboratory, Radiological and Environmental Research Division Report ANL-8060-I, pp. 14−23.
Koller, H. H. (1969): Doctoral Thesis, Universität Zürich.
Korchevoï, Yu. P., Przhonskiï, A. (1966): Zh. Eksp. Teor. Fiz. 50, 315−319. [English translation: Sov. Phys. JETP 23, 208−211.]
Korchevoï, Yu. P., Przhonskiï, A. M. (1967): Zh. Eksp. Teor. Fiz. 51, 1617−1621. [English translation: Sov. Phys. JETP 24, 1089−1092.]
Kurepa, M. V., Belić, D. S. (1978): J. Phys. B: Atom. Molec. Phys. 11, 3719−3729.
Kurepa, M. V., Cadez, I. M., Pejcev, V. M. (1974): Fizika (Zagreb) 6, 185−209.
Kurepa, M. V., Pejcev, V. M., Cadez, I. M. (1976): J. Phys. D: Appl. Phys., 481−484.
Langenberg, A., van Eck, J. (1976): J. Phys. B: Atom. Molec. Phys. 9, 2421−2433.
Leep, D., Gallagher, A. (1974): Phys. Rev. A 10, 1082−1090.

Liu, J. W. (1973): Phys. Rev. *A 7*, 103 – 109.

Long, D. R., Geballe, R. (1979): Phys. Rev. *A 1*, 260 – 265.

Lotz, W. (1968): Z. Phys. *216*, 241 – 247.

Lozier, W. W. (1934): Phys. Rev. *46*, 268 – 276.

Mann, J. B. (1967): J. Chem. Phys. *46*, 1646 – 1651.

Märk, E., Märk, T. D., Kim, Y. B., Stephan, K. (1981): J. Chem. Phys. *75*, 4446 – 4453.

Märk, T. D. (1984): In: Electron-Molecule Interactions and Their Applications (Christoporou, L. G., ed.), Vol. I chapter 3, 251 – 334. New York: Academic Press.

Märk, T. D., Egger, F. (1977): J. Chem. Phys. *67*, 2629 – 2635.

Märk, T. D., Egger, F., Cheret, M. (1977): J. Chem. Phys. *67*, 3795 – 3802.

Märk, T. D., Hille, E. (1978): J. Chem. Phys. *69*, 2492 – 2496.

Martin, S. O., Peart, B., Dolder, K. T. (1968): J. Phys. *B:* Atom. Molec. Phys. *1*, 537 – 542.

Massey, H. S. W. (1969): Electronic and Ionic Impact Phenomena (Massey, H. S. W., Burhop, E. H. S., Gilbody, H. B., eds.), Vol. II, 801 – 1064. Oxford: Clarendon Press.

Massey, H. S. W., Burhop, E. H. S. (1969): Electronic and Ionic Impact Phenomena (Massey, H. S. W., Burhop, E H. S., Gilbody, H. B., eds.), Vol. I, 97 – 168. Oxford: Clarendon Press.

McClure, G. W. (1953): Phys. Rev. *90*, 796 – 803.

McDowell, M. R. C. (1969): In: Case Studies in Atomic Collision Physics I (McDaniel, E. W., McDowell M. R. C., eds.), 47 – 97. Amsterdam-New York-Oxford: North-Holland.

McDowell, M. R. C., Meyerscough, V. P., Peach, G. (1965): Proc. Phys. Soc. (London) *85*, 703 – 707.

McFarland, R. H. (1967): Phys. Rev. *159*, 20 – 26.

McFarland, R. H., Kinney, J. D. (1965): Phys. Rev. *137*, A1058 – A1061.

McGuire, E. J. (1971): Phys. Rev. *A 3*, 267 – 279.

McGuire, E. J. (1977): Phys. Rev. *A 16*, 62 – 72.

Miles, W. T., Thompson, R., Green, A. E. S. (1972): J. Appl. Phys. *43*, 678 – 686.

Monnom, G., Gaucherel, P., Paparoditis, C. (1982): Vide, Couches Minces *212* (Suppl.), 349 – 354.

Montague, R., Harrison, M. F. A., Smith, A, C. H. (1984): J. Phys. B: Atom. Molec. Phys. *17*, 3295 – 3313.

Nagy, P., Skutlartz, A., Schmidt, V. (1980): J. Phys. B: Atom. Molec. Phys. *13*, 1249 – 1267.

Nygaard, K. J. (1968): J. Chem. Phys. *49*, 1995 – 2002.

Nygaard, K. J. (1974): In: Physics of Ionized Gases, 1974. Proceedings of Invited Lectures given at the VIIth Yugoslav Symposium and Summer School on the Physics of Ionized Gases (Vujnović, V., ed.), 229 – 264, Institute of Physics of the University of Zagreb.

Nygaard, K. J., Hahn, Y. B. (1973): J. Chem. Phys. *58*, 3493 – 3499.

Okuno, Y. (1971): J. Phys. Soc. Jpn. *31*, 1189 – 1195.

Okuno, Y., Okuno, K., Kaneko, Y., Kanomata, I. (1970): J. Phys. Soc. Jpn. *29*, 164 – 172.

Omidvar, K., Kyle, H. L., Sullivan, E. C. (1972): Phys. Rev. *A 5*, 1174 – 1187.

Otvos, J. W., Stevenson, D. P. (1956): J. Am. Chem. Soc. *78*, 546 – 551.

Pavlov, S. I., Rakhovskii, V. I., Fedorova, G. M. (1967): Zh. Eksp. Teor. Fiz. *52*, 21 – 28. [English translation: Sov. Phys. JETP *25*, 12 – 16 (1967).]

Pavlov, S. I., Stotskii, G. I. (1970): Zh. Eksp. Teor. Fiz. *58*, 108 – 114 [English translation: Sov. Phys. JETP *31*, 61 – 64 (1970).]

Peach, G. (1970): J. Phys. B: Atom. Molec. Phys. *3*, 328 – 349.

Peach, G. (1971): J. Phys. B: Atom. Molec. Phys. *4*, 1670 – 1677.

Pejcev, V. M., Kurepa, M. V., Cadez, I. M. (1979): Chem. Phys. Lett. *63*, 301 – 304.

Peterson, J. R. (1964): In: Atomic Collision Processes, The Proceedings of the IIIrd International Conference on the Physics of Electronics and Atomic Collisions, London, 1963 (McDowell, M. R. C., ed.), 465 – 473. Amsterdam-New York-Oxford: North-Holland.

Platzman, R. L. (1962): Vortex *23*, 372 – 385.

Platzman, R. L. (1963): J. Chem. Phys. *38*, 2775 – 2776.

Rapp, D., Englander-Golden, P. (1965): J. Chem. Phys. *43*, 1464 – 1479.

Rapp, D., Englander-Golden, P., Briglia, D. D. (1965): J. Chem. Phys. *42*, 4081 – 4085.

Rieke, F. F., Prepejchal, W. (1972): Phys. Rev. *A6*, 1507 – 1519.

Rothe, E. W., Marino, L. L., Neynaber, R. H., Trujillo, S. M. (1962): Phys. Rev. *125*, 582 – 583.

Samson, J. A. R. (1982): In: Handbuch der Physik (Flügge, S., ed.), Vol. 36, 123 – 213. Berlin-Heidelberg-New York: Springer.

Saxon, R. P. (1973): Phys. Rev. *A 8*, 839 – 849.

Schlachter, A. S., Tuan, Vu. N., Gautherin, G. (1973): In: Abstracts VIIIth International Conference on the Physics of Electronic and Atomic Collisions (Cobić, B. C., Kurepa, M. V., eds.), Vol. 2, 751 – 752. Beograd: Institute of Physics.

Schram, B. L., de Heer, F. J., van der Wiel, M. J., Kistemaker, J. (1965): Physica *31*, 94 – 112.

Schram, B. L., Moustafa, H. R., Schutten, J., de Heer, F. J. (1966a): Physica *32*, 734 – 740.

Schram, B. L., van der Wiel, M. J., de Heer, F. J., Moustafa, H. R. (1966b): J. Chem. Phys. *44*, 49 – 54.

Schroeer, J. A., Gunduz, D. H., Livingston, S. (1973): J. Chem. Phys. *58*, 5135 – 5140.

Schulz, G. J. (1958): Phys. Rev. *112*, 150 – 154.

Schulz, G. J. (1962): Phys. Rev. *128*, 178 – 186.

Schulz, G. J., Fox, R. E. (1957): Phys. Rev. *106*, 1179 – 1181.

Schutten, J., de Heer, F. J., Moustafa, H. R., Boerboom, A. J. H., Kistemaker, J. (1966): J. Chem. Phys. *44*, 3924 – 3928.

Seaton, M. J. (1959): Phys. Rev. *113*, 814.

Shah, M. B., Gilbody, H. B. (1981): J. Phys. B: Atom. Molec. Phys. *14*, 2831 – 2841.

Shearer-Izumi, W., Botter, R. (1974): J. Phys. B: Atom. Molec. Phys. *7*, L 125 – L 131.

Shimon, L. L., Nepiipov, É. I., Zapesochnyĭ, I. P. (1975): Zh. Tekh. Fiz. *45*, 688 – 689 [English translation: Sov. Phys. Tech. Phys. *20*, 434 (1975).]

Siegel, M. W. (1982): Intern. J. Mass Spectrom. Ion Phys. *44*, 19 – 36.

Smith, A. C. H., Caplinger, E., Neynaber, R. H., Rothe, E. W., Trujillo, S. M. (1962): Phys. Rev. *127*, 1647 – 1649.

Smith, P. T. (1930): Phys. Rev. *36*, 1293 – 1302.

Smith, P. T. (1931): Phys. Rev. *37*, 808 – 814.

Spiess, G., Valance, A., Pradel, P. (1972): Phys. Rev. *A 6*, 746 – 755.

Starace, A. F. (1982): In: Handbuch der Physik (Flügge, S., ed.), Vol. 36, 1 – 121. Berlin-Heidelberg-New York: Springer.

Stebbings, R. F. (1957): Proc. Roy. Soc. (London) *A 241*, 270 – 282.

Stephan, K., Helm, H., Kim, Y. B., Seykora, G., Ramler, J., Grössl, M., Märk, E., Märk, T. D. (1980a): J. Chem. Phys. *73*, 303 – 308.

Stephan, K., Helm, H., Märk, T. D. (1980b): J. Chem. Phys. *73*, 3763 – 3778.

Stevie, F. A., Vasile, M. J.: J. Chem. Phys. *74*, 5106 – 5110.

Tan, K. H., Brion, C. E., van der Leeuw, Ph. F., van der Wiel, M. J. (1978): Chem. Phys. *29*, 299 – 309.

Tate, J. T., Smith, P. T. (1932): Phys. Rev. *39*, 270 – 277.

Tate, J. T., Smith, P. T. (1934): Phys. Rev. *46*, 773 – 776.

Tawara, H., Harrison, K. G., de Heer, F. J. (1973): Physica *63*, 351 – 367.

Taylor, J. B., Langmuir, L. (1937): Phys. Rev. *51*, 753 – 760.

Taylor, P. O., Dolder, K. T., Kaupilla, W. E., Dunn, G. H. (1974): Rev. Sci. Instrum. *45*, 538 – 544.

Tozer, B. A. (1958): J. Electron. Control *4*, 149 – 159.

Tozer, B. A., Craggs, J. D. (1960): J. Electron. Control *8*, 103 – 107.

Vainshtein, L. A., Ochkur, V. I., Rakhovskii, V. I., Stepanov, A. M. (1971): Zh. Eksp. Teor. Fiz. *61*, 511 – 519. [English translation: Sov. Phys. JETP *34*, 271 – 275 (1972).]

van Wingerden, B., Wagenaar, R. W., de Heer F. J. (1980): J. Phys. B: Atom. Molec. Phys. *13*, 3481 – 3491.

Vriens, L., Bonsen, T. F. M., Smit, J. A. (1968): Physica *40*, 229 – 252.

Wagenaar, R. W., de Heer, F. J. (1980): J. Phys. B: Atom. Molec. Phys. *13*, 3855 – 3866.

Wiegand, W. J., Boedeker, L. R. (1982): Appl. Phys. Lett. *40*, 225 – 227.

Winters, H. F. (1975): J. Chem. Phys. *63*, 3462 – 3466.

Winters, H. F., Inokuti, M. (1982): Phys. Rev. *A 25*, 1420 – 1430.

Zapesochnyĭ, I. P., Aleksakhin, I. S. (1968): Zh. Eksp. Teor. Fiz. *55*, 76 – 85 [English translation: Sov. Phys. JETP *28*, 41 – 45 (1969).]

8
Electron-Ion Ionization

G. H. Dunn

Joint Institute for Laboratory Astrophysics, University of Colorado and National Bureau of Standards, Boulder, Colorado, U.S.A.

8.1 Introduction

Electron impact ionization of ions received little attention until the 1950's when nonequilibrium approaches to astrophysics gained favor and the international push for controlled thermonuclear fusion began. Spurred by these needs, discussed more fully in Chapter 9, and aided by technological advances in ultra-high vacuum and electronics, Dolder, Harrison and Thoneman (1961) made the first successful electron-ion ionization measurements on $e + He^+ \rightarrow He^{2+} + 2e$. Not only was this the first ever measurement of this type, but it has served as a role model of the crossed charged beams technique and of quality work for the dozens of subsequent beams experiments.

Within five years of the He^+ crossed beams measurement, Hinnov (1966) introduced techniques for measurements of rate coefficients of ionization in plasmas. Baker and Hasted (1965, 1966) introduced trapped ion techniques for ionization measurements. These two additional basic approaches to ionization rate measurements and variations thereof stood for a decade as the only ways to study multiply charged ions beyond twice charged. However, starting with measurements of Crandall *et al.* (1977, 1978) using the ORNL PIG ion source (Mallory and Crandall, 1976), measurements have now been made on ions with charges up to $q = +6$ using the crossed beams technique. Nevertheless, rate measurements using plasma and trapped ion methods are still relied upon for measurements of more highly charged species. A number of excellent reviews have been written and bibliographies compiled, and these are listed separately at the end of this chapter under "Reviews and Bibliographies". The reader is referred especially to recent ones by Dolder and Peart (1976), Dolder (1983), Salzborn (1983), Crandall (1983 a, b), and Dunn (1980).

Fig. 8-1 provides an overview of measurements which have now (1983) been made. Here the ordinate is the charge q of the target ion and the abscissa is the atomic number Z. Data are shown only for single ionization, i.e. for ions of charge q being ionized to ions of charge $q + 1$. Solid blocks represent crossed beams measurements, a diagonal line from lower left to upper right represents a trapped ion rate measurement yielding absolute cross sections, a diagonal line from upper left to lower right represents a plasma rate measurement, and an open square with a dot in the center represents measurements in traps which were normalized using beam or other data or are simply relative data. The notation is hierarchical in the order listed; i.e. a crossed beams measurements precludes listing any other measurement, and plasma rate or trapped ion rate measurements preclude showing relative trap data, but both types of rate measurements can be shown when appropriate.

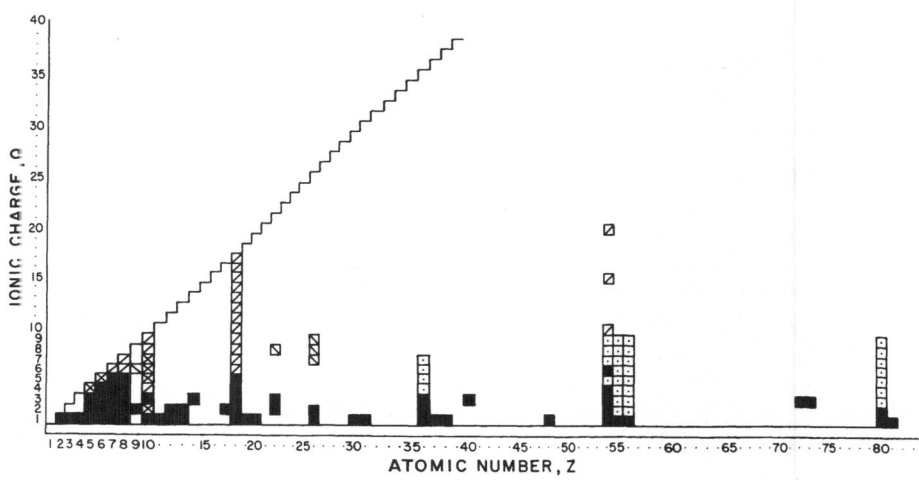

Fig. 8-1. Ions for which electron-impact single ionization cross sections or rates have been measured. The ordinate shows the charge of the target, and the abscissa the atomic number Z. Solid blocks represent crossed beams measurements, a diagonal line from lower left to upper right represents a trapped ion rate measurement yielding absolute cross sections, a diagonal line from upper left to lower right represents a plasma rate measurement, and open square with a dot represents ion trap data which were normalized to other measurements or are purely relative data. The notation is hierarchical in the order listed

Table 8-1 lists in a more detailed fashion the crossed beams measurements which have been made, including brief generic comments about the measurements. Fig. 8-1 and Table 8-1 refer exclusively to atomic ions. Crossed beams measurements have also been made for ionization of H_2^+ (Peart and Dolder, 1973) and of CO^+ (Müller et al., 1980 a), but there appear to be no measurements for any other molecular ions. The bibliographies of Takayanagi and Iwai (1978) and of Itikawa (1982) are particularly complete, well organized and easy to use for references to both experiment and theory for electron-ion ionization.

Table 8.1. *Crossed beams measurements of single ionization of positive ions by electron impact. For symbols, see key at bottom of table*

Target	Highest electron energy (eV)	Generic comments	Other comments	Reference
He⁺	1000	M, G, D, B		DHT 61
	10000	M, G, D, B		PWD 69
	74	M, G, D, B		DBCV 81
	750	N, G, D, B		AMSB 83
Li⁺	800	M, G, D, B		LHM 66
	1000	M, G, D, B		WD 67
	3000	M, G, D, B		PD 68b
	25000	M, G, D, B		PWD 69
Be⁺	1600	M, G, I⁻, B		FD 83
B⁺	400	M, E, D, B	metastables varied 50–90%	C et 83b
B²⁺	1500	M, G, I⁻, B		C et 83
B³⁺	1500	M, G, D, B		CPG 79
C⁺	800	M, E, I⁻, B	possible EA from metastable ⁴P	AHR 71
C²⁺	1000	M, E, D, B	up to 40% metastables	WHHB 78
	500	M, E, D, B	metastables varied 65–90%	F et 83b
C³⁺	500	M, G, ?, B		CPT 78
	1500	M, G, I⁺, B		CPHG 79
C⁴⁺	900	M, G, D, B		CPG 79
N⁺	500	M, E, ?, K	metastables, N₂²⁺ probably in target beam	HDT 63
N²⁺	900	M, E, I⁻, B	possible EA from metastable ⁴P	AHR 71
N³⁺	1500	M, E, D, K	metastable fraction approaching 90%	F et 83b
N⁴⁺	500	M, G, I⁺, B		CPT 78
	1500	M, G, I⁺, B		CPHG 79
N⁵⁺	1500	M, G, D, C	4 points only	CPG 79
O⁺	1000	M, E, D, B		AH 71
	350	N, ?, D, B	lower than AH 71 by ~20%	M et 80b
O²⁺	500	M, E, D, B		AH 71
O³⁺	1500	M, E, D, B	metastables ~16%	CPG 79
O⁴⁺	1500	M, E, I⁻, K	metastables ~90%	F et 83b
O⁵⁺	1400	M, G, I⁺, C		CPG 79
F²⁺	1500	M, E, D, B		M et 83
Ne⁺	1000	M, G, D, B		DHT 63
	800	N, G, D, B	lower than DHT by ~20%	M et 80b
Ne³⁺	1500	M, E, D, B		GDC 83
Na⁺	1000	M, G, D, B		HLB 66
	3500	M, G, D, B		PD 68a
Mg⁺	2000	M, G, D, B		MPD 68
	400	M, G, I⁺, B		C et 82
Mg²⁺	3000	M, G, D, B		PMD 69
Al⁺	750	M, E, D, A	ground state σ deduced	MH 83
	1000	M, G, I⁻, B		B et 82
Al²⁺	1000	M, G, I⁺, B		C et 82

Table 8.1 (*continued*)

Target	Highest electron energy (eV)	Generic comments	Other comments	Reference
Si^{3+}	1500	M, G, I^+, B		C *et* 82
Cl^{2+}	1500	M, E, I^+, B		M *et* 83
Ar^+	1000	M, G, I^+, B		WHH 78
	700	N, G, I^+, B		M *et* 80b
Ar^{2+}	750	N, ?, I^+, B		M *et* 80b
	1000	M, E, I^+, B		M *et* 83
	1000	M, E, I^+, C	efficiency of MCP assumed to be unity	MO *et* 83
Ar^{3+}	700	N, E, I^+, B		M *et* 80b
	1500	M, E, I^+, B		GDC 83
Ar^{4+}	1400	M, G, D, B		
	850	N, G, D, B		M *et* 80b
Ar^{5+}	850	N, ?, D, B		M *et* 80b
K^+	1000	M, G, D, B		HLB 66
	3000	M, G, D, B		PD 68a
Ca^+	800	M, G, I^+, B	extraordinary indirect contributions	PD 75
Ti^{2+}	1500	M, G, I^+, B	extraordinary indirect contributions	M *et* 83
Ti^{3+}	1000	M, G, I^+, B	extraordinary indirect contributions	F *et* 81 F *et* 83a
Fe^+	750	M, E, ?, B	ground state σ deduced	MDHS 83
Fe^{2+}	1500	M, E, D, B		M *et* 83
Cu^{2+}		M, E, ?, B		G 83
Zn^+	2000	M, G, I^+, B	clear autoionization onsets	R *et* 82
Ga^+	2000	M, G, I^+, B	clear autoionization onsets	R *et* 82
Kr^+	400	N, E, D, C	metastables strong function of ion source conditions	LK 68
Kr^{2+}	1500	M, G, D, B		G 83
	1000	M, E, ?, C		MO *et* 83
Kr^{3+}	1000	M, E, I^+, B	sharp fall off of indirect processes	GDC 83
Rb^+	500	M, G, D, B		PD 75
	2000	M, G, D, B		FSD 78
Sr^+	700	M, G, I^+, B	extraordinary EA contributions	PD 75
Zr^{3+}	1000	M, G, I^+, B	extraordinary EA contributions	F *et* 81 F *et* 83a
Cd^+	2000	M, G, D, B		B *et* 82
Xe^+	800	N, ?, ?, B	about 25% lower than later data	M *et* 80b
	700	N, E, D, B		AMSB 83
Xe^{2+}	700	N, E, I^+, B	extraordinary EA contributions	AMSB 82
Xe^{3+}	1500	M, E, I^+, B	extraordinary EA contributions possible REDA. Very rapid falloff of indirect contributions	GDC 83
	700	N, E, I^+, B	same comments as above $15-30\%$ larger values	AMSB 83

Table 8.1 (*continued*)

Target	Highest electron energy (eV)	Generic comments	Other comments	Reference
Xe^{4+}	700	N, E, I^+, B	extraordinary EA contributions $\sigma_{4\to5}$ becomes larger than $\sigma_{3\to4}$ above	AMSB 83
Xe^{6+}	1500	M, G, I^+, B	very high EA and probable REDA contributions	GC 83
Cs^-	400	M, G, D, B		PD 75
	500	M, G, D, B		HFHS 82
Ba^+	2000	M, G, I^+, B	extraordinary EA contributions	PD 68 b
	1000	M, G, I^+, B		FHE 72
	25	M, G, I^+, B	energy resolved electrons	PSD 73
Hf^{3+}	1000	M, G, I^+, B	extraordinary EA and probable REDA contributions	F et 81 F et 83 a
Ta^{3+}	300	M, G, I^+, B	high indirect effects	F et 83 a
Hg^+	2000	M, G, D, B		B et 82
Hg^{2+}	150	O, ?, ?, K	one point only — unexpectedly high, possible indirect processes	LKT 64
Tl^+	2000	M, G, D, B		DFSH 76

Highest electron energies are to nearest 100 eV.

Generic comments: First column refers to form factor measurements; M = measured in the experiment; N = form factor inferred by normalizing measurements for a given ion to results of another author who measured form factor, then assuming the same for other ions; O = no evidence that form factor was measured; P = electron current density measured during set-up tests, and assumed reproducible so form factor could be inferred during experiments. *Second column* refers to whether target ion beam was in ground state (G), excited ions were also present in beam (E), or quite undeterminable (K). *Third column* refers to whether ionization was due to direct ionization D; whether indirect processes of excitation-autoionization (EA) and resonant-excitation-double-autoionization (REDA) were obviously observed, but at less than a 10% level (I^-), or whether indirect processes were clearly present at greater than a 10% level (I^+). If the issue was not clear, a (D) was entered. *Fourth column* refers to accuracy of data, showing uncertainties u: A, $u < 5\%$; B, $5\% < u < 15\%$; C, $u < 30\%$; D, $u < 50\%$; K, unknown and not readily evaluated.

Other comments are to emphasize certain points about the experiments and data.

References

AHR 71, Aitken *et al.* (1971); AH 71, Aitken and Harrison (1971); AMSB 83, Achenbach *et al.* (1983); B et 82, Belić *et al.* (1982); C et 82, Crandall *et al.* (1982); CPT 78, Crandall *et al.* (1978); C et 83, Crandall *et al.* (1983); CPG 79, Crandall *et al.* (1979 a); CPHG 79, Crandall *et al.* (1979 b); DBCV 81, Defrance *et al.* (1981); DHT 61, Dolder *et al.* (1961); DHT 63, Dolder *et al.* (1963); DFSH 76, Divine *et al.* (1976); FD 83, Falk and Dunn (1983); F et 83 a, Falk *et al.* (1983 a); F et 83 b, Falk *et al.* (1983 b); FSD 78, Feeney *et al.* (1978); FHE 72, Feeney *et al.* (1972); GDC 83, Gregory *et al.* (1983); G 83, Gregory (1983); GC 83, Gregory and Crandall (1983); HDT 63, Harrison *et al.* (1963); HFHS 82, Hertling *et al.* (1982); HLB 66, Hooper *et al.* (1966); LHM 66, Lineberger *et al.* (1966); LK 68, Latypov and Kuprianov (1968); LKT 64, Latypov *et al.* (1964); MPD 68, Martin *et al.* (1968); M et 83, Mueller *et al.* (1983); M et 80 b, Müller *et al.* (1980 b); MH 83, Montague and Harrison (1983); MO et 83, Matsumoto *et al.* (1983); MDHS 83, Montague *et al.* (1983); PD 68 a, Peart and Dolder (1968 a); PD 68 b, Peart and Dolder (1968 b); PWD 69, Peart, Walton and Dolder (1969); PMD 69, Peart, Martin, Dolder (1969); PD 75, Peart and Dolder (1975); PSD 73, Peart *et al.* (1973); R et 82, Rogers *et al.* (1982); WD 67, Wareing and Dolder (1967); WHHB 78, Woodruff *et al.* (1978 b); WHH, Woodruff *et al.* (1978 a).

Mechanisms leading to ionization have been described in previous chapters. The discussions in Sections 1.2.3 and 1.3.5 are of notable significance to this chapter. The so-called "excitation-autoionization" (EA) process seems to be of particular importance. Simplistically, one can view this as a two-step process involving, first, excitation of an inner shell electron, followed by autoionization. Excitation of ions by electrons is characterized by a finite cross section at threshold, as opposed to excitation of neutrals in which the cross section rises from zero as $k_1^{2l_1+1}$ (k_1 and l_1 are respectively the wave number and angular momentum quantum number of the scattered electron). Thus, the EA process for ions gives rise to step function increases in the cross section as shown by the dashed lines in Fig. 1-1. As noted in Section 1.2.3, and as will be discussed later, this mechanism can result in enhancements of the total ionization cross section of more than an order of magnitude.

As described in Section 1.2.3, resonances in the incident electron channel can also be of importance in electron-ion ionization. Consider an electron which at infinity has ε less than Δ, the energy needed to excite a level nl of the ion. As the electron approaches the ion, it gains kinetic energy, and at near distances will have enough kinetic energy to excite the level nl. As the electron tries to leave, however, it finds itself bound with energy ε. The process is resonant, since ε must be just the energy to leave the electron in a well-defined orbital $n'l'$. This process is called dielectronic capture, and if radiation of the state nl occurs before autoionization, dielectronic recombination has taken place. In the case of interest for the purposes of this chapter, however, the target electron bound to the ion is an inner-shell electron and the states nl are the same as those leading to EA. Before any radiative stabilization can take place, both the nl and $n'l'$ electrons are ejected in autoionization. This "resonant-excitation-double-autoionization" (REDA) process was suggested by La Gattuta and Hahn (1981) and gives rise to resonant spikes in the net ionization cross section located below the step rises associated with EA. The combined effect of a number of these resonances as calculated by La Gattuta and Hahn is shown by the solid curve in Fig. 1-1.

Henry and Msezane (1982) distinguish, upon suggestion of A. R. P. Rau, between whether the doubly excited state described above decays in two steps or a single step. Thus, as an example, for a sodium-like ion the REDA process would be represented

$$
\begin{aligned}
e^- + 2p^6\,3s &\rightarrow 2p^5\,3s\,3p\;nl \\
&\rightarrow e^- + 2p^5\,3s^2 \\
&\rightarrow e^- + 2p^6 + e^-.
\end{aligned}
\tag{8-1}
$$

The suggestion of Rau was that one must also consider "resonant-excitation-auto-double-ionization" (READI) which would be represented

$$
e^- + 2p^6\,3s \rightarrow 2p^5\,3s\,3p\;nl \rightarrow 2p^6 + 2e^-.
\tag{8-2}
$$

Currently, the experimental methods will not distinguish these processes, but both should be kept in mind as further studies are made.

Multiple ionization (MI), described by Eq. (5-3), has had little attention for ions. It has come to be conventional wisdom that the cross sections for MI are typically at most only a few percent of those for single ionization, and thus not so important in the applications described in Chapter 9. However, MI cross sections can be comparable to those for single ionization and it is essential that further attention be given to this area experimentally, theoretically, and in modeling.

In the remainder of this chapter we outline the experimental methods used to measure ionization cross sections and rates, paying attention to advantages, disadvantages, limitations, and projected improvements. Status of the ionization data is discussed, with emphasis of data for species where mechanisms and issues are most readily highlighted.

8.2 Experimental Methods

8.2.1 Crossed Beams

The crossed-beams technique has been discussed in Section 7.1.6 with emphasis on measurements of ionization of neutral particles. Reviews by Harrison (1966, 1968), Dunn (1969), Dolder (1969) and Dolder and Peart (1976) give good discussions of the method and the problems associated.

The notion of colliding beams is conceptually simple as represented in the block diagram of Fig. 7-4. Beams of electrons and ions are generated and caused to intersect at some angle θ. Incident and product particles are separated, and their intensities measured with suitable instrumentation.

The ion signal R_s (number of ions formed per second) per unit volume resulting from ionizing collisions between electrons and target ions can be obtained from the relationship

$$\frac{dR_s}{dV} = N_i N_e v_r \sigma \tag{8-3}$$

where N_i and N_e are the number densities of target ions and electrons respectively, v_r is the relative velocity of these colliding species, and σ is the ionization cross section. If ions and electrons move in mutually perpendicular beams parallel to the x and y axes respectively, number densities and relative velocity are given by

$$N_i = \Gamma_i(x, z)/v_i; \quad N_e = \Gamma_e(y, z)/v_e$$

and

$$v_r = (v_e^2 + v_i^2)^{1/2}.$$

The ion signal can be obtained from integration of Eq. (8-3)

$$R_s = \sigma \frac{(v_e^2 + v_i^2)^{1/2}}{v_e v_i} \int_v \Gamma_i(x, z) \Gamma_e(y, z) \, dx \, dy \, dz,$$

where V is the volume where nonvanishing flux densities Γ_i and Γ_e exist simultaneously. For the beam conditions assumed, the integrations over x and y can be performed at once leaving

$$R_s = \frac{\sigma}{v} \int_z j_i(z) j_e(z) \, dz, \tag{8-4}$$

where $j_i(z)$ and $j_e(z)$ are one-dimensional spatial flux densities given by

$$j_i(z) = \int_x \Gamma_i(x, z)\, dx; \; j_e(z) = \int_y \Gamma_e(y, z)\, dy, \tag{8-5}$$

and

$$v = \frac{v_i v_e}{(v_i^2 + v_e^2)^{1/2}}.$$

Eq. (8-4) is the fundamental equation from which cross sections can be extracted for *crossed* beams and Eqs. (8-4) and (8-5) are essentially the same as Eqs. (7-14) and (7-15). The generalization to inclined beams at angle θ is obvious, and merely results in replacing the velocity factor v in Eq. (8-4) by

$$v_e v_i \sin\theta/(v_i^2 + v_e^2 - 2 v_i v_e \cos\theta).$$

Though inclined and merged beams have been used for a variety of collision processes, ionization work has been done predominantly using crossed beams, so the form of Eq. (8-4) is as given. Proceeding operationally to obtain quantities to use in Eq. (8-4) to get cross sections has taken two major routes.

The most widely used to date is that introduced by Dolder, Harrison, and Thoneman (1961). A thin slit (parallel to x for ions or parallel to y for electrons) performs the integrations of Eq. (8-5) automatically if the respective currents passing through the slit are measured. If the slit is scanned along z, the functions $j_i(z)$ and $j_e(z)$ can then be mapped. Noting that

$$I_i = q e \int_z j_i(z)\, dz, \; I_e = e \int_z j_e(z)\, dz,$$

one can then write Eq. (8-4) as

$$R_s = \sigma \frac{I_i I_e}{q e^2 v} \frac{1}{\mathscr{F}} \tag{8-6}$$

where

$$\mathscr{F} = \frac{\int_z j_i(z)\, dz \int_z j_e(z)\, dz}{\int_z j_i(z) j_e(z)\, dz}. \tag{8-7}$$

Use of this method, then, involves scanning the two beams to get the distributions. and performing the numerical integrals of Eq. (8-7). With modern computer controlled stepping motors, data collection, and reduction, this is a straightforward. rapid, and accurate procedure; but in the first experiments, much tedious effort was involved. When one beam is larger, is uniform, and of height h, then $\mathscr{F} \equiv h$. Most experiments have been designed to satisfy this condition, hence reducing the sensitivity of the experiment to beam energy and other variables. Nevertheless, it has been shown that measuring $\mathscr{F}$ is important, and when discrepancies arise between experiments in which $\mathscr{F}$ was and was not measured, those in which $\mathscr{F}$ was not measured are most suspect.

An alternative procedure has been introduced by Defrance *et al.* (1981). In their method one beam is moved through the other while keeping it parallel to its initial axis. Let $Z(t)$ be the position of the moving beam. If the beam sweep is over a large enough distance that at the extreme positions on either side the beams no longer overlap, and if the sweep takes place at a constant speed u, then the total number of signal ions during a sweep is

$$N_s = \int_{-\infty}^{\infty} R_s \, dt = \int_{-\infty}^{\infty} \frac{R_s}{u} \, dZ = \frac{\sigma}{v} \int_{-\infty}^{\infty} \int_{-\infty}^{\infty} j_i(z) j_e(z - Z) \, dz \, dZ$$

$$= \frac{\sigma I_i I_e}{q e^2 v} \frac{1}{u}. \tag{8-8}$$

The similar structure of Eqs. (8-6) and (8-8) is obvious, and one notes that $R_s \mathscr{F} = N_s u$

The advantage of the Defrance *et al.* method for collisions with neutral beams is obvious, since one does not then need to worry about measuring the differential distribution of a neutral beam. For charged beams it is not obvious that either method has a particular advantage over the other except when taken in the context of a particular experimental arrangement.

It was emphasized that the crossed beams approach is conceptually simple. From Eqs. (8-6) and (8-8) it would appear that the measurements are also operationally simple as well, provided the apparatus is initially configured to measure R_s (or N_s), I_i, I_e, v and $\mathscr{F}$ (or u). However, there are a number of effects which can lead to false signals, incorrect measurements, etc. Such effects must be recognized, tested for, then eliminated or accounted for. These problems have been discussed in some detail in the reviews cited at the beginning of this section; however, we mention the major effects here.

1. Accompanying every target ion beam of X^{N+} ions which are being ionized to form X^{n+} ions there is a "beam" of X^{n+} ions which have been formed by collisions of the X^{N+} with ambient gas, slit edges, etc. Trajectories are not as well defined for these X^{n+} ions due to possible scattering on collision and to the fact that upstream focusing elements have a different effect on these ions than the parent beam. These ions give rise to a background at the signal detector which is normally linear with background pressure (but may have a pressure independent component from stripping at slits, surfaces, etc.). The background can, in principle, be separated from the signal simply by turning the electrons on and off — modulating the electrons — and taking the difference in detected particles. However, due to the non-optimum trajectories of the X^{n+} ions noted above, some may not be getting into the collector, and the space charge fields due to the electron beam may cause some of these ions to now enter the collector (or, cause some which had been barely getting into the collector to now miss) thus giving rise to "false signal". Harrison (1966) has shown that this false signal can be approximated by

$$\Delta I_b^{(n)} = I_b^{(n)} C I_e / T^{1/2} E \qquad\qquad (8\text{-}9)$$

where $I_b^{(n)}$ is the current of background X^{n+} ions, C is a geometric parameter of the apparatus, T is electron energy and E is ion energy. The linear dependences upon ion and electron currents make $\Delta I_b^{(n)}$ very difficult to distinguish from legitimate signal. The best way to do this is to look for apparent signal at electron energies below the ionization threshold. When metastable ions may be in the target beam, it becomes a more difficult problem to sort out, since they also give rise to "signal" below threshold. Clearly, the best solution is to minimize C, and, by making the vacuum as good as possible, also to minimize $I_b^{(n)}$. The experiment of Crandall, Phaneuf and Taylor (1978) on C^{3+} and N^{4+} is a good example of recent work in which this issue was dealt with.

2. As noted above, the background of X^{n+} ions can, in principle, be separated from the signal by modulating the electrons and taking the difference in detected particles when electrons are on and when they are off. However, the electron beam may desorb gas from the collector or other surfaces intercepting portions of the beam. The background X^{n+} ions formed by gas collisions would then also be modulated synchronously with the electrons. The effect may be modeled as a modulation of the equilibrium vacuum chamber pressure, or as a modulated "beam" of gas passing through the interaction volume. In either case the interaction of the ion beam with the synchronously modulated neutral gas density may lead to a component of background count rate which cannot be separated from the true signal count rate by the chopping scheme. There is an important difference between two models for pressure modulation.

In the case of the equilibrium pressure model it can readily be shown that for $\omega\tau \gg 1$ the time-dependence of the chamber pressure is

$$P(t) = \frac{\rho_w}{s} + e^{-t/\tau}\left[P_0 - \frac{\rho_w}{s} + \frac{\rho_b}{s}\frac{1}{\omega\tau}\right] + \frac{\rho_b}{s}\frac{1}{\omega\tau}\sin(\omega\tau - \pi/2) \qquad (8\text{-}10)$$

where s is the pumping speed on the chamber, $\tau = (V/s)$ is the pumping time constant ($V=$ chamber volume), ρ_w and ρ_b are source terms due to desorption from chamber walls and desorption due to the beam, respectively, P_0 is the initial pressure, and ω is the fundamental beam modulation frequency. The third term in Eq. (8-10) is the modulation term, and two points should be made. First, the pressure modulation is phase-shifted by the high-frequency limiting value of $-\pi/2$ from the phase of the beam, so the spurious contribution is in general negative with respect to the true signal. Second, the pressure modulation is attenuated by $(\omega\tau)^{-1}$, so it should be observable by varying the chopping frequency.

In the gas beam model it is assumed that there is a straight path from the source of gas to the interaction region so that the phase lag $\omega\tau_f$ between the electron beam and the desorption-produced gas beam depends upon the flight time τ_f of the gas. Thus, if the beam model is operative, the *sign* of the spurious signal should be frequency-dependent.

3. It was mentioned above that ion species which have metastable states may be particularly difficult to study in an unambiguous way, since normally the target beam will contain mixtures of states in unknown fractions. In recent work of Falk

et al. (1983 b) on Be-like ions, metastable fractions up to 90% were found. In those experiments it was shown that when the metastable fraction was appreciable and could be varied, the cross section for ionization of the ground state could be extracted in the following way. The measured cross section σ for the mixed-state beam can be related to the ground-state cross section σ_g and the metastable-state cross section σ_*, below and above the ground-state threshold energy E_g^{th}, by the respective equations

and
$$\sigma(E_b) = f \, \sigma_*(E_b)$$

$$\sigma(E_a) = (1-f) \, \sigma_g(E_a) + f \, \sigma_*(E_a)$$

where f is the metastable fraction and $E_b < E_g^{th} < E_a$. If the metastable fraction can be changed, then a second set of equations can be obtained with σ' and f'. Solving the four equations, we obtain

$$\sigma_g(E_a) = \frac{\sigma(E_a) - \kappa \, \sigma'(E_a)}{1 - \kappa}, \tag{8-11}$$

where

$$\kappa = \frac{\sigma(E_b)}{\sigma'(E_b)} = \frac{f}{f'}. \tag{8-12}$$

Varying the metastable fraction in a way which is reproducible and long-lasting enough to get data is often difficult, and use of this procedure met with mixed success — depending on species — in the work of Falk *et al.* One can make assumptions (e.g. from the Lotz or classical formulations) about the ratio of ground state to metastable cross sections and proceed from that to deduce the metastable fraction f. This has been done by Aitken and Harrison (1971) and by Montague and Harrison (1983). Most often, when metastables are present, it has been the practice of workers to simply cite the results, note the probable presence of metastables, and leave it at that. As dissatisfying as this may be, it is often the only choice, since ionization data are of practical use (Chapter 9) and some data are better than no data.

8.2.2 Plasma Rate Measurements

As noted in the Introduction, just a few years after the first cross section measurements using crossed beams, Hinnov (1966) introduced a method for extracting rate coefficients from measurements of plasma parameters. The method has been successfully applied by several investigators since then, and is one of the only ways for making measurements on highly charged ions. Hinnov (1966, 1967) measured ionization rate coefficients for $T_e \sim 10 - 15$ eV for Ne ions from Ne^+ to Ne^{7+} in a stellerator discharge. Measurements were made in θ-pinch discharges for C^{4+} ions for $T_e \sim 205 - 240$ eV by Kunze *et al.* (1968); for C^{3+}, N^{4+}, O^{4+}, O^{5+}, Ne^{6+} at $T_e \sim 100 - 260$ eV by Kunze (1971); for Ar^{7+} at $T_e \sim 62$ and 260 eV by Datla *et al.* (1972); for Fe^{7+}, Fe^{8+}, Fe^{9+} at $T_e \sim 50$, 95, and 142 eV by Datla *et al.* (1975); for B^{3+} and C^{4+} at $T_e \sim 220$ eV by Datla *et al.* (1976); for Ne^{5+}, Ne^{6+}, Ne^{7+} at $T_e \sim 120 - 400$ eV by Jones *et al.* (1977); for N^{4+} and O^{5+} at $T_e \sim 80 - 150$ eV by

Källne and Jones (1977); for N^{4+} and O^{5+} and 80 eV by Rowan and Roberts (1979); for B^{3+} at 140 and 175 eV, B^{4+} at 230 eV, C^{4+} at 160, 210 eV, C^{5+} at 230 eV, N^{4+} at 110 and 120 eV, N^{5+} at 235 eV, O^{5+} at 135 and 160 eV, F^{6+} at 160 and 180 eV, and Ne^{7+} at 160 and 220 eV by Greve et al. (1981); for Ti^{8+}, Ne^{5+}, Ne^{6+}, and O^{5+} at 55 eV by Datla and Roberts (1983), and for N^{3+} at 32 and 43 eV and N^{4+} at 68 and 75 eV by Brown et al. (1983). Breton et al. (1978) measured rates in a Tokamak discharge for Mo ions about 30 times charged for T_e in the neighborhood of 2 keV.

In general, the method depends on low-density plasmas so that a coronal model holds; i.e. the steady state population of an excited state is determined by a balance between the sum of collisional transitions into that state and the sum of all radiative decay rates. The intensity of radiation from ionic transitions is observed as a function of time, and this leads to the density of ions in the upper level. Rate coefficients are deduced from the observed densities by the use of appropriate models.

Normally, the plasma is modeled with the set of linear equations

$$\frac{dN_z}{dt} = N_e N_{z-1} S_{z-1} - N_e N_z (S_z + \alpha_{z-1} + N_e \gamma_{z-1}) + N_e N_{z+1} (\alpha_z + N_e \gamma_z)$$

$$(8\text{-}13)$$

where N_z is the number density of the zth ion, N_e the electron density, S_z the ionization rate coefficient for the zth ion, α_z the recombination rate coefficient, and γ_z the collisional recombination rate coefficient. In essentially all cases, a set of rate coefficients based upon theory or semi-empirical formulas is put into the set of Eqs. (8-13), and the equations solved. The time-dependent ion populations that result from this are then used to calculate specific spectral line intensities using the coronal model for level populations within an ion. These simulated time-dependent spectral line intensities are in turn compared with the observed line intensities. The coefficients are then iteratively adjusted until suitable agreement with observations

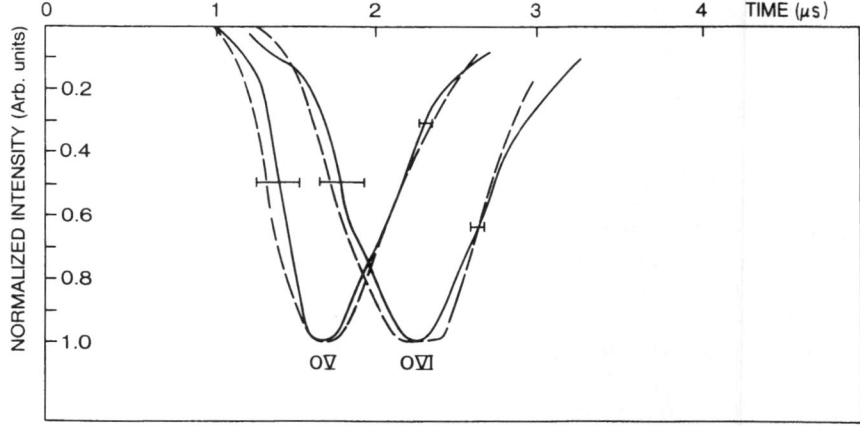

Fig. 8-2. Intensities of emission lines of O V and O VI normalized to their peak values as a function of time from the initiation of the main bank. Experiment (———); theoretical simulation (– – –). From Datla and Roberts (1983)

is obtained. Quite often, recombination is judged to be negligible and ignored in the solutions to (8-13). Fig. 8-2 is an example of observed and simulated line intensities for O^{4+} and O^{5+} from a paper by Datla and Roberts (1983).

Fig. 8-3 schematically demonstrates a typical experimental setup (Greve *et al.* 1981). The ruby laser and detector combination are for measurement of 90° Thomson scattering of the laser light by the plasma. The device can be scanned through the plasma, thus yielding electron density and temperature anywhere in the plasma. Both visible and uv spectroscopy are done, and often a high-speed framing camera is used to observe gross dynamics of plasma behavior. Observations of time histories of line intensities are made both end-on and side-on.

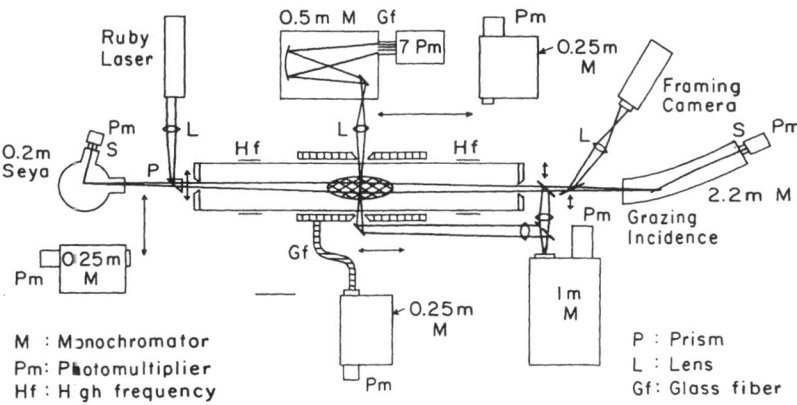

Fig. 8-3. Schematic overview of a typical experimental setup for plasma ionization rate measurements from Greve *et al.* (1981)

For the method to be applicable, the time evolution of the line intensities should be more sensitive to the ionization rates than to variations in electron density, electron temperature and plasma length. Careful diagnoses of the plasmas are made, and observations are carried out under conditions that the above plasma parameters are relatively constant.

It must be recognized that the output of such observations are *effective* ionization rates. It is typical to include in the model contributions to ionizations from at least one effective metastable level (Kunze *et al.* 1968; Datla *et al.* 1976; Greve *et al.* 1981). More recently, Datla and Roberts (1983) have modeled contributions from excitation-ionization, i.e. contributions from states above the collision limit, which is the lowest principal quantum number for which further collisional excitation is as high as radiative decay. Such a collision limit exists even in low-density plasmas, since the radiative decay rate *decreases* with increasing quantum number, and the collisional excitation to higher levels *increases* due to a decreasing energy difference between levels. Datla and Roberts (1983) found contributions from these levels to be comparable in magnitude to those from the ground state for a number of cases. Typically, multiple ionization has been ignored in the analyses and modeling.

However, it can now be said that this can be a serious shortcoming for some species.

Experimental uncertainties or "errors" are typically quoted in the range 20% to 40% depending on species.

Measured ionization rates generally agree with the modeled effective rates for H-like and He-like ions, but are lower than predicted for Li-like and other high-z ions by amounts outside the stated uncertainties.

8.2.3 Trapped Ions

Ionization work with trapped ions has taken several different forms, following many workers' preliminary experiments which will not be discussed here. Baker and Hasted (1966) obtained information on ionization of Ne^+, Ar^+, Kr^+, and Xe^+ by observing sequential ionization from an ion source, recognizing trapping of the target ions by the electron space charge. Further development of the space charge trapping was carried out by Hasted and Awad (1972) who developed a cylindrically symmetric hollow electron beam for space charge trapping and primary ionization (target preparation) and a concentric variable energy electron beam to further ionize the ion targets. Absolute cross sections were obtained by calibrating the apparatus using known cross sections from crossed beams experiments. With this method Hasted and Awad (1972) obtained cross sections up to 500 eV for Ne^{2+}, Ne^{3+}, Ar^+, Ar^{2+}, and Ar^{3+}. Hamdan et al. (1978) used both this method and a variation of it wherein the hollow beam was pulsed to obtain cross sections up to about 500 eV for Ar^+, Ar^{2+}, Ar^{3+}, Ne^+, Ne^{2+}, O^+, O^{2+}, C^+ and C^{2+}. Uncertainties were estimated to be in the vicinity of 30%.

Redhead (1967) also used space charge of the electron beam of his mass spectrometer ion source to trap ions as a target for further sequential ionization. By applying trapping potentials at the ends of his source, he was able (operating at pressures of 1.3×10^{-6} Pa [or 10^{-8} Torr]) to get trapping times the order of one second. He cautioned that charge discrimination of the ion source was not readily evaluated. Relative ionization functions were obtained using this method for He^+, Ne^{n+} ($1 \leq n \leq 5$), Ar^{n+} ($1 \leq n \leq 6$), Kr^{n+} ($1 \leq n \leq 7$) and Xe^{n+} ($1 \leq n \leq 10$) by Redhead (1967); for Hg^{n+} ($1 \leq n \leq 9$) by Redhead and Feser (1968); for C^{n+} ($1 \leq n \leq 4$) and O^{n+} ($1 \leq n \leq 6$) by Redhead (1969); and for Cs^{n+} ($1 \leq n \leq 9$) and Ba^{n+} ($1 \leq n \leq 9$) by Redhead and Gopalaraman (1971). Redhead (1970, 1971) modeled the pressure dependence in his source to give estimates of absolute cross sections for Ne^{n+} ($1 \leq n \leq 4$) and Ar^{n+} ($1 \leq n \leq 6$), but the method has not been developed to obtain absolute cross sections comparable in accuracy with the colliding beams method.

Perhaps the most productive use of traps to date for electron-impact ionization studies is represented in the work of Donets and Ovsyannikov (1981). Using an EBIS (electron beam ion source) trap, and analyzing time histories of charge state evolution, they have deduced high-energy (keV range) cross sections for ionization of ions over a range of charge states up to very highly charged (e.g. Ar^{17+}, Kr^{33+}, Xe^{47+}).

A schematic representation of the CRYEBIS-I apparatus of Arianer and Geller (1981) is shown in Fig. 8-4. A superconducting, solenoid generates a longitudinal

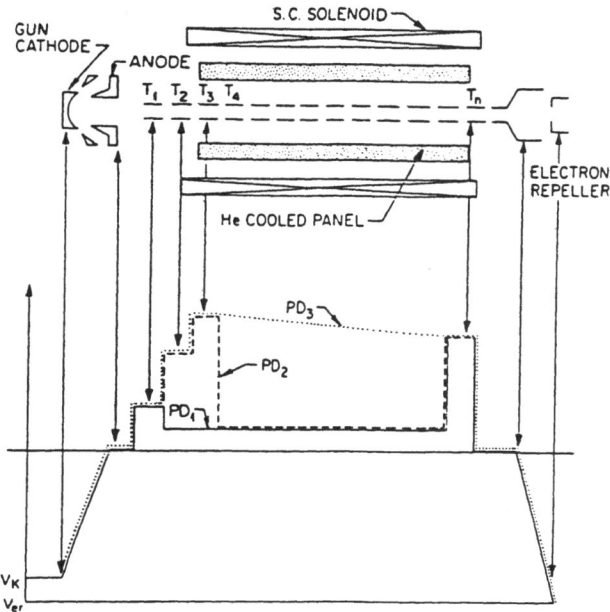

Fig. 8-4. Schematic arrangement of an Electron Beam Ion Source (EBIS) from Arianer and Geller (1981)

magnetic field (~ 2 Tesla), liquid He cooled cryopanels pump the region to ultra-high vacuum ($\sim 10^{-12}$ Torr), and an electron beam from the cathode at energies $0.5 - 18$ keV traverses the axial space. The potential distributions, PD_1, PD_2, PD_3, in the figure are applied one after the other to "create" ions, trap and bombard the ions to sequentially make high charge states, and to eject the ions in a pulsed beam. While in the distribution PD_1, a small amount of cold gas is "puffed" in at position T_3 to make low charge state ions which are then ionized to higher states during PD_2, with the final charge level depending on the energy and the time before "dumping" the ions with PD_3. Thus, the measurement cycle can be summarized: 1. turn on electron beam; 2. inject low-charge ions of the investigated element into the electron beam (PD_1 for $0.5 - 1$ ms); 3. trap ions in the beam and ionize to higher charge states (PD_2 for $0.1 - 500$ ms); 4. extract ions in the longitudinal direction (PD_3 for $10 - 50\,\mu$s); 5. determine charge states of ions by time-of-flight analysis. The charge states are determined immediately after injection and at later times (up to 500 ms). The kinetic equation for the number of ions of charge q has the form

$$\frac{dn_q}{d(jt)} = -\sum_{f=1}^{f_{max}} n_q\, \sigma_{q\to q+f} + \sum_{r=1}^{r_{max}} n_{q-r}\, \sigma_{q-r\to q} \tag{8-14}$$

where j is the electron current density, f is the number of electrons simultaneously removed from the ion of charge q, and $\sigma_{q\to q+f}$ is the cross section for such multiple ionization. Thus, in principle, multiple ionization is allowed for, and Donets and Ovsyannikov (1981) refer to some efforts and programs for doing this. However, in practice, the published data have been analyzed assuming sequential ionization, in which case Eq. (8-14) takes the simple form:

$$\frac{dn_q}{d(jt)} = -n_q \sigma_{q \to q+1} + n_{q-1} \sigma_{q-1 \to q}. \qquad (8\text{-}15)$$

There is a clear resemblance to Eq. (8-13) and the method of plasma rate measurements.

Quoted uncertainties on the deduced cross sections are typically in the range $10-20\%$, but for the heavier ions ignoring multiple ionization may be serious, and uncertainties may be much larger.

8.3 Experimental Cross Sections — Single Ionization

In this section the term "cross section" will often be used to refer to "total single ionization cross section".

8.3.1 Low Z Ions $(Z \leq 8)$

Inspection of Fig. 8-1 and Table 8-1 shows that for this range of Z, experiments have been done to measure cross sections for nearly all stages of ionization (exceptions are Li^{2+}, Be^{2+}, Be^{3+}). However, only for He^+ have crossed beams measurements been made on a hydrogen-like ion, and crossed beams measurements have not been made for helium-like oxygen (O^{6+}). There are a number of features of the cross sections for these low Z ions which are worthy of discussion. Ions in the lithium isoelectronic sequence show excitation-autoionization contributions increasing monotonically with Z. Very large metastable state populations are encountered for Be-like ions. Because of the basic importance of hydrogen-like ions, some attention will also be given these as well as some other cases.

Bell et al. (1983) have carefully reviewed the data for all ionization stages of hydrogen through oxygen, and have assembled a recommended set of cross sections. Their recommended values are in the form of a formula with coefficients chosen to give fits to the data. The recommended cross sections are represented by the equation

$$\sigma(T) = \frac{1}{IT} \left\{ A \ln(T/I) + \sum_{i=1}^{N} B_i \left(1 - \frac{I}{T} \right)^i \right\}, \qquad (8\text{-}16)$$

where T is the incident electron energy, I is the ionization potential, and the coefficients B_i are determined by a least-squares fitting procedure. The quantity A is a Bethe coefficient determined from the Born approximation or from using the photoionization cross section as described by Bell et al. For N^{4+} and O^{5+} where excitation-autoionization becomes a large fraction of the total, they made two different fits for each ion over different energy ranges to attempt to accomodate this. The parameters for Eq. (8-16) recommended by Bell et al. are shown in Table 8-2.

Table 8-2. *Parameters for use in Eq.* (8-16) *to obtain cross sections for ionization of species shown and as recommended by Bell et al.* (1983). *Spectroscopic notation is used to designate the ion, wherein the Roman numeral is one higher than the ion charge, e.g.* $C IV \equiv C^{3+}$

Species	I (eV)	A	B_1	B_2	B_3	B_4	B_5	Reliability (%)
H I	13.50	0.1845	−0.0186	0.1231	−0.1901	0.9525		± 7
He I	24.50	0.5720	−0.3440	−0.5230	3.4450	−6.8210	5.5780	± 5
He II	54.42	0.1845	0.0887	0.1315	0.3877	−1.0910	1.3541	±10
Li I	5.39	0.0854	−0.0040	0.7573	−0.1779			±10
Li II	75.54	0.7220	−0.1492	−1.3007	1.9443			±12
Li III	122.45	0.4000						±10
Be I	9.32	0.9239	−0.7697	0.3619				±20
Be II	18.21	0.7542	−0.0189	−2.9618	7.5182	−8.5431	3.1108	±20
Be III	153.39	0.7960	−0.5004	0.8836				±20
Be IV	217.71	0.4000						±10
B I	8.30	1.1063	−1.0694	−0.0879				±20
B II[a]	25.15	0.9070	−0.4770	0.1970				±20
B III[b]	37.93	0.7542	−0.0189	−2.9618	7.5182	−8.5431	3.1108	±20
B IV	259.37	0.7960	−0.5004	0.8836				±20
B V	340.22	0.4000						±10
C I	11.26	2.1143	−1.9647	−0.6084				± 5
C II	24.38	1.0824	−0.1611	−0.8563	0.9062			±10
C III	47.39	0.7150	−0.0410	0.1754				±10
C III[a]	41.38	0.6910	−0.5081	0.6993	0.0142	−0.4325		±10
C IV[b]	64.49	0.4667	−0.1298	0.2577	−0.9561	0.6441		±20
C V	392.08	0.7960	−0.5004	0.8836				±20
C VI	489.98	0.4000						±10
N I	14.53	2.2648	−1.7100	−2.3220	1.7324			± 5
N II	29.50	1.0755	−0.8287	0.8724	−0.1618	1.5331		±10
N III	47.45	0.5004	0.2234	2.2074	−4.1555	3.7686		±10
N IV	77.47	0.8125	−0.0066	−0.0459				±10
N IV[a]	69.13	0.3270	0.3570	−0.0420	−0.8740	2.1670		±10
N V	97.39	0.2182	0.2376	−0.2201	−0.4463	2.5227	−1.9021	±10 for $E \leq 4.2 I$
	97.39	0.8368	−0.2135	−2.5377	7.4882	−11.0056	5.5226	±10 for $E > 4.2 I$
N VI	552.006	0.7960	−0.5004	0.8836				±20
N VII	667.03	0.4000						±10
O I	13.52	2.4554	−2.1811	−1.5701				± 5
O II	35.12	1.5257	−0.5935	−0.3994	−0.5833	3.2355		±10
O III	54.93	1.0657	0.4420	0.4751	−2.9613	4.4700		± 7
O IV	77.41	1.0446	−0.6519	1.2988				±10
O IV[a]	68.50	0.5605	−0.6091	4.6523	−8.9404	6.7354		± 5
O V	113.90	0.7268	0.0911	0.0220				±10
O V[a]	103.58	0.3287	0.6097	−2.1048	5.9130	−3.0004		±10
O VI	138.12	0.3362	0.0803	0.1432	−0.7309	1.3363	−0.7846	±20 for $E < 4.0 I$
	138.12	0.6362	−0.3127	1.3187	−5.2457	6.3866	−2.4415	±20 for $E \geq 4.0 I$
O VII	739.32	0.7960	−0.5004	0.8836				±20
O VIII	871.39	0.4000						±10

[a] Includes contributions from ions in metastable states (see text).

[b] In making recommendations for B^{2+}, Bell et al. did not have access to experimental data. The parameters given here give a cross section more than 30% greater than the experiment and Crandall et al. (1984). For C^{3+}, Bell et al. chose to be guided by theory rather than experiment, and their parameters here give a cross section about 25% greater than experiments. We recommend the experimental values rather than those yielded by the above parameters.

They used a combination of experimental data (where available), theoretical values, and judgment to arrive at their recommendations. We here generally commend their cross sections to data users. The cases of C^{3+} and B^{2+} are exceptions. These and some other qualifications are discussed later.

Obviously, one of the most important classes of experiments in electron-ion ionization is that dealing with hydrogen-like ions. The theory must deal with only one target electron, so data on hydrogenic ions seem ideal for making comparisons between experiment and theory to test for understanding the direct ionization process. It was thus no accident that He^+ was the object of the first crossed charged beams experiment of Dolder et al. (1961). Fig. 8-5 shows the data of Peart, Walton, and Dolder (1969) as points, and cross sections calculated using distorted wave exchange (DWE) methods (see Chapter 1 and Younger 1980 a). The other measurements noted in Table 8-1 are not shown in this figure for the sake of clarity — they are the same within experimental uncertainties. The distorted wave calculations of Younger are in excellent agreement with the measurements. Born approximation calculations (Omidvar 1969, Mott and Massey 1965) are substantially higher than experiment for $u < 20$, but in good agreement at the higher energies of the Bethe-Born limit.

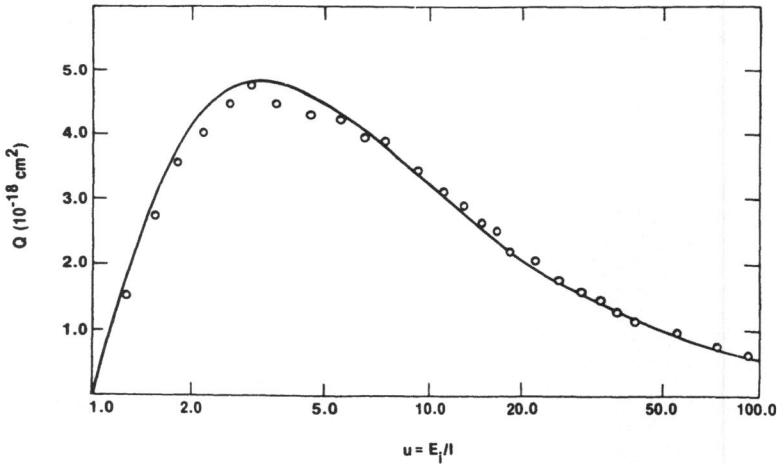

Fig. 8-5. Electron impact ionization cross section for He^+ versus energy in threshold units. Solid line, Distorted Wave Exchange (Younger, 1980 a); points are cross beams measurements (Dolder et al., 1961)

More highly charged ions in the hydrogen isoelectronic sequence have been studied only with trapped ion and plasma rate methods. Fig. 8-6 shows the scaled cross sections ($Z^4 \sigma$) versus energy u in threshold units (T/I) for He^+, C^{5+}, N^{6+}, O^{7+}, Ne^{9+}, and Ar^{17+}. Also shown is a curve for $Z = 128$ using scaling methods recommended by Younger (1981 a) based on the DWE approximation. Coulomb-Born calculations of Rudge and Schwartz (1966) for $Z = 128$ are essentially identical to the curve shown. The helium data are from Peart, Walton, and Dolder (1969) as in Fig. 8-5. The data for more highly charged ions are from Donets and Ovsyannikov

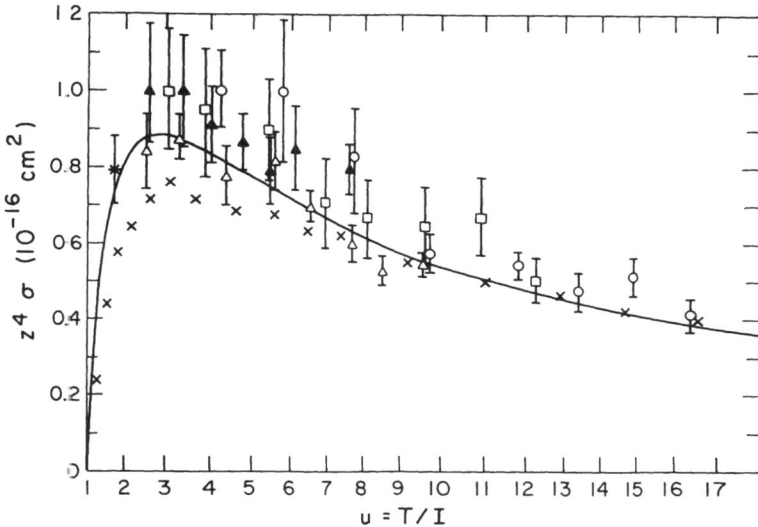

Fig. 8-6. Scaled electron impact ionization cross sections of hydrogen-like ions versus energy in threshold units. $\times$, He$^+$ (Dolder *et al.*, 1961); $\bigcirc$, C^{5+}; $\square$, N^{6+}; $\triangle$, O^{7+}; $\blacktriangle$, Ne^{9+}; *, Ar^{17+} (Donets and Ovsyannikov, 1981). Solid curve is DWE for $Z=128$ (Younger, 1981 a)

(1981). There is clearly a close correspondence between the data for highly charged ions and the DWE curve, though, as Donets and Ovsyannikov point out, at low values of u the data are, on the average, about one standard deviation above the curve.

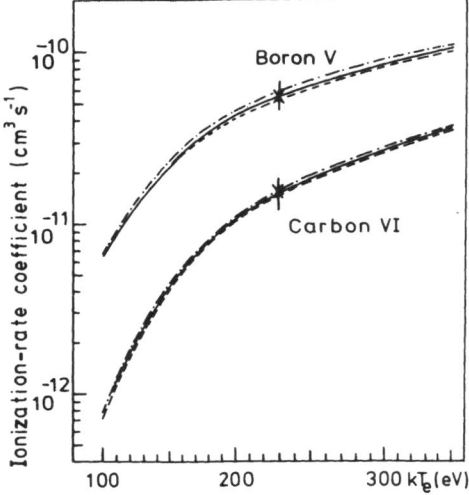

Fig. 8-7. Ionization-rate coefficients for B V and C VI versus temperature; *, experimental data (Greve *et al.*, 1981); solid curve, ECIP (Burgess, 1964); dot-dashed, SCB (Golden and Sampson, 1977); dashed curve, from modified Lotz calculation

Ionization rate coefficient measurements on the hydrogen-like ions B^{4+} and C^{5+} were made by Greve *et al.* (1981) for temperatures T_e such that $k\,T_e \sim 230\,eV$ and for electron densities $N_e \sim 3 \times 10^{16}\,cm^{-3}$. Results are shown in Fig. 8-7 along with predictions based on the semiclassical (ECIP) method of Burgess (1964), the Coulomb-Born method of Golden and Sampson (1977), and the empirical Lotz formula (Lotz 1967 a, b). The agreement between measured and predicted values is excellent, speaking for the ability to model and to measure with the plasma method for these ions of such simple structure.

Crossed beams measurements on He-like ions are extremely difficult, because the two 1 *s* electrons are very tightly bound leading to small cross sections, and because of the difficulty of obtaining intense ion beams. Thus, the data of Crandall *et al.* (1979) are sparse and have large associated uncertainties for the species investigated (B^{3+}, C^{4+}, N^{5+}). Fig. 8-8 shows data for Li^+ which, as shown in Table 8-1, has been investigated several times (Lineberger *et al.* 1966; Peart and Doldinger 1968 b; Peart, Walton, and Dolder 1969). For comparison, the calculated results of Younger (1980 b) from using the DWE approximation are shown by the solid curve. Agreement among the measurements and between the measurements and the DWE calculations must be considered good.

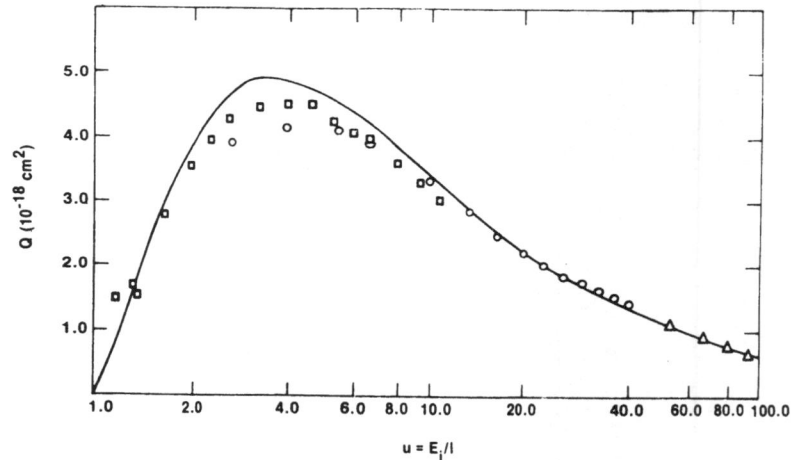

Fig. 8-8. Electron impact ionization cross section for Li^+. Solid line **DWE** (Younger, 1980 b). Crossed beams measurements; ○ (Peart and Dolder, 1968); △ (Peart *et al.*, 1969); □ (Lineberger *et al.*, 1966)

Plasma rate measurements for He-like ions (e.g. Greve *et al.* 1981) are generally in good agreement with predictions provided care is taken to properly account for ionization from metastables in the plasma. Such ionizations account for as much as 50% of the total in some measurements.

The lithium iso-electronic sequence is perhaps the most extensively studied series both experimentally and theoretically. As noted in Table 8-1, crossed beams measurements have been made for the first five members of the sequence (Be^+, B^{2+}, C^{3+}, N^{4+}, and O^{5+}). Jacubowicz and Moores (1981), followed up earlier work by Moores (1978) and calculated cross sections for C^{3+}, N^{4+}, O^{5+} and Ne^{7+} and explored several approximations varying from distorted wave with exchange (DWE) to Coulomb-Born and included calculations of the excitation autoionization contributions. Younger (1980 a) also used several approximations including DWE, and calculated direct ionization cross sections for Be^+, O^{5+}, M^{9+}, and for the hypothetical species with $Z = \infty$. From his results he deduced scaling parameters for all other Z (Younger, 1981 a). Sampson and Golden (1979) made calculations for lithium-like ions using the scaled Coulomb-Born-exchange method they have developed so fully, and also included calculations of excitation-autoionization. Henry (1979) calculated cross sections for inner shell excitation of C^{3+}, N^{4+} and O^{5+} using a six-state-close-coupling approximation, and thus deduced the excitation autoionization contributions to ionization of these species. There are many other calculations covering the classical, semiclassical, and semiempirical approaches, and references to these can be found in the bibliographies of Takayanagi and Iwai (1978) and Itikawa (1982).

Table 8-3. Be^+ scaled electron-impact ionization cross section $U\,I^2\,Q$, where $U = E/I$, E is electron energy, and $I = 1.339\,Ry$ is the threshold ionization energy. Units are $\pi\,a_0^2\,(Ry)^2$. Theoretical values owing to Younger ($\mu\chi\varphi\psi$ a)

U	DWE[a]	DWT[a]	CBE[a]	CBT[a]	PWB[a]	Experimental data
1.125	0.269	0.237	0.310	0.234	0.145	0.31 ± 0.02
1.25	0.493	0.467	0.564	0.462	0.334	0.55 ± 0.04
1.50	0.866	0.903	0.973	0.891	0.752	0.96 ± 0.07
2.25	1.67	1.94	1.80	1.92	1.79	1.98 ± 0.14

[a] DWE = distorted wave with exchange; DWT = distorted wave truncated; CBE = Coulomb-Born with exchange; CBT = Coulomb-Born truncated; PWB = plane-wave Born.

Table 8-3 shows selected experimental results (Falk and Dunn, 1983) for direct single ionization cross sections for Be^+ compared to theoretical cross sections resulting from various approximations (Younger, 1980a; see also Chapter 1). The agreement between experiment and theory is good — particularly the Coulomb-Born with exchange (CBE) approximation; though one may normally expect best agreement with the DWE approximation.

Scaled experimental single ionization cross sections for all lithium-like targets (including Li) for which measurements have been made are shown im Fig. 8-9 where the curves have been drawn by eye through the experimental points. The cross sections are plotted only up to the threshold energy for excitation-autoionization. Thus, the plots are for direct ionization only. It is observed that one could draw a single curve which would quite closely represent all of the data — except for B^{2+} and C^{3+} which both lie substantially below the other curves. This, along with consideration of the theoretical values helped lead Bell et al. (1983) to recommend cross sections for C^{3+} and B^{2+} well above the experimental values. However, the B^{2+} data (Crandall et al., 1983) have come forth since the Bell et al. paper, and a separate (third) check on the value of the C^{3+} cross section at the peak has been carried out by Howald (1984) confirming the two earlier results (Crandall et al.,

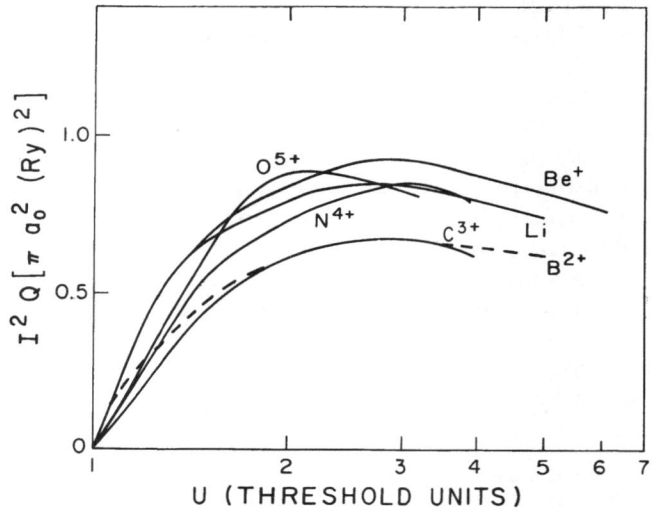

Fig. 8-9. Scaled electron impact ionization cross sections for the lithium-like isoelectronic sequence of ions versus energy in threshold units

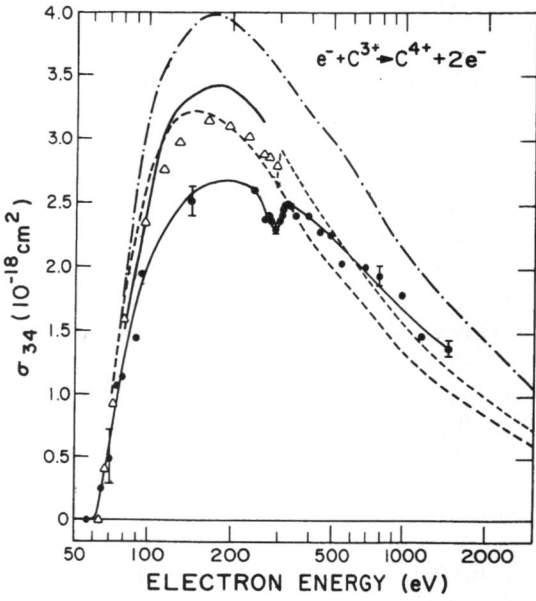

Fig. 8-10. Cross sections for electron impact ionization of C^{3+}. Experimental data, ● (Crandall et al., 1979 a, b); △, DWE (Jakubowicz and Moores, 1981); dashed, SCB (Golden and Sampson, 1977), past 294 eV addition of summed $1s^2\,2s \rightarrow 1s^2\,2s\,nl$ excitation (Magee et al., 1977); bold solid, CB (Moores, 1978); dot-dashed (Lotz, 1968)

1978, 1979). One set of data (Crandall *et al.*, 1979 a, b) for C^{3+} is shown in Fig. 8-10 along with several theoretical and semiempirical curves for comparison. The recommended values (Table 8-2) of Bell *et al.* are very close to the theoretical ones using DWE of Jacubowicz and Moores (1981) shown as triangles in the figure.

At 294 eV in Fig. 8-10, one notes a sudden increase in the cross section. This is attributable to excitation-autoionization. In this case the process can be represented by excitation of an inner shell $1s$ electron to an nl level: $1s^2 2s \rightarrow 1 s 2 s n l$. The doubly excited ion then autoionizes leaving the heliumlike ions C^{4+} ($1 s^2$) and a continuum electron. If the process can be thought of as approximated by the two-step process described above (excitation followed by autoionization) the cross section should have a step-function increase at the threshold for excitation, since that is characteristic of electron impact excitation of ions. That interpretation is consistent with the observation.

The same phenomenon is observed for all of the Li-like ions for which measurements have been made. The fractional contribution of cross section due to excitation-autoionization to the total cross section increases with Z, the nuclear charge, increasing from about a 3 or 4% effect for Be^+ to about a 24% effect for O^{5+}. This increase with Z can be anticipated theoretically and has been discussed by Henry (1979) and by Sampson and Golden (1979). Crandall *et al.* (1983) have used the DWE results of Younger (1980 a, 1981 a) to model the direct ionization cross sections in the vicinity of excitation-autoionization onset and subtracted those from the experimental total cross sections, leaving in principle only the contributions from excitation-autoionization. In all cases the resultant excitation-autoionization cross sections are consistent with the close-coupling excitation calculations of Henry (1979) for the $1 s 2 s n l$ states (Henry did not make calculations for all values of Z mentioned here, and extrapolations along Z of his calculations were used in the comparisons).

Table 8-4. *Collisional ionization-rate coefficients for lithium-like ions in units of 10^{-10} cm^3 s^{-1}*

Ion	kT_e (eV)	kT_e/E_i	N_e (10^{16} cm^3)	ECIP	GS	Lotz	Y	I_{expt}^{Cr}	$I_{eff expt}$	I_{eff}/I_1	$I_{1 expt}$	$I_{1 expt}/GS$
N V	110	1.12	0.8	3.98	5.76	7.77	6.21	6.65	4.40	1.12	3.93	0.68
N V	120	1.23	1.1	4.39	6.23	8.50	6.77	7.27	4.00	1.13	3.54	0.57
O VI	135	0.98	1.2	2.01	3.10	3.96	3.30	3.77	2.24	1.09	2.06	0.66
O VI	160	1.16	1.3	2.47	3.69	4.80	3.92	4.57	2.70	1.09	2.48	0.67
F VII	160	0.86	1.5	1.09	1.78	2.20	1.83		1.43	1.06	1.35	0.76
F VII	180	0.97	1.5	1.29	2.05	2.56	2.12		1.43	1.06	1.35	0.66
Ne VIII	160	0.67	1.8	0.49	0.86	1.10	0.86		0.51	1.05	0.49	0.57
Ne VIII	220	0.92	2.2	0.82	1.34	1.62	1.37		0.76	1.05	0.72	0.54

Just as for crossed beams measurements, plasma rate measurements have been carried out most extensively so far for lithium-like ions, as can be seen from the listing in Section 8.2.2. Table 8-4, taken from Greve *et al.* (1981) shows some of their experimental rates in comparison with calculated rates based upon various methods of estimating the cross sections — including cross sections from crossed beams measurements. The column labeled ECIP is from use of the semiclassical approximation of Burgess (1964), GS from the scaled Coulomb-Born method of Golden and Sampson (1977), Lotz from the semiempirical formula of Lotz (1967a, b; 1968, 1969), Y from the distorted wave calculations of Younger (1980a), and I_{expt}^{Cr} from the crossed beams measurements of Crandall *et al.* (1979a,b). The effective ionization rates measured are listed as $I_{eff\,expt}$. Since the theoretical and crossed beams rates are from the ground state only, the results are corrected for ionization from the $2p$ level to give results in the column labeled $I_{1\,expt}$. It is seen from the last column that the experimental rates are on the average only 0.64 of the Coulomb-Born values (roughly the same disparity for all cases except ECIP). Källne and Jones (1977) obtain reasonable agreement between measured and calculated rates at low temperatures when account is taken of the depression of the ionization limit and ionization from the $2p$ level. They made measurements for C^{3+}, N^{4+}, and O^{5+}. At high temperatures their measured values were about 40% below calculated ones. Datla and Roberts (1983) obtained good agreement between their measured and calculated rates at low temperatures (55 eV) for O^{5+} when account was taken of ionization of excited states and excitation ionization. There seem to be a number of interesting issues to resolve in order to get full consistency between crossed beams measurements (or theory) and plasma rate measurements. It seems important to discover and resolve these issues if the beams measurements (and/or theory) are to be useful in the applications areas (Chapter 9).

For Be-like ions the lowest excited states $2s\,2p\ ^3P_{0,1,2}$ are metastable. These levels lie near in energy to the ground state $2s^2\ ^1S$, and they are favored over the ground state by a 9 to 1 ratio of statistical weights. The experiments on these ions, as noted in Table 8-1 have all involved ion beams with metastable populations in the range $40-90\%$. For B^+ and C^{2+} (Falk *et al.*, 1983b) it was possible to vary the metastable populations and use the technique described in Section 8.2.1 to deduce the cross sections for ionization from the ground state. Cross sections for B^+ are shown in Fig. 8-11, where open circles result from about 90% metastable population in the beam, and solid circles result from about 50% metastables. The solid curve is the ground state cross section deduced according to Eq. (8-11). No strong evidence for a meaningful contribution to the cross section from excitation-autoionization was seen in the cross section measurements for Be-like ions except for O^{4+}. An enhancement of the cross section of about 10% is found near 550 eV, agreeing reasonably well with a prediction of Sampson and Golden (1981) of $5-7\%$ enhancement at 543 eV.

There are no real surprises nor serious issues for the B-, C-, and N-like ions. The recommendations of Bell *et al.* (1983) generally follow closely the experimental data. Predictions of the Lotz formula and those based on the Coulomb-Born approximation seem somewhat low at low energies (on the initially rising portion of the cross section curve). Excitation-autoionization does not seem to contribute significantly to the observed cross sections. However, in the boron-like ions, C^+ and N^{2+}, EA from the ^4P metastable state contributes to a small pre-threshold signal.

Fig. 8-12 shows a so-called Bethe plot for nitrogen-like ions O^+, F^{2+}, Ne^{3+}, where the product of energy (scaled) times cross section (scaled) is plotted versus logarithm of the energy. The experimental cross sections for F^{2+} and Ne^{3+} scale to almost

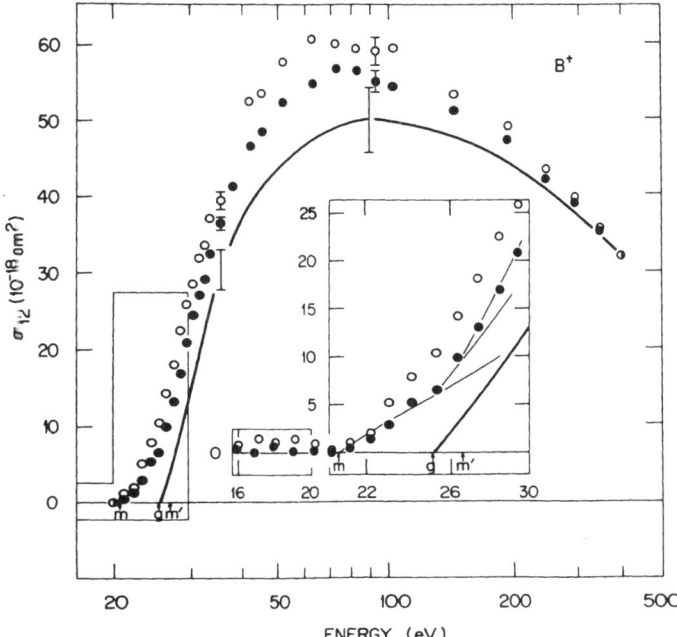

Fig. 8-11. Electron-impact ionization of B^+. Solid points are low metastable data and open circles are high metastable. Solid curve is the deduced experimental ground-state cross section. Inserts shows the near-threshold region expanded including the $16-20\,eV$ region where the nonzero measured cross sections were determined and then subtracted from each data set. Arrows indicate threshold energies for removal of metastable (m and m') and ground state (g) outer electrons. Uncertainties due to counting statistics are roughly the size of data points. Total relative uncertainties are shown on the data points at 36 and 95 eV and the total absolute uncertainties in the deduced ground-state cross sections are shown on the solid curve at 36 and 95 eV

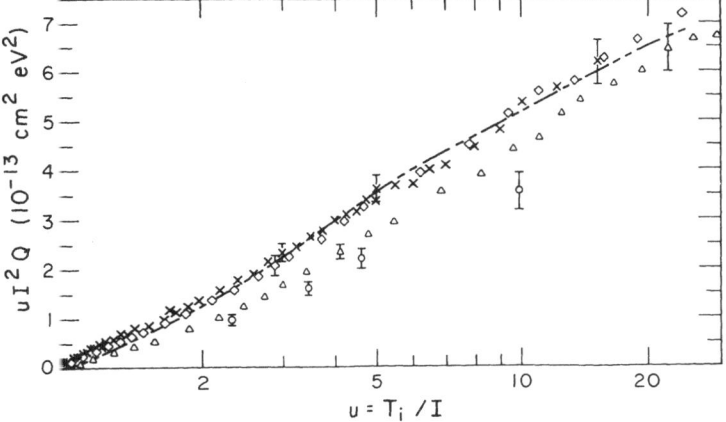

Fig. 8-12. Bethe plot of product of scaled electron impact ionizations cross sections times reduced energy, $u I^2 Q$, for N-like ions versus log u. Crossed beams measurements: $\Diamond$, F^{2+} (Mueller et al., 1983); $\times$, Ne^{3+} (Gregory et al., 1983); $\triangle$, O^+ (Aitken and Harrison, 1971); $\bigcirc$, O^+ (Müller et al., 1980 b). Chain Curve, DWE for F^{2+} (Younger, 1984)

identical functions and are described well by the DWE calculations of Younger (1984) for F^{2+} (shown) or Ne^{3+} (not shown). The O^+ data of Aitkin and Harrison (1971), however, are $15-30\%$ lower and the O^+ data of Müller et al. (1980 b) are lower than that. The F^{2+} data are from Mueller et al. (1983), and the Ne^{3+} data are from Gregory et al. (1983).

8.3.2 Intermediate Z ions $(9 \leq Z \leq 19)$

For this range of Z, there is again a significant number of cross section measurements as can be seen from Fig. 8-1. Members of the nitrogen isoelectronic sequence F^{2+} and Ne^{3+} have already been discussed in the last section. The only members of the Ne sequence for which measurements have been done are Na^+ and Mg^{2+} – and results are lower than theory; the data will be discussed. Three members of the sodium isoelectronic sequence, Mg^+, Al^{2+}, and Si^{3+}, have beeen studied. As with the lithium-like ions, excitation-autoionization (EA) plays a significant role which increases with Z. Some anomalies with the excitation-autoionization feature have led to hypothesizing resonant-excitation-double-autoionization (REDA) and this has been modeled. Excitation-autoionization also contributes significantly to the cross sections for the phosphorous-like ions Cl^{2+} and Ar^{3+}. Five members of the argon isonuclear sequence have been studied, and there are interesting systematics in the data.

The neon isolelectronic sequence was discussed at some length in Chapter 1 and is interesting for study from several points of view. First, as pointed out by Younger (1981 b) the closed-shell target is a many-electron system which is represented fairly accurately by independent particle Hartree-Fock wave functions. It is not likely that target configuration-interaction effects will substantially affect calculations of ionization cross sections. Second, the contribution from excitation-autoionization should be small compared to other atomic systems. Inner-shell ionization will occur at high energies and there should be a small contribution from the two $2s$ electrons compared to the six $2p$ electrons. The neon sequence is thus a good choice for study of direct ionization of a many-electron ion with relatively simple structure.

Measurements on Na^+ were made by Hooper et al. (1966) and by Peart and Dolder (1968 a), and measurements on Mg^{2+} by Peart et al. (1969). The Na^+ measurements were in very close agreement with each other. However, the scaled cross sections $(I^2\sigma)$ as a function of reduced energy $(u = T_e/I)$ were not even close to those of Ne. Fig. 8-13 is a Bethe plot from Younger (1981 b) for Na^+ showing points from both experiments as well as from several calculations including one based on the Lotz (1968) semiempirical formula, Coulomb-Born without exchange (Moores, 1972), distorted wave with exchange (Younger), and distorted wave with the addition of a semiclassical exchange potential (Younger). Agreement between the experiments and theories is not good – worse than generally found for the light isoelectronic sequences of the last section. There is similar poor agreement for Mg^{2+}. Younger (1981 b) concludes there is a need both for improved representation of the ejected partial waves and a more rigorous formulation of the theory of electron-impact ionization itself. The problems of theory relative to the neon isoelectronic sequence have been discussed in Chapter 1, and Fig. 1-2 presents all the data.

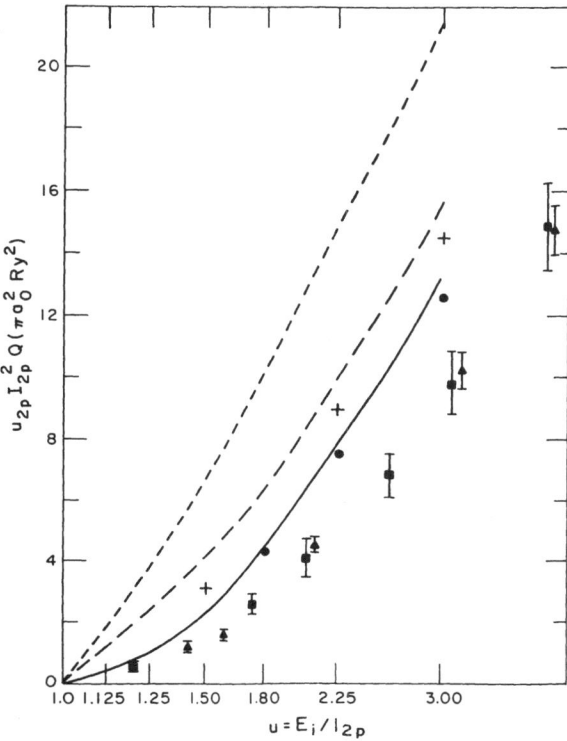

Fig. 8-13. Bethe plot of the scaled electron impact ionization cross section, uI^2Q, for Na^+ versus $\log u$. —— distorted wave Born exchange including semiclassical exchange in the distorted potentials; – – – distorted wave Born exchange without semiclassical exchange in the distorted potentials; ● same as solid line but omitting inner shell ionization from $2s^2$, all from Younger (1981 c). + Coulomb-Born (Moores, 1972): – – – Lotz (Lotz, 1968); crossed beams measurements: ■ (Hooper *et al.*, 1966), ▲ (Peart and Dolder, 1968 a)

Fig. 1-1 shows calculated cross sections for the sodium-like ion Fe^{15+}, and the figure and associated discussion in Chapter 1 demonstrate the possible role of different resonance mechanisms in the ionization process. Martin *et al.* (1968) made the first crossed beams cross section measurement for a sodium-like ion (Mg^+), but the results did not reveal the expected excitation-autoionization contribution (Bely, 1968; Moores and Nussbaumer, 1970). Crandall *et al.* (1982) made measurements on the sodium-like ions Mg^+, Al^{2+}, and Si^{3+}. Fig. 8-14 shows their experimental results and results of calculations by Griffin *et al.* (1982 a) for Al^{2+}. The energies associated with orbital promotions leading to autoionization are also shown on the figure. The distorted wave calculations for EA shown are seen to give cross section contributions about a factor of two greater than observed. Also, the theoretically dominant $2p-3p$ contribution appears to be missing in the experiment, and the experimental values are larger than theory in the energy region between the $2p-3s$ and $2p-3p$ transitions.

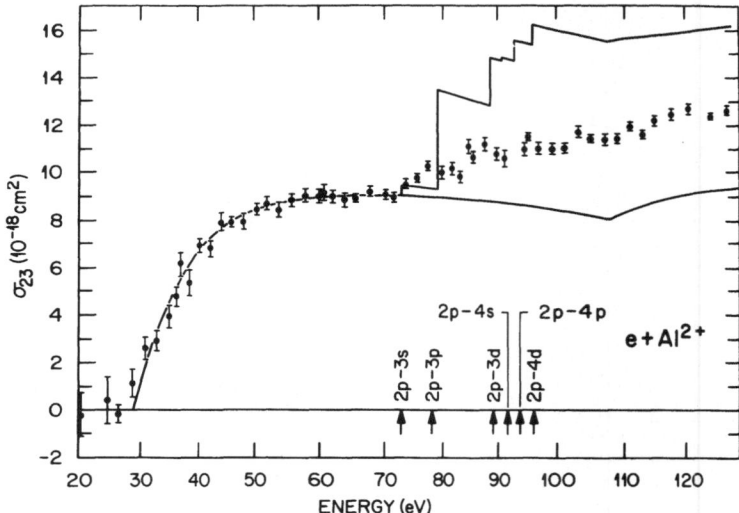

Fig. 8-14. Electron-impact ionization of Al^{2+} near threshold, ●, crossed beams measurement (Crandall et al., 1982). The solid curve is distorted-wave calculation of direct ionization by Younger (1981 a, c). normalized to the experiment at 70 eV by multiplying Younger's results by 0.65. The distorted-wave excitation of Griffin et al. (1982 a) is added to the Younger direct-ionization results with arrows indicating center-of-gravity energies for excitation of a $2p$ electron to final orbital nl

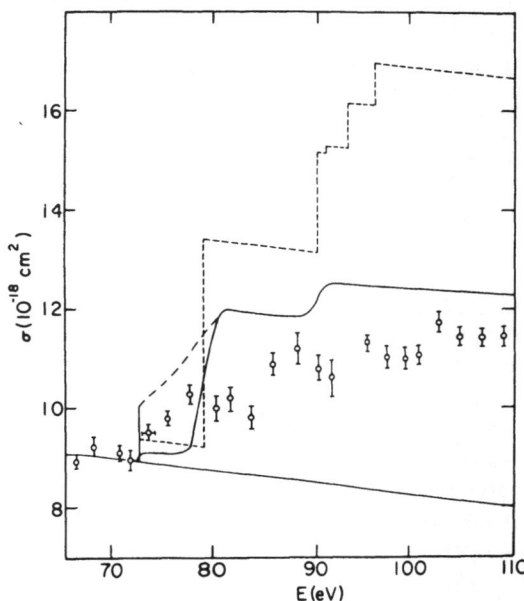

Fig. 8-15. Points, dashed curve, and curve below 70 eV, same as in Fig. 8-14 for Al^{2+}. Solid curve is two-state close coupling for EA including only $2p-3l$ excitations, and folded with a 2 eV electron energy spread; dot-dashed curve between 73 and 80 eV is estimated close-coupling results including both EA and REDA (Henry and Msezane, 1982)

Henry and Msezane (1982) made two-state close-coupling calculations of the EA contributions to the ionization cross sections for the Na-like ions noted above. They also estimated the contributions of resonant-excitation-double-autoionization from a three-state close-coupling approximation with the aid of quantum-defect-theory analysis. As shown in Fig. 8-15, it appears that these contributions explain the absence of the $2p-3p$ step which was predicted to be dominant. This coupled-state calculation still overestimates the $2p-3p$ EA contribution somewhat, but the estimated effect of the REDA resonances, just below this excitation threshold, leads to a curve consistent in shape with the observations. Predictions for Mg^+ and Si^{3+} are also in comparable agreement with the experiments.

Various theoretical values of cross sections for direct ionization of the sodium-like ions were much larger than experiment. For example, in the Al^{2+} case discussed above, the DWE values of Younger (1981 a, c) had to be multiplied by 0.65 to normalize to the experimental values. The Lotz formula gave values almost a factor of 2.5 larger than experiment at energies just below the EA threshold. In the case of Si^{3+}, the value from Younger agreed with experiment for direct ionization. Scaled plane wave Born calculations of McGuire (1982) give quite reasonable agreement ($\sim 20\%$) with the experiments when account is taken of EA.

The only Mg-like ion for which ionization cross sections have been measured is Al^+. The experimental cross sections of Belić et al. (1982) and of Montague and Harrison (1983) are in excellent agreement with each other. Predictions using the Lotz (1968) formula give values too large by about 25% at 100 eV, and predictions using the Coulomb-Born (Moores et al., 1980) are too small by about 25% at 100 eV. The scaled plane wave Born method of McGuire (1982) gives cross section values in excellent agreement with the measurements at all energies. A small ($\sim 4\%$) EA contribution due to excitation of the $2p$ electrons shows up in the data of Belić et al. and is predicted by McGuire (1982) at about 70 eV. The EA feature was not observed by Montague and Harrison. Montague and Harrison found a small ($\sim 9\%$) population of metastable ions in their beam, and made corrections to obtain ground state cross sections.

Müller et al. (1980 b) made a systematic study of ionization of Ar^{n+} ions ($n = 1, ..., 5$). Their measured single ionization cross sections are shown in the log-log plot of Fig. 8-16. They found that a single Bethe-like expression describes the data for all five ionization stages

$$\sigma = 1.4 \times 10^{-13} (T/I) \ln (T/I).$$

This expression is represented by the solid curves in Fig. 8-16, and generally falls within 20% of the measurements. Furthermore, for $n = 4$ and $n = 5$, the expression gives 2.9×10^{-18} cm^2 and 2.05×10^{-18} cm^2 respectively at 2500 eV. This can be compared to the respective values reported by Donets (1976) of 3.1×10^{-18} cm^2 and 2.05×10^{-18} cm^2. Such a simple representation would, of course, be very valuable to data users (Chapter 9) if more generally valid. However, it seems quite improbable that other isonuclear sequences will lend themselves to such descriptions. That such an expression works at all for more than one charge state seems incredible; since one is used to thinking of $A(T/I) \ln(T/I)$ as representing the cross section contribution per nl electron and since each charge state would have a correspondingly different multiplier to this expression. In fact, the formula of Müller et al. does not seem to be

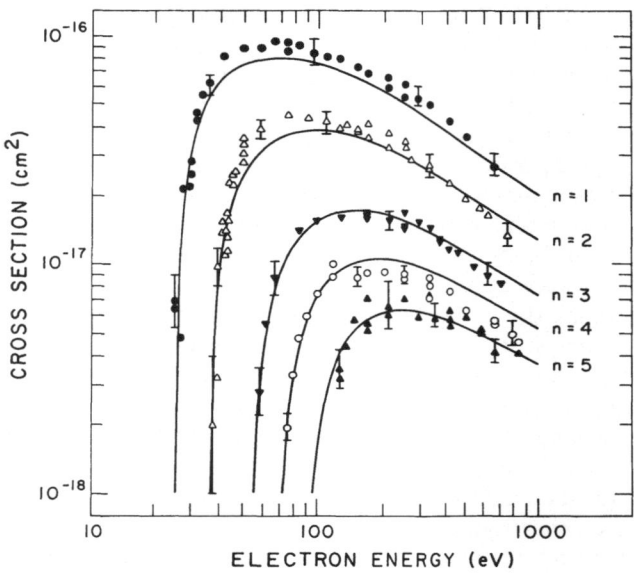

Fig. 8-16. Cross sections for $Ar^{n+} + e \rightarrow Ar^{(n+1)+} + 2e$ with $n = 1, 2, ..., 5$. Full curves are $\sigma = 1.4 \times 10^{-13}$ $(T/I)\ln(T/I)$. Müller et al. (1980 b)

applicable for the higher charge states investigated by Donets and Pikin (1976) and Donets and Ovsyannikov (1977 a, b).

A more detailed plot taken from Gregory et al. (1983) of data for Ar^{3+} is shown in Fig. 8-17. Agreement between Gregory et al. and Müller et al. is within the uncertainties. The signal below threshold for ground state ionization indicates

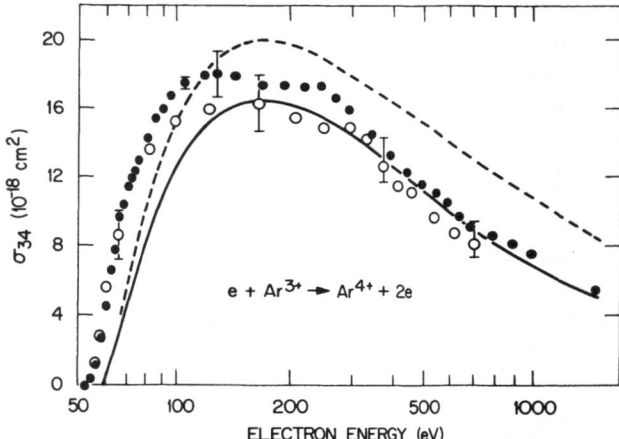

Fig. 8-17. Electron-impact single ionization cross sections for Ar^{3+}. Crossed beams measurements: solid points (Gregory et al., 1983), open circles (Müller et al., 1980 b). Solid curve, DWE (Younger, 1981 a); dashed curve, Lotz formula (Lotz, 1968)

metastable ions in the target beam, and steeper than theory rise at threshold signals possible contributions of EA involving $3s$ electrons. The feature near 180 eV may be attributed to EA associated with excitation of $2p$ electrons. In the measurements of Mueller *et al.*(1983), almost identical features are seen for the isoelectronic ion Cl^{2+}. Plots of scaled cross sections versus scaled energy for Cl^{2+} and Ar^{3+} are in close agreement, except that the features associated with EA of $2p$ electrons are shifted in energy. Calculations using the Lotz formula are not in particularly good agreement with the measurements, and the DWE calculations of Younger (1981 a) are in agreement at higher energies, but do not agree in the rising portion of the cross section where EA may play a role.

8.3.3 Ions of $Z \geq 20$

There are relatively fewer measurements per range of Z which have been made for these heavier species. However, these "heavy" ions provide remarkable examples of indirect mechanisms dominating the ionization process.

The heavy alkali-like ions provide perhaps the clearest examples. It has been reasonably supposed that when there are only a few electrons outside a many-electron closed shell or subshell which lies close in energy to the valence shell, excitation-autoionization will be prominent. The clear dominance of EA over direct ionization was first demonstrated in the experiments (Peart and Dolder, 1968 b; Feeney *et al.*, 1972; Peart *et al.*, 1973) on Ba^+ and the experiments (Peart and Dolder, 1975) on Ca^+ and Sr^+. They found excitation-autoionization cross sections which were larger than those for direct ionization by factors ranging from 1.5 for Ca^+ to 4 for Ba^+. A composite of their results is presented in Fig. 8-18.

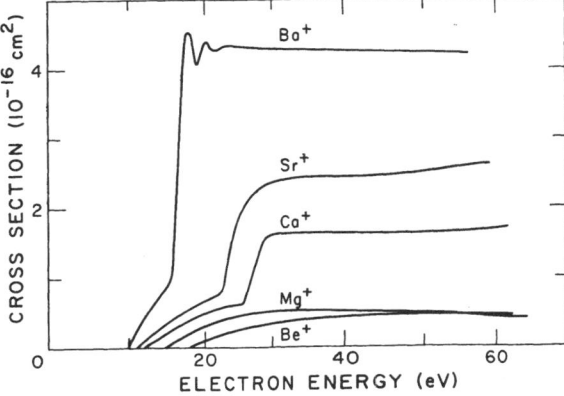

Fig. 8-18. Electron-impact single ionization cross sections for the singly ionized alkaline-earth metals. Lines are smooth curves drawn through crossed beams measurements as referenced in the text and Table 8-1

In each case the sharp increase can be attributed primarily to $np^6(n+1)s \to np^5\, nd(n+1)s$ transitions — an inner-shell p electron is excited to a d orbital with $\Delta n = 0$. Bely *et al.* (1971) calculated the effect for Ba^+ with reasonable success, finding agreement in general shape and in magnitude at the 25% level. Recently Burke *et al.* (1983) calculated the electron impact excitation of Ca^+ from the ground state to four terms of the autoionizing configurations $3p^5\,3d\,4s$ and $3p^5\,4s^2$ using the R-matrix method. They found good agreement with the Peart and Dolder (1975) results over a wide energy range, except just above threshold for EA. Here they predict a large resonance rising to twice the experimental cross section. The width of the resonance is several electron volts, and would readily have been resolved in the experiment. Griffin *et al.* (1984 b) have recently used the distorted wave approximation to make calculations for Ca^+ and Ba^+. Their results are quite similar to Burke *et al.* for Ca^+, including the large resonance not observed experimentally. For Ba^+ they predict resonance structure similar to that observed, except the resonance occurring at about 22 eV is calculated to be many times larger than observed.

As one proceeds to more highly charged isoelectronic analogs of Ca^+, Sr^+, and Ba^+, the single valence electron occupies a d rather than an s orbital. However, one may still expect strong EA effects from transitions of the type $np^6\,nd \to np^5\,nd^2$. Fig. 8-19

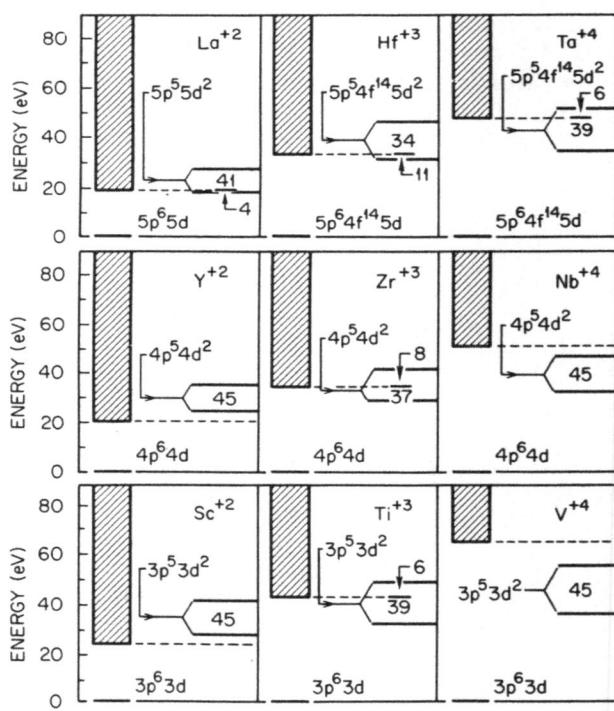

Fig. 8-19. Energy level structure of $np^5\,nd^2$ relative to the ionization limit for several transition metals. The number of levels above and below the respective ionization threshold is indicated for each case. Hatched region indicates the ionization continuum for the next higher ionization state (Falk *et al.*, 1981)

shows the calculated energy level structure (Falk *et al.*, 1981: Griffin *et al.*, 1982 b)
for the next three members of the sequences (note that the structures called
"isoelectronic" to Ba$^+$ have a filled f subshell after La^{2+}). Most of the $np^5 nd^2$ levels
are indeed autoionizing for two and three times charged ions, but become mostly
bound for four times charged.

Measurements of cross sections for ionization of the three times charged ions (Ti^{3+},
Zr^{3+}, and Hf^{3+}) yielded dramatic results. Measured cross sections (Falk *et al.*, 1981,
1983 a) along with calculated ones (Falk *et al.*, 1981: Griffin *et al.*, 1982 b) are shown
in Fig. 8-20. The dashed curves toward the bottom in each case represents the direct
ionization cross section (probably an upper limit) calculated using the Lotz (1967 a,
1967 b, 1968, 1969) formula (see Chapter 2). The solid curve was calculated using a
distorted wave dipole approximation for the excitation cross section and that was
added to the dashed curve. However, the calculations for excitation had to be
reduced a factor of 2.5 to get reasonable agreement with experiment, and it is the
scaled theory values which are plotted as the solid curve. The chain curves are
convolutions of the electron energy distribution with the scaled predictions of
theory. Ignoring the factor of 2.5 the agreement is quite acceptable except for Hf^{3+}.
In further calculations by Bottcher *et al.* (1983), attention was given to the
contribution of excitations other than $np \rightarrow nd$ and to the effect on cross sections of
correlation in the target wave function. Though agreement between experiment and
theory was improved in all cases, significant differences remain. For Hf^{3+} the
"missing" cross section below 45 eV may possibly be due in part to the REDA
process.

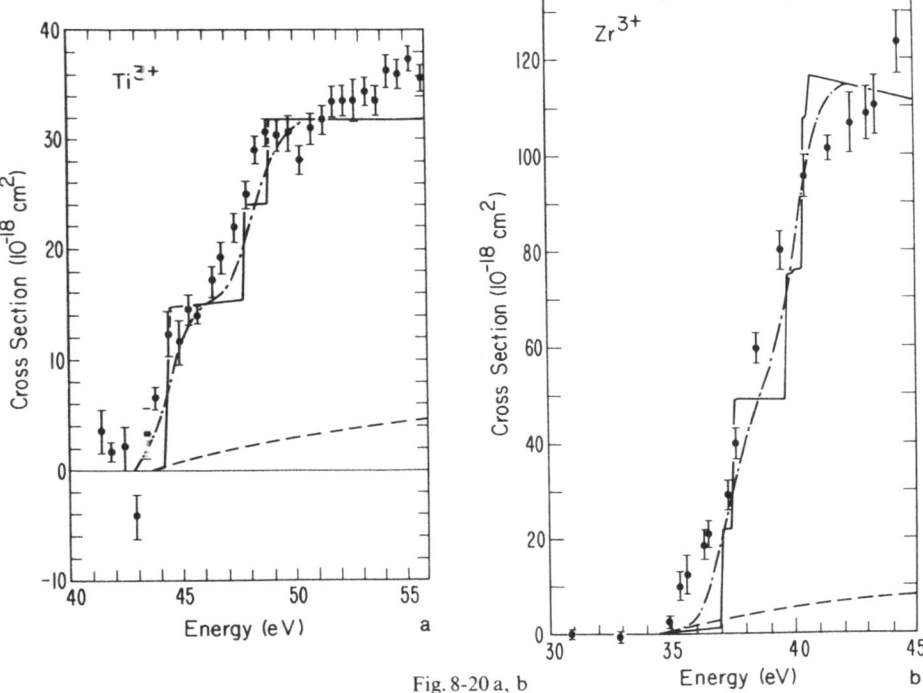

Fig. 8-20 a, b

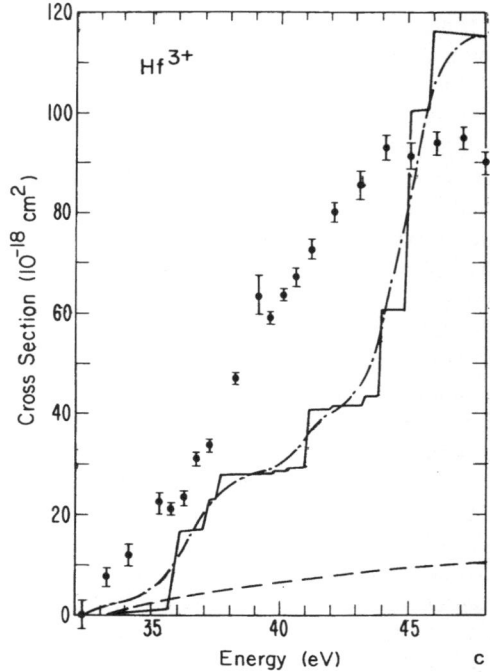

Fig. 8-20. Cross section in the threshold region for electron-impact ionization of: a) Ti^{3+}, b) Zr^{3+}. c) Hf^{3+}. Points, experimental measurements (Falk *et al.*, 1981, 1983 a); solid line, DW, dipole approximation calculations (Griffin *et al.*, 1982 b) added to dashed curve; dot-dashed, solid curve convoluted with a 2-eV-FWHM Gaussian to simulate electron-energy distribution; dashed curve. calculated direct-ionization cross section using the Lotz formula. Bars represent counting statistical uncertainties at one standard deviation (1σ)

In a discussion of the data for Hf^{3+} at higher energies than shown in Fig. 8-20, Falk *et al.* (1983 a) conclude that the data show f subshell electrons ionize much less efficiently than electrons from other subshells as is predicted by the Scaled Plane Wave Born method of McGuire (1977, 1979).

The structure of Ti^{2+} is $3p^6\,3d^2$, and one may again expect that transitions to $3p^5\,3d^2\,nl$ may give strong EA contributions with $3p^5\,3d^3$ being most prominent. Measurements of Mueller *et al.* (1983) shown in Fig. 8-21 indicate large EA cross sections. The narrow structure near 45 eV may be due to a buildup of resonances of the REDA type, though there is no clear means for identifying the cause at this point. A similar structure Ta^{3+} ($5p^6\,4f^{14}\,5d^2$) was also observed by Falk *et al.* (1983 a) to be dominated by EA, but the data were not precise enough to identify structures such as the one at 45 eV in Ti^{2+}. From the point of view of applications in controlled thermonuclear fusion and in astrophysics, it is probable that ions in the iron isonuclear sequence are as important as any one can think of. Yet there are difficulties working with this ion, and it is only recently that any measurements have been made for any charge state. After one solves the difficulty of making ions, there is the difficulty of assuring the beam purity, since q/m for iron in its various charge states seems to match q/m for various charge states of common possible impurities.

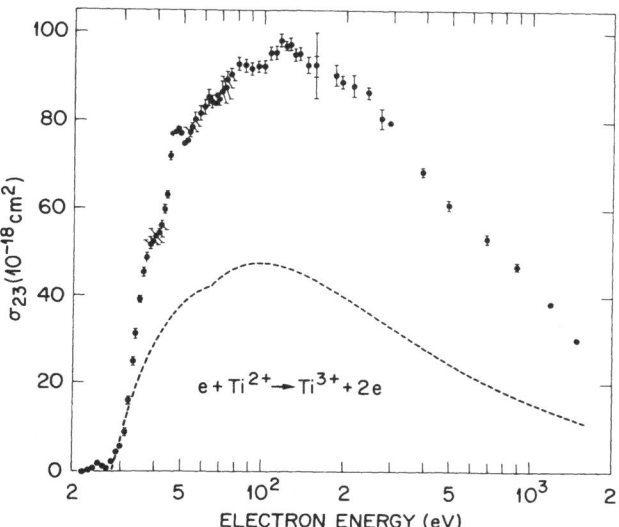

Fig. 8-21. Cross sections for electron impact single ionization of Ti^{2+}: points, crossed beams measurements (Mueller *et al.*, 1983); dashed, from Lotz formula

Furthermore, most low charge states will probably be contaminated with ions excited to metastable states. Nevertheless, some progress has been made in overcoming these problems, and Montague *et al.* (1983) reported measurements of cross sections for Fe^+, and Mueller *et al.* reported cross sections for Fe^{2+}. For Fe^+, the contribution from metastable ions in the beam was identified, and a ground state cross section was deduced. For Fe^{2+} there is a substantial apparent cross section below the threshold for ground state ionization, and there is evidence from the shape there that EA plays a strong role in ionization of the metastable. Expected

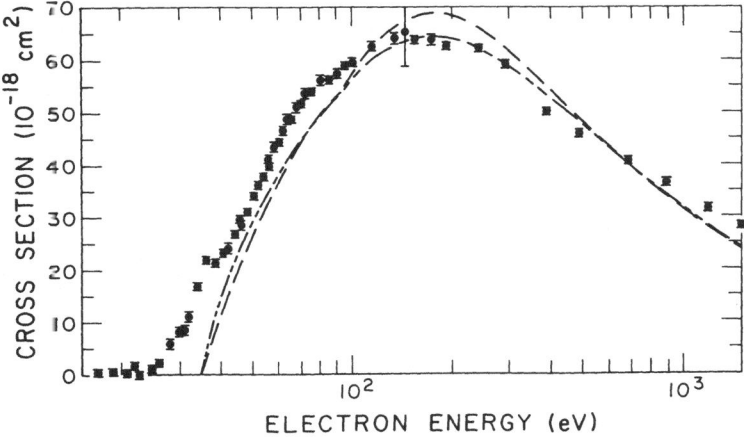

Fig. 8-22. Cross sections for electron impact single ionization of Fe^{2+}: points, crossed beams measurements (Mueller *et al.*, 1983); dashed, from Lotz formula; chain, DWE (Younger, 1981 a)

contributions from EA to ground state ionization of the type $3p^6 3d^6 \rightarrow 3p^5 3d^7$ should be seen at about 57 eV, but are not easily distinguished in the data presented in Fig. 8-22.

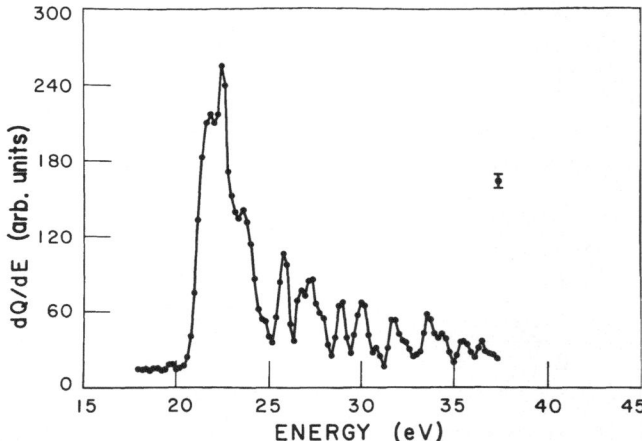

Fig. 8-23. Near threshold derivative with respect to energy of the ionization cross section of Ga^+ (Rogers *et al.*, 1982). Error bar at the right/above represents rms reproducibility of derivatives from individual data runs

Rogers *et al.* (1982) carried out high precision cross section measurements for Zn^+ and Ga^+ with ground state structures $3d^{10} 4s$ and $3d^{10} 4s^2$ respectively. The derivative of the measured cross section with respect to energy is shown near threshold for Ga^+ in Fig. 8-23, and these were tentatively identified with EA transitions to $3d^9 4s^2 nl$. Pindzola *et al.* (1982) calculated the cross section contributions from excitation to 12 states of the type $3d^9 4s^2 4p$ using the distorted wave approximation. The results were added to cross sections for direct ionization calculated using either the Lotz (1968) or the SPWB method of McGuire (1977), and are shown in Fig. 8-24 along with the experimental measurements of Rogers *et al.* for the energy range shown. The agreement seems good, but as Pindzola *et al.* point out, this may be fortuitous, since the distorted wave calculations are believed to overestimate the $3d^{10} 4s^2 \rightarrow 3d^9 4s^2 4p$ cross sections, and since final configurations of the type $3d^9 4s^2 nl$ with $nl \neq 4p$ were not included. Indeed, this assessment is probably accurate, since all of the states considered by Pindzola *et al.* lie in the first large peak of Fig. 8-23. It is readily seen from this figure that other states contribute with thresholds in the energy range $25 - 35$ eV. Two-state close-coupling was used to calculate EA contributions for Zn^+ (Rogers *et al.*, 1982), and when added to a calculated direct ionization cross section, quite reasonable agreement was obtained with experimental cross sections.

Analogous to the study of Ga^+, belonging to the zinc isoelectronic sequence, are studies on the cadmium sequence which has structure $4d^{10} 5s^2$. Measurements were carried out on Xe^{6+} by Gregory and Crandall (1983) and calculations in the distorted wave approximation were done for In^+, Sb^{3+}, and Xe^{6+} and also in the

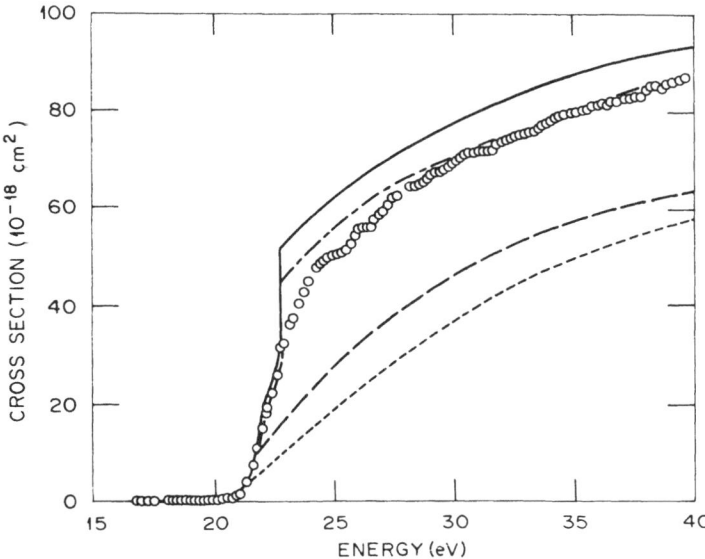

Fig. 8-24. Total ionization cross section for Ga⁻. ◯, crossed beams measurements (Rogers *et al.*, 1982): (— — —), direct-ionization cross section calculated from the Lotz equation: — — — direct ionization calculated from scaled PWB: ———, distorted-wave excitation cross section for $3d^{10} 4s^2 \to 3d^9 4s^2 4p$ plus Lotz: — — — — —, distorted-wave excitation cross section for $3d^{10} 4s^2 \to 3d^9 4s^2 4p$ plus scaled PWB (Pindzola *et al.*, 1982)

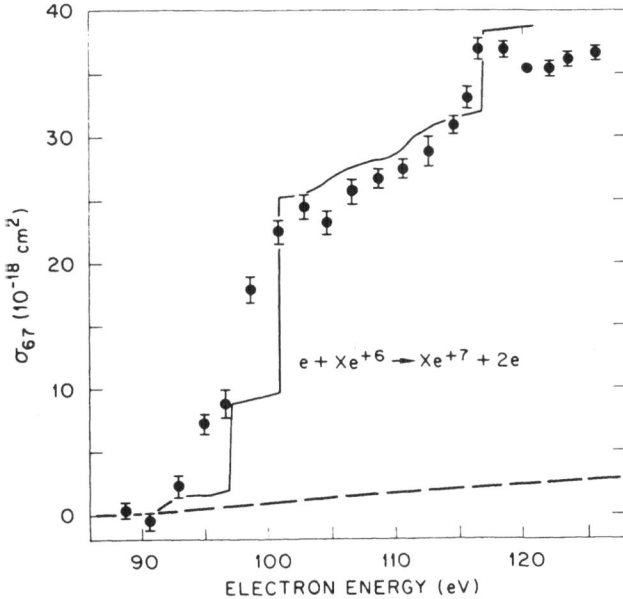

Fig. 8-25. Threshold region of the Xe⁶⁺ ionization cross section. Relative uncertainties are 1 s.d. Dashed curve. Lotz prediction: solid curve, distorted-wave excitation-autoionization theory (Pindzola *et al.*, 1983) added to Lotz

close coupling approximation for Xe^{6+} by Pindzola *et al.* (1983). Results of both measurements and calculations for Xe^{6+} are presented in Fig. 8-25. The direct ionization mechanism is seen to account for only about 5% of the total. Excitation-autoionization transitions of the type $4d^{10}4s^2 \rightarrow 4d^9 4s^2 nf$, $4d^9 4s^2 5d$, and $4d^9 5s^2 6p$ seem to account for the bulk of the cross section. There is good agreement between experiment and theory, except the step structure is not seen in the experiments — there is "snow on the stairsteps". Hahn and La Gattuta (1984) attribute this to the REDA process, and their estimates of the effect bring the predicted curve into good agreement with the data. Griffin *et al.* (1984a), on the other hand, also calculated the effect of REDA using an average branching ratio for double ionization of 0.05, and found the contribution much too small to explain the data. They speculated on the possible importance of resonant excitation auto double ionization, but concluded only that there was a significant disagreement between theory and experiment for this detail.

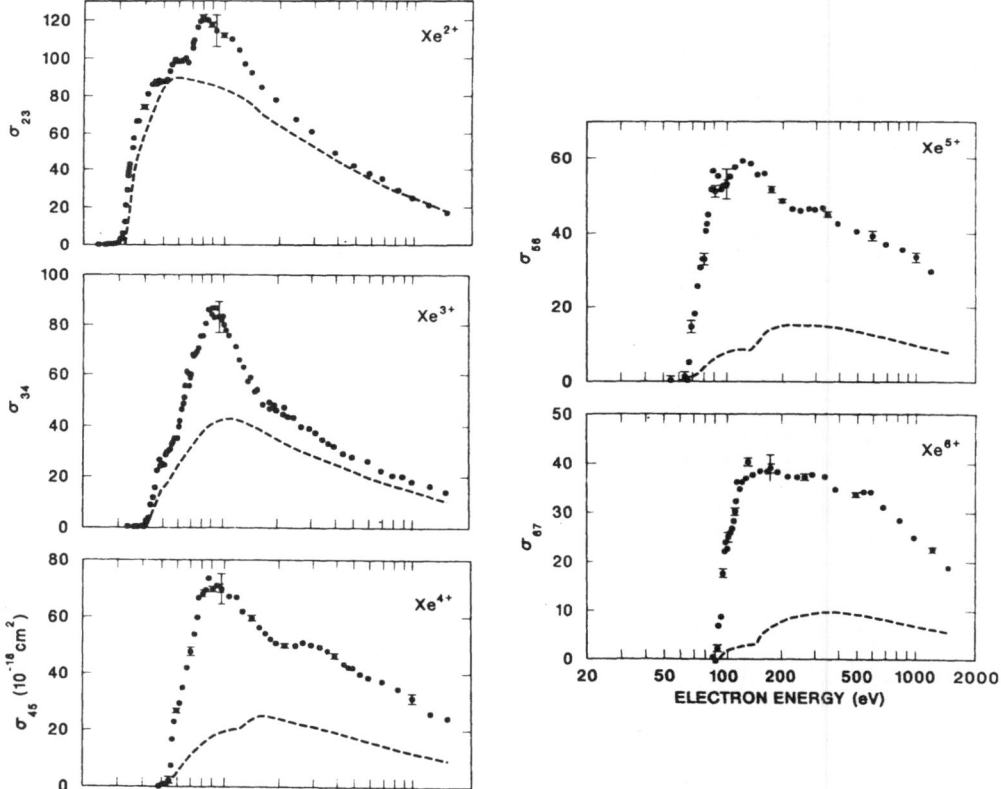

Fig. 8-26. Single ionization of Xe^{2+} through Xe^{6+}. Relative uncertainties in the measurements at the 1 s.d. level are shown at representative points. Absolute uncertainties ($\pm 7.5\%$ of the peak cross section) at good confidence level are plotted near 100 eV (175 eV on the Xe^{6+} curve). Dashed curves are distorted-wave calculations of direct ionization. Figure taken from Griffin *et al.* (1984a)

Indeed, single ionization of Xe^{n+} has been studied for $n = 1 - 4$ by Achenbach *et al.* (1983) and for $n = 2 - 6$ by Griffin *et al.* (1984 a), and the data for all stages of ionization for $n \geq 2$ are strongly affected by indirect processes. Fig. 8-26 taken from Griffin *et al.* demonstrates the data and estimates of direct ionization. For $n = 2 - 4$, where both groups made measurements, there is reasonable agreement between them, except the data of Griffin *et al.* are generally about $20 - 30\%$ lower at the peak. Griffin *et al.* also made calculations of the indirect effects, obtaining moderate agreement with experiment generally, and quite good agreement for the Xe^{6+} case shown in Fig. 8-25. Generally, they concluded that the ionization cross sections are reasonably well explained by an independent process approximation if contributions from both EA and REDA are included. They found the excitation-autoionization contributions from the transitions $4d^{10} 5s^2 5p^{6-q} \rightarrow 4d^9 5s^2 5p^{6-q} nf$ to be highly term dependent, and noted the need to employ proper radial wave functions to obtain accurate results. Agreement between theory and experiment should be considered good, given the complexities of the ions.

8.3.4 General Observations for Single Ionization

A number of conclusions can be drawn from the observations discussed in this section which can be applied with only a few exceptions.

1. Electron impact ionization cross sections of ions for $Z \leq 8$ have been thoroughly studied and are quite well understood both experimentally and theoretically. The cross sections are — with two exceptions — well characterized by the formula and fitted parameters of Bell *et al.* (1983).

2. Direct ionization is quite well represented by theory, though one must use caution and recognize exceptions as, for example, with Ne-like and Na-like ions.

3. Indirect processes involving inner-shell electrons can contribute varying amounts to the total ionization cross section, with observed contributions ranging from a few percent to more than 95%. Thus, these processes may be totally insignificant or they may dominate by more than an order of magnitude.

4. It is probable that the processes noted in Item 3 above include recombination resonances with subsequent double autoionization or auto-double ionization (REDA and READI), though the evidence for these processes is so far indirect.

5. The indirect ionization mechanisms and their contributions depend heavily on atomic structure and configuration, but usually, along an isoelectronic sequence, the fractional contribution of indirect ionization to the total increases with ionic charge.

6. Significant contributions to total ionization from indirect effects become more likely as the complexity of the ion increases for a given ionic charge.

Reviews and Bibliographies

Bazylev, V. A., Chibisov, M. I. (1981): Sov. Phys. Usp. *24*, 276.

Bell, K. L., Gilbody, H. B., Hughes, J. G., Kingston, A. E., Smith, F. J. (1983): J. Phys. Chem. Ref. Data *12*, 891; also see report: Cullham Laboratory Report CLM-R216 (1982).

Burgess, A., Summers, H. P., Cochrane, D. M., McWhirter, R. W. P. (1977): Monthly Notices Roy. Astron. Soc. *179*, 275.

Crandall, D. H. (1981): Physica Scr. *23*, 153.

Crandall, D. H. (1982): Physics of Electronic and Atomic Collisions (Datz, S., ed.), 595. Amsterdam: North-Holland.

Crandall, D. H. (1983a): Atomic Physics of Highly Ionized Atoms (Marrus, R., ed.), 399–453. New York: Plenum.

Crandall, D. H. (1983b): Nucl. Inst. Meth. *214*, 129.

Dolder, K. T. (1969): Case Studies in Atomic Collision Physics 1 (McDaniel, E. W., McDowell. M. R. C., eds.), Chapter 5, 249. Amsterdam: North-Holland.

Dolder, K. T. (1980): Atomic and Molecular Processes in Controlled Thermonuclear Fusion (McDowell, M. R. C., Fernedeci, A. M., eds.), 313–349. New York: Plenum.

Dolder, K. T. (1983): Physics of Ion-Ion and Electron-Ion Collisions (Brouillard, F., McGowan, J. W., eds.), 373–413. New York: Plenum.

Dolder, K. T., Peart, B. (1976): Rep. Prog. Phys. *39*, 693.

Dunn, G. H. (1969): Atomic Physics, Vol. I (Hughes, V. W., Cohen, V. W., Pichanick, F. M. J., eds.). 417. New York: Plenum.

Dunn, G. H. (1976): IEEE Trans. Nucl. Sci. *NS-23*, 929.

Dunn, G. H. (1979): Atomic Processes in Fusion Plasmas, IPPJ-AM-13 (Itikawa, Y., Kato, T., eds.). 57. Nagoya: Institute of Plasma Physics.

Dunn, G. H. (1980): Physics of Ionized Gases (Matić, H., Kidrić, B., eds.), 49. Belgrade: Inst. Nucl. Sci.

Gabriel, A. H., Jordan, C. (1972): Case Studies in Atomic Collision Physics 2 (McDaniel, E. W., McDowell, M. R. C., eds.), 209. Amsterdam: North-Holland.

Harrison, M. F. A. (1966): Brit. J. Appl. Phys. *17*, 371.

Harrison, M. F. A. (1968): Methods in Experimental Physics, Vol. 7B, (Fite, W. L., Bederson, B., eds.). 95–115. New York: Academic Press.

Itikawa, Y. (1982): Report No. IPPJ-AM-24. Nagoya: Institute of Plasma Physics. Bibliography 1978–1982, a supplement to IPPJ-AM-7.

Itikawa, Y., Kato, T. (1981): Report No. IPPJ-AM-17. Nagoya: Institute of Plasma Physics.

Jakubowicz, H., Moores, D. L. (1980): Comments At. Mol. Phys. *9*, 55.

Kieffer, L. J., Dunn, G. H. (1966): Rev. Mod. Phys. *38*, 1.

McGowan, J. W. (1980): Electronic and Atomic Collisions (Oda, N., Takayanagi, K., eds.), 237. Amsterdam: North-Holland.

Moores, D. L. (1982): Physics of Electronic and Atomic Collisions (Datz, S., ed.), 623. Amsterdam: North-Holland.

Moores, D. L., Nussbaumer, H. (1981): Space Sci. Rev. *29*, 379.

Salzborn, E. (1983): Physics of Ion-Ion and Electron-Ion Collisions (Brouillard, F., McGowan, J. W., eds.), 239–277. New York: Plenum.

Takayanagi, K., Iwai, T. (1978): Report No. IPPJ-AM-7. Nagoya: Institute of Plasma Physics. (Bibliography 1940 through 1977.)

Younger, S. M. (1982): Comments At. Mol. Phys. *11*, 192.

References

Achenbach, C., Müller, A., Salzborn, E., Becker, R. (1984): J. Phys. B *17*, 1405.

Aitken, K. L., Harrison, M. F. A. (1971): J. Phys. B *4*, 1176.

Aitken, K. L., Harrison, M. F. A., Rundel, R. D. (1971): J. Phys. B *4*, 1189.

Arianer, J., Geller, R. (1981): Ann. Rev. Nucl. Part. Sci. *31*, 19.

Baker, F. A., Hasted, J. B. (1965): 4th ICPEAC 447.
Baker, F. A., Hasted, J. B. (1966): Phil. Trans. Roy. Soc. A 261, 33.
Belić, D. S., Falk, R. A., Timmer, C., Dunn, G. H. (1982): Private Communication.
Bely, O. (1968): J. Phys. B 1, 23.
Bely, O., Schwartz, S. B., Val, J. L. (1971): J. Phys. B 4, 1482.
Bottcher, C., Griffin, D. C., Pindzola, M. S. (1983): J. Phys. B 16, L65.
Breton, C., DeMichelis, C., Finkenthal, M., Mattioli, M. (1978): Phys. Rev. Lett. 41, 110.
Brown, R., Deuchars, W. M., Kidd, D. E., Summers, H. P., Wood, L. (1983): J. Phys. B 16, 2053.
Burgess, A. (1964): Proceedings of the Symposium on Atomic Collision Processes in Plasmas, Culliam AERE Report No. 4818, 63.
Burke, P. G., Kingston, A. E., Thompson, A. (1983): J. Phys. B 16, L385.
Crandall, D. H., Phaneuf, R. A., Falk, R. A., Belić, D. S., Dunn, G. H. (1982): Phys. Rev. A 25, 143.
Crandall, D. H., Phaneuf, R. A., Gregory, D. C. (1979a): ORNL/TM-7020, Oak Ridge National Laboratory.
Crandall, D. H., Phaneuf, R. A., Gregory, D. C., Mueller, D., Morgan, T., Dunn, G. H. (1983): Private Communication.
Crandall, D. H., Phaneuf, R. A., Hasselquist, B. E., Gregory, D. C. (1979b): J. Phys. B 12, L249.
Crandall, D. H., Phaneuf, R. A., Taylor, P. O. (1978): Phys. Rev. A 18, 1911.
Crandall, D. H., Taylor, P. O., Phaneuf, R. A. (1977): 10th ICPEAC, 1086.
Datla, R. U., Blaha, M., Kunze, H.-J. (1975): Phys. Rev. A 12, 1076.
Datla, R. U., Kunze, H.-J., Petrini, D. (1972): Phys. Rev. A 6, 38.
Datla, R. U., Nugent, L. J., Griem, H. R. (1976): Phys. Rev. A 14, 979.
Datla, R. U., Roberts, J. R. (1983): Phys. Rev. A 28, 2201.
Defrance, P., Brouillard, F., Claeys, W., van Wassenhove, G. (1981): J. Phys. B 14, 103.
Divine, T. F., Feeney, R. F., Sayle, W. E., II, Hooper, J. W. (1976): Phys. Rev. A 13, 54.
Dolder, K. T., Harrison, M. F. A., Thonemann, P. C. (1961): Proc. Roy. Soc. A 264, 367.
Dolder, K. T., Harrison, M. F. A., Thonemann, P. C. (1963): Proc. Roy. Soc. A 274, 546.
Donets, E. D. (1976): Int. Conf. on Heavy Ion Sources, Gatlinburg. IEEE Trans. Nucl. Sci. NS-23, 897.
Donets, E. D., Ovsyannikov, V. P. (1977a): 10th ICPEAC, 1088.
Donets, E. D., Ovsyannikov, V. P. (1977b): JINR, R-10780, Dubna.
Donets, E. D., Ovsyannikov, V. P. (1981): Sov. Phys. JETP 53, 466.
Donets, E. D., Pikin, A. I. (1975 1976): Zh. Tekh. Fiz. 45, 2373 (1975); Sov. Phys. Tech. Phys. 20, 1477.
Donets, E. D., Pikin, A. I. (1976): JETP USSR 70, 2025; Sov. Phys.-JETP 43, 1057.
Falk, R. A., Dunn, G. H. (1983): Phys. Rev. A 27, 754.
Falk, R. A., Dunn, G. H., Gregory, D. C., Crandall, D. H. (1983a): Phys. Rev. A 27, 762.
Falk, R. A., Dunn, G. H., Griffin, D. C., Bottcher, C., Gregory, D. C., Crandall, D. H., Pindzola, M. S. (1981): Phys. Rev. Lett. 47, 494.
Falk, R. A., Stefani, G., Camilloni, R., Dunn, G. H., Phaneuf, R. A., Gregory, D. C., Crandall, D. H. (1983b): Phys. Rev. A 28, 91.
Feeney, R. K., Hooper, J. W., Elford, M. T. (1972): Phys. Rev. A 6, 1469.
Feeney, R. K., Sayle, W. E., II, Divine, T. F. (1978): Phys. Rev. A 18, 82.
Golden, L. B., Sampson, D. H. (1977): J. Phys. B 10, 2229.
Gregory, D. C. (1983): Private Communication.
Gregory, D. C., Crandall, D. H. (1983): Phys. Rev. A 27, 2338.
Gregory, D. C., Dittner, P. F., Crandall, D. H. (1983): Phys. Rev. A 27, 724.
Greve, P., Kato, M., Kunze, H.-J., Hornady, F. S. (1981): Phys. Rev. A 24, 429.
Griffin, D. C., Bottcher, C., Pindzola, M. S. (1982a): Phys. Rev. A 25, 154.
Griffin, D. C., Bottcher, C., Pindzola, M. S. (1982b): Phys. Rev. A 25, 1374.
Griffin, D. C., Bottcher, C., Pindzola, M. S., Younger, S. M., Gregory, D. C., Crandall, D. H. (1984a): Phys. Rev. A 29, 1729.
Griffin, D. C., Pindzola, M. S., Bottcher, C. (1984b): J. Phys. B 17, 3183.
Hahn, Y., LaGattuta, K. J. (1984): Private Communication.
Hamdan, M., Birkinshaw, K., Hasted, J. B. (1978): J. Phys. B 11, 331.
Harrison, M. F. A., Dolder, K. T., Thonemann, P. C. (1963): Proc. Phys. Soc. 82, 368.
Hasted, J. B., Awad, G. L. (1972): J. Phys. B 5, 1719.
Henry, R. J. W. (1979): J. Phys. B 12, L309.
Henry, R. J. W., Msezane, A. W. (1982): Phys. Rev. A 26, 2545.

Hertling, D. R., Feeney, R. K., Hughes, D. W., Sayle II, E. W. (1982): J. Appl. Phys. *53*, 5427.

Hinnov, E. (1966/1967): J. Opt. Soc. Am. *56*, 1179. [See also revision in J. Opt. Soc. Am. *57*, 1392 (1967).]

Hooper, J. W., Lineberger, W. C., Bacon, F. M. (1966): Phys. Rev. *141*, 165.

Howald, A. M. (1984): Private Communication.

Hughes, D. W., Feeney, R. K. (1981): Phys. Rev. *A 23*, 2241.

Jakubowicz, H., Moores, D. L. (1981): J. Phys. *B 14*, 3733.

Jones, L. A., Källne, E., Thomson, D. B. (1977): J. Phys. *B 10*, 187.

Källne, E., Jones, L. A. (1977): J. Phys. *B 10*, 3637.

Kunze, H.-J. (1971): Phys. Rev. *A 3*, 937.

Kunze, H.-J., Gabriel, A. H., Griem, H. R. (1968): Phys. Rev. *165*, 267.

Kupriyanov, S. E., Latypov, Z. Z. (1963/1964): JETP USSR *45*, 815 (1963); Sov. Phys.-JETP *18*, 558 (1964).

La Gattuta, K. J., Hahn, Y. (1981): Phys. Rev. *A 24*, 2273.

Latypov, Z. Z., Kupriyanov, S. E. (1968): Sov. Phys.-Tech. Phys. *13*, 811.

Latypov, Z. Z., Kupriyanov, S. E., Tunitskii, N. N. (1964): JETP USSR *46*, 833; Sov. Phys.-JETP *19*, 570.

Lineberger, W. C., Hooper, J. W., McDaniel, E. W. (1966): Phys. Rev. *141*, 151.

Lotz, W. (1967a): Astrophys. J. Suppl. *14*, 207.

Lotz, W. (1967b): Z. Phys. *206*, 205.

Lotz, W. (1968): Z. Phys. *216*, 241.

Lotz, W. (1969): Z. Phys. *220*, 466.

Magee, N. H., Mann, J. B., Merts, A. L., Robb, W. D. (1977): Los Alamos Scientific Laboratory Report LA6691 M.S., 109 – 112.

Mallory, M. L., Crandall, D. H. (1976): IEEE Trans. Nucl. Sci. *23*, 1069.

Martin, S. O., Peart, B., Dolder, K. T. (1968): J. Phys. *B 1*, 537.

Matsumoto, A., Ohtani, S., Danjo, A., Hanashiro, H., Hino, T., Kondo, Y., Suzuki, H., Tawara, H., Wakiya, K., Yoshino, M. (1983): XIII ICPEAC Abstracts, 198.

McGuire, E. J. (1977): Phys. Rev. *A 16*, 73.

McGuire, E. J. (1979): Phys. Rev. *A 20*, 445.

McGuire, E. J. (1982): Phys. Rev. *26*, 125.

Montague, R. G., Diserens, M. J., Harrison, M. F. A., Smith, A. C. H. (1983): XIII ICPEAC Abstracts. 205.

Montague, R. G., Harrison, M. F. A. (1983): J. Phys. *B 16*, 3045.

Moores, D. L. (1972): J. Phys. *B 5*, 286.

Moores, D. L. (1978): J. Phys. *B 11*, 1403.

Moores, D. L., Golden, L. B., Sampson, D. H. (1980): J. Phys. *B 13*, 385.

Moores, D. L., Nussbaumer, H. (1970): J. Phys. *B 3*, 161.

Mott, N. F., Massey, H. S. W. (1965): The Theory of Atomic Collisions, 3rd ed. Oxford: Clarendon.

Mueller, D. W., Morgan, T. J., Dunn, G. H., Gregory, D. C., Crandall, D. C. (1983): Bull. Am. Phys. Soc. *28*, 931; XIII ICPEAC Abstracts 1983, 204.

Müller, A., Frodl, R. (1980): Phys. Rev. Lett. *44*, 29.

Müller, A., Salzborn, E., Frodl, R., Becker, R., Klein, H. (1980a): J. Phys. *B 13*, L221.

Müller, A., Salzborn, E., Frodl, R., Becker, R., Klein, H., Winter, H. (1980b): J. Phys. *B 13*, 1877.

Omidvar, K. (1969): Phys. Rev. *177*, 212.

Peart, B., Dolder, K. T. (1968a): J. Phys. *B 1*, 240.

Peart, B., Dolder, K. T. (1968b): J. Phys. *B 1*, 872.

Peart, B., Dolder, K. T. (1969): J. Phys. *B 2*, 1169.

Peart, B., Dolder, K. T. (1973): J. Phys. *B 6*, 2409.

Peart, B., Dolder, K. T. (1975): J. Phys. *B 8*, 56.

Peart, B., Martin, S. O., Dolder, K. T. (1969): J. Phys. *B 2*, 1176.

Peart, B., Stevenson, J. G., Dolder, K. T. (1973): J. Phys. *B 6*, 146.

Peart, B., Walton, D. S., Dolder, K. T. (1969): J. Phys. *B 2*, 1347.

Pindzola, M. S., Griffin, D. C., Bottcher, C. (1982): Phys. Rev. *A 25*, 211.

Pindzola, M. S., Griffin, D. C., Bottcher, C. (1983): Phys. Rev. *A 27*, 2331.

Pindzola, M. S., Griffin, D. C., Bottcher, C., Crandall, D. H., Phaneuf, R. A., Gregory, D. C. (1984): Phys. Rev. *A 29*, 1749.

Redhead, P. A. (1967): Can. J. Phys. *45*, 1791.

Redhead, P. A. (1969): Can. J. Phys. *47*, 2449.
Redhead, P. A. (1970): Can. J. Phys. *48*, 1906.
Redhead, P. A. (1971): Can. J. Phys. *49*, 3059.
Redhead, P. A., Feser, S. (1968): Can. J. Phys. *46*, 1905.
Redhead, P. A., Gopalaraman, C. P. (1971): Can. J. Phys. *49*, 585.
Rogers, W. T., Stefani, G., Camilloni, R., Dunn, G. H., Msezane, A. Z., Henry, R. J. W. (1982): Phys. Rev. *A 25*, 737.
Rowan, W. L., Roberts, J. R. (1979): Phys. Rev. *A 19*, 90.
Rudge, M. R. H., Schwartz, S. B. (1966): Proc. Phys. Soc. *88*, 579.
Sampson, D. H., Golden, L. B. (1979): J. Phys. *B 12*, L785.
Sampson, D. H., Golden, L. B. (1981): J. Phys. *B 14*, 903.
Wareing, J. B., Dolder, K. T. (1967): Proc. Phys. Soc. *91*, 887.
Woodruff, P. R., Hublet, M.-C., Harrison, M. F. A. (1978a): J. Phys. *B 11*, L305.
Woodruff, P. R., Hublet, M.-C., Harrison, M. F. A., Brook, E. (1978b): J. Phys. *B 11*, L679.
Younger, S. M. (1980a): Phys. Rev. *A 22*, 111.
Younger, S. M. (1980b): Phys. Rev. *A 22*, 1425.
Younger, S. M. (1981a): Atomic Data for Fusion 7, 190. (A bulletin from the Controlled Fusion Atomic Data Center of Oak Ridge National Laboratory and the National Bureau of Standards.)
Younger, S. M. (1981b): Phys. Rev. *A 23*, 1138.
Younger, S. M. (1981c): Phys. Rev. *A 24*, 1272.
Younger, S. M. (1984): Private Communication.

9

Applications

9.1 Application of Electron Impact Ionization to Plasma Diagnostics

W. Lindinger and *F. Howorka*

Institut für Experimentalphysik, Leopold-Franzens-Universität,
Innsbruck, Austria

9.1.1 Introduction

Information on plasma densities (total charge carrier densities as well as densities of individual ion species), on plasma temperatures and on the excitation of various plasma components is obtained by a variety of diagnostic methods, involving photon spectroscopy, probe measurements as well as mass spectrometric plasma sampling. The latter usually requires the investigation of ions emerging through orifices in walls confining the plasma. So called hole probes (Lindinger, 1971, 1973: Howorka *et al.*, 1973, 1974; Pahl *et al.*, 1972) have proven to be a well suited tool for both ion and electron sampling from plasmas. Such a device consists (Fig. 9.1-1) of a thin metal plate, covered with a non conducting material (e.g. glass), leaving only an area of $\leq 1 \, mm^2$ of the metal exposed to the plasma. Located in the center of this area is an aperture of about 50 μm in diameter, the metal plate being only 10 μm thick at this point. Changing the potential of the metal plate allows the hole probe to be operated as a planar Langmuir probe, and at the same time ions and electrons leaving the plasma via the aperture can be investigated. By means of mass spectrometric analysis of the ions sampled from cylindrical hollow cathode discharges, which were movable with respect to the hole probe, radial density profiles of the various ion types present in the negative glow were monitored (Lindinger, 1973; Howorka *et al.*, 1973, 1974). Valuable diagnostic data on plasmas are obtained, when mass spectrometric ion sampling is combined with existing information on electron impact ionization. In the following we will discuss results

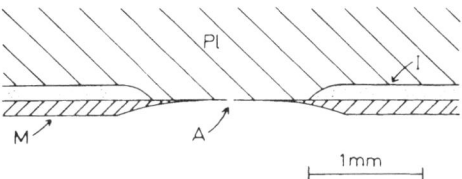

Fig. 9.1-1. Hole probe used for ion sampling from plasmas. M molybdenum, I insulating material (usually glass). Pl plasma, A aperture

obtained in this way on the densities of fast electrons, which are responsible for primary ion production within plasmas, such as the negative glows of hollow cathode discharges. Also we will show the application of electron impact ionization occurring in and behind the orifice of a hole probe to determine the densities of neutral trace constituents admixed to the discharge gas.

9.1.2 Densities of Fast Electrons in Hollow Cathode Discharges

Hollow cathode discharges (and any negative glow plasmas) are typical representatives of electron beam stabilized plasmas, in that electrons released from the walls by the impact of ions or by photons (Fig. 9.1-2) are accelerated in the cathode fall to energies of typically one to a few hundred eV, entering thereafter the negative glow, where electron impact ionization is the main source of ion production to sustain the discharge. The density of these fast electrons $[e_f]$ usually is several orders of magnitude lower than the density of the bulk of slow plasma electrons, which do not contribute significantly in the generation of positive ions (Howorka and Pahl, 1972). Obtaining information on $[e_f]$ from mass spectrometric ion sampling by means of hole probes requires some understanding of the probe characteristics and properties of hole probes, which will be discussed briefly in the following.

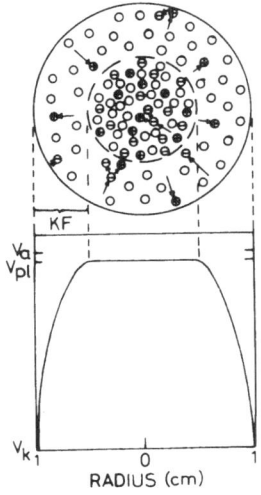

Fig. 9.1-2. Section through a cylindrical hollow cathode discharge and potential distribution, V_a anode potential, V_{pl} space potential, V_k cathode potential, KF cathode fall region

9.1.2.1 Hole Probe Characteristics

Typical hole probe characteristics for the ions Ar^+ and Ar^{++} sampled from the negative glow of a hollow cathode discharge in argon is shown schematically in Fig. 9.1-3 (Handle et al., 1984). When the hole probe potential V_p is kept negative with respect to the plasma potential, V_{pl}, part of the ions Ar^+ and Ar^{++} respectively, diffusing from the negative glow towards the probe are accelerated through the orifice. Thus, ions detected, when $V_p < V_{pl}$, originate from the negative glow region. Changing the probe potential towards positive values results in a decrease of the detected ion current $i(Ar^+)$ and $i(Ar^{++})$ due to repulsion of these ions by the positively biased probe. At the same time, slow plasma electrons, e_s, are accelerated towards the probe, gaining more energy the more positive V_p becomes. As soon as the energy of these electrons exceeds the ionization energy of Ar, ionization of Ar occurs in and behind the orifice of the hole probe resulting in the rapid increase of $i(Ar^+)$ at $V_p > 10$ Volts (region B in Fig. 9.1-3). Only electrons accelerated to more than 43.38 eV however are able to produce Ar^{++} ions, thus causing the increase of $i(Ar^{++})$ in region C of Fig. 9.1-3. Note that neither $i(Ar^+)$ nor $i(Ar^{++})$ goes to zero at any value V_p. These "residual" currents $i(Ar^{++})_{min}$ and $i(Ar^+)_{min}$ originate from Ar^{++} and Ar^+ ions produced in and behind the orifice by *fast* electrons, e_f, which are scattered towards the hole probe, and due to their high energy their scattering pattern is not significantly influenced by the relatively slight variation of V_p in the range covered in Fig. 9.1-3. Thus especially the current $i(Ar^{++})_{min}$, showing a broad minimum, can be used as an indicator of the density $[e_f]$ at the position of the hole probe.

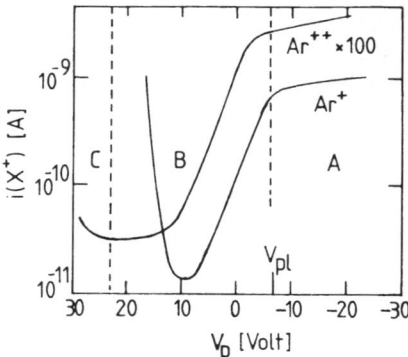

Fig. 9.1-3. Ion currents $i(Ar^+)$ and $i(Ar^{++})$ monitored as a function of the potential of the hole probe. V_p

9.1.2.2 Density $[e_f]$ Derived from Electron Impact Ionization within the Orifice of a Hole Probe

In the following we will discuss the quantitative analysis of $i(X^{n+})_{min}$ to obtain the density of fast electrons in negative glow plasmas. A condition to make this possible is that $i(X^{n+})_{min}$ consists neither of ions effusing from the plasma nor of ions produced by accelerated slow plasma electrons, e_s. This condition surely is met for $i(\mathrm{Ar}^{++})_{\min}$ in the example of Fig. 9.1-3, but not necessarily for $i(\mathrm{Ar}^{+})_{\min}$. In general, such broad minima, as observed for $i(\mathrm{Ar}^{++})$ occur for primary ions with high recombination energies as are typical for doubly charged ions, which are therefore best suited to serve as an indicator for $[e_f]$.

The current $i(X^{n+})_{\min}$ is related to the density of the neutral discharge gas and to the density $[e_f]$ by the equation (Handle *et al.*, 1984),

$$i(X^{n+})_{\min} = n \cdot \bar{\sigma}_{X^{n-}} \cdot i(e_f) \cdot \int_{-L}^{+x} [N](X)\, dX, \qquad (9.1\text{-}1)$$

with

$$i(e_f) = \frac{[e_f] \cdot \bar{v}}{4} \cdot R^2 \, \Pi, \qquad (9.1\text{-}2)$$

where R is the radius of the orifice in the hole probe, and $\bar{\sigma}_{X^{n-}}$ is the mean ionization cross section for X^{n+} production by the fast electrons, $[e_f]$ (for hollow cathode discharges the energy values for fast electrons are typically $300\,\mathrm{eV} \geq KE_{e_f} \geq 50\,\mathrm{eV}$). $\bar{v}$ is the average velocity of the electrons, e_f. According to Walcher (1944), for the neutral gas density as dependent on the distance from the hole probe, $[N](X)$, the following relation holds,

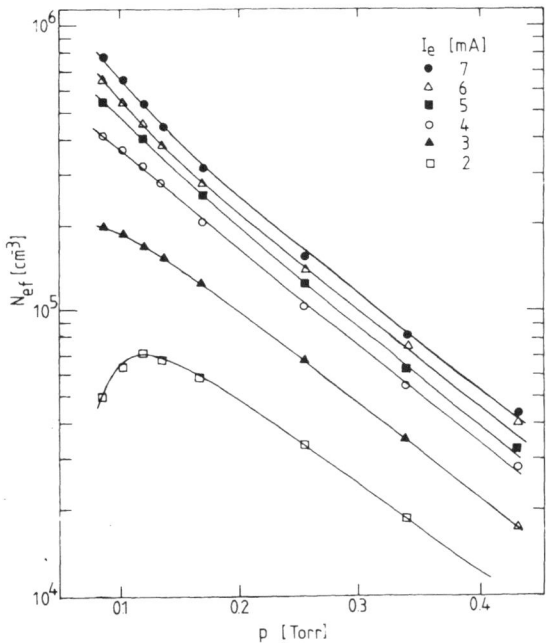

Fig. 9.1-4. Densities of fast electrons, $[e_f]$ (on the axis of a zylindrical hollow cathode discharge) as dependent on the discharge pressure, with the discharge current, I_e, being the parameter

$$\int\limits_{-L}^{+\infty} [N](X) = \frac{1}{2} [N_0](L+R), \tag{9.1-3}$$

with $[N_0]$ being the neutral gas density within the hollow cathode. L is the thickness of the metal sheet of the hole probe at the location of the orifice. Using measured values of $i(Ar^{++})_{min}$, the density $[e_f]$ on the axis of a hollow cathode discharge in argon ($\emptyset = 2$ cm) was recently obtained (Handle et al., 1984) over a range of discharge gas pressures 0.08 Torr $\leq p \leq 0.45$ Torr and discharge currents 2 mA $\leq I \leq 7$ mA, the results of which are shown in Fig. 9.1-4.

9.1.2.3 Density $[e_f]$ Derived from Measured Ion Density Profiles

A completely different approach for obtaining the densities of fast electrons in the negative glows of hollow cathode discharges utilizes measured radial density profiles of ions, an example of which is shown in Fig. 9.1-5 (Lindinger, 1973). As the potential within the negative glow is nearly constant (Howorka and Pahl, 1972), and no dependence of the total charge carrier density along the axial direction of the negative glow is observed (Lindinger, 1971), the motion of ions within the negative glow is governed solely by ambipolar diffusion and thus the steady state equation

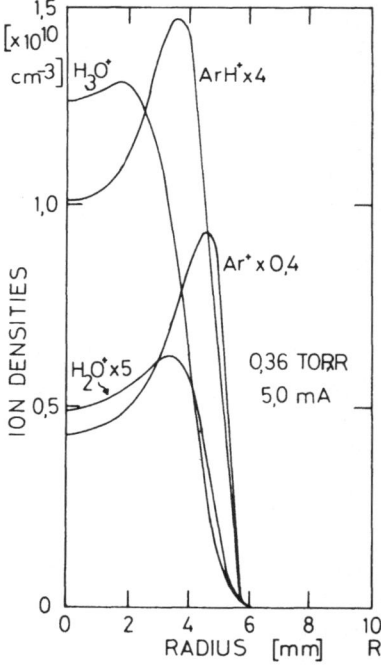

Fig. 9.1-5. Radial density profiles of the ions Ar^+, ArH^+, H_2O^+ and H_3O^+ in a hollow cathode discharge obtained by mass spectrometric sampling by means of a hole probe. The discharge gas is Ar with traces of H_2O

containing radial components only applies to the various ion species, present in the negative glcw, Y_i^+, in the form

$$\frac{d[Y_i^+]}{dt} = 0 = D_{a_i} \Delta[Y_i^+] + \Sigma v_p - \Sigma v_l - \alpha[Y_i^+][e_s], \qquad (9.1\text{-}4)$$

with

$$\Delta[Y_i^+] = \frac{\partial^2[Y_i^+]}{\partial r^2} + \frac{1}{r}\frac{\partial[Y_i^+]}{\partial r}, \qquad (9.1\text{-}5)$$

where D_{a_i} is the ambipolar diffusion coefficient for the ion species Y_i^+. Σv_p is the sum of the terms describing the production of ions Y_i^+ by either electron impact ionization due to fast electrons, e_f, or due to ion molecule reactions. Σv_l and $\alpha[Y_i^+][e_s]$ describe the loss processes due to reactive collisions between Y_i^+ and neutral reactants and due to ion electron recombination respectively. In an argon discharge with traces of H_2O present (conditions of Fig. 9.1-5), Ar^+ is produced by electron impact

$$e_f - Ar \overset{\sigma}{\to} Ar^+ + 2e \qquad (9.1\text{-}6)$$

and losses are due to diffusion as well as due to reactive collisions

$$Ar^- + H_2O \overset{k_l}{\to} Products, \qquad (9.1\text{-}7)$$

while recombination of Ar^+ is negligible. Thus the steady state equation for Ar^+ reads (Lindinger, 1973),

$$\frac{d[Ar^+]}{dt} = 0 = D_a \Delta[Ar^+] + k_p[e_f][Ar] - k_l[Ar^+][H_2O], \qquad (9.1\text{-}8)$$

with $k_p = \langle \sigma_{Ar^-} \cdot \bar{v}\rangle$. From the data in Fig. 9.1-5, both $\Delta[Ar^+]$ as well as $[Ar^+]$, both as a function of the radius are obtained, and with an average value for the ionization cross section $\bar{\sigma}_{Ar^-} \simeq 2 \times 10^{-16}\,cm^2$ for electrons of energies between 50 and 300 eV (Egger and Märk, 1978; Hasted, 1972), an electron density $[e_f] \simeq 8 \times 10^4\,cm^{-3}$ was obtained, in good agreement with the density for e_f obtained by Handle et al. (1984) with the completely different method described in Section 9.1.2.2.

9.1.3 Neutral Gas Densities

Contaminants in concentrations as low as a few ppm can not only influence the ion composition of plasmas considerably, but do often decrease the total ion density dramatically by converting atomic ions into molecular ions, which have recombination coefficients usually about 4 orders of magnitudes higher than the ones of atomic ions. Ways of determining the densities of impurities or of purposely added trace constituents in plasmas are therefore of great interest. The classical approach is the ionization of the effusing neutral gas in an adjacent electron impact ion source with consecutive mass spectrometric analysis of the ions produced in the source.

However, when a hole probe is used for plasma ion diagnostic purposes, it also can yield data on the neutral gas composition, when operated at positive potentials as described in Section 9.1.2.3. When plasma electrons e_s, are accelerated to energies higher than necessary for ionizing any of the neutral constituents of the plasma, ion currents $i(Z_j^+)$ are observed, being related to the various neutral densities in the plasma in the form

$$i(Z_1^+):i(Z_2^+)\ldots:i(Z_j^+)=\sigma_1[Z_1]:\sigma_2[Z_2]\ldots:\sigma_j[Z_j], \tag{9.1-9}$$

with σ_j being the average ionization cross section for the Z_j neutral component and for the electron energy distribution given at the respective probe potential used. These electron energy distributions and thus σ_j are usually not known, but the problem can easily be overcome by adding purposedly higher impurity concentrations to the discharge gas, in quantities which allow the determination of $[Z_j]:[Z_{Buffer}]$ e.g. by means of pressure gauges or by measuring the gas flow ratio. With these known ratios $[Z_J]/[Z_{Buffer}]$ and the values $i(Z_j^+)/i(Z^+_{Buffer})$, measured under the same conditions, the ratios, σ_J/σ_{Buffer} can be determined (Lindinger and Howorka, 1973) for any Z_j (when necessary by correcting for reactions occuring in the extraction orifice as discussed by Helm et al., 1974, 1980). With these experimentally obtained ratios $\bar{\sigma}_i/\bar{\sigma}_{Buffer}$ now any small impurity concentrations Z_j can be calculated from the measured current ratios using equation (9.1-9).

References

Egger, F., Märk, T. D. (1978): Z. Naturforsch. 33 a, 1111.

Handle, F., Lindinger, W., Howorka, F., Pahl, M. (1984): Beitr. Plasmaphys. 24, 407−416.

Hasted, J. B. (1972): Physics of Atomic Collisions, 2nd ed., 396, 399. London: Butterworth.

Helm, H., Howorka, F., Handle, F., Egger, F., Lindinger, W. (1974): J. Phys. B: Atom. Molec. Phys. 7, 1970.

Helm, H., Märk, T. D., Lindinger, W. (1980): Pure and Appl. Chem. 52, 1739.

Howorka, F., Pahl, M. (1972): Z. Naturforschg. 27 a, 1425−1433.

Howorka, F., Lindinger, W., Pahl, M. (1973): Int. J. Mass Spectrom. Ion Phys. 12, 67−77.

Howorka, F., Lindinger, W., Varney, R. N. (1974): J. Chem. Phys. 61, 1180−1188.

Lindinger, W. (1971): Thesis, University of Innsbruck.

Lindinger, W. (1973): Phys. Rev. A 7, 328.

Lindinger, W., Howorka, F. (1973): Rev. Sci. Instrum. 44, 1473.

Pahl, M., Lindinger, W., Howorka, F. (1972): Z. Naturforsch. 27 a, 678.

Walcher, W. (1944): Z. Physik 122, 62.

9.2 Collisional Ionization: Astrophysical and Fusion Applications

J. M. Shull

Department of Astrophysics and JILA, University of Colorado and National Bureau of Standards and Laboratory for Atmospheric and Space Physics, Boulder, Colorado, U.S.A.

9.2.1 Introduction

For illustration of the effects of electron-impact collisional ionization, there are few less vivid examples than the hot astrophysical plasmas in stellar coronae, interstellar and intergalactic gas, and supernova remnants. However, laboratory studies in plasma physics have also provided a profitable interchange of data and theoretical calculations between the controlled fusion and astrophysics communities. In this review, I shall describe how basic data on collisional ionization have been used in both areas to obtain diagnostics of hot plasmas in states of equilibrium and disequilibrium.

9.2.2 Astrophysical Plasmas

Matter in the universe is predominantly hydrogen (H) and helium (He), with small amounts (approximately 1% by mass) of heavier elements such as C, N, O, Ne, Na, Mg, Si, S, Ar, Ca, and Fe. In many astrophysical environments, H and He are almost fully ionized as a result of radiative or collisional ionization. In such situations, the electrons are donated by H and He, and partially ionized heavy elements take on key roles as coolants of the plasma, diagnostics of physical conditions, and opacity sources to the transport of radiation.

Fig. 9.2-1 shows an optical photograph of the Crab Nebula, the remnant of a supernova explosion in 1054 A.D. These 10^{51} erg explosions, which terminate the lives of massive stars, heat and ionize vast portions of interstellar gas with their blast waves (McCray and Snow, 1979). Although the ionization state in the Crab Nebula is also strongly influenced by relativistic electrons and radiation from a rapidly spinning neutron star (pulsar) at the center, conditions in other young remnants are dominated by collisional ionization in hot ($10^7 - 10^8$ K) plasma with a Maxwellian distribution of electron velocities. Fig. 9.2-2 illustrates the $0.5 - 4.5$ keV X-ray line and continuum emission from one such remnant, observed by the Solid State Spectrometer (SSS) aboard NASA's HEAO-2 satellite. Similar X-ray line emission has been detected from the low-density ($n_e \simeq 10^{-3}$ cm^{-3}) intergalactic gas in rich clusters of galaxies (Mushotzky et al., 1978; Jones et al., 1979), while ultraviolet lines characteristic of somewhat cooler gas have been seen in active stellar chromospheres (Fig. 9.2-3).

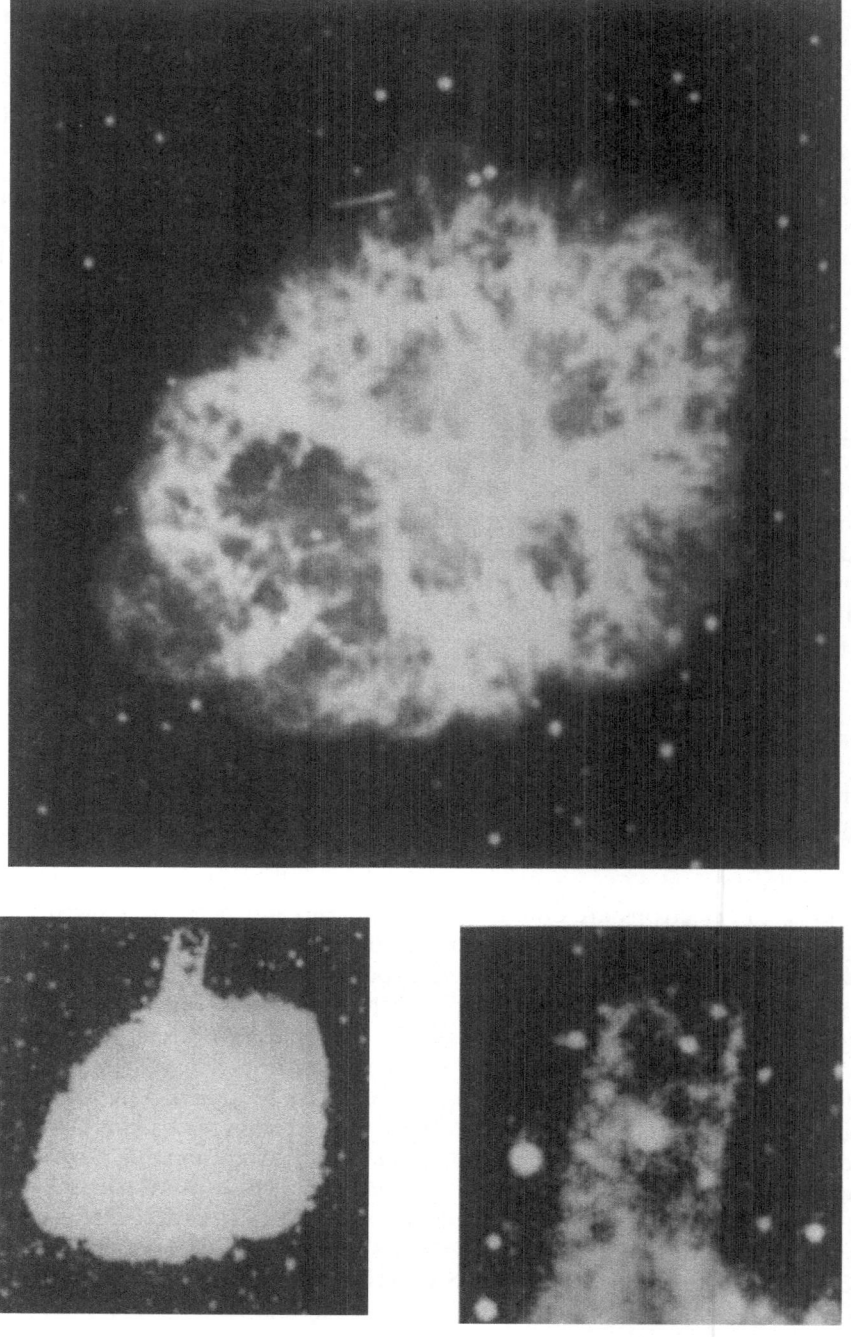

Fig. 9.2-1. Image enhancement of [O III] photograph of Crab Nebula (Gull and Fesen, 1982), showing faint jet at outer boundary and filaments resulting from 1054 A.D. supernova explosion

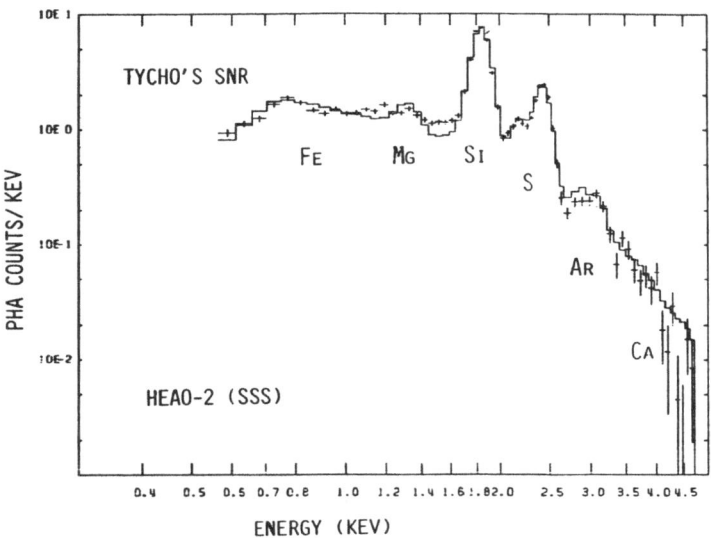

Fig. 9.2-2. X-ray spectrum of Tycho's supernova remnant (1604 A.D.), taken by the solid state spectrometer aboard the HEAO-2 X-ray observatory (Szymkowiak *et al.*, 1985). Positions of prominent He-like emission lines are indicated

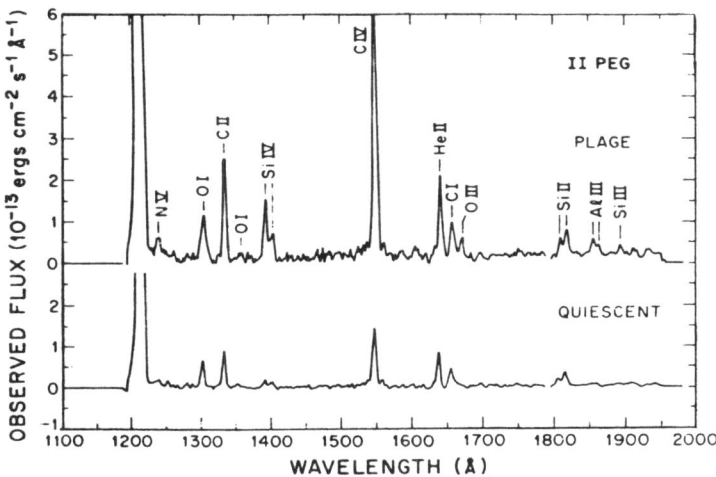

Fig. 9.2-3. Composite ultraviolet spectrum from *International Ultraviolet Explorer* satellite (Linsky, 1982) of chromospheric activity in the RS CVn-type star II Peg

9.2.2.1 Coronal Equilibrium

Following the observation that the visible surface (photosphere) of the sun is surrounded by a hot (2×10^6 K), extended corona of moderate density ($n_H \simeq 10^{10-12}$ cm^{-3}), astrophysicists developed the "coronal equilibrium" model (see e.g. Jordan, 1969, 1970) to describe the distribution of ionization states, the emission rate of electromagnetic radiation, and the transport properties of the plasma. With broad applicability to low-density plasmas in which collisional ionization dominates radiative ionization, this model is now used quite generally to describe the state of gas and emission in interstellar and intergalactic space, far from any "coronae".

In a low density plasma with electron density n_e (cm^{-3}), the rate of change of density in ion state (i) is standardly expressed as

$$1/n_e \, dn_i/dt = n_{i-1} \, C_{i-1} - n_i [C_i + \alpha_{i-1}] + n_{i+1} \, \alpha_i, \qquad (9.2\text{-}1)$$

where $C_i(T)$ and $\alpha_i(T)$ are the rate coefficients (cm^3 s^{-1}) for collisional ionization from and radiative plus dielectronic recombination to state (i). The coefficient C_i includes both direct (valence shell) ionization as well as autoionization following inner-shell excitation. Inner shell ionization is particularly important for ions in isoelectronic sequences with 1 or 2 valence electrons outside a closed shell (Cowan and Mann, 1979) — for example, the Li, Be, Na, Mg, Ca, or K-isosequences. Its relative importance to direct ionization increases with atomic charge. For large-scale plasma modeling, it is often convenient to adopt semiempirical analytic fits to $C_i(T)$ and $\alpha_i(T)$ — see Lotz (1968), Burgess et al. (1977), Crandall (1981), Younger (1982), Shull and Van Steenberg (1982) and Chapter 2.

If collisional ionization couples only adjacent ion stages, the steady state solution to equation (9.2-1) is

$$n_{i+1}/n_i = C_i(T)/\alpha_i(T), \qquad (9.2\text{-}2)$$

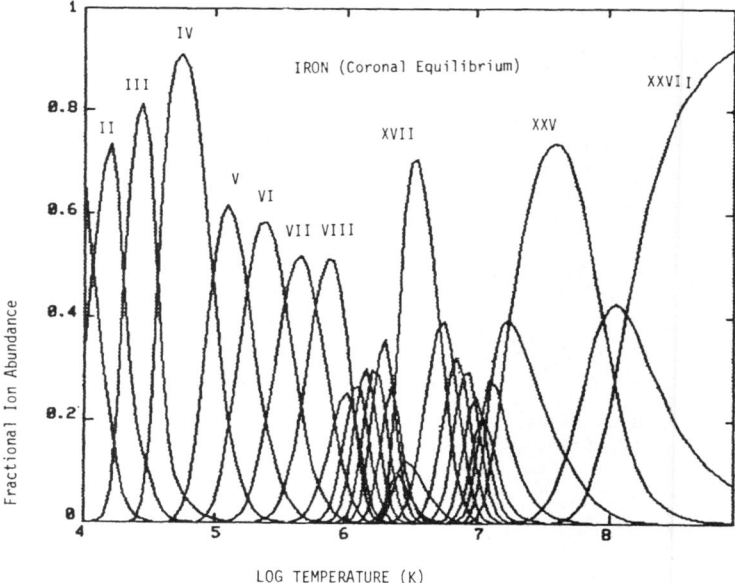

Fig. 9.2-4. Fractional abundances in various ionization stages of Fe, assuming coronal ionization equilibrium (Shull and VanSteenberg, 1982)

and equilibrium curves for the fractional ionization distribution of a given element may be computed solely as a function of electron temperature T (see Fig. 9.2-4). A primary result of such calculations is the temperature, T_{max}, at which a given ion reaches its maximum fractional abundance. Double ionization complicates this model, invalidating equation (9.2-2) and necessitating matrix inversion of the steady state rate equations (9.2-1).

Equilibrium spectral emission models of hot, optically-thin, low-density plasmas (Raymond and Smith, 1977; Shull, 1981) are now a familiar tool in X-ray astronomy for deriving the temperatures, densities, and element abundances in space. After computing the equilibrium ionization fractions (equation 9.2-1), these models generate the spectral emissivity (ergs $cm^{-3} s^{-1} eV^{-1}$) of the plasma in the X-ray continuum (bremsstrahlung, radiative recombination, and 2-photon emission from metastable states of H-like and He-like ions) and emission lines (dominated by collisionally-excited lines from heavy elements, with some contribution from radiative recombination cascade and satellite lines from dielectronic recombination). By fitting theoretical emissivity models to X-ray spectra, one may derive the plasma electron temperature, the "emission integral" (n_e^2 times the emitting volume V), and the heavy element abundances relative to H and He. This extraction is simplified in astrophysical plasmas because the continuum is dominated by electron free-free emission in the Coulomb fields of H and He ions, while the line emission is produced by the trace heavy elements. Thus, the thermal continuum shape yields kT, and the line-to-continuum ratios yield the abundances; $n_e^2 V$ follows from the total flux normalization, since the emitting processes all result from binary collisions.

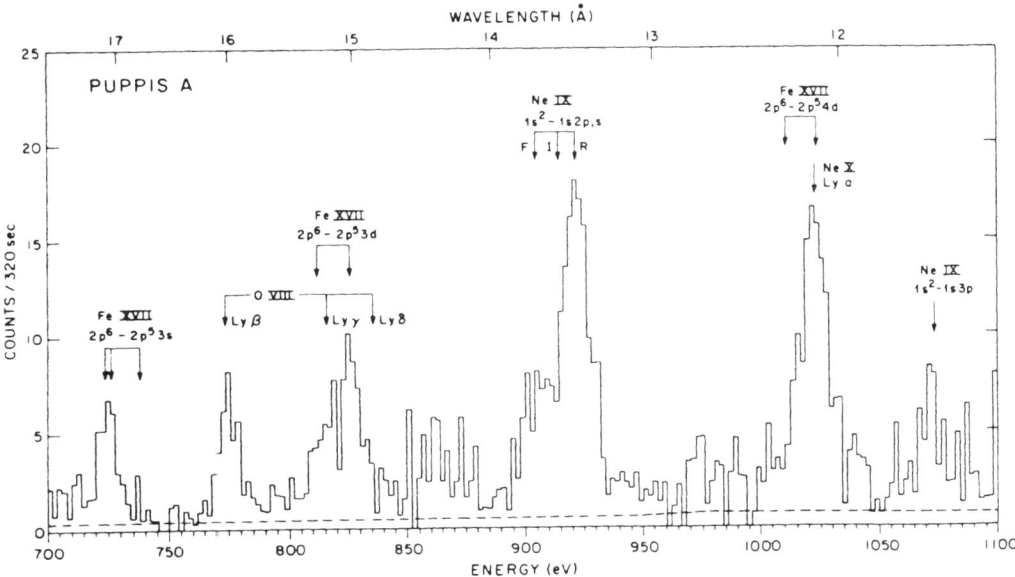

Fig. 9.2-5. X-ray spectrum of supernova remnant Puppis A (Winkler et al., 1981), taken by focal plane crystal spectrometer aboard HEAO-2 observatory. Note the 3 He-like lines of Ne IX (marked F, I, R), as well as H-like lines of Ne X and O VIII

Ratios of emission line intensities may also be used to derive n_e and T (e.g. Doschek, Feldman, and Cowan, 1981). Densities usually follow from the ratio of a forbidden to an allowed line (the forbidden line intensity is density sensitive, owing to collisional de-excitation), while temperatures are derived from lines with significantly different excitation energies (which sample the Boltzmann factor, $\exp[-E/kT]$). An elegant set of density and temperature diagnostic lines (Gabriel and Jordan, 1969, 1973; Pradhan and Shull, 1981) are the resonance $(2\,^1P-1\,^1S)$, forbidden $(2\,^3S-1\,^1S)$, and intercombination $(2\,^3P-1\,^1S)$ lines of He-like ions is seen in solar flares (Culhane et al., 1981) and supernova remnants (Winkler et al., 1981; Fig. 9.2-5). Helium-like ions are remarkably stable in high-temperature plasmas, owing to their high excitation and ionization energies. Contributions to these lines from dielectronic "satellites" and inner-shell ionization also provide diagnostics of whether the plasma is in a transient ionizing or recombining state.

The coronal model neglects several processes which occasionally become important. First, at high densities $(n_e > 10^{14}\,\mathrm{cm}^{-3})$, one must include 3-body recombination (the inverse process to electron impact ionization), as well as the suppression of dielectronic recombination by electron collisional excitation and proton-impact redistribution of high-n states (Burgess and Summers, 1969; Weisheit, 1975). Second, in partially ionized plasmas, ion charge exchange with H^0, He^0, and He^+ introducing additional parameters to the model's rate equations (equation 9.2-1), particularly for ions such as O II and Si III which exchange their charge to the ground state of the next lower ion stage (Butler, Heil, and Dalgarno, 1980; Baliunas and Butler, 1980). Third, the effects of photoionization must often be included if the gas is situated near a source of ultraviolet and/or X-ray radiation; the relative importance of such radiation is measured by the ratio, F/n_e, of ionizing photon flux to electron density (Kallman and McCray, 1982). Finally, and most importantly, the assumption of ionization equilibrium is often violated in transient situations (an ionizing or recombining plasma).

9.2.2.2 Non-Equilibrium Effects

If the characteristic timescale t (e.g. age, transport time, or heating time) of a plasma is shorter than the ionization time, $\tau_i = (n_e\,C_i)^{-1}$, or the recombination time, $\tau_r = (n_e\,\alpha_i)^{-1}$, then the plasma will be out of ionization equilibrium. As a result, a given element may be "underionized" (lower ionization stages than predicted by equilibrium; $t < \tau_i$) or "overionized" (higher ionization stages than equilibrium; $t < \tau_r$), with significant effects on the plasma emissivity and radiative cooling rate. (In general, an underionized plasma radiates far more effectively, owing to the enhancement of ion stages with more bound electrons). Departures from equilibrium may also be gauged by the He-like line ratios described above.

Non-equilibrium ionization effects are of major importance, for example, in many young supernova remnants (Itoh, 1977; Shull, 1982), solar flares (Acton and Brown, 1978; Mewe and Schrijver, 1980; Gabriel et al., 1981) and interstellar shock waves (McKee and Hollenbach, 1980). Models of non-equilibrium plasmas are far more complicated, because they require the specification of the past ionization history of every parcel of emitting gas. While this ionization history has little effect on the free-

free X-ray continuum, the emission line strengths are quite sensitive to the dominant ion stage at a given temperature, as noted above. In $10^5 - 10^6$ K gas behind $50 - 150$ km s^{-1} interstellar shock waves, rapid collisional ionization has major effects on ultraviolet emission line intensities (Raymond, 1979; Shull and McKee, 1979).

9.2.3 Fusion Plasmas

Many of the same plasma diagnostics are used in fusion research, although the trace impurities sputtered off the wall (Cr, Fe, Ni, Mo, W) are often heavier elements than those occuring in space (TFR Group, Dubau and Loulergue, 1981 for Cr; Bitter *et al.*, 1979 for Fe). However, low-mass elements (Si, S, Cl) are sometimes injected into Tokamak plasmas as spectroscopic probes of density, temperature, and transport rates. These diagnostics provide checks on the plasma confinement and fusion yield, while impurity transport rates may affect overall plasma properties such as radiative losses and plasma instabilities.

Merts, Cowan, and Magee (1976) calculated the ionization equilibrium and power output (0.8 − 10 keV) from a thin (1%) Fe-seeded plasma with $n_e = 10^{14}$ cm^{-3}. Using the collisional ionization rates of Lotz (1967, 1968), they computed diagnostic spectra for X-ray $K\alpha$ ($1s - 2p$) transitions near 6.5 keV. Källne, Källne, and

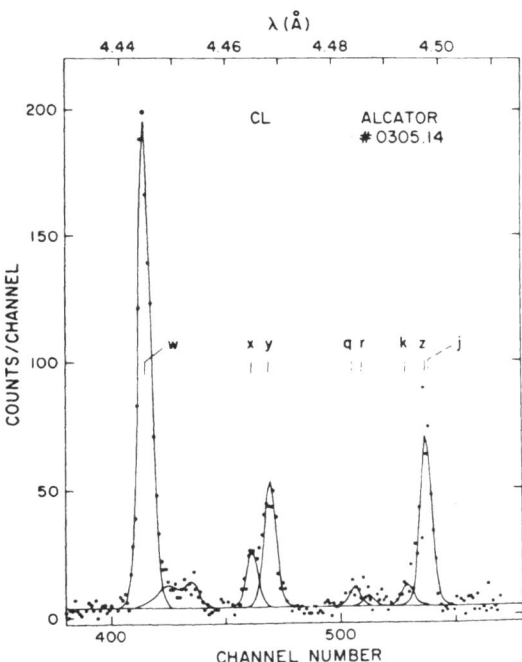

Fig. 9.2-6. Example of measured He-like spectra of Cl (Källne, Källne, and Pradhan, 1983), showing line ratios x/y larger than average. These deviations could result from impurity transport or from ionization disequilibrium

Pradhan (1983) have obtained spectra of He-like S and Cl in the Alcator C Tokamak (Fig. 9.2-6), analyzing the relative line ratios of resonance, forbidden, and intercombination lines for plasma conditions $kT = 1.0 - 1.8\,\text{keV}$ and $r_e = 1 - 7 \times 10^{14}\,\text{cm}^{-3}$. Because the He-like manifold includes contributions from dielectronic recombination from H-like ions and inner-shell excitation/auto-ionization from Li-like ions, these theoretical models rely on calculations of the relative fractional ionization (in coronal equilibrium or disequilibrium). The Alcator plasma is sustained for several hundered milliseconds, but approaches equilibrium on a shorter timescale because of the high densities. However, other Tokamak devices operate at lower densities, and impurity transport may introduce non-equilibrium effects.

More recently, Petrasso *et al.* (1982) have demonstrated a powerful diagnostic technique with spatially and temporally resolved measurements of fully stripped, H-like, and He-like Si in the Alcator C Tokamak. The fully stripped ions are detected via the X-ray continuum from ground state radiative recombination and bremsstrahlung, and the H-like and He-like ions are measured from their $n = 2 - 1$ X-ray emission lines. By assuming that the total Si density is dominated by these 3 ion stages in coronal equilibrium, they solve for all 3 densities. The calibrations are most accurate for the fully stripped Si, since the collisional ionization rate of H-like Si has been calculated to 10% (Younger, 1980) and hydrogenic radiative recombination is well understood. The central ion densities in the Tokamak were found to be close to coronal equilibrium values, but the densities of fully ionized and He-like Si exhibited large fluctuations during internal plasma disruptions. Small deviations from equilibrium can be important, since they often result in enhanced radiative losses (Bitter *et al.*, 1981), creating difficulties in achieving "break-even" controlled fusion.

References

Acton, L. W., Brown, W. A. (1978): Astrophys. J. *225*, 1065.
Baliunas, S. L., Butler, S. E. (1980): Astrophys. J. Letters *235*, L 45.
Bitter, M. *et al.* (1979): Phys. Rev. Letters *43*, 129.
Bitter, M. *et al.* (1981): Bull. Amer. Phys. Soc. *26*, 981.
Burgess, A., Summers, H. P. (1969): Astrophys. J. *157*, 1007.
Burgess, A., Summers, H. P., Culhane, J. L., McWhirter, R. W. P. (1979): M. N. R. A. S. *179*, 275.
Butler, S. E., Heil, T. G., Dalgarno, A. (1980): Astrophys. J. *241*, 442.
Cowan, R. D., Mann, J. B. (1979): Astrophys. J. *232*, 940.
Crandall, D. H. (1981): Physica Scripta *23*, 153.
Culhane. J. L. *et al.* (1981): Astrophys. J. Letters *244*, L 141.
Doschek, G. A., Feldman, U., Cowan, R. D. (1981): Astrophys. J. *245*, 315.
Gabriel, A., Jordan, C. (1969): M. N. R. A. S. *145*, 241.
Gabriel, A., Jordan, C. (1973): Astrophys. J. *186*, 327.
Gabriel, A. *et al.* (1981): Astrophys. J. *244*, L 147.
Gull, T. R., Fesen, R. A. (1982): Astrophys. J. Letters *260*, L 75.
Itoh, H. (1977): Publ. Ast. Soc. Japan *29*, 813.
Jones, C. *et al.* (1979): Astrophys. J. Letters *234*, L 21.
Jordan, C. (1969): M. N. R. A. S. *142*, 501.
Jordan, C. (1970): M. N. R. A. S. *148*, 17.
Kallman, T. R., McCray, R. (1982): Astrophys. J. Suppl. *50*, 263.

Källne, E., Källne, J., Pradhan, A. K. (1983): Phys. Rev. A. *28*, 467.
Linsky, J. (1982): Unpublished IUE data.
Lotz, W. (1967) Z. Physik *206*, 205.
Lotz, W. (1968) Z. Physik *216*, 241.
McCray, R. A., Snow, T. P. (1979): Ann. Rev. Astr. Astrophys. *17*, 213.
McKee, C. F., Hollenbach, D. J. (1980): Ann. Rev. Astr. Astrophys. *18*, 219.
Merts, A. L., Cowan, R. D., Magee, N. H. (1976): Los Alamos Scientific Laboratory Report LA-6220-MS "The Calculated Power Output from a Thin Iron-Seeded Plasma".
Mewe, R., Schrijver, J. (1980): Astr. Astrophys. *87*, 261.
Mushotzky, R. F., Serlemitsos, P. J., Smith, B. W., Boldt, E. A., Holt, S. S. (1978): Astrophys. J. *225*, 21.
Petrasso, R., Seguin, F. H., Loter, N. G., Marmar, E., Rice, J. (1982): Phys. Rev. Letters *49*, 1826.
Pradhan, A. K., Shull, J. M. (1981): Astrophys. J. *249*, 821.
Raymond, J. R. (1979): Astrophys. J. Suppl. *39*, 1.
Raymond, J. R., Smith, B. W. (1977): Astrophys. J. Suppl. *35*, 419.
Shull, J. M. (1981): Astrophys. J. Suppl. *46*, 27.
Shull, J. M. (1982): Astrophys. J. *262*, 308.
Shull, J. M., McKee, C. F. (1979): Astrophys. J. *227*, 131.
Shull, J. M., Van Steenberg, M. (1982): Astrophys. J. Suppl. *48*, 95; errata: *49*, 351.
Szymkowiak, A. Shull, J. M., Hamilton, A. J. S., Holt, S. S. (1985): Astrophys. J.: In preparation.
TFR Group, Dubau, J., Loulergue, M. (1981): J. Phys. B. *15*, 1007.
Weisheit, J. C. (1975): J. Phys. *B 8*, 2556.
Winkler, P. F., Canizares, C. R., Clark, G. W., Markert, T. H., Kalata, K., Schnopper, H. W. (1981): Astrophys. J. Letters *246*, L 27.
Younger, S. M. (1980): Phys. Rev. *A 22*, 111.
Younger, S. M. (1982): Phys. Rev. *A 26*, 3177.

9.3 Ionization in Gas Discharges: Experiment and Modeling

A. V. Phelps

Joint Institute for Laboratory Astrophysics, U.S. Bureau of Standards and University of Colorado, Boulder, Colorado, U.S.A.

9.3.1 Introduction

Electron impact ionization is the dominant source of electrons and ions in many electrical discharges in gases. This includes the ionization of atoms and molecules in excited states such as metastable states, as well as ionization of the ground state. Other processes resulting in the production of electrons in gas discharges, which we will not consider, are associative and Penning ionization, chemionization, photo-ionization, collisional and associative detachment, and collisional ionization by heavy particles. In Sections 9.3.2 and 9.3.3 we discuss the measurement and calculation of ionization coefficients in gases, while in Section 9.3.4 we discuss the production of electrons by ionization in various forms of discharges. In order to save space we will generally cite only those recent references containing enough review of the topic so as to serve as a starting point for a literature search.

9.3.2 Measurement of Ionization Coefficients in Gases

The probability of electron impact ionization by electrons moving in a moderate or high density gas is customarily described by either a Townsend ionization coefficient α_i, the number of ionization events per electron per unit distance in the direction of electron drift, or by an ionization rate coefficient k_i, the number of ionization events per unit time per electron per atom or molecule (Cherrington, 1979). Most measurements of ionization coefficients have been made using the parallel plane electrode system in which electrons are emitted photoelectrically from the cathode and drift through the gas under the action of a uniform electric field. Measurements are made of the current in the external circuit and/or of the light emitted by atoms or molecules excited by the electrons. The measurements may be steady-state or may follow the time dependence of the current or light output. Recent theory (Tagashira et al., 1977) has shown the importance of a careful distinction among various experimental techniques.

9.3.2.1 Spatial Growth Experiments

The spatial growth or steady-state Townsend experiments discussed in this subsection are conducted on a time scale which is long compared to the electron transit time across the apparatus. In the conventional Townsend experiments the growth of current is measured as the electrode separation is increased while keeping the gas density and electric field gas density ratio E/ρ constant. In the absence of secondary processes, the current measurements are fitted to an exponential of the form

$$i = i_0 \exp(\alpha_i d), \tag{9.3-1}$$

in which d is the electrode spacing and i_0 is the current leaving the cathode. Equation (9.3-1) describes the variation with distance of the current resulting from an electron avalanche in which the multiplication factor is $\exp(\alpha_i d)$. For the conditions of these experiments keeping the E/ρ value constant keeps the distribution of electron energies constant.

When the emission of electrons from the cathode due to the arrival of photons, ions or metastables is significant or when electron attachment occurs the expression for the current ratio becomes more complicated (Dutton, 1978). In order to remove the first-order dependence of the Townsend coefficient on gas density ρ it is customary to express the results of experiment in terms of the ratio α_i/ρ. The quantity α_i/ρ can be shown to be a function of E/ρ. The results of numerous experiments have been summarized by Dutton (1975) and by Gallagher et al. (1983). The points of Fig. 9.3-1 show representative experimental values of α_i/ρ for N_2 (Haydon and Williams, 1976). The curves show the results of recent theoretical calculations of α_i/ρ which will be discussed later.

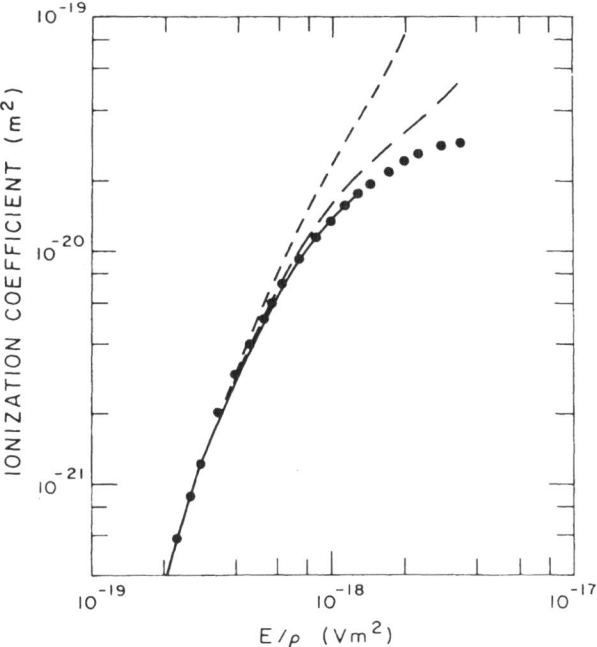

Fig. 9.3-1. Calculated and experimental ionization coefficients for N_2. Calculations: – – – neglecting new electron; — — temporal growth; ——— spatial growth. Experiment from Haydon and Williams (1976)

A problem with the conventional Townsend measurements is that it is sometimes difficult to separate the effects of secondary processes such as the release of electrons from the cathode by ions, metastables or photons from the growth of current caused by electron impact ionization. Therefore the more recent experiments make use of measurements of the current multiplication on a time scale short compared to ion transient times or metastable diffusion times (Haydon and Williams, 1976). Alternatively, measurements are sometimes made of the spatial variation of the light output (Bhattacharya, 1979). Occasionally measurements have been made of the growth of current at fixed electrode separation and gas density as the electric field is increased (Dutton, 1978).

9.3.2.2 Temporal Growth Experiments

The temporal growth or pulsed Townsend measurements are usually made on the time scale of 10 to 20 electron ionization collisions and are of the transient current flowing in the external circuit, the total light output, or of the time required for the electron current to grow to some critical value. Thus, the measurement is of a spatial integral of the electron distribution when the electrons are far from electrodes and is independent of the spatial gradients which may be of significance in other experiments (Tagashira et al., 1977).

For dc electric fields, this technique has been applied to a few gases (Raether, 1964; Dutton, 1978), to high E/ρ measurements (Byszewski et al., 1982), and to attaching gases (Fromhold, 1964). The experimental data are fitted to the function

$$i(t) = i_0 \exp(k_i \rho t),\qquad\qquad\qquad (9.3\text{-}2)$$

where $k_i \rho$ is determined from the exponential portion of a plot of $i(t)$ versus the time t. k_i is a function of E/ρ (Raether, 1964). We have not shown data from such experiments in Fig. 9.3-1 because of difficulties in the interpretation of the data at high E/ρ (Byszewski et al., 1982).

9.3.2.3 Temporal and Spatial Growth

The most difficult and, in principle, the most informative experimental measurements of ionization in gases are those in which measurements are made of both the temporal and spatial dependence of the electron density as the electrons drift and ionize under the action of a uniform electric field. The temporal distribution at a given point in space, i.e., the time-of-flight (arrival) distribution, or the spatial distribution of the electrons at a given time is generally determined by the light emitted by short-lived excited atoms or molecules (Fletcher, 1982). Thus far, these techniques have been used more for the determination of electron transport coefficients than for ionization coefficients.

9.3.2.4 High-Frequency Electric Fields

Most experimental measurements of ionization by electrons in the presence of high-frequency electric fields, e.g., microwave fields, do not fit into the previous categories since the dominant electron loss process is diffusion to the walls with its associated density gradients (Cottingham and Buchsbaum, 1963; MacDonald, 1966). However, in some cases (Felsenthal, 1966) the time scale is short enough so that the diffusive loss of electrons is not important.

9.3.2.5 Non-Equilibrium Ionization

In this section non-equilibrium ionization refers to the time and/or spatial variations in the apparent ionization coefficients which occur at short times and/or at very high values of E/ρ. For example, experiment shows that the onset of ionization occurs some distance from the cathode when the electrons are emitted with an energy below the ionization energy (Folkhard and Haydon, 1973; Hayashi, 1980). In addition. experiments at very high E/ρ have shown that the apparent α_i coefficients are not independent of position in the steady-state Townsend experiment (Folkhard and Haydon, 1973; Dutton, 1978). This phenomenon is believed to be the result of the fact that some of the electrons gain energy from the electric field more rapidly than they lose energy in collisions with the atoms and molecules, i.e. they become "runaway" electrons. This effect will be discussed further in Section 9.3.3.3.

9.3.3 Theory of Ionization Coefficients in Gases

The theory of ionization by electrons in moderate and high-pressure gases is reviewed in many texts (Dutton, 1978). We will divide this discussion into theories primarily concerned with the calculation of ionization coefficients and the associated electron transport coefficients, i.e., solutions of the Boltzmann equation, and theories concerned with the statistical aspects of ionization and the associated electrical breakdown, i.e., statistical theories and numerical simulation.

9.3.3.1 Application of the Boltzmann Equation

The application of the Boltzmann equation to the calculation of electron impact ionization and transport coefficients has received considerable attention during the past decade (Huxley and Crompton, 1974; Tagashira et al., 1977; Lin et al., 1979; Kumar et al., 1980). Ionization coefficient data have played a key role in the determination of electron collision cross sections at the higher electron energies using solutions of the two-term spherical harmonic or Lorentz approximation to the Boltzmann equation (Pitchford and Phelps, 1982). The two-term approximation has also been used to predict ionization coefficients for various gas mixtures of technological interest (Aleksandrov et al., 1981; Sakai et al., 1979; Pfau and Winkler, 1978; Kline et al., 1979; Itoh et al., 1980; Cacciatore et al., 1982). Important recent advances are the theoretical treatments of the effects of spatial density gradients (Tagashira et al., 1977; Kumar et al., 1980), electric field gradients (Alexandrov and Konchakov, 1981), anisotropic electron scattering (Haddad et al., 1981), large values of the ratio of the cross sections for inelastic collisions to the cross section for elastic scattering (Reid et al., 1980; Pitchford et al., 1981) and the production of new electrons in the ionization process (Wilhelm and Winkler, 1980; Yoshida et al., 1983). An example of the error resulting from calculations omitting the effects of spatial gradients and of the new electrons is shown by the dashed curve of Fig. 9.3-1 (Yoshida et al., 1983), which is much higher than the more accurate calculations shown by the lower two curves. The solid curve shows a calculation of α_i/p which includes spatial growth (Taniguchi, 1978), while the intermediate chain curve shows the k_i/W values appropriate to the temporal growth experiments (Yoshida et al., 1983). Here W is the calculated electron drift velocity. One important conclusion of recent investigations is that one expects a reasonably high degree of accuracy for ionization and other transport coefficients calculated using the two-term approximation under many conditions of practical importance. Investigations to define these conditions more generally are continuing (Allis, 1982; Braglia et al., 1982).

9.3.3.2 Statistical and Simulation Theories

Although they have received far less attention than the Boltzmann equation approach, the statistical and simulation theories of the growth of ionization in gases provide very useful models of early stages of discharges (Dutton, 1978; Hayashi,

1976; Tseng and Kunhardt, 1983). In addition, simulation theories using the Monte Carlo technique provide a very general method which is easily adapted to new problems. However, the high cost of Monte Carlo solutions for steady-state problems usually limits their use to testing the validity of the Boltzmann equation (Reid, 1979; Pitchford *et al.*, 1981; Braglia *et al.*, 1982).

9.3.3.3 Non-Equilibrium Theory

Both the simulation and Boltzmann approaches have been applied to the modeling of the non-equilibrium behavior of electrons. Of particular recent interest has been the modeling of the cathode fall regions of glow discharges (Boeuf and Marode, 1982). The onset of runaway behavior has been investigated by Ecker and Müller (1961) and others (Bhasavanich and Parker, 1977; Hayashi, 1976; Kunhardt and Byszewski, 1980). Non-equilibrium in the time domain has been reviewed recently by Wilhelm and Winkler (1979).

9.3.4 Ionization in Various Discharge Forms

In this section we will review studies of the role of ionization in various forms of discharges. A number of more extensive reviews are available (Pfau *et al.*, 1969; Bekefi, 1976; Hirsh and Oskam, 1978; Cherrington, 1979).

9.3.4.1 Externally Maintained Discharges

Interest in externally maintained discharges has been high because of their use for electric discharge excited lasers (Bekefi, 1976; Velikov *et al.*, 1977). Numerous papers have modeled the ionization by high-energy electrons (Bogdanov *et al.*, 1982), the distribution in energy of the secondary and plasma electrons (Konovalov and Son, 1981; Bretange *et al.*, 1981) and the criteria for stability against ionization waves, etc. (Nighan, 1976).

9.3.4.2 The Positive Column

The status of experiments and theories for the positive column of discharges in the rare gases has been thoroughly reviewed (Francis, 1956; Pfau *et al.*, 1969; Cherrington, 1979; Ingold, 1979). Very detailed models of the relevant ionization and excitation processes, including the ionization of metastable states (Hyman, 1979; Armentrout *et al.*, 1981), metastable-metastable collisional ionization and the effects of electron-electron collision, have been developed (Ligthardt and Keijser, 1980; Cernoggora *et al.*, 1981; Dothan and Kagan, 1982; Winkler *et al.*, 1983). Theory and experiment concerned with the role of ionization in the striations found in the positive column have been reviewed by Garscadden (1978). Discussions of the balance between ionization and loss of electrons by diffusion include the transition from ambipolar to free diffusion as the electron density decreases (Muller and Phelps, 1980), the effects of attachment (MacCullum, 1970; Mikhalev and Selin, 1974), and the role of excited states in ionization and recombination (Fujimoto, 1979; Devos et al., 1979).

9.3.4.3 Electrode Effects

Only a few references present recent experimental work on ionization in the cathode fall of glow discharges, while there has been much progress on the theory (Baksht and Yur'ev, 1979; Emeleus, 1981; Boeuf and Marode, 1982). Very little work has been done recently on the anode of glow discharges (Francis, 1956). A great deal of progress has been made in understanding the cathode regions of arcs (Holmes, 1978; Hantzsche, 1979), as well as the anode (Kimblin, 1974; Chen and Pfender, 1980). Ionization in the hollow cathode has been the subject of a number of recent investigations (Helm, 1979; Reshnov, 1981).

9.3.4.4 Corona and Streamers

The understanding of coronas and streamers occurring during electrical breakdown is progressing rapidly as the result of the application of improved experimental and theoretical techniques (Goldman and Goldman, 1978; Gallimberti, 1978; Craggs, 1979; Marode et al., 1979; Yoshida et al., 1979; Sigmond, 1980). In particular, detailed models of streamer development have led to reasonable agreement between experiment and theory. The effect of the non-equilibrium caused by high electric fields at the head of the streamer on the streamer growth is currently being investigated (Abbas and Bayle, 1981; Byszewski and Rheinhold, 1982; Tzeng and Kunhardt, 1983).

9.3.4.5 Arcs and Lightning

Most models of the arc column, such as found in lightning, assume that the electron, positive ion and neutral atom densities are in local thermodynamic equilibrium (LTE) at their common temperature and so do not consider the details of the ionization process (Finkenburg and Maeker, 1956). In the rare gases significant departures from LTE occur (Uhlenbush, 1974; Gleizes et al., 1982) and models of these arcs usually use some form of the collisional-radiative model of ionization and recombination (Biberman et al., 1979; Vriens and Smeets, 1980). See Sections 9.1 and 9.2. At the higher relative gas densities it may be necessary to take into account the effects of the excitation and deexcitation of highly excited atoms by neutral atoms (Devos et al., 1979; Bacri et al., 1982).

References

Abbas, I., Bayle, P. (1981): J. Phys. *D 14*, 649−674.

Aleksandrov, N. L., Konchakov, A. M. (1981): Fiz. Plasmy *7*, 185−191 [Sov. J. Plasma Phys. *7*, 103−106 (1981)].

Aleksandrov, N. L., Vysikailo, F. I., Islamov, R. Sh., Kochetov, I. V., Napartovich, A. P., Pevgo, V. G. (1981): Teplofizika Vysokikh Temperatur *19*, 22−27 [High Temperature *19*, 17−21 (1979)].

Allis, W. P. (1982): Phys. Rev. *26*, 1704−1712.

Armentrout, P. B., Tarr, S. M., Dori, A., Freund, R. S. (1981): J. Chem. Phys. *75*, 2786−2794.

Bacri, J., Lagreca, M., Medani, A. (1982): Physica *113 C*, 403−418.

Baksht, F. G., Yur'ev, V. G. (1979): Zh. Tech. Fiz. *49*, 905−944 [Sov. Phys. Tech. Phys. *24*, 535−557 (1979)].

Bekefi, G. (1976): Principles of Laser Plasmas. New York: Wiley.

Bhasavanich, D., Parker, A. B. (1977): Proc. R. Soc. Lond. *A 358*, 385−403.

Bhattacharya, A. K. (1979): J. Appl. Phys. *50*, 6207−6210.

Biberman, L. M., Vorob'ev, V. S., Yakubov, I. T. (1979): Usp. Fiz. Nauk. *128*, 233−271 [Sov. Phys. Usp. *22*, 411−432 (1979)].

Boeuf, J. P., Marode, E. (1982): J. Phys. *D 15*, 2169−2187.

Bogdanova. V. I., Burtsev, V. A., Kazachenko, N. I., Kuznetsov, V. S., Trubnikov, G. I. (1982): Fiz. Plazmy *8*, 189−192 [Sov. J. Plasma Phys. *8*, 107−109 (1982)].

Braglia, G. L., Romanò, L., Digligenti, M. (1982): Phys. Rev. *A 26*, 3689−3694.

Bretange, J., Delouya, G., Godart, J., Peuch, V. (1981): J. Phys. *D 14*, 1225−1239.

Byszewski, W. W., Reinhold, G. (1982): Phys. Rev. *A 26*, 2826−2831.

Byszewski, W. W., Enwright, M. J., Proud, J. M. (1982): IEEE Trans. Plasma Sci. *PS-10*, 281−285.

Cacciatore, M., Capitelli, M., Gorse, C. (1982): Chem. Phys. *66*, 141−151.

Cernoggora, G., Hochard, L., Touzeau, M., Ferreira, C. M. (1981): J. Phys. *B 14*, 2977−2987.

Chen, D. M., Pfender, E. (1980): IEEE Trans. Plasma Sci. *PS-8*, 252−259.

Cherrington, B. E. (1979): Gaseous Electronics and Gas Lasers. Oxford: Pergamon.

Cottingham, W. B., Buchsbaum, S. J. (1963): Phys. Rev. *130*, 1002−1006.

Craggs, J. D. (1978): Spark Channels. In: Electrical Breakdown of Gases (Meek, J. M., Craggs, J. D., eds.), 753−839. New York: Wiley.

Devos, F., Boulmer, J., Delpech, J.-F.: J. Physique *40*, 215−223 (1979).

Dothan, F., Kagan, Yu. M. (1979): J. Phys. *D 12*, 2155−2166.

Dutton, J. (1975): J. Phys. Chem. Ref. Data *4*, 577−856.

Dutton, J. (1978): Spark Breakdown in Uniform Fields. In: Electrical Breakdown of Gases (Meek, J. M., Craggs, J. D., eds), 209−318. New York: Wiley.

Ecker, G., Müller, K. G. (1961): Z. Naturforsch. *16 a*, 246−252.

Emeleus, K. G. (1981): J. Phys. *D 14*, 2179−2187.

Felsenthal, P. (1966): J. Appl. Phys. *37*, 4557−4560.

Finklenburg, W., Maecker, H. (1956): Elektrische Bögen und Thermisches Plasma. In: Handbuch der Physik, Vol. XXII (Flügge, S., ed.), 254−444. Berlin-Göttingen-Heidelberg: Springer.

Fletcher, J. (1981): Recent Measurements of Electron Transport Coefficients. In: Electron and Ion Swarms (Christophorou, L. G., ed.), 1−10. Oxford: Pergamon.

Folkhard, M. A., Haydon, S. C. (1973): J. Phys. *B 6*, 214−226.

Francis, G. (1956): The Glow Discharge at Low Pressure. In: Handbuch der Physik, Vol. XXII (Flügge, S., ed.), 53−208. Berlin-Göttingen-Heidelberg: Springer.

Fromhold, L. (1964): Fortschr. Physik. *12*, 597−643.

Fujimoto, T.: (1979): J. Quant. Spectrosc. Radiat. Transfer *21*, 439−455.

Gallagher, J. W., Beaty, E. C., Dutton, J., Pitchford, L. C. (1983): J. Phys. Chem. Ref. Data *12*, 109−152.

Gallimberti, I. (1979): J. Physique *C 7*, 193−250.

Garscadden, A. (1978): Ionization Waves in Glow Discharges. In: Gaseous Electronics, Vol. I (Hirsh, M. N., Oskam, H. J., eds.), 65−107. New York: Academic Press.

Gleizes, A., Kafrouni, H., Dang Duc, H., Maury, C. (1982): J. Phys. *D 15*, 1031−1045.

Goldman, M., Goldman, A. (1978): Corona Discharges. In: Gaseous Electronics, Vol. I (Hirsh, M. N., Oskam, H. J., eds.), 219−290. New York: Academic Press.

Haddad, G. N., Lin, S. L., Robson, R. E. (1981): Aust. J. Phys. *34*, 243 – 249.
Hantsche, E. (1979): Beitr. Plasmaphys. *19*, 59 – 79.
Hayashi, M. (1976): Proc. 4th IEE Conf. on Gas Discharges, 195 – 198. London: Inst. Electrical Engr.
Haydon, S. C., Williams, O. M. (1976): J. Phys. *D 9*, 523 – 536.
Helm, H. (1979): Beitr. Plasmaphys. *19*, 233 – 257.
Hirsh, M. N., Oskam, H. J. (1978): Gaseous Electronics, Vol. I. New York: Academic Press.
Holmes, R. (1978): Electrode Phenomena. In: Electrical Breakdown of Gases (Meek, J. M., Craggs, J. D., eds.), 839 – 867). New York: Wiley.
Huxley, L. G. H., Crompton, R. W. (1974): The Diffusion and Drift of Electrons in Gases. New York: Wiley.
Hyman, H. A. (1979): Phys. Rev. *A 20*, 855 – 859.
Ingold, J. H. (1978): Glow Discharges at DC and Low Frequencies: Anatomy of the Discharge. In: Gaseous Electronics, Vol. I (Hirsh, M. N., Oskam, H. J., eds.), 19 – 64. New York: Academic Press.
Itoh, H., Shimozu, M., Tagashira, H. (1980): J. Phys. *D 13*, 1201 – 1209.
Kimblin, C. W. (1974): IEEE Trans. Plasma Sci. *PS-2*, 310 – 319.
Kline, L. E., Davies, D. K., Chen, C. L., Chantry, P. J. (1979): J. Appl. Phys. *50*, 6789 – 6796.
Kondo, K., Ikuta, N. (1980): J. Phys. *D 13*, L 33 – L 38.
Konovalov, V. P., Son, É. E. (1981): Zh. Tekh. Fiz. *51*, 547 – 554 [Sov. Phys. Tech. Phys. *26*, 328 – 332 (1981)].
Kumar, K., Skullerud, H. R., Robson, R. E. (1980): Aust. J. Phys. *33*, 343 – 448.
Kunhardt, E. E., Byszewski, W. W. (1980): Phys. Rev. *A 21*, 2069 – 2077.
Ligthardt, F. A. S., Keijser, R. A. J. (1980): J. Appl. Phys. *51*, 5295 – 5299.
Lin, S. L., Robson, R. E., Mason, E. A. (1979): J. Chem. Phys. *71*, 3483 – 3498.
MacCullum, C. J. (1970): Plasma Phys. *12*, 143 – 148.
MacDonald, A. D. (1966): Microwave Breakdown in Gases, Chap. 4. New York: Wiley.
Marode, E., Bastien, F., Bakker, M. (1979): J. Appl. Phys. *50*, 140 – 146.
Mikhalev, L. A., Selin, L. N. (1974): Zh. Tekh. Fiz. *44*, 1095 – 1097 [Sov. Phys. Tech. Phys. *19*, 689 – 690 (1974)].
Muller III, C. H., Phelps, A. V. (1980): J. Appl. Phys. *51*, 6141 – 6148.
Nighan, W. L. (1976): Stability of High-Power Molecular Laser Discharges. In: Principles of Laser Plasmas (Bekefi, G., ed.), 257 – 314. New York: Wiley.
Pfau, S., Rutscher, A., Wojaczek, K. (1969): Beitr. Plasmaphys. *9*, 333 – 358.
Pfau, S., Winkler, R. (1978): Beitr. Plasmaphys. *18*, 113 – 118.
Pitchford, L. C., ONeil, S. V., Rumble, J. R., jr. (1981): Phys Rev. *A 23*, 294 – 304.
Pitchford, L. C., Phelps, A. V. (1982): Phys. Rev. *A 25*, 540 – 554.
Raether, H. (1964): Electron Avalanches and Breakdown in Gases. London: Butterworth.
Reid, I. D. (1979) Aust. J. Phys. *32*, 231 – 254.
Reshenov, S. P. (1981): Zh. Tech. Fiz. *51*, 1393 – 1402 [Sov. Phys. Tech. Phys. *26*, 800 – 805 (1981)].
Sakai, Y., Kanekc, S., Tagashira, H., Sakamoto, S. (1979): J. Phys. *D 12*, 23 – 31.
Sigmond, R. S. (1978): Corona Discharges. In: Electrical Breakdown in Gases (Meek, J. M., Craggs, J. D., eds.), 319 – 384. New York: Wiley.
Tagashira, H., Sakai, Y., Sakamoto, S. (1977): J. Phys. *D 10*, 1051 – 1063.
Taniguchi, T., Tagashira, H., Sakai, Y. (1978): J. Phys. *D 11*, 1757 – 1768.
Tzeng, T., Kunhardt, E. E. (1983): Bull. Am. Phys. Soc. *28*, 180.
Uhlenbusch, J. (1974): Non-Equilibrium Effects in Arc Discharges. In: Gaseous Electronics (McGowan, J. W., John, P. K., eds.), Chap. 7. Amsterdam: North-Holland.
Velikhov, E. P., Pis'mennyï, V. D., Rakhimov, A. T. (1977): Usp. Fiz. Nauk. *122*, 419 – 447 [Sov. Phys. Usp. *20*, 586 – 602 (1977)].
Vriens, L., Smeets A. H. M. (1980): Phys. Rev. *A 22*, 940 – 951.
Wilhelm, J., Winkler, R. (1979): J. Physique *C 7*, 251 – 267.
Wilhelm, J., Winkler, R. (1980): Ann. Phys. Leipzig *37*, 35 – 56.
Winkler, R. B., Wilhelm, J., Winkler, R. (1983): Beitr. Plasmaphys. *23*, 25 – 40.
Yoshida, K., Taniguchi, T., Tagashira, H. (1979): J. Phys. *D 12*, L 3 – L 7.
Yoshida, S., Phelps, A. V., Pitchford, L. C. (1983): Phys. Rev. *A 27*, 2858 – 2867.

9.4 Electron-Impact Ionization Processes in Planetary Atmospheres

E. C. Zipf

Department of Physics and Astronomy, University of Pittsburgh, Pittsburgh, Pennsylvania, U.S.A.

9.4.1 Introduction

In this section, we are concerned with the role of electron-impact ionization processes in planetary and cometary atmospheres. This is a diverse subject that deals with remarkable phenomena like the aurora borealis, with complex geophysical plasmas, like the Jupiter torus where the electron gas has an effective temperature of 10^5 K, with multi-level radiative entrapment problems, with equilibrium, non-equilibrium, and nonlinear ion and neutral chemistry, with the thermal energy balance in the upper atmosphere, and with the evolution of planetary atmospheres and escape. Electron-impact ionization processes play such an important role in all of these geophysical and cometary phenomena because they are a copious source of photons, kinetically energetic atoms and ions, and chemically active metastable species. Modeling these processes is quite difficult because of the variety of ways that energetic electrons in the 1 eV to 100 keV energy range are produced in the geophysical environment and because the steady-state electron energy distributions sustained by these sources can be remarkably different in character.

Over the past twenty years, satellite and sounding rockets have amassed a wealth of data on the terrestrial airglow and on the ion and neutral composition of the atmosphere under a wide variety of conditions. These studies now span nearly two complete solar cycles. Deep space probes are building similar data sets for the other major planets of the solar system and satellite observatories like IUE have detected aurora on planets as remote as Uranus. Stimulated by this observational database, a major effort is now underway to develop quantitative models of the physics and chemistry of planetary atmospheres. The realism of these models depends critically on the accuracy of the electron-impact excitation and ionization cross sections adopted, and it is this need that continues to stimulate and to provide a sense of direction for laboratory electron-scattering work on aeronomically important gases such as H, H_2, O, O_2, N, N_2, CO, CO_2, H_2O, and CH_4.

9.4.2 Current Trends

Laboratory studies of electron-impact ionization processes have a venerable history. The enduring work of Tate and Smith (1932), Rapp *et al.* (1965), and others on gases of aeronomic interest (Kieffer and Dunn, 1966) has provided an adequate set of cross section values for direct and dissociative ionization processes such as,

$$e + N_2 \rightarrow N_2^+ + 2e \tag{9.4-1}$$

and

$$e + N_2 \rightarrow N^+ + N + 2e. \tag{9.4-2}$$

These results allow the primary ionization rates due to electron impact in aurora or in the sunlit ionosphere to be calculated from in-situ measurements of the electron energy distribution and neutral densities. Since these observational data are seldom accurate to better than a factor of two, the absolute accuracy of the total and dissociation ionization cross sections determined in the laboratory does not compromise the aeronomic analyses at this point. The real impediment to future progress geophysically, however, comes from a different quarter: The wealth of airglow and ionospheric data show that excited ionic states produced by processes like (9.4-1) and (9.4-2) are very important in the physics and chemistry of a planetary atmosphere, but there is limited cross section data. In particular, metastable ions such as $O^+ (^2D)$ and $O^+ (^2P)$ excited by electron impact and photoionization,

$$e + O(^3P) \rightarrow O^+ (^2D) + 2e \tag{9.4-3}$$

$$h\nu - O(^3P) \rightarrow O^+ (^2D) + e \tag{9.4-4}$$

orchestrate much of the F region terrestrial ion chemistry (Torr and Torr, 1978, 1979). Similar processes involving simultaneous ionization-excitation of O^+ ions to states that relax promptly by dipole-allowed transition to the $O^+ (^4S)$ ground state, e.g.

$$e + O(^3P) \rightarrow O^+ (^4P) + 2e \tag{9.4-5}$$

$$O^+ (^4P) \rightarrow O^+ (^4S) + h\nu (\lambda 834 A) \tag{9.4-6}$$

dominate the terrestrial airglow and auroral spectra between $500 A - 1200 A$ and are involved in complex radiative entrapment problems (Meier, 1982). The trapped EUV resonance features are themselves a secondary source of ionization. For example, process (9.4-6) followed by

$$h\nu (\lambda 834 A) + O_2 \rightarrow O_2^+ + e \tag{9.4-7}$$

is simultaneously the dominant low altitude sink for these energetic photons and a significant source of O_2^+ ions in the E-region (Zipf, 1982). These second order, ionization processes further complicate the analysis of the airglow and ionospheric chemistry, and, unfortunately, are difficult to study.

There is also considerable interest in the use of these airglow features by remote-sensing satellite as a practical technique for measuring the global distribution of atomic oxygen and other species and for studying the thermal energy balance and the fluid mechanics of the terrestrial atmosphere. The feasibility of such investigations is directly related to the availability of comprehensive sets of cross sections for the direct and dissociative excitation of specific ionic states by electron impact. In addition, information is also needed on the velocity distribution of these excited states and on the linewidth profiles of ionic resonance transition in the vacuum ultraviolet wavelength region because these physical parameters affect the magnitude of many ion-molecule reaction rate coefficients and control the effective optical depth of the atmosphere. These are challenging experimental problems which suggest that laboratory work on electron-impact ionization processes has entered a new phase.

9.4.3 Energetic Electrons in the Upper Atmosphere

There are three major sources of energetic electrons in the upper atmosphere that produce distinctive airglow phenomena and related ion and neutral chemistries, and each supports a steady-state electron energy distribution with remarkably different characteristics. Satellite and sounding rocket experiments have provided a wealth of data on these processes, and a considerable effort has been made to develope physical models to account for the in-situ observations. In this section, we review briefly the general types of secondary electron distributions encountered in the terrestrial thermosphere, and we provide additional examples of unusual ionization processes sourced by inelastic electron collisions.

9.4.3.1 Photoionization

In the sunlit atmosphere, photoionization processes, e.g.

$$h\nu(\text{solar}) + N_2 \rightarrow N_2^+ + e + \text{kinetic energy} \tag{9.4-8}$$

create and sustain the ionosphere. The energetic photoelectrons produced by the absorption of solar x-rays and vacuum ultraviolet photons with wavelengths shorter than $\sim \lambda\,1300\,A$ lose their energy in inelastic collisions with the ambient gases at altitudes above 80 km and produce additional ionization and the day airglow by processes such as those listed in Table 9.4-1. The radiative livetime of the excited states produced by either direct or dissociative excitation has been noted in order to illustrate the wide variety of excited ionic states encountered in aeronomy.

Table 9.4-1. *Some simultaneous ionization-excitation processes used in studies of auroral substorms and the terrestrial cusp region*

Process	Transition	Wavelength	Lifetime
1. $e + N_2 \rightarrow N_2^+ (B^2\Sigma_u^+) + 2e$	$B^2\Sigma_u^+(v'=0) \rightarrow X^2\Sigma_g^+(v''=0)$	3914 Å	60.5 ns
2. $e + N_2 \rightarrow N^+ (^5S) + N + 2e$	$3s\,^5S \rightarrow 2p^4\,^3P$	2139.68 Å, 2143.55 Å	4 ms
3. $e + O_2 \rightarrow O_2^+ (b^4\Sigma_g^-) + 2e$	$b^4\Sigma_g^-(v'=1) \rightarrow a^4\pi_u(v''=0)$	5362 Å	1.15 μs
4. $e + O_2 \rightarrow O^+ (^4P) + O + 2e$	$2s\,2p^4\,^4P \rightarrow 2s^2\,2p^3\,^4S^0$	834 Å	7.1 ns
5. $e + O \rightarrow O^+ (^4P) + 2e$	$2s\,2p^4\,^4P \rightarrow 2s^2\,2p^3\,^4S^0$	834 Å	7.1 ns
6. $e + O \rightarrow O^+ (^2P) + 2e$	$2p^3\,^2P^0 \rightarrow 2p^3\,^4S^0$	7319 Å	5 s
7. $e + O \rightarrow O^+ (^2D) + 2e$	$2p^3\,^2D^0 \rightarrow 2p^3\,^4S^0$	3729 Å	3.6 hrs

Fig. 9.4-1 shows the terrestrial photoelectron energy spectrum at a number of different altitudes in the thermosphere (Lee *et al.*, 1980; Hernandez *et al.*, 1983). The prominent peaks in the 20 − 30 eV range are due to the photoionization of atomic

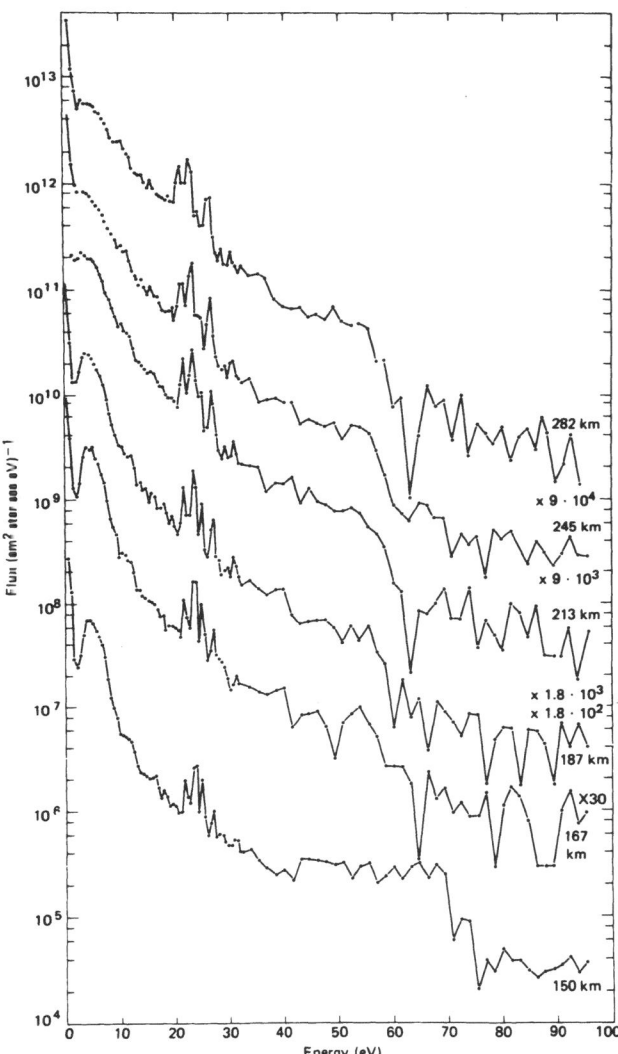

Fig. 9.4-1. The terrestrial photoelectron spectrum obtained by the Atmospheric Explorer Satellite (from Lee *et al.*, 1980; Hernandez *et al.*, 1983)

oxygen, which is the dominant neutral species above 150 km, by helium resonance radiation (λ 530 A). The steady-state photoelectron energy spectrum produced by the sun is dominated by comparatively low-energy electrons ($E < 70$ eV) so that a knowledge of the behavior of the inelastic cross section for a specific process near its energy threshold is quite important. The soft character of this electron spectrum also favors direct excitation of an atomic state versus dissociative excitation or direct ionization because of the striking differences in the magnitude and energy dependence of the pertinent excitation ionization cross sections. This point is illustrated well in Figs. 9.4-2 and 9.4-3 by the atomic oxygen cross section data for

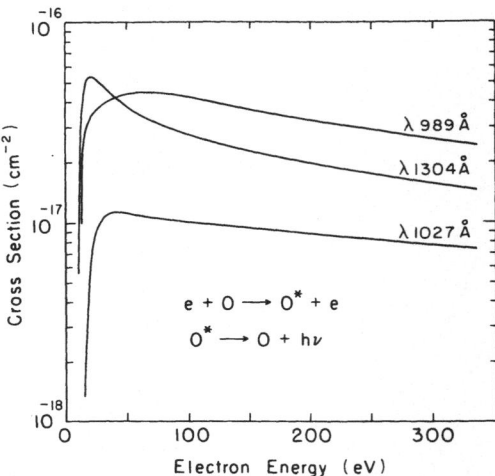

Fig. 9.4-2. The cross sections for the excitation of the O I [$2p\,^3$P $-3\,s\,^3$S^0; λ 1304 A], O I [$2p\,^3$P $-3\,s'\,^3$D^0: λ 989 A], and O I [$2\,p\,^3$P $-3\,d\,^3$D^0; λ 1027 A] VUV multiplets by electron impact on atomic oxygen (from Stone and Zipf, 1974; Kao and Zipf, 1984)

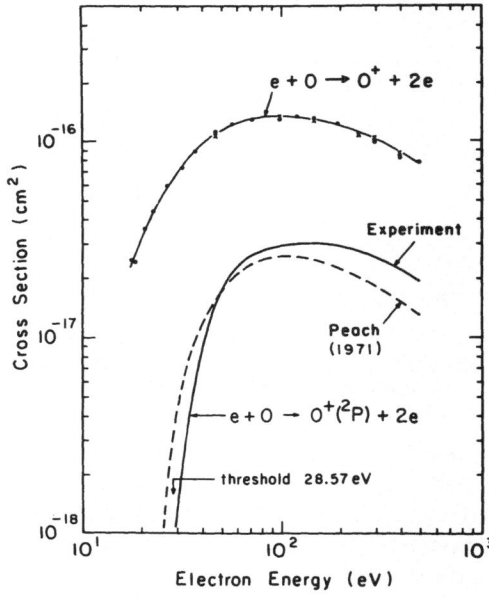

Fig. 9.4-3. The cross section for the excitation of the O II [$2p\,^3$P $-3\,s\,^2$P; λ 834 A] transition by electron impact on atomic oxygen (Kao and Zipf, 1984). The experimental results are compared with the theoretical estimate of Peach (1971) and the total O$^+$ ionization cross section (Brook *et al.*, 1978)

the O I and O II multiplets that dominate the VUV and EUV spectrum of the aurora and dayglow. In general, the direct neutral excitation processes have large cross sections ($\sim 5 \times 10^{-17}$ cm$^2 - 2 \times 10^{-16}$ cm^2) that peak at low energies (~ 20 eV) while the cross sections for the simultaneous ionization-excitation are smaller ($< 2 \times 10^{-17}$ cm^2) and peak near 100 eV (Stone and Zipf, 1973, 1974). It is also worth noting that the excitation of the OI(^{4}P $\rightarrow$ ^{4}S; λ 834 A) transition accounts for nearly 25°$_{\text{o}}$ of the total O$^+$ ionization cross section and that the ratio, $\sigma(\lambda 834 \text{A})/\sigma_{\text{total}}$ (O$^+$ + O^{++}), is nearly constant for electron energies > 50 eV. This behavior provides the basis for an optical technique for estimating the O$^+$ ionization rate from remote λ 834 A intensity measurements.

Although the photoelectron energy spectrum does not favor electron-impact ionization and dissociation processes, indirect processes do contribute to ion production in the sunlit thermosphere and in many gas discharges as well, and they complicate the task of the modelers. Two examples will serve to illustrate this point. We have already noted that the O$^+$ (λ 834 A) resonance transition is a bright feature in auroral and dayglow spectra and under many circumstances is entrapped. At altitudes below 150 km, O and O$_2$ absorption is an effective sink for these energetic photons. The absorption produces O$^+$ and O$_2^+$ ions with an efficiency of $\sim 80\%$. Analogous EUV airglow features excited by electron impact on O, O$_2$, and N$_2$ add to this locally produced photoionizing flux and make this indirect source of ionization significant in the lower F and E regions of the ionosphere (Zipf, 1982). A second indirect source of ionization is electron-impact excitation of high-lying atomic oxygen Rydberg atoms,

$$e + O \rightarrow O(\text{Rydberg}) + e. \qquad (9.4\text{-}9)$$

These atoms, with principal quantum number $n > 10$, are located energetically near the series limits converging to O II(^{4}S), O II(^{2}D), and O II(^{2}P) ions. Some 28 series are known experimentally, and estimates of the total cross section for exciting this energy-rich manifold are of the order of 10^{-16} cm^2. In an optically thick medium, the effective lifetime of these pseudo-metastable states can be many tens of microseconds. Since the collision cross section of these large atoms is typically 10^{-14} cm^2 or more and increases with the fourth power of n, quenching by the ambient gases is the likely sink for these species below 130 km, and it is quite possible that Penning ionization processes, such as

$$O(\text{Ryd}) + O_2 \rightarrow O_2^+ + O(^3\text{P}) + e \qquad (9.4\text{-}10)$$

are particularly efficient (Zipf, 1982). Thus, both of these indirect processes provide additional ways for the low-energy electrons that characterize the daytime photoelectron spectrum to serve as a source of ions in the thermosphere.

9.4.3.2 Polar Cusp Electrons

The Earth's magnetic field provides an effective shield that deflects the supersonic solar wind. There are, however, dayside cusp or cleft regions where subsonic solar wind or magnetosheath particles can penetrate deeply into the atmosphere. These cusp regions exist in both hemispheres at very high latitudes, and the airglows that result from the solar wind interaction are produced by low-energy electrons whose distribution can often be described as Maxwellian with a characteristic temperature of $\sim 40\,\text{eV}$. This electron spectrum is also comparatively soft, but unlike the photoelectron spectrum, it does have appreciable numbers of electrons in the $100-200\,\text{eV}$ range. The precipitating magnetosheath electrons do not penetrate deeply into the atmosphere. Most of their energy is deposited above 250 km where atomic oxygen is the dominant neutral species. Atomic nitrogen is also abundant in this altitude regime, and because the cross sections for exciting this species are so large (Stone and Zipf, 1974), N I contributes significantly to the observed cleft airglow. Cleft spectra are dominated by low-lying atomic and ionic transitions which are produced by direct electron-impact ionization-excitation of atomic oxygen and nitrogen. Little molecular emission is observed (Sivjee, 1983).

The cleft region of the terrestrial atmosphere provides an unusual opportunity to study inelastic electron collision with a reactive atomic gas under optically thick conditions, and it provides a particularly good illustration of how the effective cross section for ionization by electron impact under these conditions can be significantly larger than its magnitude under optically thin conditions. Furthermore, the relative proportions of ground state and metastable ions produced by these processes are also modified under optically thick conditions. This development is particularly important at F-region altitudes because the excitation of the chemically active, metastable ions, $O^+(^2D)$, $O^+(^2P)$, and $N^+(^5S)$ by cleft electrons is a major factor in the ion chemistry above 250 km, and their relative source strengths is a key parameter in ion chemistry modeling.

The excitation of the O I $(2p^3\,3s''\,^3P^0 \rightarrow 2p\,^3P; \lambda\,878\,\text{A})$ multiplet by electron impact on atomic oxygen in the cleft region,

$$e + O(2p\,^3P) \rightarrow O(2p^3\,3s''\,^3P^0) + e \qquad (9.4\text{-}11)$$

provides a good example of these complications. Atomic oxygen is quite unusual among atmospheric constituents in that ~ 28 Rydberg series arising from the ground state configuration $[O(^3P, ^1D, ^1S)]$ are known to exist above the $O^+(^4S)$ ionization limit (Jackman et al., 1977; Huffman et al., 1967). Some of these neutral O I state become O^+ ions immediately by allowed autoionizing transitions to the continuum. However, for many of the Rydberg states, autoionization is forbidden so that there is a competition between dipole-allowed radiation in the EUV and weak autoionization which results from the violation of the LS coupling selection rules (Dehmer et al., 1973). The excitation cross sections for exciting these states can be quite large particularly below 100 eV. Fig. 9.4-4 illustrates this point for the O I $(2p^3\,3s''\,^3P^0 \rightarrow 2p\,^3P)$ transition. The autoionization branching ratio for this state has been measured by Dehmer et al. (1977) who found that $\lambda\,878\,\text{A}$ emission and forbidden autoionization are equally probable. Under optically thick conditions, the recycling of the trapped $\lambda\,878\,\text{A}$ photons ultimately converts them to O^+ ions. The net effect of the entrapment process involving these Rydberg states is to increase the total O^+ ionization cross section by $\sim 30\%$ while the specific formation rate of ground state $O^+(^4S)$ atoms nearly doubles. Current aeronomy models assume that the relative proportion of ground state $O^+(^4S)$ to metastable $O^+(^2D)$ and $O^+(^2P)$ is 0.33 : 0.67. However, the most recent laboratory data suggest that under optically thick conditions, the proportion is really 0.75 : 0.25.

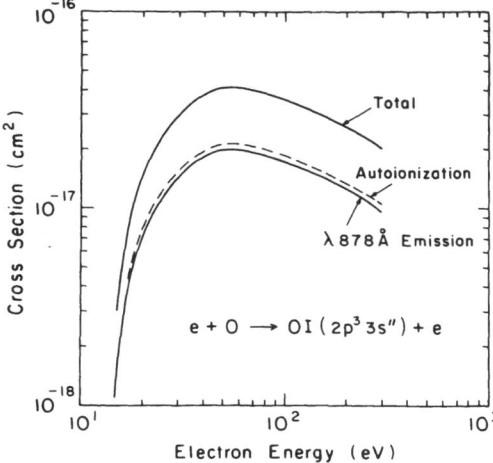

Fig. 9.4-4. The cross section for the excitation of the O I $(2 p^3 3 s'' \, ^3P^0)$ state by electron impact on atomic oxygen. This state is depopulated at nearly equal rates by radiative relaxation $[2 p^3 3 s'' \, ^3P^0 \to 2 p \, ^3P; \lambda 878 A]$ and by forbidden autoionization $[O I (2 p^3 3 s'' \, ^3P^0) \to O^+ (^4S) + e]$. Under optically thick conditions, the entrapment of $\lambda 878$ A photons enhances the $O^+ (^4S)$ ionization rate and results in the total effective ionization cross section for the O I $(2 p^3 3 s'' \, ^3P^0)$ channel shown (from Zipf, 1984)

9.4.3.3 Energetic Auroral Electrons

The aurora borealis is a spectacular visual phenomenon that is observed at high latitudes in the northern and southern hemispheres. The luminosity, for which an aurora is so noteworthy, is excited mostly by secondary electrons that are produced by the degradation of energetic primary electrons $(1 - 50 \, keV)$ that precipitate into the auroral zone from the magnetosphere (Jones and Rees, 1973; Rees and Jones, 1973; Stamnes and Rees, 1983). This process also generates substantial ionization in the lower atmosphere $(z > 90 \, km)$ in addition to the radiation which consists chiefly of molecular band emission from N_2, N_2^+, O_2, and O_2^+ and numerous O I, O II, N I, and N II multiplet transitions. High quality spectral data have now been obtained by satellites and sounding rockets and from ground-based observations over the wavelength range from $\lambda 500$ A to ~ 15 microns. Ion and neutral composition measurements also exist. Fig. 9.4-5 shows the auroral electron energy spectrum of a quiet homogeneous arc and its altitude dependence (Jones and Rees, 1973). Most of the radiation and ionization observed in auroral forms is due to electron-impact processes involving secondary electrons with energies $< 1 \, keV$. The secondary processes discussed in Section 9.4.3.1 and noted in Table 9.4-1 are also important in auroral substorms. As the figure suggests, the auroral modeling community is in need of comprehensive cross section data for the species mentioned above from threshold to $\sim 10 \, keV$.

Historically, laboratory work in support of auroral physics has evolved along two complementary lines. Firstly, in order to interpret auroral ion composition data, it is essential to have quantitative information on the specific ionization rates, $\eta(z)$, for

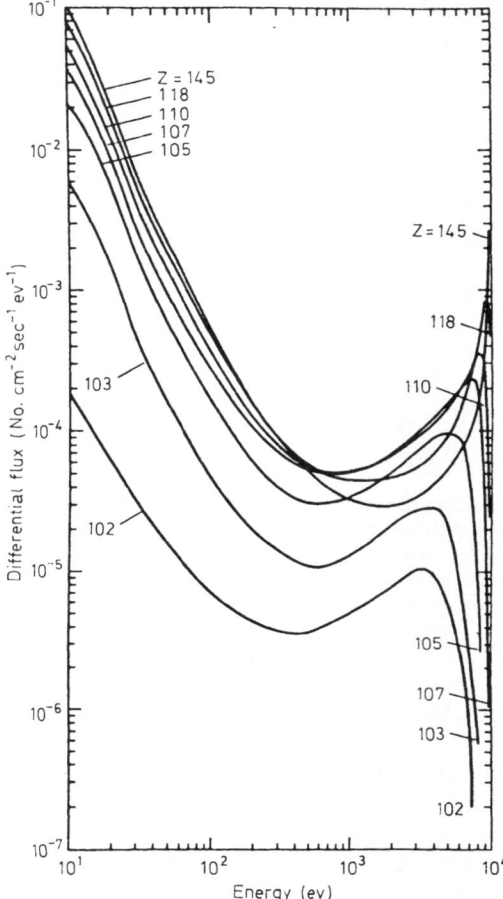

Fig. 9.4-5. The electron energy spectrum in an auroral substorm. The high energy primary electrons $[E \sim 10^4 \, \text{eV}]$ precipitating into the ionosphere from the magnetosphere can be seen in the higher altitude spectra $[z \sim 145 \, \text{km}]$. Notice the marked decrease in the magnitude of the electron flux at the lower edge of the auroral form $[z \sim 105 \, \text{km}]$ and its low energy character (from Jones and Rees, 1973)

each of the principal neutral constituents (N_2, O_2, and O), e.g.

$$e + N_2 \rightarrow N_2^+ + 2e \qquad\qquad\qquad\qquad (9.4\text{-}12)$$

and the related atomic ions produced dissociatively,

$$e + N_2 \rightarrow N^+ + N + 2e \qquad\qquad\qquad\qquad (9.4\text{-}13)$$

Conceptually, the pertinent η values could be deduced as a function of altitude, z, in an auroral substorm from simultaneous measurements of the electron energy flux distribution and the neutral composition. However, because of numerous practical difficulties, only a few such measurements actually exist. Instead, auroral modeling has relied on optical measurements to obtain needed ionic source-function data. The technique depends on measuring the N_2^+ and O_2^+ molecular band radiation and the O II λ 833 A resonance emission produced by simultaneous ionization-excitation of these targets [cf. Table 9.4-1].

Detailed studies have shown that the ratio of the emission cross section to the ionization cross section for these processes is nearly constant over the range of electron energies encountered in a typical auroral substorm (Borst and Zipf, 1970 a, b). Thus, from volume emission profiles of the N_2^+ 1 N(0,0) and O_2^+ 1 N(1,0) first negative bands, and the O I λ 833 A resonance transition, the total specific ionization rates may be inferred by noting the $\eta(N_2^+ + N^+) = 14.1 \eta(\lambda 3914 \text{A})$, $\eta(O_2^+ + O^+) = 64 \eta(\lambda 5632 \text{A})$, and $\eta(O^+) = 4.3 \eta(\lambda 833 \text{A})$. An attractive feature of this approach to the auroral ionization problem is that it makes it possible to extimate the total energy deposition in the auroral zone from remote satellite observations of these emission features. Alternatively, if primary electron flux data are available from the satellite, the optical measurements can be used to study the composition of the neutral atmosphere and its dynamic response to the large amount of energy ($\sim 3 \times 10^{12}$ watts) being deposited in the auroral ionosphere.

The second application for laboratory electron scattering work is in the analysis of auroral spectra and in estimating the production rates of the numerous chemically active metastable species formed in auroral arcs. This is a particularly challenging task because the degradation of the energetic electrons, which are found so abundantly in auroral substorms, are a rich source of unusual metastable and kinetically energetic ions that are difficult to observe in the laboratory because of wall constraints. The metastable ^{5}S state of N^+ is a good example of an excited state that has been known geophysically for more than two decades (Crosswhite et al., 1960) but has eluded study in the laboratory until recently (Erdman et al., 1981: Knight, 1982)

The forbidden doublet, which is emitted by metastable N^+(^{5}S) ions at wavelengths of λ 2139.68 A and λ 2143.55 A, dominates the near ultraviolet spectrum of an aurora (Sharp, 1978) and is excited by electron-impact dissociation of N_2,

$$e + N_2 \rightarrow N^+(^5S) + N + 2e \qquad (9.4\text{-}14)$$

(Dick, 1978) and by direct ionization-excitation of atomic nitrogen

$$e + N \rightarrow N^+(^5S) + 2e \qquad (9.4\text{-}15)$$

(Hibbert and Bates, 1981). Quantal calculations by Hibbert and Bates (1981) indicate that the radiative lifetime of the ^{5}S state is approximately 4 msec and that the λ 2143.55 A/λ 2139.68 A intensity ratio is 107/48.

In 1982, Knight obtained some evidence in support of the calculated lifetime from an ion trap experiment, and he also showed that the N^+(^{5}S) ions in the trap were rapidly quenched by N_2 with an apparent reaction rate coefficient of 2.5×10^{-9} cm^3 s^{-1}. Since the quenching of the N^+(^{5}S) state releases 20.3 eV, ion-molecule reactions involving this species can produce a wide variety of electronically excited derivative ions. For example, both charge transfer and dissociative ionization processes are energetically possible,

$$N^+(^5S) + O_2 \rightarrow O_2^+ + N \qquad (9.4\text{-}16)$$

and

$$N^+ ({}^5S) + O_2 \rightarrow O^+ + O + N \tag{9.4-17}$$

but there has been no laboratory work on either of these reactions.

Auroral analysis suggests that the cross section for process (9.4-14) must be in the range 3×10^{-18} cm^2 $- 1 \times 10^{-17}$ cm^2 at 100 eV (Sharp, 1978; Dalgarno *et al.*, 1981). By laboratory standards, this is a large excitation cross section so it would seem from a geophysical point of view that the ^5S doublet should have been easy to observe in the laboratory. But this has not been the case. Until recently, our knowledge of the $N^+ ({}^5S)$ doublet was derived entirely from spark spectra; it had never been observed in a laboratory electron-scattering experiment.

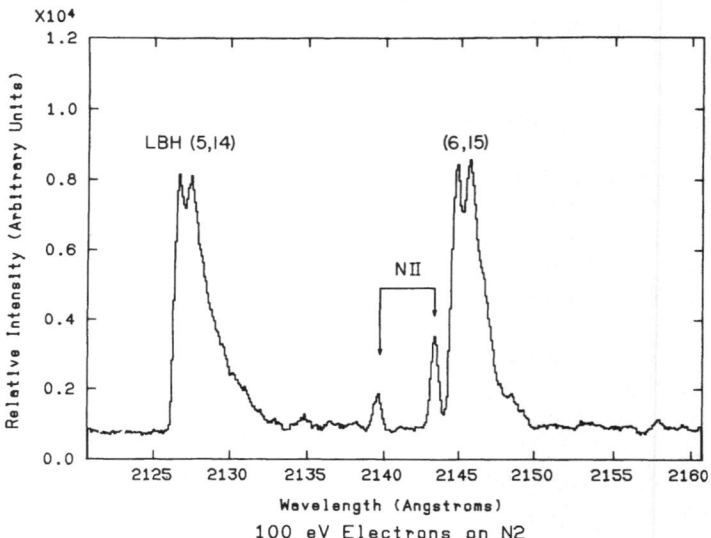

Fig. 9.4-6. Laboratory spectrum showing the λ 2143.55 and λ 2139.68 A doublet emitted by metastable $N^+ ({}^5S)$ ions produced by dissociative ionization of N_2 by electron impact. The apparent N II doublet cross section is $\sim 3 \times 10^{-21}$ cm^2. The small apparent cross section value is an artifact of the high translational energy and long lifetime of the $N^+ ({}^5S)$ state. The actual $N^+ ({}^5S)$ cross section appears to have a peak value of $\sim 3.8 \times 10^{-18}$ cm^2 at 175 eV (from Erdman *et al.*, 1981)

The laboratory impasse was due in part to the metastable character of the state and to the high velocities imparted to these ions in the dissociative ionization process. Both factors contribute to the removal of the $N^+ ({}^5S)$ ions from the field of view in the laboratory but not in the aurora where walls are not a factor. Computer-assisted laboratory experiments have now made considerable progress in overcoming these difficulties, and Fig. 9.4-6 shows the $N^+ ({}^5S)$ doublet excited by electron-impact dissociation of nitrogen. The data, which required 700 hours to acquire, were obtained at a resolution of 0.5 A. The doublet lines are clearly resolved from the nearby (6, 15) LBH band which has a cross section value of 3×10^{-20} cm^2 at 100 eV. The doublet intensity ratio is in good agreement with the theoretical prediction of Hibbert and Bates (1981). The apparent $N^+ ({}^5S)$ excitation cross section is $\sim 3 \times 10^{-21}$ cm^3 at 100 eV. The actual magnitude of the cross section when corrected for field of view, quenching, and translational energy effects, appears to be approximately 3×10^{-18} cm^2 at 100 eV: Fig. 9.4-7 shows the energy dependence of the N$^+$ (^{5}S) cross section.

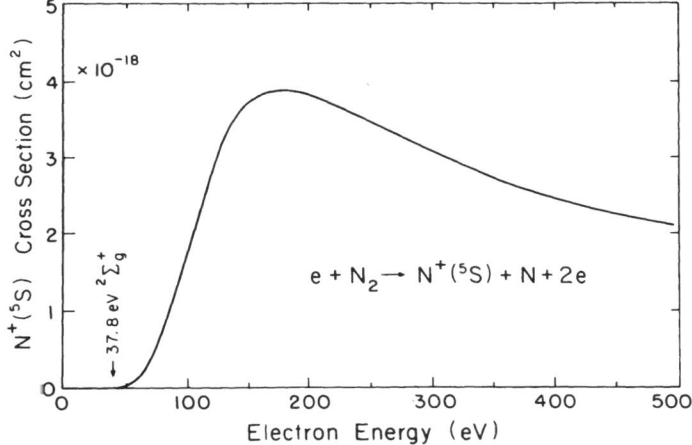

Fig. 9.4-7. The excitation function for the dissociative excitation of the metastable N$^+$ (^{5}S) state by electron impact on N$_2$. The ^{5}S state appears to form as the by-product of the dissociative autoionization of the N$_2$ $^2\Sigma_g^-$) state (from Erdman and Zipf. 1983; Dalgarno *et al.*, 1981)

9.4.4 Unfinished Business

The last twenty years has been a period marked by exceptional discoveries in planetary research. Interplanetary spacecraft have made airglow and ionospheric observations in planetary atmospheres strikingly different from our own and have raised complex and intriguing questions about the origin of our planetary atmosphere and its evolution. Auroral phenomena have now been observed in the atmospheres of Jupiter, Saturn and Uranus, and lightning is now known to occur frequently in the lower atmospheres of both Venus and Jupiter. A rich aeronomical database now exists for more detailed study. Although much progress has been made on relevant electron-impact processes, a great deal still remains to be done. For example, the cross sections for less than 25% of the known emission features observed in planetary airglow and aurora have been measured over the needed energy ranges. Laboratory work on electron-impact excitation ionization of the critical, chemically active, atomic metastable species is all but non-existent, as we have seen, and recent sounding rocket and laboratory studies of non-linear beam plasma discharges have raised additional questions about the role of electron-impact ionization of metastable species as an important element in initiating and sustaining these discharges (Gorman and Zipf, 1981; Papadopoulos, 1982; Bernstein *et al.*, 1983). Electron scattering off difficult target species like O I, N I, and S I, which are of major interest in cometary atmospheres and in the Jupiter torus, are at either a primitive stage or have yet to be attempted in the laboratory, and virtually nothing is known quantitatively about the velocity distributions of the key excited states produced in the dissociative excitation-ionization of aeronomically important molecules. This information is crucial to further progress on planetary radiation entrapment problems. Unfortunately, even the existing cross section database itself leaves much to be desired and will require additional work. Many of the excitation cross sections used routinely in aeronomic modeling were obtained with optical instrumentation with inadequate spectral purity, and, in some cases, newer studies have shown that order of magnitude errors have resulted because of this problem. The task of developing adequate calibration standards particularly in the vacuum ultraviolet continues to be a vexing problem, and grows in importance as the need for highly accurate solar flux values, absolute airglow intensity measurements, and cross section data becomes more pressing. All of these are new challenges. There is, indeed, ample work for present and future experimentalists.

References

Beitling, E. J., Feldman, P. D. (1978): Geophys. Res. Lett. *5*, 51.
Bernstein, W., McGarity, J. O., Konradi, A. (1983): Geophys. Res. Lett. *10*, 1124.
Borst, W. L., Zipf, E. C. (1970): Phys. Rev. *1 A*, 834.
Borst, W. L., Zipf, E. C. (1970): Phys. Rev. *1 A*, 1410.
Brook, E., Harrison, M. F. A., Smith, A. C. H. (1978): J. Phys. *B 11*, 3115−3132.
Crosswhite, H. M., Zipf, E. C., Fastie, W. G. (1962): J. Opt. Soc. Am. *52*, 643.
Dehmer, P. M., Berkowitz, J., Chupka, W. A. (1973): J. Chem. Phys. *59*, 5777.
Dehmer, P. M., Luken, W. L., Chupka, W. A. (1977): J. Chem. Phys. *67*, 195.
Dalgarno, A., Victor, G. A., Hartquist, T. W. (1981): Geophys. Res. Lett. *8*, 603.
Erdman, P. W., Zipf, E. C. (1983): EOS *64*, 785.
Erdman, P. W., Espy, P. J., Zipf, E. C. (1981): Geophys. Res. Lett. *8*, 1163.
Gorman, M. R., Zipf, E. C. (1981): Bull. Am. Phys. Soc. *27*, 100.
Hernandez, S. P., Doering, J. P., Abreu, V. J., Vector, G. A. (1983): Planet. Space Sci. *31*, 221.
Hibbert, A., Bates, D. R. (1981): Planet. Space Sci. *29*, 263.
Huffman, R. E., Larrabee, J. C., Tanaka, Y. (1967): J. Chem. Phys. *46*, 2213.
Jackman, C. H., Garvey, R. H., Green, A. E. S. (1977): J. Geophys. Res. *82*, 5081.
Jones, R. A., Rees, M. H. (1973): Planet. Space Sci. *21*, 537.
Kao, W. W., Zipf, E. C., Erdman, P. W. (1983): EOS *64*, 787.
Kao, W. W., Zipf, E. C. (1984): J. Chem. Phys. (in print).
Kieffer, L. J., Dunn, G. H. (1966): Rev. Mod. Phys. *38*, 1.
Knight, R. D. (1982): Phys. Rev. Lett. *48*, 792.
Lee, L. S., Doering, J. P., Potemera, T. A., Brace, L. H. (1980): Planet. Space Sci. *28*, 947.
Meier, R. R. (1982): J. Geophys. Res. *87*, 6307.
Papadopoulos, K. (1982): Theory of Beam Plasma Discharge. In: Artificial Particle Beam in Space Plasma Studies (Grandel, B., ed.), 505−523. New York: Plenum Press.
Peach, G. (1970): J. Phys. *B 3*, 328.
Rapp, D., Englander-Golden, P., Briglia, D. D. (1965): J. Chem. Phys. *42*, 4081.
Rees, M. H., Jones, R. A. (1973): Planet. Space Sci. *21*, 1213.
Sharp, W. E. (1978): Geophys. Res. Lett. *5*, 708.
Sivjee, G. G. (1983): Geophys. Res. Lett. *10*, 349.
Stamnes, K., Rees, M. H. (1983): J. Geophys. Res. *88*, 6301.
Stone, E. J., Zipf, E. C. (1973): J. Chem. Phys. *58*, 4278.
Stone, E. J., Zipf, E. C. (1974): J. Chem. Phys. *60*, 4237.
Tate, J. T., Smith, P. T. (1932): Phys. Rev. *39*, 270.
Torr, D. G., Torr, M. R. (1978): Rev. Geophys. and Space Phys. *16*, 327.
Torr, D. G., Torr, M. R. (1979): J. Atmosph. Terrest. Phys. *41*, 797.
Zipf, E. C. (1982): EOS *63*, 1050.

9.5 Applications to Radiation Physics*

Y.-K. Kim**

Argonne National Laboratory, Argonne, Illinois, U.S.A.

9.5.1 Introduction

The main objective of radiation physics is to study the interaction of ionizing radiation, such as photons and fast charged particles, with matter. The emphasis in radiation physics is different from that in conventional atomic and molecular physics. In the former, an entire chain of events, including subsequent products, is studied. In the latter, the emphasis is usually on an isolated event. The topic discussed in this book, electron-impact ionization, plays a central role in radiation physics. Regardless of the incident particles, an overwhelming number of secondary electrons are produced by primary interactions, which then proceed to interact further with the target medium. (Because of the indistinguishability of electrons, the slower electron emerging from an ionizing collision is called the secondary electron. The faster one is called the primary electron.) One of the direct applications of radiation physics is the monitoring of energy deposition by incident particles, such as radiation dose monitoring. Although basic concepts used in radiation physics are unambiguously defined in commonly used terms of atomic and molecular collision theory, theoretical radiation physics cannot fully predict an entire chain of events initiated by an incident particle. Such predictions require not only the knowledge of all relevant collision cross sections, but also full knowledge of the reaction rates of atomic and molecular species that are produced by primary as well as by all subsequent interactions. In most cases, therefore, experimental data are used for applications. Fortunately, those quantities essential to radiation monitoring and medical applications can be experimentally determined with far better reliability than ab initio calculations or straightforward modeling can provide. In this section, we will introduce some basic quantities used in radiation physics and provide their theoretical background.

9.5.2 Stopping Power

The stopping power of a material is defined as the amount of energy lost per unit path length by an incident particle while it passes through the material. The numerical value of stopping power is not only dependent on the target material but also on the type of incident particle and its kinetic energy.

The stopping power is defined in terms of collision cross sections as follows:

$$-dE/dx = \rho \, \Sigma_n E_n \sigma_n \qquad (9.5\text{-}1)$$

---

* Work performed under the auspices of the U.S. Department of Energy.
** Present address: National Bureau of Standards, Gaithersburg, Md., U.S.A.

where ρ is the number density of target particles, the summation covers both discrete excitations (referred to as excitations hereafter) and ionization, E_n is the energy required for excitation or ionization to the n-th state measured from the initial state of the target particle, and σ_n is the corresponding cross section for excitation or ionization, which also depends on the kinetic energy of the projectile.

In practice, it is common to use mass stopping power defined as

$$-dE/\mu\,dx = (\rho/\mu)\,\Sigma_n\,E_n\,\sigma_n \tag{9.5-2}$$

where μ is the mass density of the target medium. Note that the dimension of the mass stopping power is energy $\times$ area/mass, which is convenient in evaluating the energy loss per unit area.

It is obvious from Eqs. (9.5-1) and (9.5-2) that accurate determination of the stopping power requires complete knowledge of cross sections for all excitations and ionizations, preferably at all incident energies. Unfortunately, this is not possible at present even for electron-hydrogen or proton-hydrogen collision. On the other hand, one can measure stopping powers directly, thus avoiding the necessity of actually performing the summation in Eq. (9.5-1). In principle, the measured values of stopping power provide a test of relevant cross sections, but only in a gross manner since stopping power represents a weighted average of such cross sections.

Nevertheless, we can derive some qualitative conclusions on stopping power by using cross sections based on a specific theoretical model. Since most applications of radiation physics involve high-energy incident particles, it is appropriate to use cross sections based on the first Born approximation. When Born cross sections are used, a sum rule for them can be applied. Then, the stopping power for a fast projectile of nonrelativistic speed v and charge ze can be expressed in terms of only one number $\bar{I}$, *the mean excitation energy*, that depends on target properties (Bethe and Ashkin, 1953):

$$-dE/dx = (4\,\pi\,a_0^2\,z^2\,R^2/T)\,\rho\,B, \tag{9.5-3}$$

with

$$B = 2\,N\,\ln(4\,T/\bar{I}), \tag{9.5-4}$$

where a_0 is the Bohr radius (0.529 Å), R is the Rydberg energy (13.6 eV), $T = mv^2/2$ with the electron mass m, and N is the total number of electrons in the target particle. Eqations (9.5-3) and (9.5-4) are valid for heavy projectiles such as protons. For incident electrons, allowance must be made for the electron exchange effect, and Eq. (9.5-4) is replaced by:

$$B_e = 2\,N\,[\ln(T/\bar{I}) + (1 - \ln 2)/2]. \tag{9.5-5}$$

Within the context of the Born approximation, the mean excitation energy $\bar{I}$ in Eqs. (9.5-4) and (9.5-5) is defined by

$$\ln(\bar{I}/R) = N^{-1}\,\Sigma_n\,f_n\,\ln(E_n/R), \tag{9.5-6}$$

where f_n is the dipole oscillator strength for the transition to the n-th state from the initial state.

For relativistic projectiles ($T > 10$ keV), Eqs. (9.5-3) to (9.5-5) must be modified further:

$$-dE/dx = (4 \pi a_0^2 z^2 \alpha^2 R/\beta^2) \rho B, \tag{9.5-7}$$

$$B = z N \ln[2 m c^2 \beta^2 (1 - \beta^2)/\bar{I}] - \beta^2, \tag{9.5-8}$$

for heavy projectiles, and

$$B_e = \ln(m c^2 \beta^2 T/\bar{I}^2 \gamma^2) - (2/\gamma + \beta^2) \ln 2 + \gamma^{-2} + (1 - \gamma^{-1})^2/8 \tag{9.5-9}$$

for electrons. In Eqs. (9.5-7) to (9.5-9), α is the fine structure constant ($\sim 1/137$), $\beta = v/c$ with the speed of light c.

$$\gamma = (1 - \beta^2)^{-1/2}. \tag{9.5-10}$$

and

$$T = m c^2 (\gamma - 1) \tag{9.5-11}$$

is the relativistic kinetic energy of the incident electron.

It is more customary to combine factor 2 on the right-hand sides of Eqs. (9.5-4), (9.5-5), and (9.5-8) with dimensional factors on the right-hand sides of Eqs. (9.5-3) and (9.5-7). However, we kept it with B to remind the reader that one half of B, say $N \ln(4 T/\bar{I})$ in Eq. (9.5-4), is the contribution from close collisions with large energy transfers while the other half comes from distant collisions with small energy transfers. Distant collisions dominate the collision cross sections themselves, but close collisions become as important as distant collisions in stopping power, because close collisions are accompanied by large energy transfers.

Table 9.5-1. *Mean excitation energies $\bar{I}$ for noble gases recommended by International Commission on Radiation Units and Measurements (ICRU, 1983). For comparison, the lowest ionization potentials (Moore, 1970), I, are also presented*

Target gas	$\bar{I}$(eV)	I(eV)
He	41.8	24.59
Ne	137	21.56
Ar	188	15.76
Kr	352	14.00
Xe	482	12.13

Although the Bethe formulas, Eqs. (9.5-3) to (9.5-11), adequately describe the stopping power of fast projectiles, additional corrections are necessary for both slow and extremely fast projectiles. For instance, corrections are required to account for the failure of the Born approximation with slow incident electrons. In addition, for positive, heavy, and slow ions, the cross section for electron pick-up is substantial and the change in effective charge states must be taken into consideration. Similarly, when the projectile speed is extremely relativistic, corrections must be made for the change in apparent target density due to the Lorentz contraction, polarization of the target, and, for electrons, energy loss through bremsstrahlung. Tables of recommended stopping power values, which include these and other corrections, of many gaseous and solid materials are available in the literature (Anderson, 1977; ICRU, 1983). Values of the mean excitation energies for noble gases are listed in Table 9.5-1, along with the lowest ionization potential, I.

For atoms with many inner-shell electrons, $\bar{I}/I$ is large because inner-shell ionizations involve large energy transfers.

Although each collision of an incident electron with a target atom will result in an energy loss of a definite amount, ΔE, the actual value of ΔE cannot be predicted; we can only assign a probability ($=$ cross section) associated with each value of ΔE. Hence, in principle, it is impossible to predict the exact amount of energy lost by a particular projectile when it passes through matter; we can only predict an average value, which is the meaning of $-dE/dx$. In other words, the stopping power is the expected average value if we repeated energy-loss experiments with many electrons under identical conditions. Actual values of ΔE are stochastic in nature so that experimental values of $\Delta E/\Delta x$ will have a distribution around the mean value given by Eqs. (9.5-7) to (9.5-9).

In radiation research, absorbed dose is often used instead of stopping power. The absorbed dose is equivalent to the mass stopping power, Eq. (9.5-2), integrated over a unit path length. The unit of absorbed dose is the "rad" defined by

$$1 \text{ rad} = 10^{-2} \text{ J/kg}. \tag{9.5-12}$$

9.5.3 Range

Once the stopping power is known, one can evaluate the average range $R(E_1 \rightarrow E_2)$ for reducing the projectile energy from E_1 to E_2:

$$R(E_1 \rightarrow E_2) = \int_{E_1}^{E_2} \frac{dE}{-dE/dx}. \tag{9.5-13}$$

Again, since $-dE/dx$ is only a mean energy loss and the actual energy loss has a distribution around the mean value, the values of the range also fluctuate around their mean value given by Eq. (9.5-13). Fluctuations in energy loss and range values around their mean values are known as *straggling*. Since electrons are easily deflected by elastic collisions with target atoms, more straggling is observed in the range of electrons than in that of heavy projectiles.

9.5.4 Average Energy per Ion Pair

Another important quantity in radiation physics is the average energy required to produce an ion pair (ion + electron) in a target medium. This quantity is commonly called the w *value*, and it is defined as

$$w = \varepsilon/v \tag{9.5-14}$$

where ε is the total energy lost by the projectile while passing through the target medium, and v is the total number of ion pairs created by the projectile as well as by energetic secondary electrons (often referred to as delta rays) generated by the projectile.

In principle, w values depend on the type and energy of the projectile, the type of target medium, and its phase – whether it is a gas, liquid, or solid. In reality, the w values are insensitive to some of these parameters, except when the projectile is very slow. This property makes the w value a convenient quantity for practical applications in radiation dose monitoring.

Reliable experimental values of w for electron-impact ionization are available for many gas targets (ICRU, 1979). Theoretical calculations of the w values, in principle, require knowledge of all cross sections for ionization by the incident (primary) and ejected (secondary) electrons. A simple example shown below, however, qualitatively demonstrates why the values of w are insensitive to details of collision cross sections.

Let us denote cross sections and excitation energies for discrete excitations by σ_d and E_d: those for ionizations that generate secondary electrons whose kinetic energies are lower than the lowest ionization potential by $d\sigma_{i1}/dE$; and those for ionizations that generate secondary electrons whose kinetic energies are higher than $2 \times I$ by $d\sigma_{i2}/dE$. An energetic secondary electron will continue to ionize target atoms until its kinetic energy becomes lower than I. This means energies spent in creating delta rays generate more than one ion. Hence, the w value is qualitatively given by

$$w = \frac{\sum \sigma_d E_d + \displaystyle\int_I^{2I} E(d\sigma_{i1}/dE)\,dE + \displaystyle\int_{2I}^{T/2} E(d\sigma_{i2}/dE)\,dE}{\displaystyle\int_I^{T/2} (d\sigma_{i1}/dE + 2\,d\sigma_{i2}/dE)\,dE} \qquad (9.5\text{-}15)$$

Equation (9.5-15) accounts for only the first generation of secondary electrons produced, and $T/2$ is the (approximate) upper limit of the secondary electron energy. For most gas targets, one-half to one-fourth of all ionizing events produce secondary electrons fast enough to cause more ionization. Let σ_t be the total ionization cross section,

$$\sigma_t = \int_I^{T/2} (d\sigma_{i1}/dE + d\sigma_{i2}/dE)\,dE. \qquad (9.5\text{-}16)$$

Then,

$$w \sim \frac{\sum \sigma_d E_d + \sigma_t(2/3) \times 1.5 \times I + \sigma_t(1/3) \times 2.5 \times I}{\sigma_t(2/3 + 2/3)}, \qquad (9.5\text{-}17)$$

where we have assumed the ratio $(d\sigma_{i2}/dE)/\sigma_t = 1/3$, the average energy loss for $d\sigma_{i1}/dE$ to be $1.5 \times I$, and that for $d\sigma_{i2}/dE$ to be $2.5 \times I$. In most atoms, discrete energy levels are concentrated in a narrow band below I, and we can take an average value, $0.8 \times I$, for E_d's. If we denote

$$\sigma_{ex} = \Sigma \, \sigma_d \qquad\qquad (9.5\text{-}18)$$

as the sum of all cross sections for discrete excitations, then Eq. (9.5-17) can be simplified:

$$w \sim \frac{[0.8\,(\sigma_{ex}/\sigma_t) + 1.0 + 0.8]}{4/3} \times I. \qquad\qquad (9.5\text{-}19)$$

For helium, $\sigma_d/\sigma_t \sim 0.5$ (Kim and Inokuti, 1971), resulting in

$$w \approx 1.7 \times I = 42\,\text{eV}, \qquad\qquad (9.5\text{-}20)$$

which compares well with the average experimental value $41.3 \pm 1.0\,\text{eV}$ as is shown in Table 9.5-2. One can see from Eq. (9.5-19) that the coefficient of I on the right-hand side is sensitive to the ratio σ_d/σ_t. For atoms and molecules with inner shells this coefficient will be higher because inner-shell ionizations require much more energy than the lowest ionization potential. On the other hand, for such targets σ_d/σ_t is far smaller than helium, amounting to $0.1-0.2$. The net result is that the w value for electron-impact ionization remains $1.6-1.8 \times I$ for most atoms. A list of the w values for electron-impact ionization for rare gases and some molecular gases are listed in Table 9.5-2.

Table 9.5-2. *Average experimental values of w in eV for electron-impact ionization of various gases (ICRU, 1979)*

gas	w value	gas	w value
He	41.3 ± 1.0	H_2	36.5 ± 0.3
Ne	35.4 ± 0.9	N_2	34.8 ± 0.2
Ar	26.4 ± 0.5	O_2	30.8 ± 0.4
Kr	24.2 ± 0.3	H_2O	29.6 ± 0.3
Xe	22.1 ± 0.1	Air	33.85 ± 0.15

9.5.5 Track Structure

In addition to dose monitoring in medical applications, radiation physics plays an important role in the identification of tracks in particle detectors and emulsion plates (Fleischer et al., 1975). Here again, observable characteristics are dependent on radiation physics concepts, such as the stopping power and range, but the theory of track structure has not advanced enough to explain all observed facts or to predict anticipated features.

Primary interactions of fast charged particles occur along a sharply defined track. Its identification by physical or chemical methods, however, must await diffusion of the collision by-products through large enough dimensions to be observable by optical or other means. This requires a waiting time of 10^{-11} to 10^{-3} second, while an electron of $\frac{1}{2}$ Rydberg in kinetic energy will pass a region of molecular size (~ 2 Å) in 10^{-16} second. Hence, to understand observed radiation effects, one must follow the time development of each primary event, a task easier said than done. With the advent of high-speed computers with large memory, one can simulate radiation interaction by the Monte Carlo or equivalent methods. However, such studies (Turner et al., 1982) are still affected by the incomplete knowledge of collision cross sections, particularly those that occur in the 10^{-15} to 10^{-11} second time range. All secondary electrons are slowed down to very low kinetic energies (< 1 eV) during this time range, and move away from the original track by diffusion. Important electron collision processes in this energy range are rotational and vibrational excitations, transient negative ion formations, and for condensed media, simultaneous interaction with more than one molecule since the de Broglie wavelengths of these slow electrons can span dozens of molecules.

To complement the lack of precise knowledge on collision cross sections and reaction rates in the $10^{-15} - 10^{-11}$ second time range, less rigorous yet useful concepts have been introduced in the study of track structure as well as in general applications to radiation chemistry. To describe initial conditions in the study of charged-particle interaction with liquids, Mozumder and Magee (1966) introduced terms such as *spurs*, *blobs*, and *short tracks*. A spur refers to all events initiated by $6 - 100$ eV of energy deposited by the primary particles, a blob stands for those initiated by $100 - 500$ eV of energy deposited, and a short track includes those initiated by $0.5 - 5$ keV of energy deposited. In addition, radiation chemists often use a quantity known as the g *value*. It is defined as the number of a chemical species produced by an initial energy deposition of 100 eV.

The definition of the spur is based on the fact that most of the "glancing" or distant collisions involve energy transfers of roughly $6 - 100$ eV for which corresponding dipole oscillator strengths are appreciable. Blobs represent events that involve secondary electrons of moderate speed, comparable to soft x-rays in energy. Short tracks involve δ rays, or fast electrons with energies equivalent to hard x-rays. The g value is a convenient quantity when details preceding the production of the species of interest are unknown or immaterial.

In order to understand the interaction of fast electrons with condensed matter, it is necessary to reexamine some basic concepts taken for granted in electron-gas interactions. For instance, periodic potentials in solids will lower ionization potentials, and an ion may not remain charged long if "free" electrons — those not bound to any particular atom or molecule — are available nearby (e.g., metals) for recombination. Also, slow electrons may be trapped into an interstitial position surrounded by polar molecules and form bound states as is known to happen in some liquids. There are collective excitations in condensed matter, usually identified with excitation energies slightly above the first ionization potential. Details of these and other events in condensed matter will pose new challenges, both theoretical and experimental, as we learn more about the electron-impact ionization of gases. For instance, one way to bridge the gap between the gas phase and the condensed matter is to study the properties of electron-cluster interactions (Märk and Castleman, 1984).

References

Anderson, H. H. (1977): Bibliography and Index of Experimental Range and Stopping Power Data (organized by Ziegler, J. F.), Vol. 2. New York: Pergamon Press.

Bethe, H. A., Ashkin, J. (1953): In: Experimental Nuclear Physics (Segré, E., ed.), 166 – 304. New York: J. Wiley.

Fleischer, R. L., Price, P. B., Walker, R. M. (1975): Nuclear Tracks in Solids, 39 – 42. Berkeley: University of California Press.

International Commission on Radiation Units and Measurements (1979): Average Energy Required to Produce an Ion Pair, ICRU Report No. 31.

International Commission on Radiation Units and Measurements (1983): Stopping Powers for Electrons and Positrons. ICRU Report No. 36.

Kim, Y.-K., Inokuti, M. (1971): Phys. Rev. *A 3*, 665 – 678.

Märk, T. D., Castleman, A. W., jr. (1984): Adv. Atom. Molec. Phys. *20*, 65 – 172.

Moore, C. E. (1970): Ionization Potentials and Ionization Limits Derived from the Analyses of Optical Spectra. NSRDS-NBS 34 (U.S. Government Printing Office).

Mozumder, A., Magee, J. L. (1966): Radiat. Res. *28*, 203 – 214; *28*, 215 – 231.

Turner, J. E., Paretzke, H. G., Hamm, R. N., Wright, H. A., Ritchie, R. H. (1982): Radiat. Res. *92*, 47 – 60.

9.6 Mass Spectrometry Applications

J. H. Futrell

Department of Chemistry, University of Utah, Salt Lake City, Utah, U.S.A.

9.6.1 Introduction

In the interplay between issues of molecular energetics and mass spectrometric fragmentation patterns as effected by electron impact ionization it is useful to distinguish three phases of research activity: The first phase was the discovery that appearance energies of molecular and fragment ions frequently correlated well with known or estimated spectroscopic and thermochemical data. Further work revealed a number of discrepancies, leading ultimately to the second phase of improved understanding of the processes involved and improved methods for obtaining relatively precise data. Many of these developments and recent data are described in Chapter 5 of this volume.

For the mass spectrometry of polyatomic ions, the improved understanding of the physical processes involved in the generation of mass spectra evolved within the framework of the quasi-equilibrium theory (QET) of mass spectra (Rosenstock *et al.*, 1952). The advent of photoionization and photoelectron spectroscopy, particularly the combined technique of threshold photoelectron photoionization mass spectroscopy (Stockbauer, 1973; Stockbauer and Inghram, 1976; Werner and Baer, 1975) has led to ever more stringent tests for the theory and to refinements in all important details of QET calculations. A recent review announced the beginning of the "era of studies of reaction dynamics" for polyatomic ion mass spectrometry (Lifshitz, 1978).

In turn, these developments have ushered in the third phase in which highly specialized instruments (see Chapter 5) are used for investigating most fundamental aspects of mass spectrometry. Electron impact is used primarily as a convenient means for ionizing and detecting molecules and for varying the extent of fragmentation by changing electron energy. (For cogent recent examples of this application of electron ionization see Drowart and Goldfinger, 1976 and Murad 1980, 1981.) No longer used extensively as quantitative analysis devices (except through use of internal standards, particularly isotopically labelled molecules), most recently constructed mass spectrometer ion sources and analyzers of the present era do not lend themselves naturally to the kind of quantitative studies described elsewhere in this volume.

At the same time the qualitative uses of mass spectral data have been codified in the development of extensive computerized data banks of standardized spectra (Heller et al., 1978) along with algorithms detailing for small and medium sized molecules correlations between molecular structure and fragmentation patterns (Venkataraghavan et al., 1978). Together with prefractionation methods of high resolution gas chromatography and liquid chromatography for resolving mixtures prior to introducing them into spectrometers, these methods have contributed to the evolution of electron impact mass spectrometry from an active research area into that of a sophisticated but routine analytical tool.

The quantitation of GC-MS data usually is accomplished using calibrated standard detectors for gas chromatographs, coupled either in series or parallel to the mass spectrometer, with the spectrometer serving primarily as a qualitative detector. An alternative procedure is the integration of ion signals to produce a total ion current chromatogram which serves exactly the same purpose when calibration samples are available. In the absence of authentic calibration standards, however, one can also calculate concentrations by using semi-empirical methods to estimate 70 eV ionization cross-sections for the detected molecules. Quite similar methods are used to estimate cross-sections for species evolving from Knudsen cells used in thermodynamic studies. These two topics will be reviewed briefly before turning to the main subject of this chapter.

In view of the fact that it has not been reviewed extensively elsewhere in this volume and that it underlies all phenomena observed in the mass spectra of polyatomic ions, it seems appropriate to use the limited space available to review the quasi-equilibrium theory and to show a few recent examples supporting essential principles. The reader is also referred to a reference book on the subject (Forst, 1973) and recent reviews for additional details (Laidler and King, 1983; Truhlar et al., 1983). Several important topics relating to the QET — dissociations from isolated electronic states, vibrational energy randomization, isomerization, kinetic energy release distributions and isotope effects — have been discussed in a recent article concerned with intramolecular energy redistribution in polyatomic ions (Lifshitz, 1983).

9.6.2 Applications in GC-MS and High Temperature Thermodynamics

In both of these areas investigators are often concerned with species for which authentic calibration standards are unavailable. In terms of sensitivity and specificity for gas phase systems, mass spectrometric detection and approximate quantitation of resulting data are unequalled. The measurement of minor constituents and the ability to discriminate against contaminants are additional important advantages of this technique. For this reason a lot of effort has been expended in developing a detailed understanding of the limitations of mass sepctrometric methods.

For Knudsen cell sampling (Knudsen, 1915) a collimated, effusive molecular beam of vapor enters the ion source of the mass spectrometer through a collimating hole which is part of a heat shield separating the differentially-pumped Knudsen cell from the ionization region. Inside the source the beam of neutral molecules (and atoms) is crossed by an electron beam of variable energy orthogonal to the neutral beam. Characteristic molecular and fragment positive ions are generated by electron impact. Weak electric fields extract a (representative) fraction of the ions which are further accelerated and analyzed by an appropriate mass analyzer. Mechanical chopping, or modulation, of the neutral beam is a useful means for distinguishing constituents of the neutral beam from background gases. Identification of the ions follows from their mass to charge ratios; characteristic isotopic distributions in mass clusters are a very useful confirmation, especially for inorganic species. Comparison of appearance potentials usually distinguishes primary from fragment ions. Velocity and angular profiles and the dilution of metastable decays in multiple sector instruments (see Chapter 5) are useful ancillary tools for distinguishing fragmentation processes and related kinetic phenomena.

Partial pressures of individual species present in the vapor are estimated from the expression

$$P_i = I_i^+ \, T/[S_i \, \sigma_i \, (E) \, \gamma_i] = k_i \, I_i^+ \, T \tag{9.6-1}$$

where T is the absolute temperature, $\sigma_i(E)$ the energy dependent electron impact ionization cross section, γ_i the electron multiplier gain factor, and S_i the instrument sensitivity factor (Gingerich, 1980). The latter factor includes the overlap of the electron and molecular beams, the extraction efficiency, and transmission of the ion optical and mass analyzer systems.

The sensitivity factor can be calibrated by the evaporation of a weighed amount of some element, such as Ag or Au, which gives predominantly monomeric species (Margrave, 1967) for which partial pressure as a function of temperature is well-established. The multiplier gain factor can be estimated for atomic systems from an empirical equation

$$\gamma_i = C \, M_i^{-0.4} \tag{9.6-2}$$

with an average deviation of $\pm 16\%$ over the entire periodic table (Pottie et al., 1973). Molecular species are known to exhibit higher multiplier gains because of breakup of the species on impacting the first dynode (Stanton, Chupka and Inghram, 1956), introducing uncertainties of the order of a factor of 2 into results deduced from eq. (9.6-2).

Experimental absolute or relative ionization cross-sections $\sigma_i(E)$ are known for only a limited number of molecules of interest in high temperature thermodynamics. Calculations based upon an empirical modification of the Bethe-Born approximation (Mann, 1970) are estimated to be reliable within a factor of two for atomic systems (Stafford, 1971). Molecular cross sections are obtained assuming additivity of atomic cross-sections. However, this approximation is known to be inaccurate by $20-30\%$ (Gingerich, 1980). Clearly there is an identifiable need for further experimental and theoretical work on ionization cross-sections of simple inorganic species and on the contributions of fragmentation processes involving higher clusters to the observed mass spectral patterns.

The problem of quantitating molecules in GC effluents contains all the factors discussed above, but is characterized by several simplifying features. Firstly, molecular beam sampling is seldom employed. Molecules make several collisions with ion source surfaces and may be considered to be at ion source temperature when ionized. Therefore, the absolute temperature may be suppressed in eq. (9.6-1). Moreover, molecules achieve uniform density in the ion source under these conditions, and the ionization volume is defined by the electron beam alone. Only molecular species are involved, and a substantial variation in multiplier gain with mass number is not expected. Since most fragment ions are also molecular species, multiplier gain corrections are further reduced in importance in comparison with Knudsen cell work.

The questions of mass (and kinetic energy) discrimination in ion extraction and mass analysis and estimation of the ionization cross-section of the species detected obviously remain as the most significant factors in the GC-MS quantitation problem. The former issue is addressed by the availability of a host of surrogate molecules of comparable mass, volatility and chemical functionality for calibrating the mass spectrometer. The latter may be addressed the same way; alternatively, relative electron impact total ionization cross-sections may be estimated within several percent by empirical methods (Harrison *et al.* 1966; Otvos and Stevenson, 1956; Lampe, Franklin and Field, 1957; Fitch and Sauter, 1983).

The higher precision in estimating ionization cross-sections for organic molecules probably results from the relative constancy of chemical bond properties in homologous series of compounds. The availability of a wide variety of compounds for a calibration basis set and the availability of several literature reports on ionization cross-sections for comparison purposes further reduce the uncertainties in this approach for organic molecules. It was recently demonstrated in a multiple linear regression study (Fitch and Sauter, 1983) that 179 experimentally determined total ionization cross-sections were fit with a correlation coefficient of 0.996 by the equation

$$Q = 0.082 + \sum_{1}^{8} \alpha_i n_i \qquad (9.6\text{-}3)$$

where Q is the total ionization cross-section and α_i and n_i are the coefficient and number of each type of atom (H, C, N, O, F, Cl, Br, and I) in the molecule examined. Among other things, these workers also investigated the effect of bonding character for the C, N, and O atoms in the molecule (e.g., sp, sp^2, and sp^3 hybridization) using the same statistical methods. No significant improvement was noted, supporting the concept that a simple additivity relationship of atomic ionization cross-sections is an adequate representation of the data. However, the fact that both the intercept constant and individual atom cross-section regression coefficients changed in this analysis clearly demonstrates that the regression coefficients deduced are not to be identified with atomic ionization cross-sections *per se* but rather with a measure of the average atomic contribution to the *molecular* cross-section in a particular basis set suite of molecules. This point is further confirmed by comparison of the atomic ionization regression coefficients (Fitch and Sauter, 1983) with Mann's absolute calculations (Mann, 1970) for the same elements. The latter work was based on ground state atomic orbital configurations and it is interesting to speculate that a comparison with calculations based on bonding orbital configurations (e.g., sp^3 hybridized carbon rather than $s^2 p^2$ ground state) might prove to be a more fruitful approach.

9.6.3 The Theory of Mass Spectra

There are two central issues which distinguish mass spectra and mass spectroscopy from other areas of kinetics and spectroscopy, respectively. The first is that the ensemble of parent molecular ions is an isolated system. Each ion is formed (e.g., by electron collision) with some specific amount of internal energy and angular momentum. These are conserved independently for each ion in all its subsequent dissociations. It does not communicate with other molecules or with the walls of the ion source. Therefore it is characterized by average energy content rather than a temperature: in the language of statistical mechanics it is characterized as a microcanonical ensemble rather than the usual canonical ensemble of microstates.

The second feature is that the mass spectrum of a molecule is a time dependent phenomenon rather than sampling of stationary states. Thus the mass spectrum of a molecule is a function of the elapsed time between ion formation and mass analysis. The utility of standardized libraries of mass spectral data therefore depends upon two important points: (1) very rapid reactions (rate constants exceeding $10^7 s^{-1}$) dominate the mass spectrum and (2) most spectrometers have characteristic sampling times of several microseconds. The use of very high extraction fields (as in field ionization mass spectrometry (Beckey, 1971) or long time constant mass analyzer, as in ion cyclotron resonance mass spectrometry (Futrell, 1971), produces characteristic shifts in mass spectral fragmentation patterns (Futrell, 1971: Wisniewski and Futrell, 1970).

The development of the quasi-equilibrium theory (QET) by Rosenstock, Wallenstein, Wahrhaftig, and Eyring (Rosenstock *et al.*, 1952) was the first satisfactory formalization of mass spectral theory. In addition to the original paper, detailed discussion of the QET have been presented by several authors (Rosenstock and Kraus, 1963; Rosenstock, 1968; Vestal, 1968; Wahrhaftig, 1964, 1972). A simplified description, with worked examples applying both the simplified and a more rigorous version of the theory are available in the text by Kiser (Kiser, 1966). Non-QET theories have been reviewed by Wahrhaftig (Wahrhaftig, 1972).

An important recent development is the application of microscopic reversibility to the QET formalism with the explicit inclusion of conservation of angular momentum (Klots, 1964, 1971, 1976). This has the important consequences of recasting the calculation of rate constants in terms of the properties of reaction products rather than properties of an "activated complex" in the Eyring theory of absolute reaction rates (Glasstone et al., 1941) and of demonstrating the equivalence of QET (in this formulation) to phase space theory (Pechukas and Light, 1965: Light, 1967).

The basic postulate of the QET is that the rates of dissociation of the excited molecular ion are slow compared to the time of interaction with the exciting electron and also slow compared with the rates of energy re-distribution among the internal degrees of freedom of the ion, both electronic and vibrational. A second assumption is that each dissociation process may be described as a motion along a reaction coordinate separable from other internal coordinates through a critical "activated complex" configuration. The latter assumption is removed by the alternative phase space theory formulation. In the latter case the Langevin collision limit model is conventionally used to estimate the required cross sections for the reverse bimolecular reactions.

The first step in computing mass spectra is determining the relevant mechanisms for unimolecular dissociation of the excited molecular ion. The investigation of metastable ions and collision-induced metastables usually provides the required set of ion decomposition reactions (Beynon et al., 1971: Hills et al., 1970: Hills, 1970). Isotopically labelled molecules are especially useful probes for unexpected rearrangement reactions (Vestal and Futrell, 1970: Stockbauer and Inghram, 1976).

The second step is the computation of rate constants as a function of internal energy in the molecular ion. The simplest equation for computing these rate constants is

$$k_i(E) = \frac{\sigma}{h} \frac{W_i^{\neq}(E - \varepsilon_i)}{\rho(E)} \tag{9.6-4}$$

where $W_i^{\neq}(E - \varepsilon_i)$ is the number of states with energy less than E but greater than the activation energy ε_i, $\rho(E)$ is the density of states for the system with energy E, σ is the number of equivalent reaction paths and h is Planck's constant. A more complex expression (but no more cumbersome once explicit methods for counting $W^{\#}$ are introduced) for $k(E)$ expressly formulated in terms of reactant and product properties rather than the transition state is given in the rigorous phase space theory of Chesnavich and Bowers (Chesnavich and Bowers, 1977a, 1977b).

The next step is the construction of a "breakdown graph" which expresses the relative abundances of parent and fragment ions as a function of the energy deposited in the molecular ion. This graph depends critically upon the time constant of the spectrometer and the ratios of rate constants deduced theoretically. If the time constant of the instrument is 10 μs, rates larger than 10^5 s^{-1} appear as product ions while those less than 10^5 s^{-1} appear as reactant ions. Those borderline processes with rates about equal to the apparatus time constant constitute the class of metastable ions.

The final step is the convolution of the breakdown graph over the energy distribution for the excited molecular ion. For mass spectrometry this is usually the distribution imparted by 70 eV electron impact. The general shape of this distribution is now known from photoelectron spectroscopy.

The result of this calculation procedure is to produce calculated mass spectra exhibiting generally good agreement with experimental data. However, it was realized very early that the several uncertainties in input data and the convolution over several steps in the calculation made this agreement a much less than definitive test of the theory. Many more rigorous tests referred to in 9.6.1 — direct measurements of breakdown graphs, ion lifetimes (decomposition rates), energy disposal, isotope effects in normal and metastable mass spectra, temperature and ion preparation effects — support the general applicability of the theory for complex ions.

9.6.4 Comparison of Some Experimental Results with QET Predictions

An important application of electron impact mass spectrometry to testing the QET is the use of the electron space charge ion trapping method devised by Harrison for ion-molecule reaction studies (Herod and Harrison, 1970) and later applied to the unimolecular problem (Dymerski and Harrison, 1976). An elegant recent example is provided by the work of Lifshitz and Gefen (Lifshitz and Gefen, 1980).

In these experiments a trapping beam of electrons of approximately 10 μamp. at 5 eV in an otherwise field-free ion source is used to retain ions generated by a 1 μs ionizing pulse for up to 2 ms. A repeller pulse applied after a variable time delay then sweeps out the ion source. Ionization efficiency curves for 1.5 hexadiyne at 13 and 1550 μs are shown in Fig. 5-10 of Chapter 5. The striking time dependence of the relative intensities of m/z 77 and 78 is evident from these graphs.

As shown by Chupka and Kaminsky (Chupka and Kaminsky, 1961), the threshold law for direct ionization leads to the conclusion that the normalized second derivative of the ionization efficiency curve is the breakdown graph. (Alternatively the first derivative of the photoionization curve gives the same information.) Breakdown graphs at 12 μs and 1 ms delay times are shown in Figs. 9.6-1 a and 9.6-1 b. The striking result is that the dissociation reaction

$$C_6H_6^+ \rightarrow C_6H_5^+ + H \tag{9.6-5}$$

is dominated by the reactant ion at times $\leq 12 \mu$s but by the product ion at times ≥ 1 ms.

A second noteworthy feature is that m/z 78 surviving in the energy range $11 < E < 17.5$ eV represents a kinetically distinct moiety which may be interpreted as a non-inter-converting state of $C_6H_6^+$, possibly the benzene molecular ion. This appears as a shoulder on the low energy molecular ion curve in Fig. 9.6-1, but is not clearly distinguished kinetically at the shorter time. Analogous isolated states have been discovered for benzene itself (Andlauer and Ottinger, 1971: Eland, 1974) and for the small molecules C_2F_6 (Eland et al., 1976), CH_3X (X = F, Cl, Br, I) (Simm et al., 1973, 1974) and C_2H_5Cl (Baer et al., 1976 a, 1976 b) using other techniques. We note parenthetically that appearance energy measurements for m/z 78 would detect the ground state ion for hexadiyne at 12 μs but only the stable "excited" state at $t > 1$ ms.

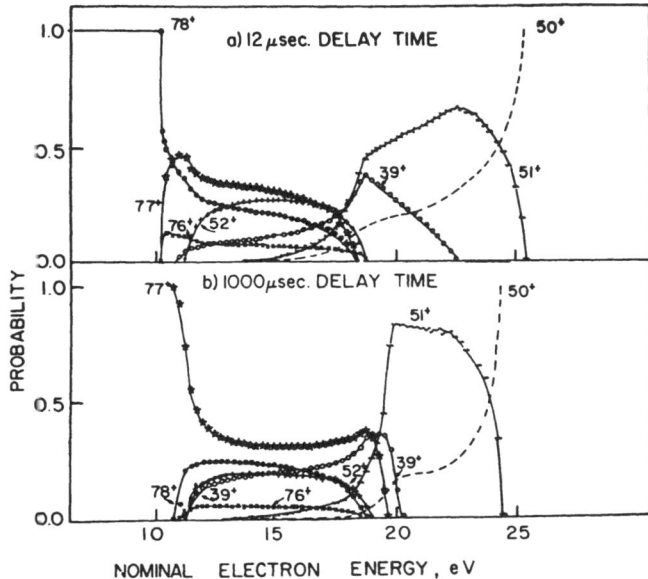

Fig. 9.6-1. *a* (Upper Curve) Breakdown graph for 1.5 hexadiyne at 12 µs delay time. *b* (Lower Curve) Breakdown graph for 1.5 hexadiyne at 1 ms delay time (Lifshitz and Gefen, 1980)

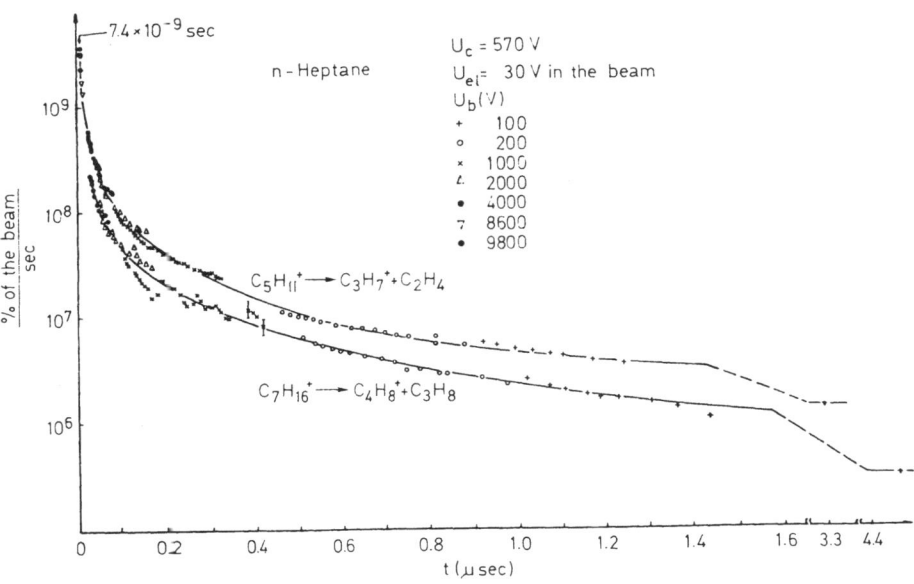

Fig. 9.6-2. Intensity of the indicated ions from *n*-heptane as a function of time after ion formation. A continuum of rate constants is indicated by non-linear shape of the semi-logarithmic curves (Ottinger, 1967)

An even closer examination of the QET prediction of a continuum of rates is provided by the molecular beam method of Ottinger and coworkers (Andlauer and Ottinger, 1971; Ottinger, 1967; Hertel and Ottinger, 1967). These workers utilized a moderate electric field to accelerate ions whose region of formation was defined by the intersection of an electron gun with an annular molecular beam. Momentum analysis of the ions was then mapped directly as the rate constant as a function of time after ion formation. Fig. 9.6-2 gives an example of the continuous distribution usually found. (An exception noted above was the $C_6H_6^+$ ion (Andlauer and Ottinger, 1971).)

The most direct measurement of the breakdown graph of the molecular ion is provided by the technique of threshold photoelectron/coincidence photoion mass spectrometry (Simm et al., 1973, 1974; Eland et al., 1976; Baer et al., 1976a, 1976b) (called either TPE-CPI MS or PIPECO, for photoion-photoelectron coincidence). Here photoionization is used to generate ions in a moderate electric field which ejects the electrons and ions in opposite directions. Electrons pass through a collimated hole structure which transmits only low energy ($<0.5\,eV$) electrons and are counted by an electron multiplier which initializes the ion detector. Mass analysis is by time-of-flight of the ion. This technique therefore displays directly the mass spectrum as a function of the deposited energy.

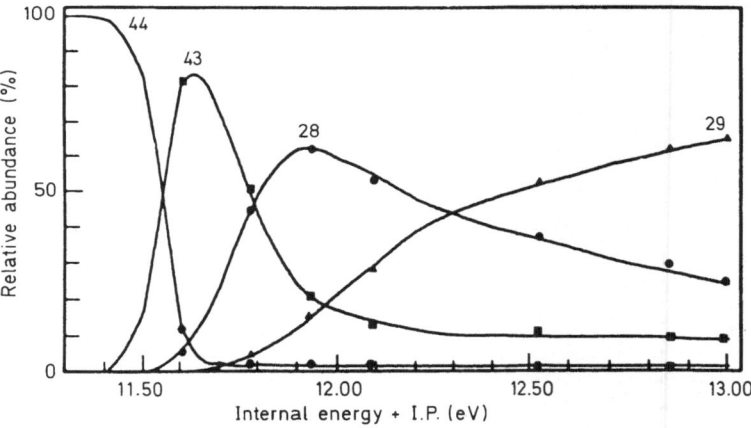

Fig. 9.6-3. Experimental breakdown graph for the propane molecular ion (Gilman, Hsieh and Meisels, 1982)

Fig. 9.6-3 is the breakdown graph for propane recently obtained by this method by Gilman, Hsieh, and Meisels (Gilman et al., 1982). It is in excellent qualitative agreement with an earlier theoretical calculation (Vestal and Futrell, 1970). Modest shifts of threshold energies and frequencies would bring theory and experiment into very good quantitative agreement.

However, the same study (Gilman et al., 1982) used ^{13}C labelling of the central carbon of propane to investigate skeletal rearrangement in the generation of $C_2H_4^+$ (and CH_4) in the mass spectrum of propane. They concluded that some 11% of methane elimination expelled the central carbon. Analogous non-specific skeletal rearrangements have been demonstrated for hexane and larger alkanes (Lavanchy, Houriet, and Gaumann, 1978, 1979; Liardon and Gaumann, 1969). Future QET calculations must include these mechanisms. For propane this implies that a skeletal rearrangement mechanism must be added to the established (Vestal and Futrell, 1970; Stockbauer and Inghram, 1976) 1.3 and 1.2 elimination mechanisms for formation of $C_2H_4^+$.

9.6.5 Some General Conclusions

Although the literature relating to QET tends to dwell upon differences between prediction and experiment rather than upon its success, it must be concluded that the QET provides a satisfactory framework for discussing organic mass spectrometry. In particular, the QET explicitly connects mass spectrometry with conventional reaction kinetics and dynamics in a satisfactory manner. The integration of QET rate expressions over a Boltzmann distribution function gives the usual expressions for unimolecular rates. Under these conditions differences in activation energies, $\varepsilon_i - \varepsilon_j$, may be identified with $\Delta H_i^{\neq} - \Delta H_j^{\neq}$. Analogously, $W_i^{\#}$ is related to $\Delta S^{\neq}$. Accordingly, under the restricted conditions that there are no complications resulting from competing reactions and no irregularities in the parent ion energy distribution function, it is expected that ion yields will correlate with the same kind of energy- and entropy-related parameters as thermal reaction kinetics. Many such correlations are known.

In other cases the special nature of mass spectrometer dissociations must be considered, particularly details of dependence of the energy distribution function on substituent groups. As discussed by McLafferty (McLafferty, 1970), metastables exhibit a large positive temperature coefficient whenever the $P(E)$ vs. E curve has a negative slope at the appearance energy of the fragmentation process and a large negative temperature coefficient where the $P(E)$ curve has a positive slope. The existence of isolated electronic states, unexpected rearrangement mechanisms and angular momentum restraints on the decomposition rates of isolated systems are similarly manifested in mass spectrometry in ways not anticipated from thermal reaction kinetics considerations.

To the instrumental conditions and constants already described in Chapter 5 are necessarily added the dynamical features of unimolecular decompositions. Fortunately the general features of energy distribution between translation and internal energy and energy distribution between fragments appear to be described adequately by the QET, especially in its phase space formulation.

Accordingly the means are at hand for characterizing the energies and kinetics of ion decomposition processes, correcting for kinetic energy release in fragmentation and extrapolating data between instruments with different characteristic sampling times.

References

Andlauer, B., Ottinger, C. (1971): J. Chem. Phys. 55, 1471.
Baer, T, Werner, A. S., Tsai, B. P. (1975): J. Chem. Phys. 62, 2497.
Beckey, H. D. (1971): Field Ionization Mass Spectrometry. New York: Pergamon Press.
Beynon, J. H., Caprioli, R. M., Ast, T. (1971): Int. J. Mass Spectrom. Ion Phys. 7, 92.
Chesnavich, W. J., Bowers, M. T. (1977): J. Chem. Phys. 66, 2306.
Chesnavich, W. J., Bowers, M. T. (1977): J. Am. Chem. Soc. 99, 1705.
Chupka, W. A., Kaminsky, M. J. (1961): Chem. Phys. 35, 1991.
Drowart, J., Goldfinger, P. (1967): Angew. Chemie 6, 581.
Dymerski, P. P., Harrison, A. G. (1976): J. Phys. Chem. 80, 2825.
Eland, J. H. D. (1974): Int. J. Mass Spectrom. Ion Phys. 13, 457.
Eland, J. H. D., Frey, R., Kuestler, A., Schulte, H., Brehm, B. (1976): Int. J. Mass. Spectrom. Ion Phys. 22, 155.
Fitch, W., L., Sauter, A. D. (1983): Anal. Chem. 55, 832.
Forst, W. (1973): Theory of Unimolecular Reactions. New York: Academic Press.
Futrell, J. H. (1971): Ion Cyclotron Resonance. In: Dynamic Mass Spectrometry, Vol. 2 (Price, D., ed.), 97 – 135. London: Heyden and Son Ltd.
Gilman, J. P., Hsieh, T., Meisels, G. G. (1982): J. Chem. Phys. 76, 3497.
Gingerich, K. A. (1980): Molecular Species in High Temperature Vaporization. In: Current Topics in Materials Science, Vol. 6 (Kaldis, E., ed.), 135 – 447. Amsterdam: North Publishing Co.
Glasstone, J., Laidler, K. J., Eyring, H. (1941): Theory of Rate Processes. New York: McGraw-Hill.
Harrison, A. G., Jones, E. G., Gupta, S. K., Nagy, G. P. (1966): Can. J. Chem. 44, 1967.
Heller, S. R., Heller, R. S., McCormick, A., Maxwell, D. C., Milne, G. W. A. (1967): In: Advances in Mass Spectrometry 7 B (Daly, N. R., ed.), 985. London: Heyden and Son Ltd.
Herod, A. A., Harrison, A. G. (1970): Int. J. Mass Spectrom. Ion Phys. 4, 415.
Hertel, J., Ottinger, C. (1967): Z. Naturforsch. 22 a, 1141.
Hills, L. P., Vestal, M. L., Futrell, J. H. (1971): J. Chem. Phys. 54, 3834.
Hills, L. P. (1970): Metastable Ions. Ph. D. thesis, Univ. of Utah.
Kiser, R. W. (1966): Introduction to Mass Spectrometry, Chapter 7. New York: Prentice-Hall.
Klots, C. E. (1964): J. Chem. Phys. 41, 117.
Klots, C. E. (1971): J. Phys. Chem. 75, 1526.
Klots, C. E. (1976): J. Chem. Phys. 64, 4269.
Knudsen, M. (1915): Ann. Physik IV Folge 47, 697.
Laidler, K. J., King, M. C. (1983): J. Phys. Chem. 87, 2657.
Lampe, F. W., Franklin, J. L., Field, F. H. (1957): J. Am. Chem. Soc. 79, 6129.
Lavanchy, A., Houriet, R., Gaumann, T. (1978): Org. Mass Spectrom. 13, 410.
Lavanchy, A., Houriet, R., Gaumann, T. (1979): Org. Mass Spectrom. 14, 79.
Liardon, R., Gaumann, T. (1969): Helv. Chim. Acta 52, 1042.
Lifshitz, C. (1978): In: Advances in Mass Spectrometry, 7A (Daly, N. R., ed.), 3 – 18. London: Heyden

Lifshitz, C., Gefen, S. (1980): Int. J. Mass Spectrom. Ion Phys. 35, 31.
Lifshitz, C. (1983): J. Phys. Chem. 87, 2304.
Light, J. C. (1967): Disc. Farad. Soc. 44, 14.
Mann, J. B. (1970): In: Recent Development in Mass Spectrometry (Ogata, K., Hayakawa, T., eds.), 814 – 819. Baltimore: University Park Press.
Margrave, J. L., ed. (1967): The Characterization of High Temperature Vapors. New York: Wiley.
McLafferty, F. W. (1970): In: Recent Developments in Mass Spectroscopy (Ogata, K., Hayakawa, T., eds.), 70. Baltimore: University Park Press.
Murad, E. (1980): J. Chem. Phys. 73, 1381.
Murad, E. (1981): J. Chem. Phys. 75, 4080.
Ottinger, C. (1967): Z. Naturforsch. 22 a, 20.
Otvos, J. W., Stevenson, D. P. (1956): J. Am. Chem. Soc. 78, 546.
Pechukas, P., Light, J. C. (1965): J. Chem. Phys. 42, 3281.
Pottie, R. F., Cocke, D. L., Gingerich, K. A. (1973): Int. J. Mass Spectrom. Ion Phys. 11, 41.
Rosenstock, H. M., Wallenstein, M. B., Wahrhaftig, A. L., Eyring, H. (1952): Proc. Nat. Acad. Sciences, USA 38, 667.

Rosenstock, H. M., Krauss, M. (1963): Quasi-Equilibrium Theory of Mass Spectra. In: Mass Spectrometry of Organic Ions (McLafferty, F. W., ed.), 1. New York: Academic Press.

Rosenstock, H. M. (1968): In: Advances in Mass Spectrometry, Vol. 4, 523. London: Inst. of Petroleum.

Simm, I. G., Danby, C. J., Eland, J. H. D. (1974): Int. J. Mass Spectrom. Ion Phys. *14*, 285.

Stafford, F. E. (1971): High Temperatures High Pressures *3*, 213.

Stanton, H. E. Chupka, W. A., Inghram, M. G. (1956): Rev. Sci. Instr. *27*, 109.

Stockbauer, R. (1973): J. Chem. Phys. *58*, 3800.

Stockbauer, R., Inghram, M. J. (1976): J. Chem. Phys. *65*, 4081.

Truhlar, D. G., Hase, W. L., Hynes, J. T. (1983): J. Phys. Chem. *87*, 2664.

Venkataraghavan, R., Dayringer, H. E., Atwater, B. L., Pesyna, G. L., McLafferty, F. W. (1978): In: Advances in Mass Spectrometry, 7B (Daly, N. R., ed.), 989–992. London: Heyden and Son Ltd.

Vestal, M. L. (1968): In: Fundamental Processes in Radiation Chemistry (Ausloos, P., ed.), Chapter 2. New York: Wiley Interscience.

Vestal, M. L., Futrell, J. H. (1970): J. Chem. Phys. *52*, 978.

Wahrhaftig, A. L. (1964): The Theory of Mass Spectra and the Interpretation of Ionization Efficiency Curves. In: NATO Advanced Study Institute on Mass Spectrometry (Reed, R. I., ed.). New York: Academic Press.

Wahrhaftig, A. L. (1972): Theory of Mass Spectra. In: Mass Spectrometry, MTP International Review of Science, Vol. 5 (Maccoll, A., ed.), 1. London: Butterworths.

Werner, A. S., Baer, T. (1975): J. Chem. Phys. *62*, 2900.

Wisniewski, S. G., Clow, R. P., Futrell, J. H. (1970): J. Phys. Chem. *74*, 2234.

Subject Index

Analysis and Simulation of Semiconductor Devices

By Dipl.-Ing. Dr. **Siegfried Selberherr,**
Institut für Allgemeine Elektrotechnik und Elektronik,
Technische Universität Wien, Austria

1984. 126 figures. XIV, 294 pages.
ISBN 3-211-81800-6

Contents: Introduction. — Some Fundamental Properties. — Process Modeling. — The Physical Parameters. — Analytical Investigations About the Basic Semiconductor Equations. — The Discretization of the Basic Semiconductor Equations. — The Solution of Systems of Nonlinear Algebraic Equations. — The Solution of Sparse Systems of Linear Equations. — A Glimpse on Results. — References. — Author Index. — Subject Index.

Numerical analysis and simulation has become a basic methodology in device research and development. This book satisfies the demand for a thorough review and judgement of the various physical and mathematical models which are in use all over the world today. A compact and critical reference with many citations is provided, which is particularly relevant to authors of device simulation programs. The physical properties of carrier transport in semiconductors are explained, great emphasis being laid on the direct applicability of all considerations. An introduction to the mathematical background of semiconductor device simulation clarifies the basis of all device simulation programs. Semiconductor device engineers will gain a more fundamental understanding of the applicability of device simulation programs. A very detailed treatment of the state-of-the-art and highly specialized numerical methods for device simulation serves in an hierarchical manner both as an introduction for newcomers and a worthwhile reference for the experienced reader.

Springer-Verlag Wien New York